Numerical Simulation Algorithm of Electromagnetic Field for Grounding Problems in Power System Substation Grounding Grids

Zhong-Xin Li

Numerical Simulation Algorithm of Electromagnetic Field for Grounding Problems in Power System Substation Grounding Grids

Springer

Zhong-Xin Li
High Voltage Institute
China Electric Power Research Institute
Beijing, China

ISBN 978-981-96-0903-1 ISBN 978-981-96-0904-8 (eBook)
https://doi.org/10.1007/978-981-96-0904-8

This Springer imprint is published by the registered company Springer Nature Singapore Pte Ltd.
The registered company address is: 152 Beach Road, #21-01/04 Gateway East, Singapore 189721, Singapore

Preface

This book focuses on grounding problems of the substation grounding grid, with the help of computer numerical simulation methods, numerical simulation of substation grounding grid grounding phenomena under various operating conditions.

This book numerically simulates the distribution of electromagnetic field around the grounding grid and the distribution of current and voltage on the grounding grid of a substation under steady state and transient operating conditions. The numerical simulation methods in this book are all based on the quasi-static electromagnetic field principle, and the Galerkin method of moments (MM), the Galerkin boundary element method (BEM), and the Galerkin hybrid algorithm based on various basis functions are developed for numerical simulation of substation grounding grid grounding problems. This includes mathematical models of equipotential and inequipotential. Meanwhile, this book also innovatively further develops the frequency-domain numerical simulation algorithms into time-domain numerical simulation algorithms. Therefore, this book not only numerically simulates the three-dimensional (3-d) distributions of electromagnetic field quantities, currents, and voltages of substation grounding grids at each frequency in the frequency domain, but also obtains the 3-d distributions of electromagnetic field quantities, currents, and voltages at each moment in the time domain.

The book begins with an in-depth discussion of one of the mathematical foundations of the numerical methods developed throughout the book, i.e., the quasi-static electromagnetic field complex image method for a point current source in a multilayer horizontally layered soil mathematical model. The method analytically calculates the infinite integral contained in the Green's function of the point current source in a multilayer horizontally layered soil.

The book goes on to discuss in depth the second mathematical basis for the numerical methods developed throughout the book, i.e., the analytical formulas for the calculation of mutual impedance and mutual inductance between discrete conductor segments of substation grounding grids. Since the numerical methods developed in this book are based on the Galerkin numerical algorithm, and the Galerkin numerical algorithm contains two-dimensional (2-d) spatial line integrals for the mutual impedance and mutual inductance between discrete conductor segments of a substation grounding grid, replacing numerical integration with analytical calculations is the mathematical foundation of this book.

The book goes on to discuss in depth the frequency-domain scalar Galerkin MM and the Galerkin BEM based on the equipotential mathematical model, and then further discusses in depth the time-domain scalar Galerkin MM and the Galerkin BEM based on the equipotential mathematical model. The advantages and disadvantages of this equipotential mathematical model and its practicality are also discussed.

This book further discusses in depth the frequency-domain vectorial Galerkin MM and Galerkin BEM based on mathematical modeling of unequal potentials, as well as Galerkin hybrid algorithms based on various basis functions. This focuses on the 3-d distribution of electromagnetic field quantities, currents, and voltages at various frequencies in the frequency domain for numerical simulations of substation grounding grids under steady state operating conditions.

The book goes on to discuss in depth the time-domain vectorial Galerkin MM and the Galerkin BEM based on mathematical modeling of unequal potentials, as well as Galerkin hybrid algorithms based on various basis functions. This focuses on the 3-d distribution of electromagnetic field quantities, currents, and voltages at various moments in the time domain for numerical simulations of substation grounding grids under transient operating conditions.

The authors first express their gratitude to their Ph.D. advisors at Tsinghua University, Prof. Xin-Shan Ma and Prof. Jian-Sheng Yuan, for guiding the authors into the issue of grounding in power system substations. The authors would also like to thank Prof. Wei-Jiang Chen, Academician of the Chinese Academy of Sciences, former Director of China Electric Power Research Institute (CEPRI), who led the authors to engage in the field of lightning protection and grounding of power system substations at CEPRI. The author would also like to express his gratitude to Prof. Jian-Bin Fan, President of Beijing Chuangtuo International Standard Technology Research Institute, for his support and understanding of my research on lightning protection and grounding of power systems. The authors would also like to express their gratitude to Prof. Peng Li, the current Director of CEPRI, for his interest and support of my research work. I would also like to thank the relevant leaders of the High Voltage Research Institute of CEPRI for their support and understanding of my research work. Finally, the authors would also like to express their gratitude to my colleagues in the Lightning Protection and Surge Arrestor Technology Research Laboratory of the High Voltage Research Institute of the CEPRI, Ltd. for their understanding and support.

This book was completed with the financial support of the CEPRI, as well as the support of the National Engineering Research Center of UHV Technology and the support of the National Natural Science Foundation of China under Grant 51177153.

The authors hope that the publication of this book will support the scientific research on lightning protection and grounding of power system substations. It will also serve as an inspiration and a backdrop for the scientific work of colleagues.

Beijing, China
September 2024

Zhong-Xin Li

Contents

Chapter 1
Progresses of Grounding Problem

Background and research progress on power system substation grounding problems.

1.1 Problems in Steady State Fault Current Grounding of Substation Grounding Grid

The problem of steady state grounding in power systems mainly involves the analysis and treatment of grounding faults in power systems, as well as the correct application of grounding technology. Grounding is an important link in the power system, which is not only related to the safe operation of the power system, but also directly affects the protection of power equipment and personnel safety. The handling and prevention of grounding faults is the key to ensure the stable operation of the power system.

The problem of grounding in the steady state case of a power system is a scientific problem of relative simplicity and a relatively stable level of scientific theory.

Since the size of the substation grounding grid is relatively small in comparison with the wavelength of the industrial frequency electromagnetic field, the mutual inductance effect of the axial currents between the conductor segments of the substation grounding grid is relatively weak and almost negligible in the case of the industrial frequency of the power system.

The power system steady state grounding problem is made relatively simple by neglecting the effects of mutual inductance between discrete conductor segments of the substation grounding grids. And because the substation grounding grid size relative to the frequency electromagnetic wave length can be regarded as short, so that the electromagnetic wave propagation effect can be ignored, therefore, at this time, the substation grounding grids due to the injected current and the electromagnetic field can be satisfied with Poisson's formula. And the electromagnetic field at this time can be regarded as quasi-static electromagnetic field.

Z.-X. Li, *Numerical Simulation Algorithm of Electromagnetic Field for Grounding Problems in Power System Substation Grounding Grids*,
https://doi.org/10.1007/978-981-96-0904-8_1

Considering the theory of electrostatic analogy, the performance of alternating current (AC) substation grounding system at low frequency of 50 or 60 Hz and the performance of direct current (DC) field are analyzed. Mathematical models for numerical simulation to calculate the steady state grounding problem of substation grounding grids can be divided into two categories, i.e., equipotential electromagnetic field mathematical model and inequipotential electromagnetic field mathematical model.

1.1.1 The Equipotential Electromagnetic Field Mathematical Model

The current literature on steady state grounding problems for substation grounding grids is mainly based on the spatial one-dimensional (1-d) scalar Galerkin boundary element method (BEM) [1–7] and the spatial one-dimensional (1-d) scalar Galerkin method of moments (MM) [8–16], where most of the literature is based on the DC field theory [1–13] and a few of them are based on the AC field theory [14–16].

Mathematical modeling of equipotential electromagnetic fields also further develops the Galerkin MM for higher order basis functions [16].

1.1.2 The Unequipotential Electromagnetic Field Mathematical Model

Like the equipotential electromagnetic fields (EMF) mathematical model, the unequipotential EMF mathematical model is also based on various EMF mathematical models under the DC field theory and AC field theory, respectively.

The literature based on mathematical modeling of unequal electromagnetic fields under DC field theory has the following multiple articles [17–19]. Also, there are several papers that are based on mathematical modeling of unequal electromagnetic fields under AC field theory [20–25]. It is important to note here that the DC field based literature [17–19] are all hybrid algorithms based on T basis functions, while the AC field based literature [20–25] are all hybrid algorithms based on PI basis functions.

1.2 Problems in Transient Fault Grounding of Substation Grounding Grid

Grounding grids plays a very important role in the safe operation of the substation, and its grounding performance has always been one of the important performance of the design and production and operation departments. With the capacity and voltage level of China's power systems continue to improve, substation faults into the grounding grids of the inrush current increases, resulting in an abnormal rise in the grounding grids potential, bringing threats to the safety of operators, caused by the ground potential counterattacks on the secondary equipment insulation was damaged, and even lead to secondary equipment malfunction or inactivity. Therefore, the grounding grids in the substation puts forward higher requirements for current dissipation and voltage equalization, and it is of great significance to carry out research on the impact characteristics of the grounding grids.

Like the previous power system substation grounding grids steady state grounding problem, the power system substation grounding grids transient grounding problem can also be divided into equipotential electromagnetic field model as well as unequal potential electromagnetic field mathematical model.

1.2.1 The Equipotential Electromagnetic Field Mathematical Model

It is important to note that, unlike the steady state grounding problem, the presence of equipotential EMF mathematical models in the transient fault grounding problem is only a requirement for the evolution of EMF numerical analysis methods, and therefore there is no published literature on the algorithms, and the methodology is only described in this book, taking into account the needs of the evolution of numerical methods. It is also important to note here that the equipotential electromagnetic field mathematical model can also be further divided into frequency and time domain electromagnetic field numerical methods.

1.2.2 The Unequipotential Electromagnetic Field Mathematical Model

Mathematical models of unequal electromagnetic fields for transient fault grounding problems can be divided into two broad categories: frequency domain and time domain.

The frequency domain unequipotential electromagnetic field mathematical model

The literature on the mathematical modeling of unequal electromagnetic fields in the frequency domain can be further divided into the literature [26–39] on the mathematical modeling of unequal electromagnetic fields in the single-frequency case for simulating and calculating the grounding characteristics of the grounding grid at that frequency, and the literature [40–92] on the numerical simulation of the lightning impulse response of the grounding grid of a transformer substation in the full-frequency case in conjunction with the Fourier transform.

It is important to note here that the literature [26–39] on the numerical calculation of the grounding characteristics of substation grounding grids at a single frequency can be further categorized into the following numerical methods: Literatures [26, 27] are about BEM for EMF numerical methods, while Literatures [28–32] are about MM for EMF numerical methods, in addition, the rest of the literatures [33–36, 38, 39] are about hybrid algorithms for EMF numerical methods, of which Literatures [33–36, 38] are about hybrid methods based on PI basis functions, while Literature [39] is about hybrid algorithms for T basis functions.

The literature on full frequency and combined with Fourier transform for numerical simulation of lightning transient shock response of substation grounding grids [40–92] can likewise be further categorized into various algorithms related to numerical methods of electromagnetic fields. Literatures [40–91] are about BEM in EMF numerical methods, while Literatures [51–57] are about MM in EMF numerical methods, and Literatures [58–61] are about finite element method (FEM) in EMF numerical methods, while literature [62] is about moment of line in numerical methods for electromagnetic fields. And Literatures [63–75] are about hybrid algorithms in EMF numerical methods, where Literatures [52–71] are about hybrid algorithms based on the PI basis function, while Literatures [72, 73] are hybrid algorithms based on PEEC basis functions, while papers [74, 75] are hybrid algorithms based on T and TPEEC basis functions. It should be further noted that there are also many literatures [77–91] on transmission line based methods in EMF numerical methods, and finallyliterature [92] on numerical methods for EMPT in EMF numerical methods.

The time domain unequipotential electromagnetic field mathematical model

In addition to the above mentioned frequency domain EMF numerical methods that rely on the Fourier transform and can be used to numerically simulate the lightning transient impulse response of a substation grounding grid, there are also a lot of time domain EMF numerical methods in the literature, which have developed time domain algorithms that can be used to directly numerically simulate the lightning transient impulse response of a substation grounding grid in the time domain without going through the Fourier transform. These papers can be further categorized into various EMF numerical methods based on the EMF numerical methods developed.

Literature [93] is a time-domain BEM that can be used to numerically simulate the lightning transient impulse response problem for substation grounding grids, whereas Literature [94, 95] is a time-domain MM that can be used to numerically simulate the lightning transient impulse response problem for substation grounding grids, whereas Literature [96–98] is a time-domain FEM that can be used to numerically compute the lightning transient impulse response problem for substation grounding grids. These literatures [93–98] have numerically simulated the lightning impact grounding problem for substation grounding grids in uniform earth. In contrast, literature [99, 100] describes the time-domain Galerkin BEM for the numerical simulation of the lightning transient shock response of substation grounding grids in multilayer horizontally stratified soils.

There is also a large body of literature [101–120] describing the finite difference time domain (FDTD) method, which is an algorithm that can be used to numerically simulate the problem of transient processes of lightning strikes in substation grounding grids. There is also a lot of literature [121–128] describing the time-domain electrical circuit approach which can also numerically simulate the problem of lightning transient impulse response of substation grounding grids.

Here it is also necessary to introduce the hybrid algorithm which is one of the numerical EMF algorithms mostly used for the lightning transient impulse response problem of substation grounding grids in multilayer horizontally layered soils, and here there have been many literatures [129–138] on the method, among which literatures [129–136] are based on the hybrid algorithm with PI basis functions, while literatures [121, 138] are based on the hybrid algorithm with PEEC basis functions.

Some more papers [139–141] describe the problem of EMPT-based algorithms for numerical simulation of lightning impulse transient processes in substation grounding grids in a uniform earth. As well as a large body of literature [142–153] describing time-domain transmission line methods that can be used to numerically simulate the lightning transient shock response problem for substation grounding grids in a uniform earth.

Finally, we have to introduce a numerical method of electromagnetic field in recent years, which can numerically simulate the lightning transient impulse grounding problem of substation grounding grids in a uniform earth. The method first obtains the frequency domain matrix equations by the numerical methods of electromagnetic field in the frequency domain (such as the BEM, the MM, the FEM, as well as a variety of hybrid algorithms), and then further transforms the frequency domain matrix equations into the time domain by using the vector fitting method, and the corresponding transient grounding problems are analyzed by numerical computation in the time-domain with EMPT. References [154–167] here all elaborate on such electromagnetic field numerical algorithms. Among them, references [154–163] are about the conversion of frequency-domain moment method into time-domain equation through vector fitting method and its combination with EMPT. And references [164–166] are about converting the frequency domain hybrid algorithm into a time domain formula through vector fitting method and using EMPT for calculation.

1.3 The Prospect of Research Direction to the Grounding Problem of Power System Substation

Obviously, the grounding problem of the power system substation grounding grid can be divided into steady-state grounding problem and transient grounding problem.

For steady-state grounding problems, due to the mature development of relevant theories and algorithms, the development of this research direction is only focused on the trend of developing powerful and mature programs or software.

For the transient grounding problem of power system substation grounding grid, traditional frequency domain electromagnetic field numerical algorithms, such as BEM, MM, FEM, and hybrid algorithms based on various basis functions, have also developed relatively mature theories and algorithms, especially for the transient lightning grounding problem of substation grounding grid in uniform ground. However, due to the Fourier transform required by this frequency domain algorithm to calculate the electromagnetic field field of the substation grounding grid and its surroundings in the time domain, it not only increases the computational load, but also cannot consider the ionization phenomenon caused by the large amount of lightning flowing into the ground on the surface of the grounding grid conductor and the surrounding soil. Therefore, the research direction of frequency domain algorithm still needs to further develop the lightning transient impulse grounding problem of the substation grounding grid in multi-layer horizontal soil.

In addition to frequency-domain algorithms, time-domain algorithms are currently flourishing, which can be divided into field domain methods and source domain methods. Here, field domain methods refer to discretizing the entire required computational field for relevant numerical calculations, such as FDTD and time-domain FEM. The source domain method refers to discretizing only the source domain that needs to be calculated, and then numerically calculating the required computational field using known sources. For example, time-domain BEM, time-domain MM, and hybrid algorithms based on various basis functions (PI and PEEC). At present, the hybrid algorithm in the time domain has fully studied and calculated the grounding problem of lightning transient impulse response of substation grounding grid in multi-layer horizontal soil. The time-domain Galerkin BEM and MM still need to further develop the grounding problem of lightning transient impulse response of substation grounding grid in multi-layer horizontal soil.

The development of time-domain algorithms still requires further research and development of corresponding programs or software. Time domain algorithms are particularly good at simulating and calculating the ionization phenomenon caused by the grounding of the grounding grid conductor surface and surrounding soil in multi-layer horizontal substations under large lightning current impact grounding. This is also one of the advantages of time-domain algorithms. The development of numerical algorithms for time-domain electromagnetic fields not only requires the development of related algorithms based on quasi-static magnetic field theory, but also requires the development of related algorithms based on dynamic electromagnetic field theory.

The development of time-domain algorithms still requires further research and development of corresponding programs or software. Time domain algorithms are particularly good at simulating and calculating the ionization phenomenon caused by the grounding of the grounding grid conductor surface and surrounding soil in multi-layer horizontal substations under large lightning current impact grounding. This is also one of the advantages of time-domain algorithms.

References

1. V. Slavko, K. Mate, Numerical analysis of earthing grids buried in horizontal stratified multilayered earth. Int. J. Numer. Methods Eng. **41**, 1297–1319 (1998)
2. I. Colominas, F. Navarrina, M. Casteleiro, A boundary element numerical approach for earthing grid computation. Comput. Methods Appl. Mech. Eng. **174**, 73–90 (1999)
3. I. Colominas, F. Navarrina, M. Casteleiro, A numerical formula for grounding system analysis in stratified soils. IEEE Trans. Power Delivery **17**(2), 587–595 (2002)
4. F. Navarrina, I. Colominas, M. Casteleiro, Why do computer methods for grounding analysis produce anomalous results? IEEE Trans. Power Delivery **18**(4), 1192–1202 (2003)
5. U. Adriano, 0. Bottauscio, M. Zucca, Boundary element approach for the analysis and design of grounding systems in presence of non-homogeneousness. IEE Proc. Gener. Distrib. Distrib. **150**(3), 360–366 (2003)
6. I. Colominas, F. Navarrina, M. Casteleiro, Analysis of transferred earth potentials in grounding systems: A bem numerical approach. IEEE Trans. Power Delivery **20**(1), 339–345 (2005)
7. I. Colominas, F. Navarrina, M. Casteleiro, Numerical simulation of transferred potentials in earthing grids considering layered soil models. IEEE Trans. Power Delivery **22**(3), 1514–1522 (2007)
8. R.J. Heppe, Computation of potential at surface above an energized grid or other electrode, allowing for non-uniform current distribution. IEEE Trans. Power Apparatus Syst., **PAS-98**(6), 1978–1989 (1979)
9. Y.L. Chow, J.J. Yang, K.D. Srivastavaa, Complex images of a ground electrode in layered soils. J. Appl. Phys. **71**(2), 569–574 (1992)
10. A.P. Sakis Meliopoulos, F Xia, G.J. Cokkinides, An advanced computer model for grounding system analysis. IEEE Trans. Power Delivery **8**(1), 13–23 (1993)
11. Y.L. Chow, M.M. Elsherbiny, M.M.A. Salama, Efficient computation of rodbed grounding resistance in a homogenous earth by galerkin's moment method. IEE Proc. Gener. Trans. Distrib. **142**(6), 653–660 (1997)
12. Y.L. Chow, M.M. Elsherbiny, M.M.A. Salama, Surface voltages and resistance of grounding systemof grid and rods in two-layer earth by the rapid galerkin's moment method. IEEE Trans. Power Delivery **12**(1), 179–185 (1999)
13. L. Zhang, J. Yuan, Z.X. Li, The complex image method and its application in numerical simulation of substation grounding grids. Commun. Numer. Methods Eng. **15**, 835–839 (1999)
14. P.D. Rancic, L.V. Stefanovic, D.R. Djordjevic, An improved linear grounding system analysis in two-layer earth. IEEE Trans. Magn. **32**, 5179–5187 (1996)
15. Z.X. Li, J.B. Fan, W.J. Chen, Numerical simulation of substation grounding grids buried in both horizontal and vertical multilayered earth model. Int. J. Numer. Methods Eng. **69**(11), 2359–2380 (2007)
16. Z.X. Li, G.F. Li, J.B. Fan, Numerical simulation of substation grounding system in low frequency domain based on the boundary element method with linear basis function. Commun. Numer. Methods Eng. **26**(12), 1819–1914 (2010)
17. A.P.S. Meliopoulos, G.J. Cokkinides, J. Dunlap, Analysis of dc grounding systems. IEEE Trans. Power Delivery **3**(4), 1595–1604 (1988)
18. J. Yuan, H. Yang, L. Zhang, X. Cui, X. Ma, Simulation of substation grounding grids with unequal-potential. IEEE Trans. Magn. **36**(4), 1468–1471 (2000)
19. Y. Dan, Z. Zhang, H. Zhao, Y. Li, H. Ye, J. Deng, A novel segmented sampling numerical calculation method for grounding parameters in horizontally multilayered soil. Int. J. Electr. Power Energy Syst. **126**, 106586 (2021)
20. F.P. Dawalibi, R.D. Southey, Analysis of electrical interference from power lines to gas pipelines part 1: computation methods. IEEE Trans. Power Delivery **4**(3), 1840–1846 (1989)
21. A.F. Otero, J. Cidras, J.L. de Alamo, Frequency-dependant grounding system calculation bu means of a conventional nodal analysis technique. EEE Trans. Power Delivery **14**(3), 873–878, 873–878

22. L. Huang, D.G. Kasten, Modeling of ground grid and metallic structure currents in high voltage a.c. substations for the computation of electromagnetic fields. Electr. Power Syst. Res. **59**, 31–37 (2001)
23. Z.-X. Li, W. Chen, J.-B. Fan, J. Lu, A novel mathematical modeling of grounding system buried in multilayer earth. IEEE Trans. Power Delivery **21**(3), 1267–1272 (2006)
24. Z.-X. Li, W. Chen, Numerical simulation grounding system buried within horizontal multilayer earth in frequency domain. Commun. Numer. Methods Eng. **27**, 11–27 (2007)
25. Z.-X. Li, J.-B. Fan, Numerical calculation of grounding system in low-frequency domain based on the boundary element method. Int. J. Numer. Methods Eng. **73**, 685–705 (2008)
26. V. Arnautovski-Toseva, L. Grcev, Electromagnetic analysis of horizontal wire in two-layered soil. J. Comput. Appl. Mathem. **168**, 21–29 (2004)
27. V. Arnautovski-Toseva, L. Grcev, Image and exact models of a vertical wire penetrating a two-layered earth. IEEE Trans. Electromagn. Compat. **53**(4), 968–976 (2011)
28. F. Dawalibi, A. Selby, Electromagnetic fields of engergized conductors. IEEE Trans. Power Delivery **8**(3), 1505–1513 (1993)
29. R. Andolfato, L. Bernardi, L. Fellin, Aerial and grounding system analysis by the shifting complex images method. IEEE Trans. Power Delivery **15**(3), 1001–1009 (2000)
30. H. Karami, K. Sheshyekani, F. Rachidi, Mixed-potential integral equation for full-wave modeling of grounding systems buried in a lossy multilayer stratified ground. IEEE Trans. Electromagn. Compat. **59**(5), 1505–1513 (2017)
31. H. Karami, K. Sheshyekani, Harmonic impedance of grounding electrodes buried in a horizontally stratified multilayer ground: A full-wave approach. IEEE Trans. Electromagn. Compat. **60**(4), 899–906 (2018)
32. A.R.J. de Arao, S. Kurokawa, C.M. Seixas, B. Kordi, Optimization of tower-footing grounding impedance for guyed-v transmission towers. Electr. Power Syst. Res. **177**, 105947 (2004)
33. L. Huang, D.G. Kasten, Modeling of ground grid and metallic structure currents in high voltage a.c. substations for the computation of electromagnetic fields. Electr. Power Syst. Res. **59**, 31–37 (2001)
34. Z.-X. Li, W. Chen, J.-B. Fan, J. Lu, A novel mathematical modeling of grounding system buried in multilayer earth. IEEE Trans. Power Delivery **21**(3), 1267–1272 (2006)
35. S. Vujevic, P. Sarajcev, Potential distribution for a harmonic current point source in horizontally stratified multilayer medium. COMPEL Int. J. Comput. Mathem. Electr. Electron. Eng. **27**(3), 624–637 (2008)
36. P. Sarajcev, S. Vujevic, D. Lovric, Time-harmonic current distribution on conductor grid in horizontal stratified multilayer medium. Progress Electromagn. Res. B **31**, 67–87 (2011)
37. S. Vujevic, P. Sarajcev, D. Lovric, Time-harmonic analysis of grounding system in horizontally stratified multilayer medium. Electr. Power Syst. Res. **83**, 28–34 (2012)
38. Z.-X. Li, Y. Yin, C.-X. Zhang, L.-C. Zhang, Numerical simulation of currents distribution along grounding grid buried in horizontal multi-layered earth in frequency domain based dynamic state complex image method. Electr. Power Components Syst. **43**(14), 1573–1582 (2015)
39. Z.-X. Li, Y. Yin, C.-X. Zhang, L.-C. Zhang, Numerical simulation of a grounding grid buried in horizontal multilayered earth based on typical basic elements in the frequency domain. IET Sci. Measur. Technol. **9**(6), 645–653 (2015)
40. L. Grcev, Computer analysis of transient voltage in large grounding systems. IEEE Trans. Power Delivery **11**(2), 815–823 (1996)
41. M. Heimbach, L. Grcev, Grounding system analysis in transient programs applying electromagnetic field approach. IEEE Trans. Power Delivery **12**(1), 186–193 (1997)
42. D. Poljak, V. Doric, Wire antenna mopdel for transient analyais of simple grounding systems, part i: the vericak grounding electrode. Progress Electromagn. Res. **64**, 149–166 (2006)
43. D. Poljak, V. Doric, Wire antenna mopdel for transient analyais of simple grounding systems, part i: the horizontal grounding electrode. Progress Electromagn. Res. **64**, 167–189 (2006)
44. T. Topa, A. Karwowski, A. Noga, Using gpu with cuda to accelerate mom-based electromagnetic simulation of wire-grid models. IEEE Antenna Wirel. Progagation Lett. **10**, 342–345 (2011)

45. D. Cavka, B. Harrat, D. Poljak, B. Nekhoul, K. Kerroum, K.E.K. Drissi, Wire antenna versus modified transmission line approach to the transient analysis of grounding grid. Eng. Anal. Boundary Elements **35**, 1101–1108 (2011)
46. K. Sheshyekani, S.H.H. Sadeghi, R. Moini, F. Rachidi, Frequency-domain analysis of ground electrodes buried in an ionized soil when subjected to surge currents: A mom-aom approach. Electr. Power Syst. Res. **81**, 290–296 (2011)
47. D. Cavka, D. Poljak, On the evaluation of input impedance and transient impedance for grounding electrodes using antenna theory. COMPEL Int. J. Comput. Mathem. Electr. Electron. Eng. **32**(6), 2045–2062 (2013)
48. Ma. Chouki, B. Nekhoul, D. Poljak, K. Kerroum, K.E.K. Drissi, Simplified dipole concept for the assessment of transient electromagnetic field in the vicinity of grounding grid. Int. J. Numer. Modelling Electron. Network Dev. Fields **28**, 404–418 (2015)
49. B. Markovski, L. Grcev, V. Arnautovski-Toseva, Fast and accurate transient analysis of large grounding systems in multilayer soil. IEEE Trans. Power Delivery **36**(2), 598–606 (2021)
50. Z.-X. Li, Y. Yu, K.-C. Wang, Lightning response of a grounding system buried in multilayered earth based on the galerkin's boundary element and quasi-static complex image methods. IEEE Trans. Electromagn. Compat. **63**(4), 1118–1127 (2021)
51. L. Grcev, F. Dawalibi, An electromagnetic model for transient in grounding systems. IEEE Trans. Power Delivery **11**(2), 1773–1781 (1990)
52. G. Ala, M.L.D. Silvestre, A simulation model for electromagnetic transients in lightning protection systems. IEEE Trans. Electromagn. Compat. **44**(4), 539–554 (2002)
53. G. Ala, M.L.D. Silvestre, E. Francomano, A. Tortorici, Wavelet-based efficient simulation of electromagnetic transients in a lightning protection system. IEEE Trans. Magn. **39**(3), 1257–1260 (2003)
54. G. Ala, M.L.D. Silvestre, E. Francomano, A. Tortorici, An advanced numerical model in solving thin-wire integral equations by using semi-orthogonal compactly supported spline wavelets. IEEE Trans. Electromagn. Compat. **45**(2), 218–228 (2023)
55. T. Rouibaha, A. Bayadia, K. Kerroum, Accelerating the frequency-domain response calculation of complexgrounding system using wavelet based mbpe technique. Electr. Power Syst. Res. **121**, 287–294 (2015)
56. M. Nazari, R. Moini, S. Fortin, F.P. Dawalibi, F. Rachidi, Impact of frequency-dependent soil models on grounding system performance for direct and indirect lightning strikes. IEEE Trans. Electromagn. Compat. **63**(1), 134–144 (2019)
57. M. Shafieipour, M. Nazari, F.P. Dawalibi, S. Fortin, F.P. Dawalibi, S. Fortin, A. Tatematsu, J.D. Silva, P. Gez, Full-wave 3-d transient analysis with method of moments and numerical laplace transform including resistive non-linear elements. IEEE Trans. Power Delivery **38**(1), 299–308 (2023)
58. L. Qi, X. Cui, Z. Zhao, H. Li, Grounding performance analysis of the substation grounding grids by finite element method in frequency domain. IEEE Trans. Magn. **43**(4), 1773–1781 (2007)
59. M. Akbari, K. Sheshyekani, M.R. Alemi, The effect of frequency dependence of soil electrical parameters on the lightning performance of grounding systems. IEEE Trans. Electromagn. Compat. **55**(4), 739–746 (2013)
60. B. Salarieh, J.D. Silva, B. Kordi, High frequency response of grounding electrodes: effect of soil dielectric constant. IET Gener. Transmission Distrib. **14**(15), 2915–2921 (2020)
61. N. Yu, Z. Cai, R. Li, Q. Zhang, L. Gao, Z. Sun, Calculation of earth surface potential and neutral current caused by hvdc considering three-dimensional complex soil structure. IEEE Trans. Electromagn. Compat. **63**(5), 1480–1490 (2021)
62. H. Zamani, K. Sheshyekani, Method of line for modeling of grounding electrodes buried in stratified multilayer soil. IEEE Trans. Electromagn. Compat. **64**(6), 2131–2140 (2022)
63. A.F. Otero, J. Cidras, J.L.D. Alamo, Frequency-dependent grounding system calculation by means of a conventional nodal analysis technique. IEEE Trans. Power Delivery **14**(3), 837–878 (1999)

64. J. Cidras, A.F. Otero, C. Garrido, Nodal frequency analysis of grounding systems considering the soil ionization effect. IEEE Trans. Power Delivery **15**(1), 103–107 (2000)
65. S. Visacro, Jr. A. Soares, Hem: a model for simulation of lightning-related engineering problems. IEEE Trans. Power Delivery **20**(2), 1206–1208 (2005)
66. Z.-X. Li, G.-F. Li, J.-B. Fan, Y. Yin, A novel mathematical model for the lightning response of a grounding system buried in multilayered earth based on the quasi-static complex image method. Europ. Trans. Electr. Power **22**, 831–845 (2012)
67. Z.-X. Li, Y. Yin, C.-X. Zhang, L.-C. Zhang, A mathematical model for the transient lightning response from grounding systems. Progress Electromagn. Res. B **57**, 47–61 (2014)
68. R. Araneo, S. Celozzi, Transient behavior of wind towers grounding systems under lightning strikes. Int. J. Energy Environ. Eng. **7**, 235–247 (2015)
69. Ro. A.R. de Moura, M.A. de Oliveira Schroeder, A.C.S. de Lima, P.H.N. Vieira, R. Alipio, Closed-form approximation for grounding grids transient analysis. J. Control Autom. Electr. Syst. **32**, 796–806 (2021)
70. K.P. Sengar, K. Chandrasekaran, Transient behaviour of grounding systems inmultilayer soil under lightning strikes. Electr. Eng. **104**, 1205–1218 (2022)
71. K.P. Sengar, K. Chandrasekaran, Analysis of grounding system transient with consideration of soil ionization and mutual coupling effect. J. Electr. Syst. Inf. Technol. **10**(53) (2023)
72. P. Yutthagowith, A. Ametani, N. Nagaoka, Y. Baba, Application of the partial element equivalent circuit method to analysis of transient potential rises in grounding systems. IEEE Trans. Electromagn. Compat. **53**(3), 381–386 (2011)
73. Z.-X. Li, K.-L. Gao, S.-W. Rao, Lightning response of a grounding system buried in multiple layers in earth using the peec method based on the quasi-static complex image method. Electr. Power Components Syst. **48**(3), 291–303 (2020)
74. Z.-X. Li, K.-L. Gao, Y. Yin, D. Ge, Transient lightning responses of grounding systems buried in horizontal multilayered earth with a hybrid method. J. Electrost. **72**, 381–386 (2014)
75. J.C. Salari, C. Portela, Grounding systems modeling including soil ionization. IEEE Trans. Power Delivery **23**(4), 1939–1945 (2008)
76. Y. Liu, M. Zitnik, R. Thottappillil, An improved transmission-line model of grounding system. IEEE Trans. Electromagn. Compat. **43**(3), 348–355 (2001)
77. Y. Liu, N. Theethayi, R. Thottappillil, An engineering model for transient analysis of grounding system under lightning strikes: Nonuniform transmission-line approach. IEEE Trans. Power Delivery **20**(2), 722–730 (2001)
78. N. Harid, H. Griffiths, A. Haddad, A new frequency-dependent surge impedance calculation method for high-voltage towers. IEEE Trans. Power Delivery **21**(3), 1430–1437 (2006)
79. M. Al-Shaher, M. Saied, s-domain analysis of grounding electrodes impulse response. Electr. Power Components Syst. **34**, 985–999 (2006)
80. N. Theethayi, R. Thottappillil, M. Paolone, C.A. Nucci, F. Rachidi, External impedance and admittance of buried horizontal wires for transient studies using transmission line analysis. IEEE Trans. Dielectrics Electr. Insul. **14**(3), 751–761 (2007)
81. J.-H. Choi, B.-H. Lee, An analysis on the frequency-dependent grounding impedance based on the ground current dissipation of counterpoises in the two-layered soils. J. Electrost. **70**, 184–191 (2012)
82. A. Jardines, J.L. Guardado, J. Torres, J.J. Chez, M. Herndez, A multiconductor transmission line model for grounding grids. Electr. Power Energy Syst. **60**, 24–33 (2014)
83. R. Shariatinasab, J. Gholinezhad, The effect of grounding system modeling on lightning-related studies of transmission lines. J. Appl. Res. Technol. **15**, 545–554 (2017)
84. S. Yassin, A.Z. El Dein, Harmonic impedance of horizontal grounding electrode under soil ionization effect. Wseas Trans. Power Syst. **13**, 200–208 (2018)
85. S.S. Sajjadi, S.R. Ostadzadeh, Lightning response of multi-port grids buried in dispersive soils: An approximation versus full-wave methods and experiment. Adv. Electromagn. **8**(1), 43–50 (2019)
86. C.M. de Seixas, S. Kurokawa, Using circuit elements to represent the distributed parameters of a grounding system under lightning strokes. Electr. Power Syst. Res. **172**, 213–220 (2019)

87. R. Batista, J.O.S. Paulino, A practical approach to estimate grounding impedance of a vertical rod in a two-layer soil. Electr. Power Syst. Res. **177**, 105973 (2019)
88. R. Batista, C.E.F. Caetano, J.O.S. Paulino, W.C. Boaventur, I.J.S. Lopes, E.N. Cardoso, A study of grounding arrangements composed by vertical electrodes for two layered stratified soil models. Electr. Power Syst. Res. **180**, 106129 (2020)
89. S.S. Sajadi, S.R. Ostadzadeh, S.H.H. Sadeghi, Parametric dependence of lightning impulse behavior of grounding electrodes buried in a dispersive and ionized lossy soil under first- and subsequent-stroke currents. COMPEL Int. J. Comput. Mathem. Electr. Electron. Eng. **39**(4), 757–773 (2020)
90. R. Batista, J.O.S. Paulino, Computing grounding resistance and impulse impedance of horizontal electrodes parallel or perpendicular to the interface of a vertically stratified soil using transmission line theory. Electr. Power Syst. Res. **194**, 107060 (2021)
91. D. Tashiro, K. Sameshima, T. Hisakado, A.K.M.M. Islam, O. Wada, Single-conductor transmission-line model for bent wire structures. IEEE Trans. Electromagn. Compat. **65**(5), 1432–1445 (2023)
92. E. Frank, L. Grcve, Empt-based model for grounding system analysis. IEEE Trans. Power Delivery **9**(4), 1838–1849 (1994)
93. S. Sesnic, D. Poljak, S. Tkachenko, Time domain analytical modeling of a straight thin wire buried in a lossy medium. Progress Electromagn. Res. **121**, 485–504 (2011)
94. M.R. Alemi, S.H.H. Sadeghi, H. Askarian-Abyaneh, A marching-on-in-time method of moments for transient modeling of a vertical electrode buried in a lossy medium. IEEE Trans. Electromagn. Compat. **64**(6), 2170–2178 (2022)
95. M.R. Alemi, S.H.H. Sadeghi, H. Askarian-Abyaneh, A marching-on-in-time method of moments for computation of transient potential rise of grounding grids exposed to lightning strikes. IEEE Trans. Electromagn. Compat. **65**(5), 1484–1491 (2023)
96. A. Habjanic, M. Trlep, The simulation of the soil ionization phenomenon around the grounding system by the finite element method. IEEE Trans. Magn. **42**(4), 867–870 (2006)
97. J. Li, T. Yuan, Q. Yang, W. Sima, C. Sun, M. Zahn, Numerical and experimental investigation of grounding electrode impulse-current dispersal regularity considering the transient ionization phenomenon. IEEE Trans. Power Delivery **26**(4), 2647–2658 (2011)
98. R. Syafruddin, F. Hetty, A. Salamatul, D.R. Givy, H. Rahmad, Voltages transient analysis in electric grounding systems. J. Phys. Conf. Ser. **1783**, 012059 (2021)
99. Z.-X. Li, P. Li, X. Zhao, Galerkin boundary element method for simulating lightning response of grounding grid in horizontal multilayered soil model considering soil ionization effect. Int. J. Numer. Model. Electron. Network Dev. Fields **37**(5), e3285 (2024)
100. J.-B. Fan, Z.-X. Li, Lightning transient impulse problem of substation grounding grid in multilayer horizontal layered soil considering soil ionization effect and frequency characteristics. IEEE Trans. Electromagn. Compat., 1–10 (2024)
101. G. Ala, E. Francomano, E. Toscano, F. Viola, Finite difference time domain simulation of soil ionization in grounding systems under lightning surge conditions. Appl. Numer. Anal. Comput. Mathem. **1**(1), 90–103 (2004)
102. R.L. Stoll, G. Chen, N. Pilling, Comparison of two simple high-frequency earthing electrodes. IEE Proc. Gener. Transmission Distrib. **151**(2), 219–224 (2004)
103. Y. Baba, N. Nagaoka, A. Ametani, Modeling of thin wires in a lossy medium for fdtd simulations. IEEE Trans. Electromagn. Compat. **47**(1), 54–60 (2005)
104. M. Tsumura, Y. Baba, N. Nagaoka, A. Ametani, Fdtd simulation of a horizontal grounding electrode and modeling of its equivalent circuit. IEEE Trans. Electromagn. Compat. **48**(4), 817–825 (2006)
105. G. Ala, P.L. Buccheri, P. Romano, F. Viola, Finite difference time domain simulation of earth electrodes soil ionisation under lightning surge condition. IET Sci. Measur. Technol. **2**(3), 134–145 (2008)
106. G. Ala, M.L.D. Silvestre, F. Viola, Soil ionization due to high pulse transient currents leaked bu earth electrodes. Progress Electromagn. Res. B **14**, 1–21 (2009)

107. R. Xiong, B. Chen, Y.-F. Mao, W. Deng, Q. Wu, Y.-Y. Qiu, Fdtd modeling of the earthing conductor in the transient grounding resistance analysis. IEEE Antenna Wirel. Progagation Lett. **11**, 957–960 (2012)
108. A. Ametania, T. Chikaraa, H. Morii, T. Kubo, Impedance characteristics of grounding electrodes on earth surface. Electr. Power Syst. Res. **85**, 38–43 (2012)
109. R. Xiong, B. Chen, C. Gao, Y. Yi, W. Yang, Fdtd calculation model for the transient analyses of grounding systems. IEEE Trans. Electromag. Compat. **56**(5), 1155–1162 (2014)
110. M.E.M. Rizk, E.A. Badran, M.H. Abdel-Rahman, Fdtd calculation of transient electromagnetic fields in the grounding system of wind towers due to direct lightning strikes. J. Lightning Re. **6**, 1–13 (2014)
111. K. Yamamoto, K. Yoshioka, S. Sumi, S. Yanagawa, S. Sekioka, Mutual influence of a deeply buried grounding electrode and the surrounding grounding mesh. IEEE Trans. Indus. Appl. **51**(6), 1–13 (2015)
112. Y. Du, B. Li, M. Chen, The extended thin-wire model of lossy round wire structures for fdtd simulations. IEEE Trans. Power Delivery **32**(6), 2472–2480 (2017)
113. R. Alipioa, M.T.C. de Barros, M.A.O. Schroeder, K. Yamamotod, Analysis of the lightning impulse and low-frequency performance of wind farm grounding systems. Electr. Power Syst. Res. **180**, 106068 (2020)
114. C. Lia, S. Wang, W. Wang, C. Zhang, L. Jiang, X. Wen, H. Lu, Equivalent circuit model of tower based on full-wave analysis. Energy Reports **8**, 1026–1033 (2022)
115. J. Cao, Y. Du, Y. Ding, B. Li, R. Qi, Y. Zhang, Z. Li, Lightning surge analysis of transmission line towers with a hybrid fdtd-peec method. IEEE Trans. Power Delivery **37**(2), 1275–1284 (2022)
116. A. Boutadjine, B. Nekhoul, B. Harrat, M. Melit, Realistic modeling of the grounding system transient behavior with frequency dependence. Electr. Power Syst. Res. **203**, 107644 (2022)
117. B. Nekhoul, B. Harrat, A. Boutadjine, M. Melit, A simplified numerical modeling of the transient behavior of grounding systems considering soil ionization. Electr. Power Syst. Res. **211**, 108182 (2022)
118. K.A. Carvalho, F.H. Silveira, Evaluation of the lightning performance of grounding electrodes using fdtd-based computational model: Influence of absorbing boundary conditions and representation of the frequency-dependence of soil parameters. Electr. Power Syst. Res. **213**, 108728 (2022)
119. L. Jiang, X. Dong, X. Zhou, J. Wang, J. Song, Q. Ma, Propagation characteristics of lightning radiation field on three-layer vertical layered ground based on cpml absorption boundary. IEEE Trans. Electromagn. Compat. **64**(4), 1191–1201 (2023)
120. K.de A. Carvalho, F.H. Silveira, Fdtd-based computational model applied to assess the influence of the frequency dependence of soil parameters on the lightning response of grounding electrodes for typical parameters of first and subsequent return stroke currents. Electr. Power Syst. Res. **229**, 110170 (2024)
121. A.C. Liew, M. Darveniza, Dynamic model of impulse characteristics of concentrated earths. Proc. IEEE **121**(2), 123–135 (1974)
122. A. Geri, E. Garbagnati, G. Sartorio, Non-linear behaviour of ground electrodes under lightning surge currents: Computer modelling and comparison with experimental results. IEEE Trans. Magn. **28**(2), 1442–1445 (1992)
123. A. Geri, E. Garbagnati, G. Sartorio, Behaviour of grounding systems excited by high impulse currents: the model and its validation. IEEE Trans. Power Delivery **14**(3), 1008–1017 (1999)
124. J. Wang, A.C. Liew, M. Darveniza, Extension of dynamic model of impulse behavior of concentrated grounds at high currents. IEEE Trans. Power Delivery **20**(3), 2160–2165 (2005)
125. M. Mokhtari, Z. Abdul-Malek, Z. Salam, An improved circuit-based model of a grounding electrode by considering the current rate of rise and soil ionization factors. IEEE Trans. Power Delivery **30**(1), 211–219 (2015)
126. M. Mokhtari, G.B. Gharehpetian, Integration of energy balance of soil ionization in cigre grounding electrode resistance model. IEEE Trans. Electromagn. Compat. **60**(2), 402–413 (2018)

127. A.F. Andrade, E.G. da Costa, Marconni F.B.R. Gonlves, G.R.S. Lira, R.O. Teixeira, Modeling grounding systems response to current impulses considering nonlinear effects. IEEE Trans. Power Delivery **36**(6), 3858–3866 (2021)
128. Y. Zhang, A practical method for transient analysis of grounding grids. IEEJ Trans. Electr. Electron. Eng. **18**, 1120–1126 (2023)
129. M. Ramamoorty, M.M. Babu Narayanan, S. Parameswaran, D. Mukhedkar. Transient performance of grounding grids. IEEE Trans. Power Delivery **4**(4), 2053–2059 (1989)
130. Z.-X. Li, S.-W. Rao, Transient lightning response of grounding grids in a horizontal, multilayered soil model that considers soil ionization effects with time-domain quasi-static complex images. Int. J. Numer. Model **32**, e2463 (2019)
131. M. Asadpourahmadchali, M. Niasati, Y. Alinejad-Beromi, Hybrid continuous circuit-trapezoidal integration method analysis of multi-cross structure of grounding system. IET Sci. Measur. Technol. **14**(3), 292–302 (2020)
132. Z.-X. Li, P. Li, K.-C. Wang, Transient lightning response to grounding grids buried in horizontal multilayered earth model considering time domain quasi-static complex image method and soil ionization effect. COMPEL Int. J. Comput. Mathem. Electr. Electron. Eng. **40**(3), 516–534 (2021)
133. B.P. Silva, S. Visacro, F.H. Silveira, Hem-td: New time-domain electromagnetic model for calculating the lightning response of electric systems and their components. IEEE Trans. Power Delivery **37**(6), 4848–4857 (2022)
134. M. Moradi, An electrostatic approach for simulation of grounding systems lightning response in time domain. IEEE Trans. Power Delivery **37**(6), 5289–5298 (2022)
135. Z.-X. Li, W.-J. Chen, K.-C. Wang, Frequency domain and time domain analysis of the transient behavior of buried grounding grids in horizontalmultilayered earth model. Electr. Eng. **104**, 2515–529 (2022)
136. B.P. Silva, S. Visacro, F.H. Silveira, New approaches to represent the frequency dependence of soil parameters on the time-domain hybrid electromagnetic model (hem-td). IEEE Trans. Electromagn. Compat. **65**(6), 1793–1800 (2023)
137. P. Yutthagowitha, A. Ametani, F. Rachidi, N. Nagaoka, Y. Baba, Application of a partial element equivalent circuit method to lightning surge analyses. Electr. Power Syst. Res. **94**, 30–37 (2023)
138. Z.-X. Li, Lightning response transient characteristics of substation grounding grid in multilayer horizontal stratified earth model considering soil ionization effect. IEEE Trans. Electromagn. Compat. **65**(6), 1665–1673 (2023)
139. M. Heimbach, L. Grcev, Grounding system analysis in transient programs apply electromagnetic field approach. IEEE Trans. Power Delivery **12**(1), 186–193 (1997)
140. S. Piliica, I. Uglec, B. Juric, Evaluating the overvoltage performance of an overhead line taking into account the frequency-dependence of its tower grounding electrodes with high soil resistivity. Electr. Power Energy Syst. **116**, 105547 (2020)
141. L. Zhou, Z. Lin, C. Liu, L. Huang, J. Yang, D. Wang, L. Guo, A novel lightning strike response simulation method for transmission lines considering transient grounding resistance. IEEE Sensors J. **23**(19), 22803–22812 (2023)
142. M.I. Lorentzou, N.D. Hatziargyriou, B.C. Papadias, Time domain analysis of grounding electrodes impulse response. IEEE Trans. Power Delivery **18**(2), 517–524 (2003)
143. A.I. Ramirez, A. Semlyen, R. Iravani, Modeling nonuniform transmission lines for time domain simulation of electromagnetic transients. IEEE Trans. Power Delivery **18**(3), 968–974 (2003)
144. N. Theethayi, Y. Baba, F. Rachidi, R. Thottappillil, On the choice between transmission line equations and full-wave maxwell equations for transient analysis of buried wires. IEEE Trans. Electromagn. Compat. **50**(2), 347–357 (2008)
145. D. Poljak, S. Antonijevic, K. El K. Drissi, K. Kerroum, Transient response of straight thin wires located at different heights above a ground plane using antenna theory and transmission line approach. IEEE Trans. Electromag. Compat. **52**(1), 108–116 (2010)

146. D.S. Gazzana, A.S. Bretas, G.A.D. Dias, M. Tell, D.W.P. Thomas, C. Christopoulos, Transient response of straight thin wires located at different heights above a ground plane using antenna theory and transmission line approach. IEEE Trans. Magn. **50**(2), 7012404 (2014)
147. D.S. Gazzana, G.A.D. Dias, A.S. Bretas, M. Tell, D.W.P. Thomas, C. Christopoulos, Novel formulation to determine the potential on the soil surface generated by a lightning surge. IEEE Trans. Magn. **52**(3), 7012404 (2016)
148. S. Chiheb, O. Kherif, M. Teguar, A. Mekhaldi, N. Harid, Transient behaviour of grounding electrodes in uniform and in vertically stratified soil using state space representation. IET Sci. Measur. Technol. **12**(4), 427–435 (2018)
149. O. Kherif, S. Chiheb, M. Teguar, A. Mekhaldi, N. Harid, Time-domain modeling of grounding systems?impulse response incorporating nonlinear and frequency-dependent aspects. IEEE Trans. Electromagn. Compat. **60**(4), 907–916 (2018)
150. D.S. Gazzana, A.B. Tronchoni, R.C. Leborgne, A.S. Bretas, D.W.P. Thomas, C. Christopoulos, An improved soil ionization representation to numerical simulation of impulsive grounding systems. IEEE Trans. Magn. **54**(3), 7200204 (2018)
151. V. Steinsland, L.H. Sivertsen, E. Cimpan, S. Zhang, A new approach to include complex grounding system in lightning transient studies and emi evaluations. Energies **12**, 12163142 (2019)
152. M. Moradi, Analysis of transient performance of grounding system considering frequency-dependent soil parameters and ionization. IEEE Trans. Electromagn. Compat. **62**(3), 785–797 (2020)
153. S.S. Sajadi, S.R. Ostadzadeh, S.H.H. Sadeghi, Parametric dependence of lightning impulse behavior of grounding electrodes buried in a dispersive and ionized lossy soil under first- and subsequent-stroke currents. COMPEL Int. J. Comput. Mathem. Electr. Electro. Eng. **39**(4), 757–773 (2020)
154. K. Sheshyekani, M. Akbari, B. Tabei, R. Kazemi, Wideband modeling of large grounding systems to interface with electromagnetic transient solvers. IEEE Trans. Power Delivery **29**(4), 1868–1876 (2014)
155. M.R. Alemi, K. Sheshyekani, Wide-band modeling of tower-footing grounding systems for the evaluation of lightning performance of transmission lines. IEEE Trans. Electromag. Compat. **57**(6), 1627–1636 (2015)
156. R. Shariatinasab, J. Gholinezhad, K. Sheshyekani, M.R. Alemi, The effect of wide band modeling of tower-footing grounding systemon the lightning performance of transmission lines: A probabilistice valuation. Electr. Power Syst. Res. **141**, 1–10 (2016)
157. V. Mashayekhi, S.H.H. Sadeghi, R. Moini, K. Sheshyekani, An adaptive chebyshev approach for fast computation of grounding system admittance matrix. IEEE Trans. Electromagn. Compat. **59**(2), 1–10 (2017)
158. R. Shariatinasab, J. Gholinezhad, K. Sheshyekani, Estimation of energy stress of surge arresters considering the high-frequency behavior of grounding systems. IEEE Trans. Electromagn. Compat. **60**(4), 917–925 (2018)
159. J. Gholinezhad, R. Shariatinasab, Time-domain modeling of tower-footing grounding systems based on impedance matrix. IEEE Trans. Power Delivery **34**(3), 910–918 (2019)
160. J. Gholinezhad, R. Shariatinasab, Interfacing electromagnetic model of tower-footing impedance with the emtp software package. Electr. Power Energy Syst. **105**, 394–403 (2019)
161. B. Salarieh, H.M. Jeewantha De Silva, B. Kordi, Electromagnetic transient modeling of grounding electrodes buried in frequency dependent soil with variable water content. Electr. Power Syst. Res. **189**, 106595 (2020)
162. A.R.J. de Arao, J.S.L. Colqui, C.M. de Seixas, S. Kurokawa, B. Salarieh, J.P. Filho, B. Kordi, Computation of ground potential rise and grounding impedance of simple arrangement of electrodes buried in frequency-dependent stratified soil. Electr. Power Syst. Res. **198**, 107364 (2021)
163. A. Mehri, R. Ghanizadeh, M. Beiraghi, Transient analysis of grounding electrodes in a two-layer soil considering frequency dependence of soil. Electr. Eng. **104**, 1019028 (2022)

164. P. Sarajcev, S. Vujevic, D. Lovric, Interfacing harmonic electromagnetic models of grounding systems with the emtp-atp software package. Renew. Energy **68**, 163–170 (2014)
165. H. Chen, Y. Du, Lightning grounding grid model considering both the frequency-dependent behavior and ionization phenomenon. IEEE Trans. Electromagn. Compat. **61**(1), 157–165 (2019)
166. R. Alipio, D. Concei, A. De Conti, K. Yamamoto, R.N. Dias, S. Visacro, A comprehensive analysis of the effect of frequency-dependent soil electrical parameters on the lightning response of wind-turbine grounding systems. Electr. Power Syst. Res. **175**, 105927 (2019)
167. A. De Conti, R. Alipio, Single-port equivalent circuit representation of grounding systems based on impedance fitting. IEEE Trans. Electromagn. Compat. **61**(5), 1683–1685 (2019)

Chapter 2
Fundamentals of Numerical Calculation Techniques I

Spatial 1-d Galerkin Moment Method and Boundary Element Method & General Analytical Calculation Formulas for Mutual Impedance and Mutual Inductance.

2.1 Preface

In the electromagnetic field numerical simulation methods for substation grounding problems, both steady-state and transient problems, including frequency domain and time domain methods, as well as equipotential and unequal potential models, all have a terrifying problem of huge computational complexity.

Fortunately, the conductor rods that make up the substation grounding system have relatively small conductor radii. Therefore, the condition of thin wire approximation can be adopted, and the surface current flowing along the 3-d cylindrical surface in space can be approximately regarded as the line current flowing on the 1-d straight line in space, so that the spatial 3-d Galerkin MM and BEM can be reduced to the spatial 1-d Galerkin MM and BEM, which successfully reduces the calculation amount while maintaining the calculation accuracy.

It must be pointed out here that the matrix equations formed by spatial 1-d Galerkin MM and BEM include 2-d spatial line integration formulas for mutual impedance and mutual inductance. These 2-d spatial line integration formulas are undoubtedly computable numerically, but the computational cost and time are enormous. However, these 2-d spatial line integration formulas can be analytically calculated and effectively overcome this problem. This is also one of the fundamental foundations of the electromagnetic field numerical methods introduced in this book.

Z.-X. Li, *Numerical Simulation Algorithm of Electromagnetic Field for Grounding Problems in Power System Substation Grounding Grids*,
https://doi.org/10.1007/978-981-96-0904-8_2

2.2 Introduction

Due to the insufficient development of computer technology at present, the simulation calculation of grounding problems in substation grounding grids still needs to consider the issues of computational complexity and speed. Therefore, so far, most numerical simulation calculations related to the grounding problem of substation grounding grids have considered the assumption of thin wire conductor approximation method. Whether it is the mathematical models of equipotential [1–3] and unequal potential [4–6] for grounding problem in DC or low-frequency AC fields, or the frequency domain [7–9] and time domain [10, 11] mathematical models in lightning protection grounding transient numerical simulation methods. Several of these literature [1–3] extensively discuss how the assumption of thin wire conductors can reduce computational complexity.

2.3 Spatial 1-d Galerkin MM and BEM

2.3.1 *Frequency Domain Spatial 1-d Scalar Galerkin MM and BEM*

Because the thin wire approximation condition can be used, the frequency domain spatial 3-d Galerkin MM and BEM can reduce the dimension and can be approximated to the frequency domain spatial 1-d Galerkin MM and BEM, thus reducing the calculation amount and improving the calculation speed while maintaining a certain calculation accuracy.

Frequency domain spatial 1-d Scalar Galerkin Weighted Residual Method

- Frequency domain spatial 3-d Scalar Galerkin Weighted Residual Method
  In the numerical simulation of the grounding system including floating electrode buried in horizontal N_e-layer earth model, the grounding system including floating electrode is divided into small cylindrical segments. First, it can be supposed the system consists of total N_b segments of cylindrical bars, and consider the surface leakage currents density J_{s_ϑ}, $(\vartheta = 1, \ldots, N_b)$ emanating from the surface of each segment as unknowns. Next, construct an equation system, and suppose the grounding system and floating electrode are equal-potential. In this way, the currents can be obtained by solving the equation system. Finally the voltage (Reference point at infinity) at any points can be calculated by the contributions of all the leakage currents.
  The voltage with zero potential reference point at infinity on the surface of each small bar can be calculated as

$$\varphi_{s_\theta}(\bar{r}_\theta) = \sum_{\vartheta=1}^{N_b} \int_{S_\vartheta} G_{\xi\zeta}^{\varphi f}(\bar{r}_\theta, \bar{r}_\vartheta) J_{s_\vartheta}(\bar{r}_\vartheta) ds_\vartheta \tag{2.1}$$

where $\xi = 0, 1, \ldots, N_e$; $\zeta = 0, 1, \ldots, N_e$. $\theta = 1, \ldots, N_b$. S_ϑ and S_θ is the area of the ϑ-th and θ-th small cylindrical segments, respectively. $\bar{r}_\theta$ and $\bar{r}_\vartheta$ are the radius vector of the field point and the source point, respectively. $G^{\varphi f}_{\xi\zeta}(\bar{r}_\theta, \bar{r}_\vartheta)$ is frequency domain monopole's Green's function, which has been fully discussed in Chap. 3. s_ϑ is a parameter variable distributed along the surface of ϑ-th cylindrical conductor.

For a given set of M order basic functions $\{[\bar{N}^n_{s_\vartheta}(\bar{r}_\vartheta)], n = 1, \ldots, M\}$ defined on S_ϑ, and a given set of N_b 3-d boundary elements $\{S_\vartheta, \vartheta = 1, \ldots, N_b\}$, the unknown surface leakage current density $[\bar{J}_{s_\vartheta}(\bar{r}_\vartheta)]$ on each segments, and the earthing electrode surface s_ϑ can be discretized in the form

$$[\bar{J}_{s_\vartheta}(\bar{r}_\vartheta)] = \sum_{n=1}^{M} N^n_{s_\vartheta}(\bar{r}_\vartheta) J^n_{s_\vartheta} = [\bar{N}^n_{s_\vartheta}(\bar{r}_\vartheta)][\bar{J}^n_{s_\vartheta}] \tag{2.2}$$

$$S = \bigcup_{\vartheta=1}^{N_b} S_\vartheta \tag{2.3}$$

where S is the surface area of the entire grounding grid cylindrical conductor.

Further, to apply the scalar Galerkin weighted residual method (SGWRM) [12, 13], consider using test functions, which are set of M trial functions $\{[\bar{W}^m_{s_\theta}(\bar{r}_\theta)], m = 1, \ldots, M\}$ defined on S_θ, imposing Eq. (2.1) is verified in the sense of weight residuals, weaker variational form can be obtained as

$$\int_{S_\theta} [\bar{W}^m_{s_\theta}(\bar{r}_\theta)][\bar{\varphi}_{s_\theta}(\bar{r}_\theta)] ds_\theta = \sum_{\vartheta=1}^{N} \int_{S_\theta} [\bar{W}^m_{s_\theta}(\bar{r}_\theta)] \int_{S_\vartheta} G^{\varphi f}_{\xi\zeta}(\bar{r}_\vartheta, \bar{r}_\theta)[\bar{N}^n_{s_\vartheta}(\bar{r}_\vartheta)][\bar{J}^n_{s_\vartheta}] ds_\vartheta ds_\theta \tag{2.4}$$

$$\int_{S_\theta} [\bar{W}^m_{s_\theta}(\bar{r}_\theta)] ds_\theta [\bar{\varphi}_{s_\theta}(\bar{r}_\theta)] = \sum_{\vartheta=1}^{N} \int_{S_\theta} \int_{S_\vartheta} G^{\varphi f}_{\xi\zeta}(\bar{r}_\vartheta, \bar{r}_\theta)[\bar{W}^m_{s_\theta}(\bar{r})_\theta][\bar{N}^n_{s_\vartheta}(\bar{r}_\vartheta)] ds_\vartheta ds_\theta [\bar{J}^n_{s_\vartheta}] \tag{2.5}$$

here $\theta = 1, \ldots, N_b$, S_ϑ and S_θ are the surface of the ϑ-th and θ-th segments, respectively, and ds_ϑ and ds_θ are the infinite small surface elements of the cylinder of the segments, respectively.

Refer to Fig. (2.1)a, It should be pointed out that spatial 3-d discretizations required to solve the above equations in real cases (grounding grids) imply an extremely large number of degrees of freedom. Taking into account that the coefficients matrix in Eq. (2.5) is not sparse, and that spatial 3-d surface integration in Eq. (2.5) must be performed twice over the electrode surface, some reasonable simplifications must be introduced to reduce computational cost under acceptable level.

- Approximated frequency domain spatial 1-d SGWRM

The conductors are subject to the thin-wire approximation, assuming that conductors radius is much smaller than the wavelength and the wire length is much greater than the radius. This assumption enables to approximate the total current in the

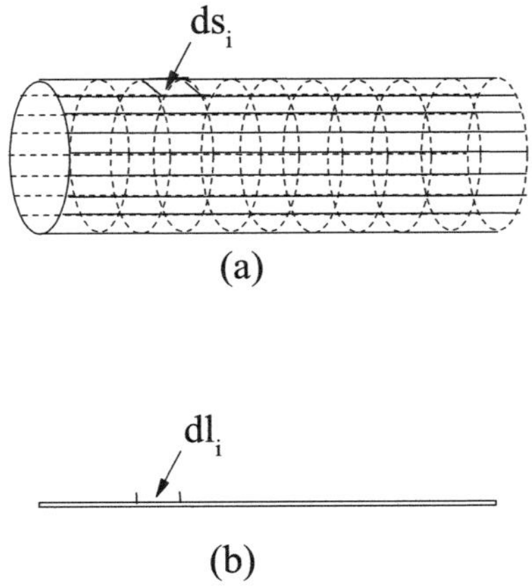

Fig. 2.1 Discrete surface element and discrete line element

conductors as filamentary line current in the conductors' axis. Also the current on open end points is assumed to be zero. It is reasonable to suppose that the leakage current density is constant around the cross section of the cylindrical electrode [3, 14], and seems not restrictive whatsoever if we take into account the real geometry of grounding grids.

Let l_ϑ, ($\vartheta = 1, 2, \ldots, N_b$) be the whole set of axial lines of the buried conductors, and suppose all diameter of conductor segments are same, and $J_{l_\vartheta}(\bar{r}_\vartheta)$ be the approximated leakage current density at this point (assumed uniform around the cross section) in each segment, we can rewrite Eq. (2.1) in the form

$$\varphi_{l_\theta}(\bar{r}_\theta) = \sum_{\vartheta=1}^{N_b} \int_{l_\vartheta} G_{\xi\zeta}^{\varphi^f}(\bar{r}_\theta, \bar{r}_\vartheta) J_{l_\vartheta}(\bar{r}_\vartheta) dr_\vartheta \tag{2.6}$$

where $\theta = 1, \ldots, N_b$. r_ϑ is a parameter variable distributed along the surface of the ϑ-th thin conductor segment

For given sets of spatial 1-d SGWRM basic functions $\{l_\vartheta; \vartheta = 1, \ldots, N_b\}$, and trial functions $\{\bar{N}_{l_\vartheta}^n(\bar{r}_\vartheta), 1, \ldots, M\}$ defined on l_ϑ, the whole set of axial lines of the buried conductors l_ϑ and the unknown leakage current density $J_{l_\vartheta}^n(\hat{r}_\vartheta)$ of each segment can be discretized as

$$[\bar{J}_{l_\vartheta}(\bar{r}_\vartheta)] = \sum_{n=1}^{M} N_{l_\vartheta}^n(\bar{r}_\vartheta) J_{l_\vartheta}^n = [\bar{N}_{l_\vartheta}^n(\bar{r}_\vartheta)][J_{l_\vartheta}^n] \tag{2.7}$$

$$l = \bigcup_{\vartheta=1}^{N_b} l_\vartheta \tag{2.8}$$

where l is the entire length of the conductor section of the thin wire grounding grid.

Finally, for a given set $\{\bar{W}_{l_\theta}^m(\bar{r}_\theta);\ m = 1, \ldots, M\}$ of test functions defined on l_θ, variational statement Eq. (2.6) is reduced to the linear equations system

$$\int_{l_\theta} [\bar{W}_{l_\theta}^m(\bar{r}_\theta)] dr_\theta [\bar{\varphi}_{l_\theta}(\bar{r}_\theta)] = \sum_{\vartheta=1}^{N_b} \int_{l_\theta} \int_{l_\vartheta} G_{\xi\zeta}^{\varphi^f}(\bar{r}_\theta, \bar{r}_\vartheta) [\bar{W}_{l_\theta}^m(\bar{r}_\theta)] [\bar{N}_{l_\vartheta}^n(\bar{r}_\theta)] dr_\vartheta dr_\theta [\bar{J}_{l_\vartheta}^n] \tag{2.9}$$

where $\theta = 1, \ldots, N_b$.

Refer to Fig. 2.1a, b compared to the spatial 3-d SGWRM, the computational complexity of the spatial 1-d SGWRM is significantly reduced, greatly improving the computational speed of the algorithm while ensuring a certain degree of accuracy.

Frequency domain spatial 1-d scalar Galerkin MM and BEM

With the spatial 1-d SGWRM, then, different test functions and basis functions will be used to generate corresponding spatial 1-d Galerkin MM and BEM. This part of the content will be introduced and discussed in detail in the corresponding chapters later.

2.3.2 *Time Domain Spatial 1-d Scalar Galerkin MM and BEM*

Same as frequency-domain Galerkin MM and BEM, because the thin wire approximation condition can also be used in time domain, the time domain spatial 3-d Galerkin MM and BEM can reduce the dimension and can be approximated to the time domain spatial 1-d Galerkin MM and BEM, thus reducing the calculation amount and improving the calculation speed while maintaining a certain calculation accuracy.

Time domain spatial 1-d SGWRM

- Time domain spatial 3-d SGWRM

  In the horizontal layered N_e-layer grounding model, in the numerical simulation of the earthing system including the floating electrode, the grounding system including the floating electrode is divided into small cylindrical segments. Like frequency case, first, it can be supposed the system consists of total N_b segments of cylindrical bars, and consider the surface leakage currents density $J_{s_\vartheta}(t)$, $(\vartheta = 1, \ldots, N_b)$ emanating from the surface of each segment as unknowns. Next, construct an equation system, and suppose the grounding system and floating electrode are equal-potential. In this way, the currents can be obtained by solving the equation system. Finally the voltage (zero potential reference point at infinity) at any points can be calculated by the contributions of all the leakage currents.

  The voltage on the surface of each small bar can be calculated as

$$\varphi_{s_\theta}(\bar{r}_\theta, t) = \sum_{\vartheta=1}^{N_b} \int_{S_\vartheta} G_{\xi\zeta}^{\varphi^t}(\bar{r}_\theta, \bar{r}_\vartheta) J_{s_\vartheta}(\bar{r}_\vartheta, t) ds_\vartheta \tag{2.10}$$

where $\theta = 1, \ldots, N_b$. S_ϑ and S_θ is the area of the ϑth and θth small cylindrical segments, respectively. $\bar{r}_\theta$ and $\bar{r}_\vartheta$ are the radius vector of the field point and the source point, respectively. $G^{\varphi^t}_{\xi\zeta}(\bar{r}_\theta, \bar{r}_\vartheta)$ is time domain monopole's Green's function, which will be introduced next chapter.
For a given set of M trial functions $\{[\bar{N}^n_{s_\vartheta}(\bar{r}_\vartheta)], n = 1, \ldots, M\}$ defined on S_ϑ, and a given set of N_b 3-d boundary elements $\{S_\vartheta, \vartheta = 1, \ldots, N_b\}$, the unknown surface leakage current density $[\bar{J}_{s_\vartheta}(\bar{r}_\vartheta, t)]$ on each segments, and the earthing electrode surface S_ϑ can be discretized in the form

$$[\bar{J}_{s_\vartheta}(\bar{r}_\vartheta, t)] = \sum_{n=1}^{M} N^n_{s_\vartheta}(\bar{r}_\vartheta) J^n_{s_\vartheta}(t) = [\bar{N}^n_{s_\vartheta}(\bar{r}_\vartheta)][\bar{J}^n_{s_\vartheta}(t)] \tag{2.11}$$

$$S = \bigcup_{\vartheta=1}^{N_b} S_\vartheta \tag{2.12}$$

Further, like frequency domain case, to apply the SGWRM, consider using test functions,, which are set of M trial functions $\{[\bar{W}^m_{s_\theta}(\bar{r}_\theta)], m = 1, \ldots, M\}$ defined on s_θ, imposing Eq. (2.10) is verified in the sense of weight residuals, weaker variational form can be obtained as

$$\int_{S_\theta} [\bar{W}^m_{s_\theta}(\bar{r}_\theta)][\bar{\varphi}_{s_\theta}(\bar{r}_\theta, t)] ds_\theta = \sum_{\vartheta=1}^{N} \int_{S_\theta} [\bar{W}^m_{s_\theta}(\bar{r}_\theta)] \int_{S_\vartheta} G^{\varphi^t}_{\xi\zeta}(\bar{r}_\vartheta, \bar{r}_\theta)[\bar{N}^n_{s_\vartheta}(\bar{r_\vartheta})][\bar{J}^n_{s_\vartheta}(t)] ds_\vartheta ds_\theta \tag{2.13}$$

$$\int_{S_\theta} [\bar{W}^m_{s_\theta}(\bar{r}_\theta)] ds_\theta [\bar{\varphi}_{s_\theta}(\bar{r}_\theta, t)] = \sum_{\vartheta=1}^{N} \int_{S_\theta} \int_{S_\vartheta} G^{\varphi^t}_{\xi\zeta}(\bar{r}_\vartheta, \bar{r}_\theta)[\bar{W}^m_{s_\theta}(\bar{r})_\theta][\bar{N}^n_{s_\vartheta}(\bar{r}_\vartheta)] ds_\vartheta ds_\theta [\bar{J}^n_{s_\vartheta}(t)] \tag{2.14}$$

here $\theta = 1, \ldots, N_b$, S_ϑ and S_θ are the surface of the ϑth and θth segments, respectively, and ds_ϑ and ds_θ are the infinite small surface elements of the cylinder of the segments, respectively.
Just like frequency case, tt could also be pointed out that 3-d discretizations required to solve the above equations in real cases (grounding grids) imply an extremely large number of degrees of freedom. Taking into account that the coefficients matrix in Eq. (2.14) is not sparse, and that 3-d surface integration in Eq. (2.14) must be performed twice over the electrode surface, some reasonable simplifications must be introduced to reduce computational cost under acceptable level.

- Approximated time domain spatial 1-d SGWRM
  Like frequency case, it is reasonable to suppose that the leakage current density is constant around the cross section of the cylindrical electrode [3, 14], and seems not restrictive whatsoever if we take into account the real geometry of grounding grids. Assuming that the radius of the conductor is much smaller than the wavelength and the length of the conductor is much larger than the radius, the conductor follows the thin line approximation. This assumption can approximate the total current in a conductor to the filamentous line current on the conductor axis. Furthermore, it is assumed that the current at the open endpoint is zero.

Let l_ϑ, $(\vartheta = 1, 2, \ldots, N_b)$ be the whole set of axial lines of the buried conductors, and suppose all diameter of conductor segments are same, and $J_{l_\vartheta}(\bar{r}_\vartheta, t)$ be the approximated line leakage current density at this point (assumed uniform around the cross section) in each segment, we can rewrite Eq. (2.1) in the form

$$\varphi_{l_\theta}(\bar{r}_\theta, t) = \sum_{\vartheta=1}^{N_b} \int_{l_\vartheta} G_{\xi\zeta}^{\varphi^t}(\bar{r}_\theta, \bar{r}_\vartheta) J_{l_\vartheta}(\bar{r}_\vartheta, t) dr_\vartheta \tag{2.15}$$

where $\theta = 1, \ldots, N_b$.

For given sets of spatial 1-d SGWRM basic functions $\{l_\vartheta; \vartheta = 1, \ldots, N_b\}$, and trial functions $\{\bar{N}_{l_\vartheta}^n(\bar{r}_\vartheta), 1, \ldots, M\}$ defined on l_ϑ, the whole set of axial lines of the buried conductors l_ϑ and the unknown leakage current density $J_{l_\vartheta}^n(\bar{r}_\vartheta)$ of each segment can be discretized as

$$[\bar{J}_{l_\vartheta}(\bar{r}_\vartheta, t)] = \sum_{n=1}^{M} N_{l_\vartheta}^n(\bar{r}_\vartheta) J_{l_\vartheta}^n = [\bar{N}_{l_\vartheta}^n(\bar{r}_\vartheta)][J_{l_\vartheta}^n(t)] \tag{2.16}$$

$$l = \bigcup_{\vartheta=1}^{N_b} l_\vartheta \tag{2.17}$$

Finally, for a given set $\{\bar{W}_{l_\theta}^m(\bar{r}_\theta); m = 1, \ldots, M\}$ of test functions defined on l_θ, variational statement Eq. (2.15) is reduced to the linear equations system

$$\int_{l_\theta} [\bar{W}_{l_\theta}^m(\bar{r}_\theta)] dr_\theta [\bar{\varphi}_{l_\theta}(\bar{r}_\theta, t)] = \sum_{\vartheta=1}^{N_b} \int_{L_\theta} \int_{l_\vartheta} G_{\xi\zeta}^{\varphi^t}(\bar{r}_\theta, \bar{r}_\vartheta) [\bar{W}_{l_\theta}^m(\bar{r}_\theta)][\bar{N}_{l_\vartheta}^n(\bar{r'})] dr_\vartheta dr_\theta [\bar{J}_{l_\vartheta}^n(t)] \tag{2.18}$$

where $\theta = 1, \ldots, N_b$.

Like frequency domain case, compared to the spatial 3-d SGWRM, the computational complexity of the spatial 1-d SGWRM is significantly reduced, greatly improving the computational speed of the algorithm while ensuring a certain degree of accuracy.

Time domain spatial 1-d scalar Galerkin MM and BEM

Like frequency domain case, with the time domain spatial 1-d SGWRM, then, different test functions and basis functions will be used to generate corresponding time domain spatial 1-d Galerkin MM and BEM. This part of the content will be introduced and discussed in detail in the corresponding chapters later.

2.3.3 Frequency Domain Spatial 1-d Vector Galerkin MM and BEM

Frequency domain spatial 1-d vector Galerkin weighted residual method

- Frequency domain spatial 3-d vector Galerkin weighted residual method
  A set of interconnected cylindrical thin conductors placed in any position or orientation makes up a grid to form the grounding system. The grounding network's conductors are assumed to be completely buried in a conductive N_e-layer media (earth) with conductivity σ_e and permittivity $\varepsilon_e = \varepsilon_{r_e} \cdot \varepsilon_0$ (here $e = 1, \ldots, N_e$). The grounding grid is divided into N_b pieces of segments that can be studied as elemental units, where the discrete grounding system has N_p nodes.
  If the complex conductivity of the source material is much greater than that of the surrounding medium, the current distribution along the generic cylindrical conductors is given by the following expression:

$$Z_s \bar{I}_s(\bar{r}) + j\omega \bar{A}(\bar{r}) + \nabla\varphi(\bar{r}) = 0 \tag{2.19}$$

where Z_s is the internal impedance of the conductor, $\bar{I}_s(\bar{r})$ is the directional vector current flowing along cylindrical conductor surface, $\bar{A}(\bar{r})$ and $\varphi(\bar{r})$ are the directional vector magnetic potential and the voltage (zero potential reference point at infinity) on surface of the cylindrical conductor respectively, j is a imaginary unit, ω is angular frequency, $\bar{r}$ is location vector along the grid conductor surface.
Generally (2.19) can be solved only using a numerical approach. Dividing the whole conductor system into segments (elementary sources), each one having a length l_ϑ, much smaller than the dimension of conductor grid, Eq. (2.19), for the segment ϑ, gives:

$$Z_{s_\vartheta} I_{s_\vartheta}(\bar{r}_\vartheta) + j\omega \hat{r}_\vartheta \cdot \bar{A}(\bar{r}_\vartheta) + \frac{\partial\varphi(\bar{r}_\vartheta)}{\partial r_\vartheta} = 0 \tag{2.20}$$

where $\vartheta = 1, 2, \ldots, N_b$, $I_{s_\vartheta}(\bar{r}_\vartheta)$ is the branch current along ϑth conductor segment surface. $\bar{r}_\vartheta$ is location vector along ϑth conductors segment; r_ϑ is parameter variable along the ϑ-th cylindrical conductor surface.

$$Z_{s_\vartheta} I_{s_\vartheta}(\bar{r}_\vartheta) + j\omega \sum_{\theta=1}^{N_b} \int_{S_\theta} \hat{r}_\vartheta \cdot \bar{\bar{G}}_{\xi\zeta}^{A^f}(\bar{r}_\vartheta, \bar{r}_\theta) \cdot \bar{I}_{s_\theta}(\bar{r}_\theta) ds_\theta + \sum_{\theta=1}^{N_b} \int_{S_\theta} \frac{\partial}{\partial r_\vartheta} G_{\xi\zeta}^{\varphi^f}(\bar{r}_\vartheta, \bar{r}_\theta) I_{s_\theta}(\bar{r}_\theta) ds_\theta = 0 \tag{2.21}$$

where $\hat{r}_\vartheta$ is the unit director vector along the axis of the cylindrical conductor. $\bar{\bar{G}}_{\xi\zeta}^{A^f}(\bar{r}_\vartheta, \bar{r}_\theta)$ is the frequency domain vector dipole's dyadic Green's function. $G_{\xi\zeta}^{\varphi^f}(\bar{r}_\vartheta, \bar{r}_\theta)$ is the frequency domain scalar monopole's Green's function. $\bar{I}_{s_\theta}(\bar{r}_\theta)$ is a directional vector branch current flowing along the surface of a cylindrical conductor. $I_{s_\theta}(\bar{r}_\theta)$ is a branch current flowing along the surface of a cylindrical conductor. $\bar{r}_\theta$ is location vector along θth conductors segment. s_θ is parameter variable along the θ-th cylindrical conductor surface.

For a given set of M basic functions $\{[\bar{N}^n_{s_\vartheta}(\bar{r}_\vartheta)], n = 1, \ldots, M\}$ defined on S_ϑ, The unknown current $\bar{I}_{s_\vartheta}(\bar{r}_\vartheta)$ along the electrode segment surface S_ϑ is expressed by the sum of a finite number of independent basis functions as below

$$\bar{I}_{s_\vartheta}(\bar{r}_\vartheta) = I_{s_\vartheta}(\bar{r}_\vartheta)\hat{r}_\vartheta = \sum_{n=1}^{M} \bar{N}^n_{s_\vartheta}(\bar{r}_\vartheta) I^n_{s_\vartheta}(\bar{r}_\vartheta)\hat{r}_\vartheta = [N^n_{s_\vartheta}(\bar{r}_\vartheta)][I^n_{s_\vartheta}(\bar{r}_\vartheta)]\hat{r}_\vartheta \quad (2.22)$$

where M is maximum value of independent basis functions, vector Garlerkin's weighted residual method (VGWRM) will be used in the model, so test function $[W^n_{s_\vartheta}(\bar{r}_\vartheta)]$ can obtained as

$$[W^n_{s_\vartheta}(\bar{r}_\vartheta)] = [N^n_{s_\theta}(\bar{r}_\theta)]^T \quad (2.23)$$

where $[W^n_{s_\vartheta}(\bar{r}_\vartheta)]$ is $M \times 1$ vector. The symbol T represents matrix transpose operation.

$$Z_{s_\vartheta} \int_{S_\vartheta} [W^n_{s_\vartheta}(\bar{r}_\vartheta)][N^n_{s_\vartheta}(\bar{r}_\vartheta)] d\tau_\vartheta [I^n_{s_\vartheta}(\bar{r}_\vartheta)]$$

$$+ \sum_{\theta=1}^{N_b} \int_{S_\vartheta} \int_{S_\theta} \hat{r}_\vartheta \cdot j\omega \bar{\bar{G}}^{Af}_{\xi\zeta}(\bar{r}_\vartheta, \bar{r}_\theta) \cdot \hat{r}_\theta [W^n_{s_\vartheta}(\bar{r}_\vartheta)][N^n_{s_\theta}(\bar{r}_\theta)] ds_\theta ds_\vartheta [I^n_{s_\theta}(\bar{r}_\theta)]$$

$$+ \sum_{\theta=1}^{N_b} \int_{S_\vartheta} \int_{S_\theta} G^{\varphi f}_{\xi\zeta}(\bar{r}_\vartheta, \bar{r}_\theta) \frac{\partial [W^n_{s_\vartheta}(\bar{r}_\vartheta, \bar{r}_\vartheta)]}{\partial r_\vartheta} \frac{d N^n_{s_\theta}(\bar{r}_\vartheta, \bar{r}_\theta)}{d r_\theta} ds_\theta ds_\vartheta [I^n_{s_\theta}(\bar{r}_\theta)] = 0 \quad (2.24)$$

- Approximated frequency domain spatial 1-d VGWRM

  Like the SGRWM case, Assuming that the radius of the conductor is much smaller than the wavelength, and that the length of the conductor is much larger than the radius, the conductor follows the fine line approximation. This assumption allows the total current in the conductor to be approximated as the filament line current on the conductor axis. In addition, assume that the current at the open endpoint is zero. If the conductivity of the source material is much greater than that of the surrounding medium, the current distribution along the generic conductors is given by the following expression:

$$\hat{r} \cdot \left[Z_l \bar{I}(\bar{r}) + j\omega \bar{A}(\bar{r}) + \nabla \varphi(\bar{r}) \right] = 0 \quad (2.25)$$

where Z_l is the internal impedance per unit length of the thin conductor, $\bar{A}(\bar{r})$ and $\varphi(\bar{r})$ are the directional vector magnetic potential and the voltage (zero potential reference point at infinity) on surface of the thin conductor respectively, $\hat{r}$ is direction unit vector along the grid thin conductor, $\bar{r}$ is location vector along the grid thin conductor, $\bar{I}(\bar{r})$ is directional vector of unknown current along the grid thin conductor. j is imaginary unit, ω is angular frequency.

Generally (2.25) can be solved only using a numerical approach. Dividing the whole conductor system into segments (elementary sources), each one having a length l_ϑ, much smaller than the dimension of conductor grid, Eq. (2.25), for the segment ϑ, gives:

$$Z_{l_\vartheta} I_{l_\vartheta}(\bar{r}_\vartheta) + \hat{r}_\vartheta \cdot j\omega \bar{A}(\bar{r}_\vartheta) + \frac{\partial \varphi(\bar{r}_\vartheta)}{\partial r_\vartheta} = 0 \tag{2.26}$$

where $\vartheta = 1, 2, \ldots, N_b$. $I_{l_\vartheta}(\bar{r}_\vartheta)$ is the branch current along ϑ-th thin conductor segment axis. $\bar{r}_\vartheta$ is location vector along ϑ-th thin conductors segment; r_ϑ is parameter variable along the ϑ-th thin conductor axis; $\hat{r}_\vartheta$ is the unit director vector along the axis of the thin conductor segment.

$$Z_{l_\vartheta} I_{l_\vartheta}(\bar{r}_\vartheta) + j\omega \sum_{\theta=1}^{N_b} \int_{l_\theta} \hat{r}_\vartheta \cdot \bar{\bar{G}}_{\xi\zeta}^{Af}(\bar{r}_\vartheta, \bar{r}_\theta) \cdot \bar{I}_{l_\theta}(\bar{r}_\theta) dr_\theta$$

$$+ \sum_{\theta=1}^{N_b} \int_{l_\theta} \frac{\partial}{\partial r_\vartheta} G_{\xi\zeta}^{\varphi f}(\bar{r}_\vartheta, \bar{r}_\theta) I_{l_\theta}(\bar{r}_\theta) dr_\theta = 0 \tag{2.27}$$

where $\bar{\bar{G}}_{\xi\zeta}^{Af}(\bar{r}_\vartheta, \bar{r}_\theta)$ is the frequency domain vector dipole's dyadic Green's function, $G_{\xi\zeta}^{\varphi f}(\bar{r}_\vartheta, \bar{r}_\theta)$ is the frequency domain scalar monopole's Green's function. $\bar{I}_{l_\theta}(\bar{r}_\theta)$ is a directional vector branch current flowing along the surface of a thin conductor segment. $I_{l_\theta}(\bar{r}_\theta)$ is the branch current along θ-th thin conductor segment. $\bar{r}_\theta$ is location vector along θ-th conductors segment.
For a given set of M order basic functions $\{[\bar{N}_{l_\theta}^n(\bar{r}_\theta)], n = 1, \ldots, M\}$ defined on l_ϑ, The unknown current $\bar{I}_{l_\theta}(\bar{r}_\theta)$ along the electrode segment surface l_θ is expressed by the sum of a finite number of independent basis functions as below

$$\bar{I}_{l_\theta}(\bar{r}_\theta) = I_{l_\theta}(\bar{r}_\theta)\hat{r}_\theta = \sum_{n=1}^{M} \bar{N}_{l_\theta}^n(\bar{r}_\theta) I_{l_\theta}^n(\bar{r}_\theta)\hat{r}_\theta = [N_{l_\theta}^n(\bar{r}_\vartheta)][I_{l_\theta}^n(\bar{r}_\theta)]\hat{r}_\theta \tag{2.28}$$

where M is maximum value of independent basis functions, VGWRM will be used in the model, so test function $[W_{l_\theta}^n(\bar{r}_\theta)]$ can obtained as

$$[W_{l_\theta}^n(\bar{r}_\theta)] = [N_{l_\theta}^n(\bar{r}_\theta)]^T \tag{2.29}$$

where $[W_{s_\theta}^n(\bar{r}_\theta)]$ is $M \times 1$ vector.

$$Z_{l_\vartheta} \int_{l_\vartheta} [W_{l_\vartheta}^n(\bar{r}_\vartheta)][N_{l_\vartheta}^n(\bar{r}_\vartheta)] dr_\vartheta [I_{l_\vartheta}^n(\bar{r}_\vartheta)]$$

$$+\sum_{\theta=1}^{N_b}\int_{l_\vartheta}\int_{l_\theta}\hat{r}_\vartheta\cdot j\omega\bar{\bar{G}}_{\xi\zeta}^{A^f}(\bar{r}_\vartheta,\bar{r}_\theta)\cdot\hat{r}_\theta[W_{l_\vartheta}^n(\bar{r}_\vartheta)][N_{l_\theta}^n(\bar{r}_\theta)]dr_\theta dr_\vartheta[I_{l_\theta}^n(\bar{r}_\theta)]$$

$$+\sum_{\theta=1}^{N_b}\int_{l_\vartheta}\int_{l_\theta}G_{\xi\zeta}^{\varphi^f}(\bar{r}_\vartheta,\bar{r}_\theta)\frac{\partial[W_{l_\vartheta}^n(\bar{r}_\vartheta,\bar{r}_\vartheta)]}{\partial r_\vartheta}\frac{dN_{l_\theta}^n(\bar{r}_\vartheta,\bar{r}_\theta)}{dr_\theta}dr_\theta dr_\vartheta[I_{l_\theta}^n(\bar{r}_\theta)]=0 \quad (2.30)$$

Compared to the frequency domain spatial 3-d VGWRM, the computational complexity of the frequency domain spatial 1-d VGWRM is significantly reduced, greatly improving the computational speed of the algorithm while ensuring a certain degree of accuracy.

Frequency domain spatial 1-d vector Galerkin MM and BEM

With the spatial 1-d VGWRM, then, different test functions and basis functions will be used to generate corresponding spatial 1-d vector Galerkin MM and BEM. This part of the content will be introduced and discussed in detail in the corresponding chapters later.

2.3.4 Time Domain Spatial 1-d Vector Galerkin MM and BEM

Time domain spatial 1-d VGWRM

- Time domain spatial 3-d VGWRM
  Like frequency domain case, a set of interconnected cylindrical thin conductors placed in any position or orientation makes up a grid to form the grounding system. The grounding grid's conductors are assumed to be completely buried in a conductive N_e-layer media (earth) with conductivity σ_e and permittivity $\varepsilon_e=\varepsilon_{r_e}\cdot\varepsilon_0$ (here $e=1,\ldots,N_e$). The grounding grid is divided into N_b pieces of segments that can be studied as elemental units, where the discrete grounding system has N_p nodes.
  If the complex conductivity of the source material is much greater than that of the surrounding medium, the current distribution along the generic cylindrical conductors is given by the following expression:

$$Z_s\bar{I}_s(\bar{r},t)+\frac{d\bar{A}(\bar{r},t)}{dt}+\frac{\partial\varphi(\bar{r},t)}{\partial r}=0 \quad (2.31)$$

where Z_s is the internal impedance of the conductor, $\bar{I}_s(\bar{r},t)$ is the directional vector current flowing along cylindrical conductor surface, $\bar{A}(\bar{r},t)$ and $\varphi(\bar{r},t)$ are the directional vector magnetic potential and the voltage (zero potential reference point at infinity) on surface of the cylindrical conductor respectively, r is parameter variable along the grid conductor surface.

Generally (2.31) can be solved only using a numerical approach. Dividing the whole conductor system into segments (elementary sources), each one having a length l_ϑ, much smaller than the dimension of conductor grid, Eq. (2.31), for the segment ϑ, gives:

$$Z_{s_\vartheta} I_{s_\vartheta}(\bar{r}_\vartheta, t) + \hat{r}_\vartheta \cdot \frac{d\bar{A}(\bar{r}_\vartheta)}{dt} + \frac{\partial \varphi(\bar{r}_\vartheta)}{\partial r_\vartheta} = 0 \tag{2.32}$$

where $\vartheta = 1, 2, \ldots, N_b$, $I_{s_\vartheta}(\bar{r}_\vartheta, t)$ is the branch current along ϑth conductor segment surface. $\bar{r}_\vartheta$ is location vector along ϑ-th conductors segment; r_ϑ is parameter variable along the ϑ-th cylindrical conductor surface.

$$Z_{s_\vartheta} I_{s_\vartheta}(\bar{r}_\vartheta, t) + \sum_{\theta=1}^{N_b} \int_{S_\theta} \hat{r}_\vartheta \cdot \frac{d\bar{\bar{G}}^{A^t}_{\xi\zeta}(\bar{r}_\vartheta, \bar{r}_\theta) \cdot \bar{I}_{s_\theta}(\bar{r}_\theta, t)}{dt} ds_\theta$$

$$+ \sum_{\theta=1}^{N_b} \int_{S_\theta} \frac{\partial}{\partial r_\vartheta} G^{\varphi^t}_{\xi\zeta}(\bar{r}_\vartheta, \bar{r}_\theta) I_{s_\theta}(\bar{r}_\theta, t) ds_\theta = 0 \tag{2.33}$$

where $\hat{r}_\vartheta$ is the unit director vector along the axis of the cylindrical conductor. $\bar{\bar{G}}^{A^t}_{\xi\zeta}(\bar{r}_\vartheta, \bar{r}_\theta)$ is the time domain vector dipole's dyadic Green's function. $G^{\varphi^t}_{\xi\zeta}(\bar{r}_\vartheta, \bar{r}_\theta)$ is the time domain scalar monopole's Green's function. $\bar{I}_{s_\theta}(\bar{r}_\theta, t)$ is a directional vector branch current flowing along the surface of a cylindrical conductor. $\bar{r}_\vartheta$ and $\bar{r}_\theta$ are location vector along ϑ-th and θ-th conductors segments, respectively.
For a given set of M basic functions $\{[\bar{N}^n_{s_\vartheta}(\bar{r}_\vartheta)], n = 1, \ldots, M\}$ defined on S_ϑ, The unknown current $\bar{I}_{s_\vartheta}(\bar{r}_\vartheta, t)$ along the electrode segment surface S_ϑ is expressed by the sum of a finite number of independent basis functions as below

$$\bar{I}_{s_\vartheta}(\bar{r}_\vartheta, t) = I_{s_\vartheta}(\bar{r}_\vartheta, t)\hat{r}_\vartheta = \sum_{n=1}^{M} \bar{N}^n_{s_\vartheta}(\bar{r}_\vartheta) I^n_{s_\vartheta}(\bar{r}_\vartheta, t)\hat{r}_\vartheta = [N^n_{s_\vartheta}(\bar{r}_\vartheta)][I^n_{s_\vartheta}(\bar{r}_\vartheta, t)]\hat{r}_\vartheta \tag{2.34}$$

where M is maximum value of independent basis functions, VGWRM will be used in the model, so test function $[W^n_{s_\vartheta}(\bar{r}_\vartheta)]$ can obtained as

$$[W^n_{s_\vartheta}(\bar{r}_\vartheta)] = [N^n_{s_\theta}(\bar{r}_\theta)]^T \tag{2.35}$$

where $[W^n_{s_\vartheta}(\bar{r}_\vartheta)]$ is $M \times 1$ vector.

$$Z_{s_\vartheta} \int_{S_\vartheta} [W^n_{s_\vartheta}(\bar{r}_\vartheta)][N^n_{s_\vartheta}(\bar{r}_\vartheta)] d\tau_\vartheta [I^n_{s_\vartheta}(\bar{r}_\vartheta, t)]$$

$$+\sum_{\theta=1}^{N_b}\int_{S_\vartheta}\int_{S_\theta}\hat{r}_\vartheta\cdot\frac{d}{dt}\left[\bar{\bar{G}}_{\xi\zeta}^{A^t}(\bar{r}_\vartheta,\bar{r}_\theta)\cdot\hat{r}_\theta[I_{s_\theta}^n(\bar{r}_\theta,t)]\right][W_{s_\vartheta}^n(\bar{r}_\vartheta)][N_{s_\theta}^n(\bar{r}_\theta)]ds_\theta ds_\vartheta$$

$$+\sum_{\theta=1}^{N_b}\int_{S_\vartheta}\int_{S_\theta}G_{\xi\zeta}^{\varphi^t}(\bar{r}_\vartheta,\bar{r}_\theta)\frac{\partial[W_{s_\vartheta}^n(\bar{r}_\vartheta,\bar{r}_\vartheta)]}{\partial r_\vartheta}\frac{dN_{s_\theta}^n(\bar{r}_\vartheta,\bar{r}_\theta)}{dr_\theta}ds_\theta ds_\vartheta[I_{s_\theta}^n(\bar{r}_\theta,t)]=0 \tag{2.36}$$

- Approximated time domain spatial 1-d VGWRM
  Like frequency domain case, assuming that the radius of the conductor is much smaller than the wavelength, and that the length of the conductor is much larger than the radius, the conductor follows the fine line approximation. This assumption allows the total current in the conductor to be approximated as the filament line current on the conductor axis. In addition, assume that the current at the open endpoint is zero.
  If the complex conductivity of the source material is much greater than that of the surrounding medium, the current distribution along the generic thin wire conductors is given by the following expression:

$$Z_l\bar{I}_l(\bar{r},t)+\hat{r}\cdot\frac{d\bar{A}(\bar{r},t)}{dt}+\frac{\partial\varphi(\bar{r},t)}{\partial r}=0 \tag{2.37}$$

where Z_l is the internal impedance of the thin conductor, $\bar{I}_l(\bar{r},t)$ is the directional vector current flowing along thin conductor axis, $\bar{A}(\bar{r},t)$ and $\varphi(\bar{r},t)$ are the directional vector magnetic potential and the voltage (zero potential reference point at infinity) on surface of the thin conductor respectively, $\hat{r}$ is unit directional vector along thin conductor axis, r is parameter variable along the grid thin conductor axis. Generally (2.37) can be solved only using a numerical approach. Dividing the whole conductor system into segments (elementary sources), each one having a length l_ϑ, much smaller than the dimension of conductor grid, Eq. (2.37), for the ϑ-th segment, gives:

$$Z_{l_\vartheta}I_{l_\vartheta}(\bar{r}_\vartheta,t)+\hat{r}_\vartheta\cdot\frac{d\bar{A}(\bar{r}_\vartheta,t)}{dt}+\frac{\partial\varphi(\bar{r}_\vartheta,t)}{\partial r_\vartheta}=0 \tag{2.38}$$

where $\vartheta=1,2,\ldots,N_b$, $I_{l_\vartheta}(\bar{r}_\vartheta,t)$ is the branch current along ϑ-th thin conductor segment axis. $\bar{r}_\vartheta$ is location vector along ϑ-th thin conductors segment; r_ϑ is parameter variable along the ϑ-th thin conductor axis; $\hat{r}_\vartheta$ is the unit director vector along the axis of the thin conductor segment.

$$Z_{l_\vartheta}I_{l_\vartheta}(\bar{r}_\vartheta,t)+\sum_{\theta=1}^{N_b}\int_{l_\theta}\hat{r}_\vartheta\cdot\frac{d\bar{\bar{G}}_{\xi\zeta}^{A^t}(\bar{r}_\vartheta,\bar{r}_\theta)\cdot\bar{I}_{l_\theta}(\bar{r}_\theta,t)}{dt}dr_\theta$$

$$+\sum_{\theta=1}^{N_b}\int_{l_\theta}\frac{\partial}{\partial r_\vartheta}G_{\xi\zeta}^{\varphi^t}(\bar{r}_\vartheta,\bar{r}_\theta)I_{l_\theta}(\bar{r}_\theta,t)dr_\theta=0 \tag{2.39}$$

where $\bar{\bar{G}}^{t}_{A}(\bar{r}_\vartheta, \bar{r}_\theta)$ is the time domain vector dipole's dyadic Green's function, $G^{t}_{\varphi}(\bar{r}_\vartheta, \bar{r}_\theta)$ is the time domain scalar monopole's Green's function. $\bar{I}_{l_\theta}(\bar{r}_\theta, t)$ is a directional vector branch current flowing along θ-th thin conductor segment. $I_{l_\theta}(\bar{r}_\theta, t)$ is the branch current along θ-th thin conductor segment. $\bar{r}_\theta$ is location vector along θ-th conductors segment.

For a given set of M basic functions $\{[\bar{N}^{n}_{l_\vartheta}(\bar{r}_\vartheta)], n = 1, \ldots, M\}$ defined on l_ϑ, The unknown current $\bar{I}_{l_\vartheta}(\bar{r}_\vartheta)$ along the electrode segment surface l_ϑ is expressed by the sum of a finite number of independent basis functions as below

$$\bar{I}_{l_\vartheta}(\bar{r}_\vartheta, t) = I_{l_\vartheta}(\bar{r}_\vartheta, t)\hat{r}_\vartheta = \sum_{n=1}^{M} \bar{N}^{n}_{l_\vartheta}(\bar{r}_\vartheta) I^{n}_{l_\vartheta}(\bar{r}_\vartheta, t)\hat{r}_\vartheta = [N^{n}_{l_\vartheta}(\bar{r}_\vartheta)][I^{n}_{l_\vartheta}(\bar{r}_\vartheta, t)]\hat{r}_\vartheta \quad (2.40)$$

where M is maximum value of independent basis functions, VGWRM will be used in the model, so test function $[W^{n}_{l_\vartheta}(\bar{r}_\vartheta)]$ can obtained as

$$[W^{n}_{l_\vartheta}(\bar{r}_\vartheta)] = [N^{n}_{l_\theta}(\bar{r}_\theta)]^{T} \quad (2.41)$$

where $[W^{n}_{s_\vartheta}(\bar{r}_\vartheta)]$ is $M \times 1$ vector.

$$Z_{l_\vartheta} \int_{l_\vartheta} [W^{n}_{l_\vartheta}(\bar{r}_\vartheta)][N^{n}_{l_\vartheta}(\bar{r}_\vartheta)] dr_\vartheta [I^{n}_{l_\vartheta}(\bar{r}_\vartheta, t)]$$

$$+ \sum_{\theta=1}^{N_b} \int_{l_\vartheta} \int_{l_\theta} \hat{r}_\vartheta \cdot \frac{d}{dt} \left[\bar{\bar{G}}^{A^t}_{\xi\zeta}(\bar{r}_\vartheta, \bar{r}_\theta) \cdot \hat{r}_\theta [I^{n}_{l_\theta}(\bar{r}_\theta, t)] \right] [W^{n}_{l_\vartheta}(\bar{r}_\vartheta)][N^{n}_{l_\theta}(\bar{r}_\theta)] dr_\theta dr_\vartheta$$

$$+ \sum_{\theta=1}^{N_b} \int_{l_\vartheta} \int_{l_\theta} G^{\varphi^t}_{\xi\zeta}(\bar{r}_\vartheta, \bar{r}_\theta, t) \frac{\partial [W^{n}_{l_\vartheta}(\bar{r}_\vartheta, \bar{r}_\vartheta)]}{\partial r_\vartheta} \frac{d N^{n}_{l_\theta}(\bar{r}_\vartheta, \bar{r}_\theta)}{d r_\theta} dr_\theta dr_\vartheta [I^{n}_{l_\theta}(\bar{r}_\theta, t)] = 0 \quad (2.42)$$

Like frequency case, compared to the time domain spatial 3-d VGWRM, the computational complexity of the time domain spatial 1-d VGWRM is significantly reduced, greatly improving the computational speed of the algorithm while ensuring a certain degree of accuracy.

Time domain spatial 1-d vector Galerkin MM and BEM

Like frequency domain case, with the time domain spatial 1-d VGWRM, then, different test functions and basis functions will be used to generate corresponding time domain spatial 1-d vector Galerkin MM and BEM. This part of the content will be introduced and discussed in detail in the corresponding chapters later.

2.4 General Analytical Calculation Formulas for Mutual Impedance and Mutual Inductance

2.4.1 Introduction

Analytical formulas for mutual resistance has been developed in [15] for zero order basis function, which can accelerate the numerical calculation and reduce the errors caused by numerical integration. The general analytical mutual resistance formulas for high order basis function has been developed several years later [3], However, the procedure offered in that paper for obtaining these formulas is simplified, and its method is also based on direct current theory. The general formulas therefore only work in real space. Since quasi-static electrical field theory is used to deal with the grounding system, which lead to application of the quasi-static complex image method (QSCIM) into the closed form of Green's function, in this way, analytical formulas for mutual impedance have been developed in complex space instead of real space in [16–19]. However, only low order (zero and one order) basis function's general formulas for mutual impedance are discussed in these papers. So high order basis function's general formulas for mutual impedance should be further developed.

2.4.2 General Analytical Calculation Formulas for Mutual Impedance and Mutual Inductance

When a substation grounding grid is buried in multi-layer horizontal layered soil, considering the need for electromagnetic field numerical simulation, the entire grounding grid is discretized according to numerical calculation requirements, consisting of N_b conductor segments and N_p endpoints.

The element $Z_{\vartheta\theta}^{nm}$ in (2.43) is the mutual impedance or mutual induction coefficient between a pair of segments with leakage currents or branch currents in the grounding system. For a general case, it has been defined in (2.43) as below

$$Z_{\vartheta\theta}^{nm} = \int_{l_\vartheta}\int_{l_\theta} \frac{t_\vartheta^m t_\theta^n}{\overline{R}} dt_\vartheta dt_\theta, \tag{2.43}$$

Here $n = 0, 1,\ m = 0, 1\ ;\ \vartheta = 1, \ldots, N_b,\ \theta = 1, \ldots, N_b$, and $\overline{R} = |\bar{r} - \bar{r'}|$. We should note that $\overline{R}$ is complex number, so all work here occurs in complex space.

For a pairs of short conductors l_ϑ and l_θ at any position, we can write their line equations as for l_ϑ: $x = x_\vartheta(t_\vartheta) = a_\vartheta^x t_\vartheta + b_\vartheta^x,\ y = y_\vartheta(t) = a_\vartheta^y t_\vartheta + b_\vartheta^y,\ z = z_\vartheta(t_\vartheta) = a_\vartheta^z t_\vartheta + b_\vartheta^z$; and l_θ: $x = x_\theta(t_\theta) = a_\theta^x t_\theta + b_\theta^x,\ y = y_\theta(t_\theta) = a_\theta^y t_\theta + b_\theta^y$, and $z = z_\theta(t_\theta) = a_\theta^z t_\theta + b_\theta^z$. So we have

$$Z_{\vartheta\theta}^{nm} = \int_{l_\vartheta}\int_{l_\theta} \frac{t_\vartheta^m t_\theta^n}{\sqrt{t_\vartheta^2 + 2\beta t_\vartheta + \gamma}} dt_\vartheta dt_\theta, \tag{2.44}$$

Here $\beta = -(a^x_\vartheta a^x_\theta + a^y_\vartheta a^y_\theta + a^z_\vartheta a^z_\theta)t_\theta + (a^x_\vartheta b^x_\vartheta + a^y_\vartheta b^y_\vartheta + a^z_\vartheta b^z_\vartheta) - (a^x_\vartheta b^x_\theta + a^y_\vartheta b^y_\theta + a^z_\vartheta b^z_\theta) = -\xi t_\theta + \zeta$; $\gamma = t_\theta^2 + 2[(a^x_\theta b^x_\theta + a^y_\theta b^y_\theta + a^z_\theta b^z_\theta) - (b^x_\vartheta a^x_\theta + b^y_\vartheta a^y_\theta + b^z_\vartheta a^z_\theta)]$ $t_\theta - 2(b^x_\vartheta b^x_\theta + b^y_\vartheta b^y_\theta + b^z_\vartheta b^z_\theta) + (b^x_\vartheta)^2 + (b^y_\vartheta)^2 + (b^z_\vartheta)^2 + (b^x_\theta)^2 + (b^y_\theta)^2 + (b^z_\theta)^2 + \frac{D}{2} = t_\theta^2 + 2\varphi t_\theta + \chi$.

Furthermore

$$Z^{nm}_{\vartheta\theta} = \sum_{k=0}^{k=n} \binom{n}{k}(-1)^k \int_0^{l_\theta} \beta^k \int_\beta^{l_\vartheta+\beta} \frac{t_\vartheta^{n-k}}{\sqrt{t_\vartheta^2 + \gamma - \beta^2}} dt_\vartheta t_\theta^m dt_\theta, \tag{2.45}$$

- (1) if $n - k \geq 2$

$$Z^{nm}_{\vartheta\theta} = \binom{n}{k} \sum_{k_1=0}^{k_1=k_{a1}} \binom{k_{a1}}{k_1} \sum_{k_2=0}^{k_2=k_{a2}} \binom{k_{a2}}{k_2} \sum_{k_3=0}^{k_3=k_{a3}} \binom{k_{a3}}{k_3} c_{11} \int_{\varphi-l_\vartheta\xi}^{l_\theta+\varphi-l_\vartheta\xi} t_\theta^{k_{a6}} \sqrt{t_\theta^2 + R_2}\, dt_\theta$$

$$-\binom{n}{k} \sum_{k_1=0}^{k_1=k_{a7}} \binom{k_{a7}}{k_1} \sum_{k_2=0}^{k_2=k_{a8}} \binom{k_{a8}}{k_2} c_{12} \int_\varphi^{l_\theta+\varphi} \tau^{k_{a3}} \sqrt{t_\theta^2 + R_1}\, dt_\theta$$

$$-\binom{n}{k} c_{13} \int_0^{l_\theta} t_\theta^m ((1-\xi^2)t_\theta^2 + 2(\varphi + \xi\zeta)t_\theta + \chi - \zeta^2) K_3(n-k-2) dt_\theta,$$

here $k_{a1} = n - k - 1$, $k_{a2} = n - 1 - k_1$, $k_{a3} = n + m - 1 - k_1 - k_2$, $k_{a4} = n - 1 + k - k_1 - k_2 + k_3$, $k_{a5} = k_{a2} - k_2$, $k_{a6} = k_{a3} - k_3$, $k_{a7} = n - 1$, $k_{a8} = n + m - 1 - k_1$, $k_{a9} = n - 1 + k - k_1 + k_2$, $c_{11} = \frac{(-1)^{k_{a4}} l_\vartheta^{k_1} \zeta^{k_2} \xi^{k_{a5}} (\varphi - l_\vartheta \xi)^{k_3}}{n-k}$, $c_{12} = \frac{(-1)^{k_{a9}} \zeta^{k_1} \xi^{k_{a2}} \varphi^{k_2}}{n-k}$ and $c_{13} = \frac{(-1)^k k_{a1}}{n-k}$.

$$K_3(n) = \int_0^{l_\vartheta} \frac{t_\vartheta^n}{\sqrt{t_\vartheta^2 + R^2_{1or2}}} dt_\vartheta, \tag{2.46}$$

$$K_3(n) = \frac{t_\vartheta^{n-1}}{n} \sqrt{t_\vartheta^2 + R^2_{1or2}}\Big|_0^{l_\vartheta} - \frac{(n-1)R^2_{1or2}}{n} K_3(n-2), \tag{2.47}$$

- (a) if $n - k - 2 = 0$

$$\int_0^{l_\theta} t_\theta^m ((1-\xi^2)t_\theta^2 + 2(\varphi + \xi\zeta)t_\theta + \chi - \zeta^2) K_3(0) dt_\theta$$

$$= -\sum_{s=1}^{s=3} \binom{n}{k} \sum_{k_1=0}^{k_1=m+s-1} \binom{m+s-1}{k_1} c_{2(4-s)} \times$$

$$\int_{\varphi-l_\vartheta\xi}^{l_\theta+\varphi-l_\vartheta\xi} t_\theta^{k_{b(4-s)}} \ln(f_2+b_2t_\theta+\zeta+\sqrt{t_\theta^2+R_2})dt_\theta$$

$$=-\binom{n}{k}\sum_{k_1=0}^{k_1=m+2}\binom{m+2}{k_1}c_{21}\int_{\varphi-l_\vartheta\xi}^{l_\theta+\varphi-l_\vartheta\xi} t_\theta^{k_{b1}} \ln(f_2+b_2t_\theta+\zeta+\sqrt{t_\theta^2+R_2})dt_\theta$$

$$-\binom{n}{k}\sum_{k_1=0}^{k_1=m+1}\binom{m+1}{k_1}c_{22}\int_{\varphi-l_\vartheta\xi}^{l_\theta+\varphi-l_\vartheta\xi} t_\theta^{k_{b2}} \ln(f_2+b_2t_\theta+\zeta+\sqrt{t_\theta^2+R_2})dt_\theta$$

$$-\binom{n}{k}\sum_{k_1=0}^{k_1=m}\binom{m}{k_1}c_{23}\int_{\varphi-l_\vartheta\xi}^{l_\theta+\varphi-l_\vartheta\xi} t_\theta^{k_{b3}} \ln(f_2+b_2t_\theta+\zeta+\sqrt{t_\theta^2+R_2})dt_\theta$$

$$+\sum_{s=1}^{s=3}\binom{n}{k}\sum_{k_1=0}^{k_1=m+s-1}\binom{m+s-1}{k_1}c_{2(7-s)}\int_{\varphi}^{l_\theta+\varphi} t_\theta^{k_{b(7-s)}} \ln(f_1+b_1t_\theta+\zeta+\sqrt{t_\theta^2+R_1})dt_\theta,$$

$$+\binom{n}{k}\sum_{k_1=0}^{k_1=m+2}\binom{m+2}{k_1}c_{24}\int_{\varphi}^{l_\theta+\varphi} t_\theta^{k_{b4}} \ln(f_1+b_1t_\theta+\zeta+\sqrt{t_\theta^2+R_1})dt_\theta$$

$$+\binom{n}{k}\sum_{k_1=0}^{k_1=m+1}\binom{m+1}{k_1}c_{25}\int_{\varphi}^{l_\theta+\varphi} t_\theta^{k_{b5}} \ln(f_1+b_1t_\theta+\sqrt{t_\theta^2+R_1})dt_\theta$$

$$+\binom{n}{k}\sum_{k_1=0}^{k_1=m}\binom{m}{k_1}c_{26}\int_{\varphi}^{l_\theta+\varphi} t_\theta^{k_{b6}} \ln(f_1+b_1t_\theta+\sqrt{t_\theta^2+R_1})dt_\theta$$

here $k_{b1}=m+2-k_1,\ k_{b2}=m+1-k_1,\ k_{b3}=m-k_1,\ k_{b4}=m+2-k_1,\ k_{b5}=m+1-k_1,\ k_{b6}=m-k_1,\ c_{21}=\frac{(-1)^{k+k_1}(n-k-1)(1-\xi^2)(\varphi-l_\vartheta\xi)^{k_1}}{n-k},\ c_{22}=\frac{(-1)^{k+k_1}(n-k-1)2(\varphi+\xi\zeta)(\varphi-l_\vartheta\xi)^{k_1}}{n-k},\ c_{23}=\frac{(-1)^{k+k_1}(n-k-1)(\chi-\zeta^2)(\varphi-l_\vartheta\xi)^{k_1}}{n-k},\ c_{24}=\frac{(-1)^{k+k_1}(n-k-1)(1-\xi^2)\varphi^{k_1}}{n-k},\ c_{25}=\frac{(-1)^{k+k_1}(n-k-1)2(\varphi+\xi\zeta)\varphi^{k_1}}{n-k},\ c_{26}=\frac{(-1)^{k+k_1}(n-k-1)(\chi-\zeta^2)\varphi^{k_1}}{n-k}$

- (b) if $n-k-2=1$

$$\int_0^{l_\theta} t_\theta^m((1-\xi^2)t_\theta^2+2(\varphi+\xi\zeta)t_\theta+R_1)K_3(1)dt_\theta$$

$$=-\sum_{s=1}^{s=3}\binom{n}{k}\sum_{k_1=0}^{k_1=m+s-1}\binom{m+s-1}{k_1}c_{3(4-s)}\int_{\varphi-l_\vartheta\xi}^{l_\theta+\varphi-l_\vartheta\xi} t_\theta^{m+s-1-k_1}\sqrt{t_\theta^2+R_2}dt_\theta$$

$$= -\binom{n}{k}\sum_{k_1=0}^{k_1=m+2}\binom{m+2}{k_1}c_{31}\int_{\varphi-l_\vartheta\xi}^{l_\theta+\varphi-l_\vartheta\xi} t_\theta^{m+2-k_1}\sqrt{t_\theta^2+R_2}dt_\theta$$

$$-\binom{n}{k}\sum_{k_1=0}^{k_1=m+1}\binom{m+1}{k_1}c_{32}\int_{\varphi-l_\vartheta\xi}^{l_\theta+\varphi-l_\vartheta\xi} t_\theta^{m+1-k_1}\sqrt{t_\theta^2+R_2}dt_\theta$$

$$-\binom{n}{k}\sum_{k_1=0}^{k_1=m}\binom{m}{k_1}c_{33}\int_{\varphi-l_\vartheta\xi}^{l_\theta+\varphi-l_\vartheta\xi} t_\theta^{m-k_1}\sqrt{t_\theta^2+R_2}dt_\theta$$

$$-\sum_{s=1}^{s=3}\binom{n}{k}\sum_{k_1=0}^{k_1=m+s-1}\binom{m+s-1}{k_1}c_{3(7-s)}\int_{\varphi}^{l_\theta+\varphi} t_\theta^{m+s-1-k_1}\sqrt{t_\theta^2+R_1}dt_\theta,$$

$$-\binom{n}{k}\sum_{k_1=0}^{k_1=m+2}\binom{m+2}{k_1}c_{34}\int_{\varphi}^{l_\theta+\varphi} t_\theta^{m+2-k_1}\sqrt{t_\theta^2+R_1}dt_\theta$$

$$-\binom{n}{k}\sum_{k_1=0}^{k_1=m+1}\binom{m+1}{k_1}c_{35}\int_{\varphi}^{l_\theta+\varphi} t_\theta^{m+1-k_1}\sqrt{t_\theta^2+R_1}dt_\theta$$

$$-\binom{n}{k}\sum_{k_1=0}^{k_1=m}\binom{m}{k_1}c_{36}\int_{\varphi}^{l_\theta+\varphi} t_\theta^{m-k_1}\sqrt{t_\theta^2+R_1}dt_\theta,$$

here $c_{31}=\frac{(-1)^{k+k_1}(n-k-1)(\varphi-l_\vartheta\xi)^{k_1}(1-\xi^2)}{n-k}$, $c_{32}=\frac{(-1)^{k+k_1}(n-k-1)(\varphi-l_\vartheta\xi)^{k_1}2(\varphi+\xi\zeta)}{n-k}$, $c_{33}=\frac{(-1)^{k+k_1}(n-k-1)(\varphi-l_\vartheta\xi)^{k_1}(\chi-\zeta^2)}{n-k}$, $c_{34}=\frac{(-1)^{k+k_1}(n-k-1)\varphi^{k_1}(1-\xi^2)}{n-k}$, $c_{35}=\frac{(-1)^{k+k_1}(n-k-1)\varphi^{k_1}2(\varphi+\xi\zeta)}{n-k}$, $c_{36}=\frac{(-1)^{k+k_1}(n-k-1)\varphi^{k_1}(\chi-\zeta^2)}{n-k}$

- (2) if $n-k=1$

$$Z_{\vartheta\theta}^{nm}=\binom{n}{k}\sum_{k_1=0}^{k_1=k}\binom{k}{k_1}\sum_{k_2=0}^{k_2=k_{d1}}\binom{k_{d1}}{k_2}c_{41}\int_{\varphi-l_\vartheta\xi}^{l_\theta+\varphi-l_\vartheta\xi} t_\theta^{k_{d2}}\sqrt{t_\theta^2+R_2}dt_\theta$$

$$-\binom{n}{k}\sum_{k_1=0}^{k_1=k}\binom{k}{k_1}\sum_{k_2=0}^{k_2=k_{d1}}\binom{k_{d1}}{k_2}c_{42}\int_{\varphi}^{l_\theta+\varphi} t_\theta^{k_{d2}}\sqrt{t_\theta^2+R_1}dt_\theta,$$

here $k_{d1}=m+k-k_1$, $k_{d2}=m+k-k_1-k_2$, $c_{41}=(-1)^{k_2-k_1}(\varphi-l_\vartheta\xi)^{k_2}\zeta^{k_1}\xi^{k-k_1}$, $c_{42}=(-1)^{k_2-k_1}\varphi^{k_2}\zeta^{k_1}\xi^{k-k_1}$

- (3) if $n - k = 0$

$$Z_{\vartheta\theta}^{nm} = \binom{n}{k}\sum_{k_1=0}^{k_1=k}\binom{k}{k_1}\sum_{k_2=0}^{k_2=m}\binom{m}{k_2}c_{51}\int_{\varphi-l_\vartheta\xi}^{l_\theta+\varphi-l_\vartheta\xi} t_\theta^{k_{e1}}\ln(f_2 + b_2t_\theta + \sqrt{t_\theta^2 + R_2})dt_\theta$$

$$-\binom{n}{k}\sum_{k_1=0}^{k_1=k}\binom{k}{k_1}\sum_{k_2=0}^{k_2=m}\binom{m}{k_2}c_{52}\int_{\varphi}^{l_\theta+\varphi} t_\theta^{k_{e1}}\ln(f_1 + b_1t_\theta + \sqrt{t_\theta^2 + R_1})dt_\theta. \quad (2.48)$$

here $k_{e1} = m + k - k_1 - k_2, c_{51} = (-1)^{k_2-k_1}(\varphi - l_\vartheta\xi)^{k_2}(\xi(\varphi - l_\vartheta\xi) + \zeta)^{k_1}\xi^{k-k_1}$, $c_{52} = (-1)^{k_2-k_1}\varphi^{k_2}(\xi\varphi + \zeta)^{k_1}\xi^{k-k_1}$

In all above formulas, there are, $R_2 = l_\vartheta^2 + 2l_\vartheta\zeta + \chi - (\varphi - l_\vartheta\xi)^2$, $R_1 = \chi - \varphi^2$, $f_2 = l_\vartheta + \xi(\varphi - l_\vartheta\xi) + \zeta$, $b_2 = -\xi$, $f_1 = \xi\varphi + \zeta$, $b_1 = -\xi$.

All above formulas have shown that $Z_{\vartheta\theta}^{nm}$ consists of two kind of pieces of basis integral formulas, they are

$$I_1(n) = \int_{l_1}^{l_2} t^n \ln(f + bt + \sqrt{t^2 + R^2})dt, \quad (2.49)$$

$$I_2(n) = \int_{l_1}^{l_2} t^n \sqrt{t^2 + R^2}dt, \quad (2.50)$$

and the two pieces of basis integral formulas can be deduced out to be analytical formulas, we can see

- (i) $I_1(n)$ case
  - (i-1) $b^2 = 1$ case;

$$I_1(n) = \frac{l_2^{n+1}}{n+1}\ln(bl_2 + \sqrt{l_2^2 + R^2}) - \frac{l_1^{n+1}}{n+1}\ln(bl_1 + \sqrt{l_1^2 + R^2})$$
$$-\frac{b}{n+1}K_3(n+1), \quad (2.51)$$

  - (i-2) $b^2 \neq 1$ case.

$$I_1(n) = \frac{l_2^{n+1}}{n+1}\ln(f + bl_2 + \sqrt{l_2^2 + R^2}) - \frac{l_1^{n+1}}{n+1}\ln(f + bl_1 + \sqrt{l_1^2 + R^2})$$
$$+\frac{bf}{n+1}K_1(n+1) - \frac{1-b^2}{n+1}K_1(n+2) - \frac{bR^2}{n+1}K_2(n+1)$$
$$+\frac{f}{n+1}K_2(n+2), \quad (2.52)$$

- (ii) $I_2(n)$ case

$$I_2(n) = K_3(n+2) + R^2 K_3(n) \tag{2.53}$$

$$K_1(n) = \frac{l_2^{n-1} - l_1^{n-1}}{n-1} + \frac{1}{1-b^2}\left[2bfK_1(n-1) - gK_1(n-2)\right], \tag{2.54}$$

$$K_1(1) = \frac{1}{2}\ln\frac{l_2^2(1-b^2) - 2fbl_2 + g}{l_1^2(1-b^2) - 2fbl_1 + g} + \frac{fb}{1-b^2}K_1(0), \tag{2.55}$$

$$K_1(0) = \frac{1-b^2}{d}\arctan\frac{(l_1 - l_2)d}{fb(l_1+l_2) - l_1 l_2(1-b^2) - g}, \tag{2.56}$$

$$K_2(n) = K_3(n-2) + \frac{2fb}{1-b^2}K_2(n-1) - \frac{g}{1-b^2}K_2(n-2), \tag{2.57}$$

- (1) $b = f = 0$

$$K_2(1) = \frac{1}{\sqrt{l_1^2 + R^2}} - \frac{1}{\sqrt{l_2^2 + R^2}}, \tag{2.58}$$

$$K_2(0) = \frac{(l_1 - l_2)(l_1 l_2 - R^2)}{R^2(l_2^2 + R^2)(l_1^2 + R^2)}, \tag{2.59}$$

- (1) $b \neq 0$ and $f \neq 0$

$$K_2(1) = \frac{f(1-b^2)}{2h}\left[\ln(\kappa_2) - \ln(\kappa_1)\right] + \frac{R^2 b(1-b^2)}{hd}\left[\arctan(\varpi_2) - \arctan(\varpi_1)\right], \tag{2.60}$$

$$K_2(0) = \frac{b(1-b^2)}{2h}\left[\ln(\kappa_2) - \ln(\kappa_1)\right] + \frac{f(1-b^2)}{hd}\left[\arctan(\varpi_2) - \arctan(\varpi_1)\right], \tag{2.61}$$

$$K_3(n) = \frac{l_2^{n-1}}{n}\sqrt{l_2^2 + R^2} - \frac{l_1^{n-1}}{n}\sqrt{l_1^2 + R^2} - \frac{(n-1)R^2}{n}K_3(n-2), \tag{2.62}$$

$$K_3(1) = \sqrt{l_2^2 + R^2} - \sqrt{l_1^2 + R^2}, \tag{2.63}$$

$$K_3(0) = \ln\frac{e_2}{e_1}. \tag{2.64}$$

Here $d = \sqrt{R^2(1-b^2) - f^2}$, $e_2 = l_2 + \sqrt{l_2^2 + R^2}$, $e_1 = l_1 + \sqrt{l_1^2 + R^2}$, $\kappa_2 = \frac{((1+b)e_2+f)^2+d^2}{((1+b)e_1+f)^2+d^2}$, $\kappa_1 = \frac{((1-b)e_2-f)^2+d^2}{((1-b)e_1-f)^2+d^2}$, $\varpi_2 = \frac{(be_2+f)d}{(fbe_2+f^2)-(1-b^2)(R^2+l_2e_2)}$, $\varpi_1 = \frac{(be_1+f)d}{(fbe_1+f^2)-(1-b^2)(R^2+l_1e_1)}$, $h = R^2b^2 + f^2$, and $g = R^2 - f^2$.

The analytical mutual impedance formulas have to work in complex space due to the application of the QSCIM.

2.5 Simulation Result and Analysis

2.5.1 *Simulation Analysis for Spatial 1-d Scalar or Vector Galerkin MM and BEM*

By observing and comparing four groups of formulas, such as, frequency domain formula (2.5) and (2.9) for SGWRM case, time domain formula (2.14) and (2.18) for SGWRM case, frequency domain formula (2.24) and (2.30) for VGWRM case, time domain formula (2.36) and (2.42) for VGWRM case, it can be seen that for formula (2.5) (2.14) (2.24) and (2.36), all spatial double 3-d surface integral need a lot of calculation, and it is difficult to have analytical formulas to apply. Therefore, the only choice is numerical integration, which will inevitably lead to lengthy calculation time. On the other hand, it can be seen that for formula (2.9) (2.18) (2.30) and (2.42), the amount of calculation required for all spatial double 1-d line integral is relatively small, and there are analytical formulas that can be applied, so this will inevitably reduce the amount of calculation and calculation time. Since the calculation amount and time can be greatly improved from the perspective of numerical formulas, simulation and comparison of numerical examples will no longer be carried out.

2.5.2 *Simulation Analysis for General Mutual Impedance and Mutual Induction Analytical Formula*

In order to show the calculation efficiency of the mutual impedance or mutual inductance matrix elements generated by the spatial 1-d Galerkin MM and BEM, the numerical integration and analytical calculation formula methods are respectively used for simulation calculation, and the calculation efficiency of the two methods is compared, of course, under the same calculation accuracy.

In this example, the grounding grid is composed of a single grid regular quadrilateral, with four sides of 20 m in length. Its material resistivity of 5.8×10^7 Ωm. It is buried in a four layer horizontal layered soil model with a depth of 0.5 m, and its soil parameters are: the four layer horizontal layered soil resistivity are 100 Ωm, 400 Ωm, 200 Ωm, and 500 Ωm, respectively. The relative dielectric constants of four horizontal layered soils are 10, 22, 22, and 22, respectively. The thickness of the three horizontal layers of soil is 1 m, 2.4 m, and 3.4 m, respectively.

The frequency domain spatial 1-d vector Galerkin MM is used to simulate and calculate the AC field problem when the calculation frequency is 1 kHz. The purpose of the simulation calculation of this example is to compare the analytical calculation formula of mutual impedance or mutual inductance with the calculation efficiency of numerical calculation. Therefore, only the calculation time required by the two calculation methods and the comparison of numerical accuracy of matrix elements are given. The two calculation methods refer to the analytical calculation formula of mutual impedance or mutual inductance and numerical integration calculation.

The time required for simulation calculation using the analytic formula of mutual impedance and mutual inductance is 2.91 s, while the time required for simulation calculation using numerical integration for mutual impedance and mutual inductance is 9.22 s. Therefore, the calculation efficiency of analytical formula for mutual impedance and mutual inductance is 3 times faster than that of numerical integration. Meanwhile, by observing Tables (2.1) and (2.2), it can be seen that the numerical accuracy is exactly the same after retaining three decimal places (Table 2.1 and 2.2).

Table 2.1 Mutual impedance matrix element coefficients

	Coefficients							
a	(23.9, −5.6*e* − 3)	(3.3, −1.7*e* − 3)	(8.1, −3.4*e* − 3)	(4.0, −2.0*e* − 3)	(6.8, −2.9*e* − 3)	(3.0, −1.6*e* − 3)	(4.0, −2.0*e* − 3)	(3.1, −1.7*e* − 3)
b	(23.9, −5.6*e* − 3)	(3.3, −1.7*e* − 3)	(8.1, −3.4*e* − 3)	(4.0, −2.0*e* − 3)	(6.8, −2.9*e* − 3)	(3.0, −1.6*e* − 3)	(4.0, −2.0*e* − 3)	(3.1, −1.7*e* − 3)
a	(3.3, −1.7*e* − 3)	(23.8, −5.6*e* − 3)	(4.0, −2.0*e* − 3)	(3.2, −1.7*e* − 3)	(3.0, −1.6*e* − 3)	(6.8, −2.9*e* − 3)	(8.1, −3.4*e* − 3)	(4.0, −2.0*e* − 3)
b	(3.3, −1.7*e* − 3)	(23.9, −5.6*e* − 3)	(4.0, −2.0*e* − 3)	(3.2, −1.7*e* − 3)	(3.0, −1.6*e* − 3)	(6.8, −2.9*e* − 3)	(8.1, −3.4*e* − 3)	(4.0, −2.0*e* − 3)
a	(8.1, −3.4*e* − 3)	(4.0, −2.0*e* − 3)	(23.9, −5.6*e* − 3)	(3.3, −1.7*e* − 3)	(4.0, −2.0*e* − 3)	(3.2, −1.7*e* − 3)	(6.8, −2.9*e* − 3)	(3.0, −1.7*e* − 3)
b	(8.1, −3.4*e* − 3)	(4.0, −2.0*e* − 3)	(23.9, −5.6*e* − 3)	(3.3, −1.7*e* − 3)	(4.0, −2.0*e* − 3)	(3.2, −1.7*e* − 3)	(6.8, −2.9*e* − 3)	(3.0, −1.7*e* − 3)
a	(4.0, −2.0*e* − 3)	(3.2, −1.7*e* − 3)	(3.3, −1.7*e* − 3)	(23.9, −5.6*e* − 3)	(8.1, −3.4*e* − 3)	(4.0, −2.0*e* − 3)	(3.0, −1.6*e* − 3)	(6.8, −2.9*e* − 3)
b	(4.0, −2.0*e* − 3)	(3.2, −1.7*e* − 3)	(3.3, −1.7*e* − 3)	(23.9, −5.6*e* − 3)	(8.1, −3.4*e* − 3)	(4.0, −2.0*e* − 3)	(3.0, −1.6*e* − 3)	(6.8, −2.9*e* − 3)
a	(6.8, −2.9*e* − 3)	(3.0, −1.6*e* − 3)	(4.0, −2.0*e* − 3)	(8.1, −3.4*e* − 3)	(23.9, −5.6*e* − 3)	(3.3, −1.7*e* − 3)	(3.2, −1.7*e* − 3)	(4.0, −2.0*e* − 3)
b	(6.8, −2.9*e* − 3)	(3.0, −1.6*e* − 3)	(4.0, −2.0*e* − 3)	(8.1, −3.4*e* − 3)	(23.9, −5.6*e* − 3)	(3.3, −1.7*e* − 3)	(3.2, −1.7*e* − 3)	(4.0, −2.0*e* − 3)
a	(3.0, −1.6*e* − 3)	(6.8, −2.9*e* − 3)	(3.2, −1.7*e* − 3)	(4.0, −2.0*e* − 3)	(3.3, −1.7*e* − 3)	(23.9, −5.6*e* − 3)	(4.0, −2.0*e* − 3)	(8.1, −3.4*e* − 3)
b	(3.0, −1.6*e* − 3)	(6.8, −2.9*e* − 3)	(3.2, −1.7*e* − 3)	(4.0, −2.0*e* − 3)	(3.3, −1.7*e* − 3)	(23.9, −5.6*e* − 3)	(4.0, −2.0*e* − 3)	(8.1, −3.4*e* − 3)
a	(4.0, −2.0*e* − 3)	(8.1, −3.4*e* − 3)	(6.8, −2.9*e* − 3)	(3.0, −1.6*e* − 3)	(3.2, −1.7*e* − 3)	(4.0, −2.0*e* − 3)	(23.9, −5.6*e* − 3)	(3.3, −1.7*e* − 3)
b	(4.0, −2.0*e* − 3)	(8.1, −3.4*e* − 3)	(6.8, −2.9*e* − 3)	(3.0, −1.6*e* − 3)	(3.2, −1.7*e* − 3)	(4.0, −2.0*e* − 3)	(23.9, −5.6*e* − 3)	(3.3, −1.7*e* − 3)
a	(3.2, −1.7*e* − 3)	(4.0, −2.0*e* − 3)	(3.0, −1.6*e* − 3)	(6.8, −2.9*e* − 3)	(4.0, −2.0*e* − 3)	(8.1, −3.4*e* − 3)	(3.3, −1.7*e* − 3)	(23.9, −5.6*e* − 3)
b	(3.2, −1.7*e* − 3)	(4.0, −2.0*e* − 3)	(3.0, −1.6*e* − 3)	(6.8, −2.9*e* − 3)	(4.0, −2.0*e* − 3)	(8.1, −3.4*e* − 3)	(3.3, −1.7*e* − 3)	(23.9, −5.6*e* − 3)

Table 2.2 Mutual inductance matrix element coefficients

	Coefficients							
a	(11.07, 0.0)	(0.39, 0.0)	(0.0, 0.0)	(0.0, 0.0)	(0.1, 0.0)	(0.35, 0.0)	(0.0, 0.0)	(0.0, 0.0)
b	(11.07, 0.0)	(0.39, 0.0)	(0.0, 0.0)	(0.0, 0.0)	(0.1, 0.0)	(0.35, 0.0)	(0.0, 0.0)	(0.0, 0.0)
a	(0.39, 0.0)	(11.07, 0.0)	(0.0, 0.0)	(0.0, 0.0)	(0.35, 0.0)	(0.11, 0.0)	(0.0, 0.0)	(0.0, 0.0)
b	(0.39, 0.0)	(11.07, 0.0)	(0.0, 0.0)	(0.0, 0.0)	(0.35, 0.0)	(0.11, 0.0)	(0.0, 0.0)	(0.0, 0.0)
a	(0.0, 0.0)	(0.0, 0.0)	(11.07, 0.0)	(0.39, 0.0)	(0.0, 0.0)	(0.0, 0.0)	(0.0, 0.0)	(0.35, 0.0)
b	(0.0, 0.0)	(0.0, 0.0)	(11.07, 0.0)	(0.39, 0.0)	(0.0, 0.0)	(0.0, 0.0)	(0.0, 0.0)	(0.35, 0.0)
a	(0.0, 0.0)	(0.0, 0.0)	(0.39, 0.0)	(11.07, 0.0)	(0.0, 0.0)	(0.0, 0.0)	(0.35, 0.0)	(0.11, 0.0)
b	(0.0, 0.0)	(0.0, 0.0)	(0.39, 0.0)	(11.07, 0.0)	(0.0, 0.0)	(0.0, 0.0)	(0.35, 0.0)	(0.11, 0.0)
a	(0.11, 0.0)	(0.35, 0.0)	(0.0, 0.0)	(0.0, 0.0)	(11.07, 0.0)	(0.39, 0.0)	(0.0, 0.0)	(0.0, 0.0)
b	(0.11, 0.0)	(0.35, 0.0)	(0.0, 0.0)	(0.0, 0.0)	(11.07, 0.0)	(0.39, 0.0)	(0.0, 0.0)	(0.0, 0.0)
a	(0.35, 0.0)	(0.11, 0.0)	(0.0, 0.0)	(0.0, 0.0)	(0.39, 0.0)	(11.07, 0.0)	(0.0, 0.0)	(0.0, 0.0)
b	(0.35, 0.0)	(0.11, 0.0)	(0.0, 0.0)	(0.0, 0.0)	(0.39, 0.0)	(11.07, 0.0)	(0.0, 0.0)	(0.0, 0.0)
a	(0.0, 0.0)	(0.0, 0.0)	(0.11, 0.0)	(0.35, 0.0)	(0.0, 0.0)	(0.0, 0.0)	(11.07, 0.0)	(0.39, 0.0)
b	(0.0, 0.0)	(0.0, 0.0)	(0.11, 0.0)	(0.35, 0.0)	(0.0, 0.0)	(0.0, 0.0)	(11.07, 0.0)	(0.39, 0.0)
a	(0.0, 0.0)	(0.0, 0.0)	(0.35, 0.0)	(0.11, 0.0)	(0.0, 0.0)	(0.0, 0.0)	(0.39, 0.0)	(11.07, 0.0)
b	(0.0, 0.0)	(0.0, 0.0)	(0.35, 0.0)	(0.11, 0.0)	(0.0, 0.0)	(0.0, 0.0)	(0.39, 0.0)	(11.07, 0.0)

2.6 Summary

Due to the characteristics of the substation grounding grid's own thin wire conductor, which meets the conditions of thin wire approximation, the spatial 3-d scalar and vector Galerkin MM and BEM can be simplified to reduce the dimension to spatial 1-d scalar and vector Galerkin MM and BEM. Under the premise of meeting certain calculation accuracy, the calculation amount is greatly reduced and the calculation efficiency is improved.

At the same time, the adoption of analytical calculation formula of mutual impedance and mutual inductance matrix elements generated by spatial 1-d scalar and vector Galerkin MM and BEM further improves the calculation efficiency.

This chapter introduces the basic foundation of electromagnetic field numerical methods and serves as the cornerstone of all subsequent chapters in this book on electromagnetic field numerical calculation methods.

References

1. I. Colominas, F. Navarrina, M. Casteleiro, A numerical formulation for grounding analysis in stratified soils. IEEE Trans. Power Delivery **17**(2), 587–595 (2002)
2. F. Navarrina, I. Colominas, M. Casteleiro, Why do computer methods for grounding produce anomalous results? IEEE Trans. Power Delivery **18**(4), 1192–1202 (2003)
3. F. Navarrina, I. Colominas, M. Casteleiro, Aboundary element numerical approach for grounding grid computation. Comput. Methods Appl. Mech. Eng. **174**, 73–90 (1999)

4. J. Yuan, H. Yang, L. Zhang, X. Cui, X. Ma, Simulation of substation grounding grids with unequal-potential. IEEE Trans. Magn. **36**(4), 1468–1471 (2000)
5. Z.-X. Li, W. Chen, J.-B. Fan, L. Jiayu, A novel mathematical modeling of grounding system buried in multilayer earth. IEEE Trans. Power Delivery **21**(3), 1267–1272 (2006)
6. Z.-X. Li, J.-B. Fan, Numerical calculation of grounding system in low-frequency domain based on the boundary element method. Int. J. Numer. Methods Eng. **73**, 685–705 (2008)
7. A.F. Otero, J. Cidras, J.L. del Alamo, Frequency-dependant grounding system calculation by means of a converntional nodal analysis technique. IEEE Trans. Power Delivery **14**(3), 873–878 (1999)
8. L. Zhong-Xin, L. Guang-Fan, F. Jian-Bin, Yu. Yin, A novel mathematical model for the lightning response of a grounding system buried in multilayered earth based on the quasi-static complex image method. Europ. Trans. Electr. Power **22**, 831–845 (2012)
9. D. Cavka, D. Poljak, On the evaluation of input impedance and transient impedance for grounding electrodes using antenna theory. COMPEL Int. J. Comput. Mathem. Electr. Electron. Eng. **32**(6), 2045–2062
10. Z.-X. Li, P. Li, K.-C. Wang, Transient lighting response to grounding grids buried in horizontal multilayered earth model considering time domain quasi-static complex image method and soil ionization effect. Int. J. Comput. Mathem. Electr. Electron. Eng. **40**(3), 516–534 (2021)
11. Z.-X. Li, W.-J. Chen, K.-C. Wang, Frequency domain and time domain analysis of the transient behavior of buried grounding grids in horizontal multilayered earth model. Electr. Eng. **104**(4), 2512–2529 (2022)
12. C.A.J. Fletcher, *Weighted Residual Methods*, pp. 98–162. Springer, Berlin, Heidelberg (1998)
13. A. Öchsner, *Weighted Residual Methods for Finite Elements*, pp. 2771–2786. Springer, Berlin, Heidelberg (2020)
14. J.G. Sverak, W.K. Dick, T.H. Dodds, R.H Heppe, Substations grounding. Part I. IEEE Trans. Power Apparaus Syst. **100**(9), 4281–4290 (1981)
15. R.J. Heppe, Computation of potential at surface above an energized grid or other electrode, allowing for non-uniform current distribution. IEEE Trans. Power Apparatus Syst. **98**(6), 1978–1989 (1979)
16. Z.X. Li, J.B. Fan, W.J. Chen, Numerical simulation of substation grounding grids buried in both horizontal and vertical multilayered earth model. Int. J. Numer. Methods Eng. **69**(11), 2359–2380 (2007)
17. Z.X. Li, G.F. Li, J.B. Fan, Numerical simulation of substation grounding system in low frequency domain based on the boundary element method with linear basis function. Commun. Numer. Methods Eng. **26**(12), 1819–1914 (2010)
18. Z.-X. Li, J.-B. Fan, Numerical calculation of grounding system in low-frequency domain based on the boundary element method. Int. J. Numer. Methods Eng. **73**(5), 685–705 (2008)
19. Z.-X. Li, K.-L. Gao, S.-W. Rao, Numerical calculation of grounding system buried in vertical earth model in low frequency domain based on the boundary element method. Europ. Trans. Electr. Power **19**(8), 1177–1190 (2009)

Chapter 3
Fundamentals of Numerical Calculation Techniques II

Frequency and Time domain Complex Image Method Based on Quasi-Static Electric Field Theory and its Closed Form of the Green's Function.

3.1 Preface

In the grounding problem of substation grounding grids, it involves the problem of how to construct a ground conductive medium model. However, in practical engineering situations, the multi-layer horizontal layered soil conductive medium model is closer to the actual geographical situation.

After adopting a multi-layer horizontal layered soil conductive medium model that is more in line with the actual situation, regardless of the frequency domain method or the time domain method, the quasi-static electromagnetic field numerical simulation method appears in the point current Green's function containing infinite integration problems. If the numerical integration method is used to calculate the infinite integral, it will undoubtedly cause a larger calculation time problem.

Fortunately, the frequency domain or time domain quasi-static complex image method (QSCIM) can be used to analytically calculate the infinite integral, improving computational efficiency. These frequency domain or time domain QSCIM are the fundamental foundations of the electromagnetic field numerical methods introduced in this book.

Z.-X. Li, *Numerical Simulation Algorithm of Electromagnetic Field for Grounding Problems in Power System Substation Grounding Grids*,
https://doi.org/10.1007/978-981-96-0904-8_3

3.2 QSCIM Based on Matrix Pencil Method in Frequency Domain and its Closed Form of the Green's Function

3.2.1 Introduction

The grounding problems under AC substation have been studied based on either the electrostatic field theory or the direct current electric field theory, with the equal potential model [1–6] or the unequal potential model [7, 8]. However, these models ignore the mutual capacitative interactions between leakage currents along a pairs of short conductors. Mutual induction interactions between branch currents along a pair of short conductors have also been overlooked in the unequal potential model. Although the hypothesis for the electrostatic field theory can be accepted under 50 Hz or 60 Hz work frequency condition, restrictions for the model will be evident for more general cases, for example, if a large resistivity of earth is observed for a buried grounding grid, the imaginary number of grounding impedance will be large [9], in the case, grounding resistance substituting for grounding impedance will be inappropriate. In other words, electrostatic field theory or direct current electric field theory is inappropriate for this type of grounding problem, and a more updated electromagnetic theory is required.

In recent years, quasi-static electric field theory has been used to solve steady-state frequency domain grounding problems in combination with equipotential model [10–17] and unequal potential model [9, 18–25]. At the same time, combined with Fourier transform, these methods are further developed to solve lightning protection and grounding problems [26–30]. In these models, only the propagation effect of electromagnetic wave has been ignored. Meanwhile, if a point source is buried in the horizontal multilayered earth, the scalar electrical potential (SEP) of a scalar monopole or the vector magnetic potential (VMP) of a vector dipole is satisfied with scalar or vector Poisson equation. After solving the scalar or vector Poisson equation for the monopole or dipole, scalar or vector Green's function for the monopole or dipole can be obtained. The infinite integral for Bessel function of the first kind, of order zero, has appeared within these Green's functions formulae. The infinite integral can be transformed into finite exponential series by 1-d Matrix Pencil (MP) method [31]. In this way, closed form of scalar or vector Green's function of monopole or dipole can be achieved. Among these derivation procedure we also discuss how to achieve a finite exponential series with 1-d MP method, called the QSCIM. It should be pointed out that references [5, 32] have also given finite exponential series for scalar Green's function of a scalar monopole based on Prony's method, and are called the complex image method. Actually, the complex image method should be called the static complex image method due to its electrostatic field theory.

In this chapter, QSCIM has been fully discussed based on the 1-d MP method. First, the procedure for vector or scalar Green's function of vector dipole or scalar monopole buried in horizontal multilayered earth model is compendiously introduced. Then the process to achieve finite exponential series for the kernel of Green's function based on 1-d MP method is discussed. Finally, numerical results for the kernel of the closed form of Green's function are carefully studied and discussed.

3.2.2 *Theory of QSCIM in Frequency Domain*

The closed form of scalar or vector Green's function of monopole or dipole in frequency domain

For a frequency domain point source (scalar monopole or vector dipole) buried in horizontal multilayered earth model, we consider that a low frequency (50 or 60 Hz and higher harmonic wave) electromagnetic field occurs around the AC substation. Meanwhile, the size of grounding system under the AC substation is limited, so the electromagnetic waves' propagation effect can be ignored. Hence, quasi-static electrical field can be regarded as a better theory for this grounding problem. Further, in this case, the SEP φ of a scalar monopole satisfies the scalar Poisson equation, and VMP $\bar{A}$ of a vector dipole satisfies the vector Poisson equation. Here we suppose that the point source current is located at the origin of the co-ordinate system (Fig. 3.1). The horizontal N_e-layer earth model is considered.

$$\nabla^2 \varphi_{\vartheta\theta}(x', y', z'; x, y, z; \omega) = -\frac{I_s(x', y', z')}{\bar{\sigma}_\vartheta} \tag{3.1}$$

where $\vartheta, \theta = 1, \ldots, N_e$, $I_s(x', y', z')$ is scalar unit leakage current for scalar monopole. The (x', y', z') and (x, y, z) are the rectangular coordinate values of source point and field point respectively. $\overline{\sigma_\vartheta} = \sigma_\vartheta + j\omega\varepsilon_\vartheta$, σ_ϑ and ε_ϑ are the soil conductivity and dielectric constant of the ϑth layer in the layered conductive medium. ω is angular frequency, and j is imaginary unit.

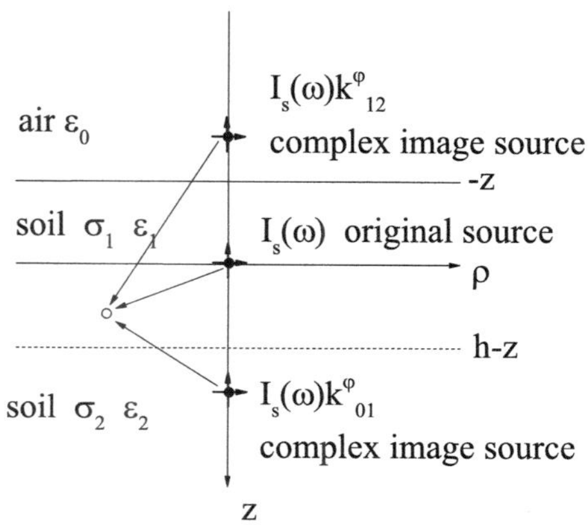

Fig. 3.1 The point source current in frequency domain

$$\nabla^2 \bar{A}_{\vartheta\theta}(x', y', z'; x, y, z; \omega) = -\mu \hat{I}_b(x', y', z') \quad (3.2)$$

where $\hat{I}_b(x', y', z')$ is vector unit branch current for vector dipole, and $\hat{I}_b(x', y', z') = I_b(x', y', z') \cos\theta_x \hat{x} + I_b(x', y', z') \cos\theta_y \hat{y} + I_b(x', y', z') \cos\theta_z \hat{z} = I_{b_x}(x', y', z')$ $\hat{x} + I_{b_y}(x', y', z')\hat{y} + I_{b_z}(x', y', z')\hat{z}$. μ is magnetoconductivity.

After solving the Poisson equations (3.1) and (3.2), we can get the scalar Green's function for monopole and vector Green's function for vector dipole. For more detail about the derivation process refer to [12, 14, 23–25] for scalar Greens' function, and [23–25] for vector Green's function. There is a special chapter at the end of this book to discuss this part in detail. Here, we will simply give, as examples, the formula for scalar Green's function $G_{11}^{\varphi f}$ of monopole and vector Green's function $G_{11}^{A_{zz}^f}$ or $G_{11}^{A_{xz}^f}$ of dipole in frequency domain, both of which are buried in horizontal two-layer earth model with source and field point at the same layer.

For monopole case, from [12, 14, 23–25], we know

$$G_{11}^{\varphi f}(x', y', z'; x, y, z; \omega) = \frac{I_s(x', y', z')}{4\pi\overline{\sigma_1}} \int_0^\infty (e^{-k_\rho|z-z'|} + f^\varphi(k_\rho)(-k_{01}^\varphi e^{-k_\rho(z+z')}$$
$$+ k_{12}^\varphi e^{-k_\rho(2h_1-z'-z)} - k_{01}^\varphi k_{12}^\varphi e^{-k_\rho(2h_1+z-z')} - k_{01}^\varphi k_{12}^\varphi e^{-k_\rho(2h_1-z+z')}) J_0(k_\rho \rho) dk_\rho \quad (3.3)$$

where $f^\varphi(k_\rho) = \frac{1}{(1+k_{01}^\varphi k_{12}^\varphi \cdot e^{-2k_\rho h_1})}$, $k_{01}^\varphi = \frac{\overline{\sigma_0}-\overline{\sigma_1}}{\overline{\sigma_0}+\overline{\sigma_1}}$, $k_{12}^\varphi = \frac{\overline{\sigma_1}-\overline{\sigma_2}}{\overline{\sigma_1}+\overline{\sigma_2}}$, $\overline{\sigma_0} = j\omega\varepsilon_0$, $\overline{\sigma_1} = \sigma_1 + j\omega\varepsilon_1$, $\overline{\sigma_2} = \sigma_2 + j\omega\varepsilon_2$, h_1 is thickness of the first layer earth. And $\rho = \sqrt{(x-x')^2 + (y-y')^2}$. It must be pointed out here that although the electrical parameters (resistivity and dielectric constant) of horizontal layered soil given here are constants, if these electrical parameters (resistivity and dielectric constant) of horizontal layered soil are functions of frequency, or the electrical parameters (resistivity and dielectric constant) of horizontal layered soil with frequency characteristics, the QSCIM developed in this chapter can also be applied.

The kernel $f^\varphi(k_\rho)$ of scalar Green's function $G_{11}^{\varphi f}$ can be expanded into finite exponential series by the 1-d MP method,

$$f^\varphi(k_\rho) = \frac{1}{(1 + k_{01}^\varphi k_{12}^\varphi \cdot e^{-2k_\rho h_1})} = 1 + \sum_{i=1}^{M} \alpha_i e^{\beta_i k_\rho} \quad (3.4)$$

where M, α_i and β_i are 1-d quasi-static complex image's coefficients.

Replacing $f^\varphi(k_\rho)$ of $G_{11}^{\varphi f}$ with the finite exponential series and employing the Lipschitz integration [23], we have

$$G_{11}^{\varphi^f}(x', y', z'; x, y, z; \omega) = \frac{I_s(x', y', z')}{4\pi\overline{\sigma_1}}[\frac{1}{R} - \frac{k_{01}^{\varphi}}{R_1} + \frac{k_{12}^{\varphi}}{R_2} - \frac{k_{01}^{\varphi}k_{12}^{\varphi}}{R_3} - \frac{k_{01}^{\varphi}k_{12}^{\varphi}}{R_4}$$

$$+ \sum_{i=1}^{M} \alpha_i(-\frac{k_{01}^{\varphi}}{R_{i_1}} + \frac{k_{12}^{\varphi}}{R_{i_2}} - \frac{k_{01}^{\varphi}k_{12}^{\varphi}}{R_{i_3}} - \frac{k_{01}^{\varphi}k_{12}^{\varphi}}{R_{i_4}})] \tag{3.5}$$

where $R = \sqrt{(x - x')^2 + (y - y')^2 + (z - z')^2}$,

$R_i = \sqrt{(x - x')^2 + (y - y')^2 + (sign_a z + sign_b z' + z_i^0)^2}$, in which $sign_a = 1$ for R_1 and R_3, $sign_b = 1$ for R_3 and R_4, $sign_a = sign_b = -1$ for others; for R_{2-4}, $z_i^0 = 2h$, for R_1, $z_i^0 = 0$.

$R_{i_{1-4}} = \sqrt{(x - x')^2 + (y - y')^2 + (sign_a z + sign_b z' + z_i^1)^2}$, in which $sign_a = 1$ for R_{i_1} and R_{i_3}, $sign_b = 1$ for R_{i_3} and R_{i_4}, $sign_a = sign_b = -1$ for others; for $R_{i_{2-4}}$ $z_i^1 = 2h - \beta_i$, for for R_{i_1} $z_i^1 = -\beta_i$.

For vertical dipole case, from [23–25], we know

$$G_{11}^{A_{zz}^f}(x', y', z'; x, y, z; \omega) = \frac{\mu I_{b_z}(x', y', z')}{4\pi} \int_0^{\infty} (e^{-k_\rho|z-z'|} + f^{TM}(k_\rho)(-k_{01}^{TM}e^{-k_\rho(z+z')}$$

$$+ k_{12}^{TM}e^{-k_\rho(2h_1-z'-z)} - k_{01}^{TM}k_{12}^{TM}e^{-k_\rho(2h_1+z-z')} - k_{01}^{TM}k_{12}^{TM}e^{-k_\rho(2h_1-z+z')})J_0(k_\rho\rho)dk_\rho \tag{3.6}$$

where $f^{TM}(k_\rho) = \frac{1}{(1+k_{01}^{TM}k_{12}^{TM}\cdot e^{-2k_\rho h_1})}$, $k_{01}^{TM} = \frac{\overline{\sigma_1}-\overline{\sigma_0}}{\overline{\sigma_1}+\overline{\sigma_0}}$, $k_{12}^{TM} = \frac{\overline{\sigma_2}-\overline{\sigma_1}}{\overline{\sigma_2}+\overline{\sigma_1}}$.

Since we know $k_{01}^{TM} = -k_{01}^{\varphi}$ and $k_{12}^{TM} = -k_{12}^{\varphi}$, and we can observe that the kernel $f(k_\rho)$ and $f^{TM}(k_\rho)$ are equal, it implies $f^{TM}(k_\rho) = f^{\varphi}(k_\rho)$ except for $k_{ij}^{TM}(k_\rho) = -k_{ij}^{\varphi}(k_\rho)$, $(i, j = 1, \ldots, N_s)$. Similar to scalar Green's function $G_{11}^{\varphi^f}$, we have

$$G_{11}^{A_{zz}^f}(x', y', z'; x, y, z; \omega) = \frac{\mu I_{b_z}(x', y', z')}{4\pi}(\frac{1}{R} - \frac{k_{01}^{TM}}{R_1} + \frac{k_{12}^{TM}}{R_2} - \frac{k_{01}^{TM}k_{12}^{TM}}{R_3} - \frac{k_{01}^{TM}k_{12}^{TM}}{R_4}$$

$$+ \sum_{i=1}^{M} \alpha_i(-\frac{k_{01}^{TM}}{R_{i_1}} + \frac{k_{12}^{TM}}{R_{i_2}} - \frac{k_{01}^{TM}k_{12}^{TM}}{R_{i_3}} - \frac{k_{01}^{TM}k_{12}^{TM}}{R_{i_4}})) \tag{3.7}$$

where R, R_{1-4}, $R_{i_{1-4}}$ are same as in scalar Green's function $G_{11}^{\varphi^f}$ in Eq. (3.5).

For horizontal dipole case, from [23–25], we know

$$G_{11}^{A_{xz}^f}(x', y', z'; x, y, z; \omega) = \frac{\mu I_{b_x}(x', y', z')}{4\pi} \frac{\partial}{\partial x} \int_0^{\infty} f^{TM}(k_\rho)(-k_{01}^{TM}e^{-k_\rho(z+z')}$$

$$+ k_{12}^{TM}e^{-k_\rho(2h_1-z'-z)} - k_{01}^{TM}k_{12}^{TM}e^{-k_\rho(2h_1+z-z')} - k_{01}^{TM}k_{12}^{TM}e^{-k_\rho(2h_1-z+z')})J_0(k_\rho\rho)\frac{dk_\rho}{k_\rho} \tag{3.8}$$

Just as in Green's function $G_{11}^{A_{zz}^f}$, considering the Lipschitz integration's varied form [23], we have

$$G_{11}^{A_{xz}^f}(x', y', z'; x, y, z; \omega) = \frac{\mu I_{b_x}(x', y', z')}{4\pi}\frac{\partial}{\partial x}(-k_{01}^{TM}\ln(z_1 + R_1)$$
$$+k_{12}^{TM}\ln(z_2 + R_2) - k_{01}^{TM}k_{12}^{TM}\ln(z_3 + R_3) - k_{01}^{TM}k_{12}^{TM}\ln(z_4 + R_4)$$
$$+\sum_{i=1}^{M}\alpha_i(-k_{01}^{TM}\ln(z_{i_1} + R_{i_1}) + k_{12}^{TM}\ln(z_{i_2} + R_{i_2}) - k_{01}^{TM}k_{12}^{TM}\ln(z_{i_3} + R_{i_3}) - k_{01}^{TM}k_{12}^{TM}\ln(z_{i_4} + R_{i_4}))) \tag{3.9}$$

where R_{1-4}, $R_{i_{1-4}}$ are same as in scalar Green's function G_{φ}^{11} in Eq. (3.5). Other parameters are $z_{1-4} = sign_a z + sign_b z' + z_i^0$, $z_{i_{1-4}} = sign_a z + sign_b z' + z_i^1$, in which $sign_a = 1$ for z_1, z_3, z_{i_1} and z_{i_3}, $sign_b = 1$ for z_3, z_4, z_{i_3} and z_{i_4}, $sign_a = sign_b = -1$ for others, for z_{2-4}, $z_i^0 = 2h$, for z_1, $z_i^0 = 0$, for $z_{i_{2-4}}$, $z_i^1 = 2h - \beta_i$, for z_{i_1}, $z_i^1 = \beta_i$.

If a dipole was lying in general orientation, the dipole can be divided into horizontal and vertical dipole components.

The Green's function of horizontal dipole

$$G_{11}^{A_{xx}^f}(x', y', z'; x, y, z) = \frac{\mu I_{b_x}(x', y', z')}{4\pi}\frac{1}{R} \tag{3.10}$$

The formula (3.10) will not be discussed since it is as simple formula.

As we observe the kernels of Green's functions for all vertical and horizontal dipole and monopole, we see $f^{TM}(k_\rho) = f^{\varphi}(k_\rho)$. So for horizontal two layer earth model, only kernel $f(k_\rho)$ can be considered as the unified kernel formula to be expanded into finite exponential series. This conclusion can apply for all other vector or scalar Green's function of vector dipole or scalar monopole buried in horizontal two-layer earth model with different field point locations, such as $G_{10}^{\varphi^f}$, $G_{10}^{A_{zz}^f}$, $G_{10}^{A_{xz}^f}$, $G_{12}^{\varphi^f}$, $G_{12}^{A_{zz}^f}$, $G_{12}^{A_{xz}^f}$, $G_{21}^{\varphi^f}$, $G_{21}^{A_{zz}^f}$, $G_{21}^{A_{xz}^f}$, $G_{22}^{\varphi^f}$, $G_{22}^{A_{zz}^f}$, $G_{22}^{A_{xz}^f}$, etc. Further, the same conclusion can be arrived at with vector or scalar Green's function of vector dipole or scalar monopole buried in the other horizontal multilayered earth model, such as 3-layer or 4-layer earth model cases.

The kernel of closed form of Green's function for both scalar monopole and vector dipole buried in horizontal three-layer and four-layer earth model will be introduced next

3-layer earth model case

$$f^{\Phi}(k_\rho) = \frac{1}{f_{sub_3}^{\Phi}(k_\rho)} \tag{3.11}$$

and

$$f_{sub_3}^{\Phi}(k_\rho) = 1 + k_{01}^{\Phi}k_{12}^{\Phi}e^{-2k_\rho h_1} + k_{12}^{\Phi}k_{23}^{\Phi}e^{-2k_\rho h_2} + k_{01}^{\Phi}k_{23}^{\Phi}e^{-2k_\rho(h_1+h_2)} \tag{3.12}$$

where $\Phi = TM$ for vector dipole case and $\Phi = \varphi$ for scalar monopole case; $k_{23}^{\varphi} = \frac{\overline{\sigma_2}-\overline{\sigma_3}}{\overline{\sigma_2}+\overline{\sigma_3}}$, $\overline{\sigma_3} = \sigma_3 + j\omega\varepsilon_3$, $k_{23}^{TM} = -k_{23}^{\varphi}$, h_2 is thickness of second layer earth.

4-layer earth model case

$$f^{\Phi}(k_\rho) = \frac{1}{f_{sub_4}^{\Phi}(k_\rho)} \tag{3.13}$$

and

$$\begin{aligned} f_{sub_4}^{\Phi}(k_\rho) = 1 &+ k_{01}^{\Phi}k_{12}^{\Phi}e^{-2k_\rho h_1} + k_{12}^{\Phi}k_{23}^{\Phi}e^{-2k_\rho h_2} + k_{23}^{\Phi}k_{34}^{\Phi}e^{-2k_\rho h_3} + k_{01}^{\Phi}k_{23}^{\Phi}e^{-2k_\rho(h_1+h_2)} \\ &+ k_{12}^{\Phi}k_{34}^{\Phi}e^{-2k_\rho(h_2+h_3)} + k_{01}^{\Phi}k_{12}^{\Phi}k_{23}^{\Phi}k_{34}^{\Phi}e^{-2k_\rho(h_1+h_3)} + k_{01}^{\Phi}k_{34}^{\Phi}e^{-2k_\rho(h_1+h_2+h_3)} \end{aligned} \tag{3.14}$$

where $\Phi = TM$ for vector dipole case and $\Phi = \varphi$ for scalar monopole case; $k_{34}^{\varphi} = \frac{\overline{\sigma_3}-\overline{\sigma_4}}{\overline{\sigma_3}+\overline{\sigma_4}}$, $\overline{\sigma_4} = \sigma_4 + j\omega\varepsilon_4$, $k_{34}^{TM} = -k_{34}^{\varphi}$, h_3 is thickness of third layer earth.

For simplified description for 1-d MP in this chapter, only scalar monopole case is discussed in below, which means that $f(k_\rho)$ is used to represent $f^{\varphi}(k_\rho)$.

Application range for the QSCIM

The maximum frequency of applicability of the method is limited by the quasi-stationary approximation of the electromagnetic fields, which means the propagation effect of the electromagnetic field around the dipole can be neglected, so

$$e^{-jk_i R} \approx 1 \tag{3.15}$$

where $k_i = j\omega\sqrt{\mu(\varepsilon_i + \frac{\sigma_i}{j\omega})}$, $i = 1, \ldots, N_s$.

3.2.3 1-d MP Method

If a function $f(x)$ can be expanded into finite exponential series, as below,

$$f(x) = f_{x\to\infty} + \sum_{i=1}^{M} \alpha_i e^{\beta_i x} \tag{3.16}$$

The three parameters to be decided are M, α_i and β_i.

To get these parameters, we first decide the maximum sample points according to the characteristic of function $f(x)$, the maximum value for x_{max} can thus be obtained according minimum $f(x)$ decided by set truncation precise. Once x_{max} is known, we can get uniformly discrete values of $f(x)$ within scope $(0 \le x \le x_{max})$. Therefore, we have uniform discrete function values of $f(x)$ with $(f(0), f(1), f(2), \ldots, f(N))$ corresponding to value of x as $(0, x_1, x_2, \ldots, x_{max})$ or $(0, \Delta x, \ldots, i\Delta x, \ldots, (N-1)\Delta x)$. Here Δx is rate of sampling. we can obtain

$$f(x) \approx f_{x\to\infty} + \sum_{i=1}^{M} \alpha_i z_i^k \qquad (k = 0, \ldots, N-1) \tag{3.17}$$

here

$$z_i^k = e^{\beta_i \Delta x} \qquad (i = 1, \ldots, M) \tag{3.18}$$

Method to decide number of M

Since we have total N number of uniform discrete function values of $f(x)$, we can get the matrix $[Y]$ from the sampling data $f(x)$ by combining $[Y_1]$ and $[Y_2]$ as

$$[Y] = \begin{bmatrix} f(0) & f(1) & \ldots & f(L-1) \\ f(1) & f(2) & \ldots & f(L) \\ \vdots & \vdots & & \vdots \\ f(N-L) & f(N-L+1) & \ldots & f(N-1) \end{bmatrix}_{(N-L+1)\times L}, \tag{3.19}$$

$$[Y_1] = \begin{bmatrix} f(0) & f(1) & \ldots & f(L-1) \\ f(1) & f(2) & \ldots & f(L) \\ \vdots & \vdots & & \vdots \\ f(N-L-1) & f(N-L) & \ldots & f(N-2) \end{bmatrix}_{(N-L)\times L}, \tag{3.20}$$

$$[Y_2] = \begin{bmatrix} f(1) & f(2) & \ldots & f(L) \\ f(2) & f(3) & \ldots & f(L+1) \\ \vdots & \vdots & & \vdots \\ f(N-L) & f(N-L+1) & \ldots & f(N-1) \end{bmatrix}_{(N-L)\times L}, \tag{3.21}$$

where L is referred as the pencil parameter [33].

Note that $[Y_1]$ is obtained from $[Y]$ by omitting the last row, and $[Y_2]$ is obtained from $[Y]$ by omitting the first row. The parameter L can be chosen between $N/3$ and $N/2$.

Next, a singular-value decomposition (SVD) of the matrix $[Y]$ can be implemented out as

$$[Y] = [U][\Sigma][V]^H, \tag{3.22}$$

here, $[U]$ and $[V]$ are unitary matrices, comprised of the eigenvectors of $[Y][Y]^H$ and $[Y]^H[Y]$, respectively, and $[\Sigma]$ is a diagonal matrix including the singular values of $[Y]$, i. e. where the superscript "H" denotes the conjugate transpose.

$$[U]^H[Y][V] = [\Sigma], \tag{3.23}$$

The choice of the parameter M can be achieved at this stage by studying the ratios of various singular values to the largest one. Typically, the singular values beyond M

are set equal to zero. The way which M is chosen is as follows. Observe the singular value σ_c such that

$$\frac{\sigma_c}{\sigma_{max}} \approx 10^{-p}, \tag{3.24}$$

where p is the number of significant decimal digits in the data. For example, if the sampling data is accurate up to three significant digits, the singular values for which the ratio in Eq. (3.24) is below 10^{-3} are essentially useless singular values, and they should not be used in the reconstruction of the sampling data.

We next introduce the "filtered" matrix, $[U']$, and $[V']$, constructed so that it contains only M predominant right-singular vectors of $[U]$ and $[V]$;

$$[U'] = [u_1, u_2, \ldots, u_M], \tag{3.25}$$

$$[V'] = [v_1, v_2, \ldots, v_M], \tag{3.26}$$

The right-singular vectors from $M + 1$ to L, corresponding to the small singular values, are omitted. Therefore,

$$[Y_1] = [U_1'][\Sigma'][V']^H, \tag{3.27}$$

$$[Y_2] = [U_2'][\Sigma'][V']^H, \tag{3.28}$$

where $[U_1']$ is obtained from $[U']$ with the last row of $[U']$ omitted, $[U_2']$ is obtained by deleting the first row of $[U']$; and $[\Sigma']$ is obtained from the M columns of $[\Sigma]$ corresponding to the M predominant singular values.

Method to decide β_i

To motivate the 1-d MP method, we can use the two $(N - L) \times L$ matrices, Y_1 and Y_2. We can rewrite

$$[Y_2] = [Z_1][Z_0][R][Z_2], \tag{3.29}$$

$$[Y_1] = [Z_1][R][Z_2], \tag{3.30}$$

where

$$[Z_1] = \begin{bmatrix} 1 & 1 & \ldots & 1 \\ z_1 & z_2 & \ldots & z_M \\ \vdots & \vdots & & \vdots \\ z_1^{N-L-1} & z_2^{N-L-1} & \ldots & z_M^{N-L-1} \end{bmatrix}_{(N-L)\times M}, \tag{3.31}$$

$$[Z_2] = \begin{bmatrix} 1 & z_1 & \ldots & z_1^{L-1} \\ 1 & z_2 & \ldots & z_2^{L-1} \\ \vdots & \vdots & & \vdots \\ 1 & z_M & \ldots & z_M^{L-1} \end{bmatrix}_{M\times L}, \tag{3.32}$$

$$[Z_0] = diag[z_1, z_2, \ldots, z_M], \tag{3.33}$$

$$[R] = diag[\alpha_1, \alpha_2, \ldots, \alpha_M], \tag{3.34}$$

where $diag[\bullet]$ represents a $M \times M$ diagonal matrix.

Now we introduce the matrix pencil

$$[Y_2] - \lambda[Y_1] = [Z_1]\{[Z_0] - \lambda[I]\}[R][Z_2], \tag{3.35}$$

where $[I]$ is the $M \times M$ identity matrix. We can demonstrate that, in general, the rank of $\{[Y_2] - \lambda[Y_1]\}$ will be M, provided $M \le L \le N - M$. However, if $\lambda = z_i, i = 1, 2, \ldots, M$, the ith row of $\{[Z_0] - \lambda[I]\}$ is zero, and the rank of this matrix is $M - 1$. Hence, the parameters z_i, may be found as generalized eigenvalues of the matrix pair $\{[Y_2]; [Y_1]\}$. Equivalently, the problem of solving for z_i can be transformed into an ordinary eigenvalue problem,

$$\{[Y_1]^+[Y_2] - \lambda[I]\}, \tag{3.36}$$

where $[Y_1]^+$ is Moore-Penrose pseudo-inverse of $[Y_1]$, and is defined as

$$[Y_1]^+ = \{[Y_1]^H[Y_1]\}^{-1}[Y_1]^H, \tag{3.37}$$

where the superscript "H" denotes the conjugate transpose.

The eigenvalues of the matrix

$$\{[Y_2] - \lambda[Y_1]\}_{L\times M} \Longrightarrow \{[Y_1]^+[Y_2] - \lambda[I]\}_{M\times M} \tag{3.38}$$

are equivalent to the eigenvalues of the matrix

$$\{[U_2'] - \lambda[U_1']\} \Longrightarrow \{[U_1']\}^+\{[U_2']\} - \lambda[I] \tag{3.39}$$

This methodology can be used to solve for z_i.

Lastly, we point out that in this case z_i represent β_i according to Eq. (3.18).

Method to decide α_i

Once M and the z_i are known, α_i, are solved with the help of the following linear least-squares problem:

$$\begin{bmatrix} f(0) \\ f(1) \\ \vdots \\ f(N-1) \end{bmatrix} = \begin{bmatrix} 1 & 1 & \ldots & 1 \\ z_1 & z_2 & \ldots & z_M \\ \vdots & \vdots & & \vdots \\ z_1^{N-1} & z_2^{N-1} & \ldots & z_M^{N-1} \end{bmatrix} \begin{bmatrix} \alpha_1 \\ \alpha_2 \\ \vdots \\ \alpha_M \end{bmatrix}, \tag{3.40}$$

3.2.4 Closed Form of the Green's Function of a Point Source in Horizontal Multilayered Earth Model in Frequency Domain

The derivation process of the closed form Green's function of the point current source in a two-layer soil model in frequency domain was introduced earlier. Using the same method, the closed form Green's function of a point current source in any horizontally layered soil in frequency domain can be derived. Here, the closed form Green's function of a point current source in a four layer horizontally layered soil model in frequency domain is directly presented.

The closed form Green's function of a monopole in a four layer horizontally layered soil model in frequency domain is firstly presented as below.

$$
\begin{aligned}
G_{11}^{\varphi f}(x', y', z'; x, y, z; \omega) = {} & \frac{I_s(x', y', z')}{4\pi\bar{\sigma}_1}\Big[\frac{1}{R} + \Big(-\frac{k_{01}^{\varphi}}{R_1} + \frac{k_{12}^{\varphi}}{R_2} - \frac{k_{01}^{\varphi}k_{12}^{\varphi}}{R_3} - \frac{k_{01}^{\varphi}k_{12}^{\varphi}}{R_4} + \frac{k_{23}^{\varphi}}{R_5} - \frac{k_{01}^{\varphi}k_{23}^{\varphi}}{R_6} - \frac{k_{01}^{\varphi}k_{23}^{\varphi}}{R_7} \\
& - \frac{k_{01}^{\varphi}k_{12}^{\varphi}k_{23}^{\varphi}}{R_8} + \frac{k_{34}^{\varphi}}{R_9} - \frac{k_{01}^{\varphi}k_{34}^{\varphi}}{R_{10}} - \frac{k_{01}^{\varphi}k_{34}^{\varphi}}{R_{11}} - \frac{k_{01}^{\varphi}k_{23}^{\varphi}k_{34}^{\varphi}}{R_{12}} - \frac{k_{01}^{\varphi}k_{12}^{\varphi}k_{34}^{\varphi}}{R_{13}} + \frac{k_{12}^{\varphi}k_{23}^{\varphi}k_{34}^{\varphi}}{R_{14}} - \frac{k_{01}^{\varphi}k_{12}^{\varphi}k_{23}^{\varphi}k_{34}^{\varphi}}{R_{15}}\Big) \\
& - \frac{k_{01}^{\varphi}k_{12}^{\varphi}k_{23}^{\varphi}k_{34}^{\varphi}}{R_{16}} + \sum_{i=1}^{M}\alpha_i\Big(-\frac{k_{01}^{\varphi}}{R_{i_1}} + \frac{k_{12}^{\varphi}}{R_{i_2}} - \frac{k_{01}^{\varphi}k_{12}^{\varphi}}{R_{i_3}} - \frac{k_{01}^{\varphi}k_{12}^{\varphi}}{R_{i_4}} + \frac{k_{23}^{\varphi}}{R_{i_5}} - \frac{k_{01}^{\varphi}k_{23}^{\varphi}}{R_{i_6}} - \frac{k_{01}^{\varphi}k_{23}^{\varphi}}{R_{i_7}} - \frac{k_{01}^{\varphi}k_{12}^{\varphi}k_{23}}{R_{i_8}} \\
& + \frac{k_{34}^{\varphi}}{R_{i_9}} - \frac{k_{01}^{\varphi}k_{34}^{\varphi}}{R_{i_{10}}} - \frac{k_{01}^{\varphi}k_{34}^{\varphi}}{R_{i_{11}}} - \frac{k_{01}^{\varphi}k_{23}^{\varphi}k_{34}^{\varphi}}{R_{i_{12}}} - \frac{k_{01}^{\varphi}k_{12}^{\varphi}k_{34}^{\varphi}}{R_{i_{13}}} + \frac{k_{12}^{\varphi}k_{23}^{\varphi}k_{34}^{\varphi}}{R_{i_{14}}} - \frac{k_{01}^{\varphi}k_{12}^{\varphi}k_{23}^{\varphi}k_{34}^{\varphi}}{R_{i_{15}}} - \frac{k_{01}^{\varphi}k_{12}^{\varphi}k_{23}^{\varphi}k_{34}^{\varphi}}{R_{i_{16}}}\Big)\Big]
\end{aligned} \tag{3.41}
$$

where $R = \sqrt{(x - x')^2 + (y - y')^2 + (z - z')^2}$,

$R_i = \sqrt{(x - x')^2 + (y - y')^2 + (sign_a z + sign_b z' + z_i^0)^2}$, in which $sign_a = 1$ for R_1, R_3, R_6, R_8, R_{10}, R_{12}, R_{13} and R_{16}, $sign_b = 1$ for R_1, R_4, R_7, R_8, R_{11}, R_{12}, R_{13} and R_{15}, $sign_a = sign_b = -1$ for others; for R_{2-4}, $z_i^0 = 2h_1$, for R_{5-7}, $z_i^0 = 2(h_1 + h_2)$, for R_{9-11}, $z_i^0 = 2(h_1 + h_2 + h_3)$, for R_{14-16}, $z_i^0 = 2(h_1 + h_3)$, for R_8, $z_i^0 = 2h_2$, for R_{12}, $z_i^0 = 2(h_2 + h_3)$, for R_{13}, $z_i^0 = 2h_3$, for R_1, $z_i^0 = 0$.

$R_{i_{1-16}} = \sqrt{(x - x')^2 + (y - y')^2 + (sign_a z + sign_b z' + z_i^1)^2}$, in which $sign_a = 1$ for R_{i_1}, R_{i_3}, R_{i_6}, R_{i_8}, $R_{i_{10}}$, $R_{i_{12}}$, $R_{i_{13}}$ and $R_{i_{16}}$, $sign_b = 1$ for R_1, R_4, R_7, R_8, $R_{i_{11}}$, $R_{i_{12}}$, $R_{i_{13}}$ and $R_{i_{15}}$, $sign_a = sign_b = -1$ for others; for $R_{i_2 - i_4}$, $z_i^1 = 2h_1 - \beta_i$, for $R_{i_5 - i_7}$, $z_i^1 = 2(h_1 + h_2) - \beta_i$, for $R_{i_9 - i_{11}}$, $z_i^1 = 2(h_1 + h_2 + h_3) - \beta_i$, for $R_{i_{14} - i_{16}}$, $z_i^1 = 2(h_1 + h_3) - \beta_i$, for R_{i_8}, $z_i^1 = 2h_2 - \beta_i$, for $R_{i_{12}}$, $z_i^1 = 2(h_2 + h_3) - \beta_i$, for $R_{i_{13}}$, $z_i^1 = 2h_3 - \beta_i$, for R_1, $z_i^1 = -\beta_i$.

Formula (3.41) can be further written into two parts, namely the part without frequency factor and the part with frequency factor, as shown in the following equation.

$$G_{11}^{\varphi^f}(x',y',z';x,y,z;\omega) = G_{11}^{\varphi_s^f}(x',y',z';x,y,z) + G_{11}^{\varphi_q^f}(x',y',z';x,y,z;\omega) \tag{3.42}$$

$$\begin{aligned}
G_{11}^{\varphi^f}(x',y',z';x,y,z;\omega) = {} & \frac{I_s(x',y',z')}{4\pi\bar{\sigma}_1}\frac{1}{R} + \frac{I_s(x',y',z')}{4\pi\bar{\sigma}_1}[(-\frac{k_{01}^{\varphi}}{R_1} + \frac{k_{12}^{\varphi}}{R_2} - \frac{k_{01}^{\varphi}k_{12}^{\varphi}}{R_3} - \frac{k_{01}^{\varphi}k_{12}^{\varphi}}{R_4} + \frac{k_{23}^{\varphi}}{R_5} - \frac{k_{01}^{\varphi}k_{23}^{\varphi}}{R_6} \\
& - \frac{k_{01}^{\varphi}k_{23}^{\varphi}}{R_7} - \frac{k_{01}^{\varphi}k_{12}^{\varphi}k_{23}^{\varphi}}{R_8} + \frac{k_{34}^{\varphi}}{R_9} - \frac{k_{01}^{\varphi}k_{34}^{\varphi}}{R_{10}} - \frac{k_{01}^{\varphi}k_{34}^{\varphi}}{R_{11}} - \frac{k_{01}^{\varphi}k_{23}^{\varphi}k_{34}^{\varphi}}{R_{12}} - \frac{k_{01}^{\varphi}k_{12}^{\varphi}k_{34}^{\varphi}}{R_{13}} + \frac{k_{12}^{\varphi}k_{23}^{\varphi}k_{34}^{\varphi}}{R_{14}} - \frac{k_{01}^{\varphi}k_{12}^{\varphi}k_{23}^{\varphi}k_{34}^{\varphi}}{R_{15}} \\
& - \frac{k_{01}^{\varphi}k_{12}^{\varphi}k_{23}^{\varphi}k_{34}^{\varphi}}{R_{16}}) + \sum_{i=1}^{M}\alpha_i(-\frac{k_{01}^{\varphi}}{R_{i_1}} + \frac{k_{12}^{\varphi}}{R_{i_2}} - \frac{k_{01}^{\varphi}k_{12}^{\varphi}}{R_{i_3}} - \frac{k_{01}^{\varphi}k_{12}^{\varphi}}{R_{i_4}} + \frac{k_{23}^{\varphi}}{R_{i_5}} - \frac{k_{01}^{\varphi}k_{23}^{\varphi}}{R_{i_6}} - \frac{k_{01}^{\varphi}k_{23}^{\varphi}}{R_{i_7}} - \frac{k_{01}^{\varphi}k_{12}^{\varphi}k_{23}}{R_{i_8}} \\
& + \frac{k_{34}^{\varphi}}{R_{i_9}} - \frac{k_{01}^{\varphi}k_{34}^{\varphi}}{R_{i_{10}}} - \frac{k_{01}^{\varphi}k_{34}^{\varphi}}{R_{i_{11}}} - \frac{k_{01}^{\varphi}k_{23}^{\varphi}k_{34}^{\varphi}}{R_{i_{12}}} - \frac{k_{01}^{\varphi}k_{12}^{\varphi}k_{34}^{\varphi}}{R_{i_{13}}} + \frac{k_{12}^{\varphi}k_{23}^{\varphi}k_{34}^{\varphi}}{R_{i_{14}}} - \frac{k_{01}^{\varphi}k_{12}^{\varphi}k_{23}^{\varphi}k_{34}^{\varphi}}{R_{i_{15}}} - \frac{k_{01}^{\varphi}k_{12}^{\varphi}k_{23}^{\varphi}k_{34}^{\varphi}}{R_{i_{16}}})]
\end{aligned} \tag{3.43}$$

so

$$G_{11}^{\varphi_s^f}(x',y',z';x,y,z) = \frac{I_s(x',y',z')}{4\pi\bar{\sigma}_1}\frac{1}{R} \tag{3.44}$$

$$\begin{aligned}
G_{11}^{\varphi_q^f}(x',y',z';x,y,z;\omega) = {} & \frac{I_s(x',y',z')}{4\pi\bar{\sigma}_1}[(-\frac{k_{01}^{\varphi}}{R_1} + \frac{k_{12}^{\varphi}}{R_2} - \frac{k_{01}^{\varphi}k_{12}^{\varphi}}{R_3} - \frac{k_{01}^{\varphi}k_{12}^{\varphi}}{R_4} + \frac{k_{23}^{\varphi}}{R_5} - \frac{k_{01}^{\varphi}k_{23}^{\varphi}}{R_6} \\
& - \frac{k_{01}^{\varphi}k_{23}^{\varphi}}{R_7} - \frac{k_{01}^{\varphi}k_{12}^{\varphi}k_{23}^{\varphi}}{R_8} + \frac{k_{34}^{\varphi}}{R_9} - \frac{k_{01}^{\varphi}k_{34}^{\varphi}}{R_{10}} - \frac{k_{01}^{\varphi}k_{34}^{\varphi}}{R_{11}} - \frac{k_{01}^{\varphi}k_{23}^{\varphi}k_{34}^{\varphi}}{R_{12}} - \frac{k_{01}^{\varphi}k_{12}^{\varphi}k_{34}^{\varphi}}{R_{13}} + \frac{k_{12}^{\varphi}k_{23}^{\varphi}k_{34}^{\varphi}}{R_{14}} - \frac{k_{01}^{\varphi}k_{12}^{\varphi}k_{23}^{\varphi}k_{34}^{\varphi}}{R_{15}} \\
& - \frac{k_{01}^{\varphi}k_{12}^{\varphi}k_{23}^{\varphi}k_{34}^{\varphi}}{R_{16}}) + \sum_{i=1}^{M}\alpha_i(-\frac{k_{01}^{\varphi}}{R_{i_1}} + \frac{k_{12}^{\varphi}}{R_{i_2}} - \frac{k_{01}^{\varphi}k_{12}^{\varphi}}{R_{i_3}} - \frac{k_{01}^{\varphi}k_{12}^{\varphi}}{R_{i_4}} + \frac{k_{23}^{\varphi}}{R_{i_5}} - \frac{k_{01}^{\varphi}k_{23}^{\varphi}}{R_{i_6}} - \frac{k_{01}^{\varphi}k_{23}^{\varphi}}{R_{i_7}} - \frac{k_{01}^{\varphi}k_{12}^{\varphi}k_{23}}{R_{i_8}} \\
& + \frac{k_{34}^{\varphi}}{R_{i_9}} - \frac{k_{01}^{\varphi}k_{34}^{\varphi}}{R_{i_{10}}} - \frac{k_{01}^{\varphi}k_{34}^{\varphi}}{R_{i_{11}}} - \frac{k_{01}^{\varphi}k_{23}^{\varphi}k_{34}^{\varphi}}{R_{i_{12}}} - \frac{k_{01}^{\varphi}k_{12}^{\varphi}k_{34}^{\varphi}}{R_{i_{13}}} + \frac{k_{12}^{\varphi}k_{23}^{\varphi}k_{34}^{\varphi}}{R_{i_{14}}} - \frac{k_{01}^{\varphi}k_{12}^{\varphi}k_{23}^{\varphi}k_{34}^{\varphi}}{R_{i_{15}}} - \frac{k_{01}^{\varphi}k_{12}^{\varphi}k_{23}^{\varphi}k_{34}^{\varphi}}{R_{i_{16}}})]
\end{aligned} \tag{3.45}$$

Formula (3.42) can be generalized to the scalar Green's function where the scalar monopole is located in the ξ-th layer of multi-layer horizontal layered soil and the field point is located in the ζ-th layer.

$$G_{\xi\zeta}^{\varphi^f}(x', y', z'; x, y, z; \omega) = G_{\xi\zeta}^{\varphi_s^f}(x', y', z'; x, y, z) + G_{\xi\zeta}^{\varphi_q^f}(x', y', z'; x, y, z; \omega) \tag{3.46}$$

In addition, since $\bar{r}_s$ is the position vector of the source point, the position coordinates in the Cartesian coordinate system are (x', y', z'); and $\bar{r}_f$ is the position vector of the field point, and the position coordinates in the Cartesian coordinate system are (x, y, z). Therefore, it is shown in the following equation.

$$G_{\xi\zeta}^{\varphi^f}(\bar{r}_s, \bar{r}_f; \omega) = G_{\xi\zeta}^{\varphi_s^f}(\bar{r}_s, \bar{r}_f) + G_{\xi\zeta}^{\varphi_q^f}(\bar{r}_s, \bar{r}_f; \omega) \tag{3.47}$$

The closed form Green's function of a dipole in a four layer horizontally layered soil model in frequency domain is then presented as below.

- Green's function of vertical dipole in multilayered horizontal layered soil.

$$G_{11}^{A_{zz}^f}(x', y', z'; x, y, z; \omega) = \frac{\mu I_{b_z}(x', y', z')}{4\pi}\Big[\frac{1}{R} + \Big(-\frac{k_{01}^{TM}}{R_1} + \frac{k_{12}^{TM}}{R_2} - \frac{k_{01}^{TM}k_{12}^{TM}}{R_3}$$

$$-\frac{k_{01}^{TM}k_{12}^{TM}}{R_4} + \frac{k_{23}^{TM}}{R_5} - \frac{k_{01}^{TM}k_{23}^{TM}}{R_6} - \frac{k_{01}^{TM}k_{23}^{TM}}{R_7} - \frac{k_{01}^{TM}k_{12}^{TM}k_{23}^{TM}}{R_8} + \frac{k_{34}^{TM}}{R_9} - \frac{k_{01}^{TM}k_{34}^{TM}}{R_{10}}$$

$$-\frac{k_{01}^{TM}k_{34}^{TM}}{R_{11}} - \frac{k_{01}^{TM}k_{23}^{TM}k_{34}^{TM}}{R_{12}} - \frac{k_{01}^{TM}k_{12}^{TM}k_{34}^{TM}}{R_{13}} + \frac{k_{12}^{TM}k_{23}^{TM}k_{34}^{TM}}{R_{14}} - \frac{k_{01}^{TM}k_{12}^{TM}k_{23}^{TM}k_{34}^{TM}}{R_{15}}$$

$$-\frac{k_{01}^{TM}k_{12}^{TM}k_{23}^{TM}k_{34}^{TM}}{R_{16}}\Big) + \sum_{i=1}^{M}\alpha_i\Big(-\frac{k_{01}^{TM}}{R_{i_1}} + \frac{k_{12}^{TM}}{R_{i_2}} - \frac{k_{01}^{TM}k_{12}^{TM}}{R_{i_3}} - \frac{k_{01}^{TM}k_{12}^{TM}}{R_{i_4}} + \frac{k_{23}^{TM}}{R_{i_5}} - \frac{k_{01}^{TM}k_{23}^{TM}}{R_{i_6}}$$

$$-\frac{k_{01}^{TM}k_{23}^{TM}}{R_{i_7}} - \frac{k_{01}^{TM}k_{12}^{TM}k_{23}}{R_{i_8}} + \frac{k_{34}^{TM}}{R_{i_9}} - \frac{k_{01}^{TM}k_{34}^{TM}}{R_{i_{10}}} - \frac{k_{01}^{TM}k_{34}^{TM}}{R_{i_{11}}} - \frac{k_{01}^{TM}k_{23}^{TM}k_{34}^{TM}}{R_{i_{12}}} - \frac{k_{01}^{TM}k_{12}^{TM}k_{34}^{TM}}{R_{i_{13}}}$$

$$+\frac{k_{12}^{TM}k_{23}^{TM}k_{34}^{TM}}{R_{i_{14}}} - \frac{k_{01}^{TM}k_{12}^{TM}k_{23}^{TM}k_{34}^{TM}}{R_{i_{15}}} - \frac{k_{01}^{TM}k_{12}^{TM}k_{23}^{TM}k_{34}^{TM}}{R_{i_{16}}}\Big)\Big] \tag{3.48}$$

Formula (3.48) can be further written into two parts, namely the part without frequency factor and the part with frequency factor, as shown in the following equation.

$$G_{11}^{A_{zz}^f}(x', y', z'; x, y, z; \omega) = G_{11}^{A_s^f\,zz}(x', y', z'; x, y, z) + G_{11}^{A_q^f\,zz}(x', y', z'; x, y, z; \omega) \tag{3.49}$$

$$G_{11}^{A_{zz}^f}(x', y', z'; x, y, z; \omega) = \frac{\mu I_{b_z}(x', y', z')}{4\pi}\frac{1}{R} + \frac{\mu I_{b_z}(x', y', z')}{4\pi}\Big[\Big(-\frac{k_{01}^{TM}}{R_1} + \frac{k_{12}^{TM}}{R_2} - \frac{k_{01}^{TM}k_{12}^{TM}}{R_3}$$

$$-\frac{k_{01}^{TM}k_{12}^{TM}}{R_4}+\frac{k_{23}^{TM}}{R_5}-\frac{k_{01}^{TM}k_{23}^{TM}}{R_6}-\frac{k_{01}^{TM}k_{23}^{TM}}{R_7}-\frac{k_{01}^{TM}k_{12}^{TM}k_{23}^{TM}}{R_8}+\frac{k_{34}^{TM}}{R_9}-\frac{k_{01}^{TM}k_{34}^{TM}}{R_{10}}-\frac{k_{01}^{TM}k_{34}^{TM}}{R_{11}}$$

$$-\frac{k_{01}^{TM}k_{23}^{TM}k_{34}^{TM}}{R_{12}}-\frac{k_{01}^{TM}k_{12}^{TM}k_{34}^{TM}}{R_{13}}+\frac{k_{12}^{TM}k_{23}^{TM}k_{34}^{TM}}{R_{14}}-\frac{k_{01}^{TM}k_{12}^{TM}k_{23}^{TM}k_{34}^{TM}}{R_{15}}-\frac{k_{01}^{TM}k_{12}^{TM}k_{23}^{TM}k_{34}^{TM}}{R_{16}})$$

$$+\sum_{i=1}^{M}\alpha_i(-\frac{k_{01}^{TM}}{R_{i_1}}+\frac{k_{12}^{TM}}{R_{i_2}}-\frac{k_{01}^{TM}k_{12}^{TM}}{R_{i_3}}-\frac{k_{01}^{TM}k_{12}^{TM}}{R_{i_4}}+\frac{k_{23}^{TM}}{R_{i_5}}-\frac{k_{01}^{TM}k_{23}^{TM}}{R_{i_6}}-\frac{k_{01}^{TM}k_{23}^{TM}}{R_{i_7}}-\frac{k_{01}^{TM}k_{12}^{TM}k_{23}}{R_{i_8}}$$

$$+\frac{k_{34}^{TM}}{R_{i_9}}-\frac{k_{01}^{TM}k_{34}^{TM}}{R_{i_{10}}}-\frac{k_{01}^{TM}k_{34}^{TM}}{R_{i_{11}}}-\frac{k_{01}^{TM}k_{23}^{TM}k_{34}^{TM}}{R_{i_{12}}}-\frac{k_{01}^{TM}k_{12}^{TM}k_{34}^{TM}}{R_{i_{13}}}+\frac{k_{12}^{TM}k_{23}^{TM}k_{34}^{TM}}{R_{i_{14}}}$$

$$-\frac{k_{01}^{TM}k_{12}^{TM}k_{23}^{TM}k_{34}^{TM}}{R_{i_{15}}}-\frac{k_{01}^{TM}k_{12}^{TM}k_{23}^{TM}k_{34}^{TM}}{R_{i_{16}}})] \tag{3.50}$$

so

$$G_{11}^{A_s^f zz}(x',y',z';x,y,z)=\frac{\mu I_{b_z}(x',y',z')}{4\pi}\frac{1}{R} \tag{3.51}$$

$$G_{11}^{A_q^f zz}(x',y',z';x,y,z;\omega)=\frac{\mu I_{b_z}(x',y',z')}{4\pi}[(-\frac{k_{01}^{TM}}{R_1}+\frac{k_{12}^{TM}}{R_2}-\frac{k_{01}^{TM}k_{12}^{TM}}{R_3}-\frac{k_{01}^{TM}k_{12}^{TM}}{R_4}$$

$$+\frac{k_{23}^{TM}}{R_5}-\frac{k_{01}^{TM}k_{23}^{TM}}{R_6}-\frac{k_{01}^{TM}k_{23}^{TM}}{R_7}-\frac{k_{01}^{TM}k_{12}^{TM}k_{23}^{TM}}{R_8}+\frac{k_{34}^{TM}}{R_9}-\frac{k_{01}^{TM}k_{34}^{TM}}{R_{10}}-\frac{k_{01}^{TM}k_{34}^{TM}}{R_{11}}$$

$$-\frac{k_{01}^{TM}k_{23}^{TM}k_{34}^{TM}}{R_{12}}-\frac{k_{01}^{TM}k_{12}^{TM}k_{34}^{TM}}{R_{13}}+\frac{k_{12}^{TM}k_{23}^{TM}k_{34}^{TM}}{R_{14}}-\frac{k_{01}^{TM}k_{12}^{TM}k_{23}^{TM}k_{34}^{TM}}{R_{15}}-\frac{k_{01}^{TM}k_{12}^{TM}k_{23}^{TM}k_{34}^{TM}}{R_{16}})$$

$$+\sum_{i=1}^{M}\alpha_i(-\frac{k_{01}^{TM}}{R_{i_1}}+\frac{k_{12}^{TM}}{R_{i_2}}-\frac{k_{01}^{TM}k_{12}^{TM}}{R_{i_3}}-\frac{k_{01}^{TM}k_{12}^{TM}}{R_{i_4}}+\frac{k_{23}^{TM}}{R_{i_5}}-\frac{k_{01}^{TM}k_{23}^{TM}}{R_{i_6}}-\frac{k_{01}^{TM}k_{23}^{TM}}{R_{i_7}}-\frac{k_{01}^{TM}k_{12}^{TM}k_{23}}{R_{i_8}}$$

$$+\frac{k_{34}^{TM}}{R_{i_9}}-\frac{k_{01}^{TM}k_{34}^{TM}}{R_{i_{10}}}-\frac{k_{01}^{TM}k_{34}^{TM}}{R_{i_{11}}}-\frac{k_{01}^{TM}k_{23}^{TM}k_{34}^{TM}}{R_{i_{12}}}-\frac{k_{01}^{TM}k_{12}^{TM}k_{34}^{TM}}{R_{i_{13}}}+\frac{k_{12}^{TM}k_{23}^{TM}k_{34}^{TM}}{R_{i_{14}}}$$

$$-\frac{k_{01}^{TM}k_{12}^{TM}k_{23}^{TM}k_{34}^{TM}}{R_{i_{15}}}-\frac{k_{01}^{TM}k_{12}^{TM}k_{23}^{TM}k_{34}^{TM}}{R_{i_{16}}})] \tag{3.52}$$

- Green's function of horizontal dipole in multilayered horizontal layered soil.
  - Green's function of the vertical component of the horizontal dipole in multilayered horizontal layered soil.

$$G_{11}^{A_{xz}^f}(x',y',z';x,y,z;\omega)=\frac{\mu I_{b_x}(x',y',z')}{4\pi}\frac{\partial}{\partial x}[+(-k_{01}^{TM}\ln(z_1+R_1)+k_{12}^{TM}\ln(z_2+R_2)$$

$$-k_{01}^{TM}k_{12}^{TM}\ln(z_3+R_3)-k_{01}^{TM}k_{12}^{TM}\ln(z_4+R_4)+k_{23}^{TM}\ln(z_5+R_5)$$

$$-k_{01}^{TM}k_{23}^{TM}\ln(z_6+R_6)-k_{01}^{TM}k_{23}^{TM}\ln(z_7+R_7)-k_{01}^{TM}k_{12}^{TM}k_{23}^{TM}\ln(z_8+R_8)$$

$$+k_{34}^{TM}\ln(z_9+R_9)-k_{01}^{TM}k_{34}^{TM}\ln(z_{10}+R_{10})-k_{01}^{TM}k_{34}^{TM}\ln(z_{11}+R_{11})$$

$$-k_{01}^{TM}k_{23}^{TM}k_{34}^{TM}\ln(z_{12}+R_{12})-k_{01}^{TM}k_{12}^{TM}k_{34}^{TM}\ln(z_{13}+R_{13})+k_{12}^{TM}k_{23}^{TM}k_{34}^{TM}\ln(z_{14}+R_{14})$$

$$-k_{01}^{TM}k_{12}^{TM}k_{23}^{TM}k_{34}^{TM}\ln(z_{15}+R_{15})-k_{01}^{TM}k_{12}^{TM}k_{23}^{TM}k_{34}^{TM}\ln(z_{16}+R_{16}))$$

$$+\sum_{i=1}^{M}\alpha_i(-k_{01}^{TM}\ln(z_{i_1}+R_{i_1})+k_{12}^{TM}\ln(z_{i_2}+R_{i_2})-k_{01}^{TM}k_{12}^{TM}\ln(z_{i_3}+R_{i_3})$$

$$-k_{01}^{TM}k_{12}^{TM}\ln(z_{i_4}+R_{i_4})+k_{23}^{TM}\ln(z_{i_5}+R_{i_5})-k_{01}^{TM}k_{23}^{TM}\ln(z_{i_6}+R_{i_6})$$

$$-k_{01}^{TM}k_{23}^{TM}\ln(z_{i_7}+R_{i_7})-k_{01}^{TM}k_{12}^{TM}k_{23}\ln(z_{i_8}+R_{i_8})+k_{34}^{TM}\ln(z_{i_9}+R_{i_9})$$

$$-k_{01}^{TM}k_{34}^{TM}\ln(z_{i_{10}}+R_{i_{10}})-k_{01}^{TM}k_{34}^{TM}\ln(z_{i_{11}}+R_{i_{11}})-k_{01}^{TM}k_{23}^{TM}k_{34}^{TM}\ln(z_{i_{12}}+R_{i_{12}})$$

$$-k_{01}^{TM}k_{12}^{TM}k_{34}^{TM}\ln(z_{i_{13}}+R_{i_{13}})+k_{12}^{TM}k_{23}^{TM}k_{34}^{TM}\ln(z_{i_{14}}+R_{i_{14}})$$

$$-k_{01}^{TM}k_{12}^{TM}k_{23}^{TM}k_{34}^{TM}\ln(z_{i_{15}}+R_{i_{15}})-k_{01}^{TM}k_{12}^{TM}k_{23}^{TM}k_{34}^{TM}\ln(z_{i_{16}}+R_{i_{16}}))] \tag{3.53}$$

Formula (3.53) can also be theoretically divided into two parts: the part without frequency and the part with frequency, however, observing formula (3.53), it can be seen that this formula only contains the frequency part, therefore, the non frequency part is zero.

$$G_{11}^{A_{xz}^f}(x',y',z';x,y,z;\omega)=G_{11}^{A_s^f xz}(x',y',z';x,y,z)+G_{11}^{A_q^f xz}(x',y',z';x,y,z;\omega) \tag{3.54}$$

$$G_{11}^{A_s^f xz}(x',y',z';x,y,z)=0 \tag{3.55}$$

$$G_{11}^{A_q^f xz}(x',y',z';x,y,z;\omega)=\frac{\mu I_{b_x}(x',y',z')}{4\pi}\frac{\partial}{\partial x}[(-k_{01}^{TM}\ln(z_1+R_1)+k_{12}^{TM}\ln(z_2+R_2)$$

$$-k_{01}^{TM}k_{12}^{TM}\ln(z_3+R_3)-k_{01}^{TM}k_{12}^{TM}\ln(z_4+R_4)+k_{23}^{TM}\ln(z_5+R_5)$$

$$-k_{01}^{TM}k_{23}^{TM}\ln(z_6+R_6)-k_{01}^{TM}k_{23}^{TM}\ln(z_7+R_7)-k_{01}^{TM}k_{12}^{TM}k_{23}^{TM}\ln(z_8+R_8)$$

$$+k_{34}^{TM}\ln(z_9+R_9)-k_{01}^{TM}k_{34}^{TM}\ln(z_{10}+R_{10})-k_{01}^{TM}k_{34}^{TM}\ln(z_{11}+R_{11})$$

$$-k_{01}^{TM}k_{23}^{TM}k_{34}^{TM}\ln(z_{12}+R_{12})-k_{01}^{TM}k_{12}^{TM}k_{34}^{TM}\ln(z_{13}+R_{13})+k_{12}^{TM}k_{23}^{TM}k_{34}^{TM}\ln(z_{14}+R_{14})$$

$$-k_{01}^{TM}k_{12}^{TM}k_{23}^{TM}k_{34}^{TM}\ln(z_{15}+R_{15})-k_{01}^{TM}k_{12}^{TM}k_{23}^{TM}k_{34}^{TM}\ln(z_{16}+R_{16}))$$

$$+\sum_{i=1}^{M}\alpha_i(-k_{01}^{TM}\ln(z_{i_1}+R_{i_1})+k_{12}^{TM}\ln(z_{i_2}+R_{i_2})-k_{01}^{TM}k_{12}^{TM}\ln(z_{i_3}+R_{i_3})$$

$$-k_{01}^{TM}k_{12}^{TM}\ln(z_{i_4}+R_{i_4})+k_{23}^{TM}\ln(z_{i_5}+R_{i_5})-k_{01}^{TM}k_{23}^{TM}\ln(z_{i_6}+R_{i_6})$$

$$-k_{01}^{TM}k_{23}^{TM}\ln(z_{i_7}+R_{i_7})-k_{01}^{TM}k_{12}^{TM}k_{23}\ln(z_{i_8}+R_{i_8})+k_{34}^{TM}\ln(z_{i_9}+R_{i_9})$$

$$-k_{01}^{TM}k_{34}^{TM}\ln(z_{i10}+R_{i10})-k_{01}^{TM}k_{34}^{TM}\ln(z_{i_{11}}+R_{i_{11}})-k_{01}^{TM}k_{23}^{TM}k_{34}^{TM}\ln(z_{i_{12}}+R_{i_{12}})$$

$$-k_{01}^{TM}k_{12}^{TM}k_{34}^{TM}\ln(z_{i_{13}}+R_{i_{13}})+k_{12}^{TM}k_{23}^{TM}k_{34}^{TM}\ln(z_{i_{14}}+R_{i_{14}})$$

$$-k_{01}^{TM}k_{12}^{TM}k_{23}^{TM}k_{34}^{TM}\ln(z_{i15}+R_{i_{15}})-k_{01}^{TM}k_{12}^{TM}k_{23}^{TM}k_{34}^{TM}\ln(z_{i_{16}}+R_{i_{16}}))] \tag{3.56}$$

- Green's function of the horizontal component of the horizontal dipole in multi-layered horizontal layered soil.

$$G_{11}^{A_{xx}^f}(x',y',z';x,y,z;\omega)=\frac{\mu I_{b_x}(x',y',z')}{4\pi}\frac{1}{R} \tag{3.57}$$

Similarly, by observing formula (3.57), it can be seen that this formula only contains the non frequency part, so the frequency part is zero.

$$G_{11}^{A_{xx}^f}(x',y',z';x,y,z;\omega)=G_{11}^{A_s^f\,xx}(x',y',z';x,y,z)+G_{11}^{A_q^f\,xx}(x',y',z';x,y,z;\omega) \tag{3.58}$$

$$G_{11}^{A_s^f\,xx}(x',y',z';x,y,z)=\frac{\mu I_{b_x}(x',y',z')}{4\pi}\frac{1}{R} \tag{3.59}$$

$$G_{11}^{A_q^f\,xx}(x',y',z';x,y,z;\omega)=0 \tag{3.60}$$

Obviously, for the three-layer horizontally layered soil model, the $k_{34}^{\varphi}=0$ and $k_{34}^{TM}=0$ in Eqs. (3.41), (3.48) and (3.53) will degenerate into the closed form Green's function of the point current source in the three-layer horizontally layered soil model.

For a two-layer horizontally layered soil model, the $k_{23}^{\varphi} = 0$, $k_{23}^{TM} = 0$, $k_{34}^{\varphi} = 0$ and $k_{34}^{TM} = 0$ in Eqs. (3.41), (3.48) and (3.53) degenerate into the closed form Green's function of the point current source in the two-layer horizontally layered soil model.

The dyadic Green's function for arbitrarily placed vector dipoles in the first layer of soil in the multilayer horizontally stratified soil model can be simply written as follows.

$$\bar{\bar{G}}_{11}^{A^f}(x', y', z'; x, y, z; \omega) = \begin{bmatrix} G_{11}^{A_{xx}^f}(x', y', z'; x, y, z; \omega) & 0 & G_{11}^{A_{xz}^f}(x', y', z'; x, y, z; \omega) \\ 0 & G_{11}^{A_{yy}^f}(x', y', z'; x, y, z; \omega) & G_{11}^{A_{yz}^f}(x', y', z'; x, y, z; \omega) \\ 0 & 0 & G_{11}^{A_{zz}^f}(x', y', z'; x, y, z; \omega) \end{bmatrix} \tag{3.61}$$

$$\bar{\bar{G}}_{11}^{A^f}(x', y', z'; x, y, z; \omega) = \bar{\bar{G}}_{11}^{A_s^f}(x', y', z'; x, y, z) + \bar{\bar{G}}_{11}^{A_q^f}(x', y', z'; x, y, z; \omega) \tag{3.62}$$

$$\bar{\bar{G}}_{11}^{A_s^f}(x', y', z'; x, y, z; \omega) = \begin{bmatrix} G_{11}^{A_s^f xx}(x', y', z'; x, y, z) & 0 & \\ 0 & G_{11}^{A_s^f yy}(x', y', z'; x, y, z) & \\ 0 & 0 & G_{11}^{A_s^f zz}(x', y', z'; x, y, z) \end{bmatrix} \tag{3.63}$$

$$\bar{\bar{G}}_{11}^{A_q^f}(x', y', z'; x, y, z; \omega) = \begin{bmatrix} 0 \; 0 \; G_{11}^{A_q^f xz}(x', y', z'; x, y, z; \omega) \\ 0 \; 0 \; G_{11}^{A_q^f yz}(x', y', z'; x, y, z; \omega) \\ 0 \; 0 \; G_{11}^{A_q^f zz}(x', y', z'; x, y, z; \omega) \end{bmatrix} \tag{3.64}$$

Formula (3.62) can be generalized to the dyadic Green's function where the vector dipole is located in the ξ-th layer of multi-layer horizontal layered soil and the field point is located in the ζ-th layer.

$$\bar{\bar{G}}_{\xi\zeta}^{A^f}(x', y', z'; x, y, z; \omega) = \bar{\bar{G}}_{\xi\zeta}^{A_s^f}(x', y', z'; x, y, z) + \bar{\bar{G}}_{\xi\zeta}^{A_q^f}(x', y', z'; x, y, z; \omega) \tag{3.65}$$

In addition, there is a vector position expression as follows

$$\bar{\bar{G}}_{\xi\zeta}^{A^f}(\bar{r}_s; \bar{r}_f; \omega) = \bar{\bar{G}}_{\xi\zeta}^{A_s^f}(\bar{r}_s; \bar{r}_f) + \bar{\bar{G}}_{\xi\zeta}^{A_q^f}(\bar{r}_s; \bar{r}_f; \omega) \tag{3.66}$$

3.2.5 Closed Form of the Green's Function of a Point Source in Horizontal Multilayered Earth Model in s Domain

In the previous section, the frequency domain Green's function of point sources in multilayered horizontal layered soil is introduced. In this section, the frequency domain Green's function is further extended to the s domain.

Referring to the frequency domain Green's function formulas (3.46) and (3.66) of point sources in multilayered horizontal layered soil, it can be seen that the s-domain Green's function formulas 1 and 2 of point sources in multilayered horizontal layered soil can be obtained by simply replacing $j\omega$ with s.

$$G_{ij}^{\varphi^s}(x', y', z'; x, y, z; s) = G_{ij}^{\varphi_s^s}(x', y', z'; x, y, z) + G_{ij}^{\varphi_q^s}(x', y', z'; x, y, z; s) \tag{3.67}$$

$$\bar{\bar{G}}_{ij}^{A^s}(x', y', z'; x, y, z; s) = \bar{\bar{G}}_{ij}^{A_s^s}(x', y', z'; x, y, z) + \bar{\bar{G}}_{ij}^{A_q^s}(x', y', z'; x, y, z; s) \tag{3.68}$$

In addition, there is a vector position expression as follows

$$G_{ij}^{\varphi^s}(\bar{r}_s; \bar{r}_f; s) = G_{ij}^{\varphi_s^s}(\bar{r}_s; \bar{r}_f) + G_{ij}^{\varphi_q^s}(\bar{r}_s; \bar{r}_f; s) \tag{3.69}$$

$$\bar{\bar{G}}_{ij}^{A^s}(\bar{r}_s; \bar{r}_f; s) = \bar{\bar{G}}_{ij}^{A_s^s}(\bar{r}_s; \bar{r}_f) + \bar{\bar{G}}_{ij}^{A_q^s}(\bar{r}_s; \bar{r}_f; s) \tag{3.70}$$

3.2.6 Simulation Result Analysis in Frequency Domain

Numerical results for QSCIM in frequency domain

For two-layer earth model, we give conductivities $300^{-1}\ S/m$, $100^{-1}\ S/m$ and permittivities $10\ \varepsilon_0$, $12\ \varepsilon_0$ to the first and second layer of the earth, respectively. The thickness of first layer earth is 5 m, and frequency is 50 Hz. Applying the 1-d MP approach, the kernel of Green's function will arrive at relative error 0.0031% with only three terms of quasi-static complex images. Figure 3.2 show this monotone function and superposition situation of two spatial curves from function and simulation and calculation error along with the distribution of k_ρ variables. The coefficients of quasi-static complex images are given in Table 3.1.

It can be seen from the figure that the two spatial curves fully agree. These demonstrate the accuracy and rationality of the frequency domain 1-d QSCIM developed in this chapter. From Table 3.1, we know that there are a pair of conjugate complex number, α_n and β_n $(n = 2, 3)$, which occurred in electrostatic field complex images [32]. Reference [32] explains that "It is to be pointed out that the complex images, in general, have complex locations and amplitudes. In electrostatics, the complex images come in conjugate pairs to give the real potential field value". Here, although quasi-static electrical field is chosen for the mathematical model, conjugate complex images also occur.

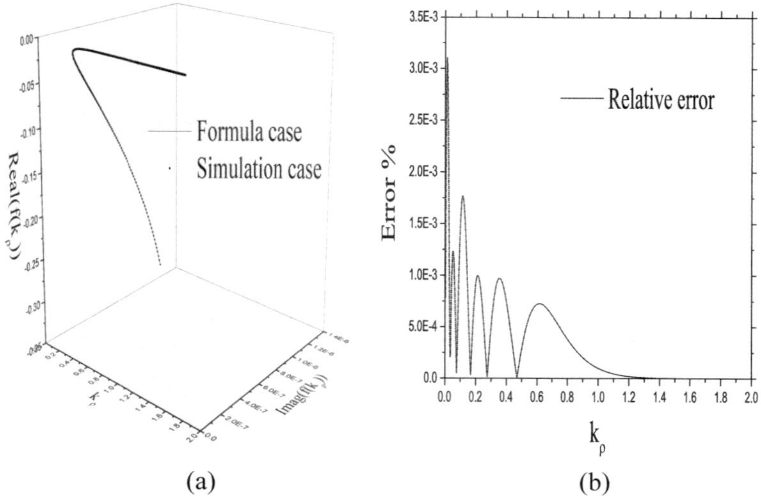

Fig. 3.2 Two spatial curves from function $f(k_\rho)$ and its simulation for 2-layer earth case

Table 3.1 Coefficients of Quasi-static complex images for a two-layer earth model

n	α_n	β_n
1	$0.489\angle+180$	$9.964\angle-180$
2	$9.045\angle-30.382$	$20.137\angle+167.082$
3	$9.045\angle+30.382$	$20.137\angle-167.082$

In the previous example, due to the fact that the resistivity of the upper layer of soil is greater than the resistivity of the lower layer of earth, a conjugate complex image coefficient appears. In order to verify whether this phenomenon is common, the upper and lower resistivities of the following calculation examples are interchanged. So for the two-layer earth model, the conductivities 100^{-1} S/m, 300^{-1} S/m and permittivities 10 ε_0, 12 ε_0 are given to the first and second layer of the earth, respectively. The thickness of first layer earth is 5 m, and frequency is also 50 Hz. Applying the 1-d MP approach, the kernel of Green's function will arrive at relative error 0.0121% with only three terms of quasi-static complex images. Figure 3.3 show this monotone function and superposition situation of two spatial curves from function and simulation and calculation error along with the distribution of k_ρ variables. The coefficients of quasi-static complex images are given in Table 3.2.

It can also be seen from the figure that the two spatial curves fully agree. These also demonstrate the accuracy and rationality of the frequency domain 1-d QSCIM developed in this chapter. As can be seen from Table 3.2, the conjugate complex number no longer appears, indicating that whether the conjugate complex image coefficient occurs depends on the refractive index of the resistivity.

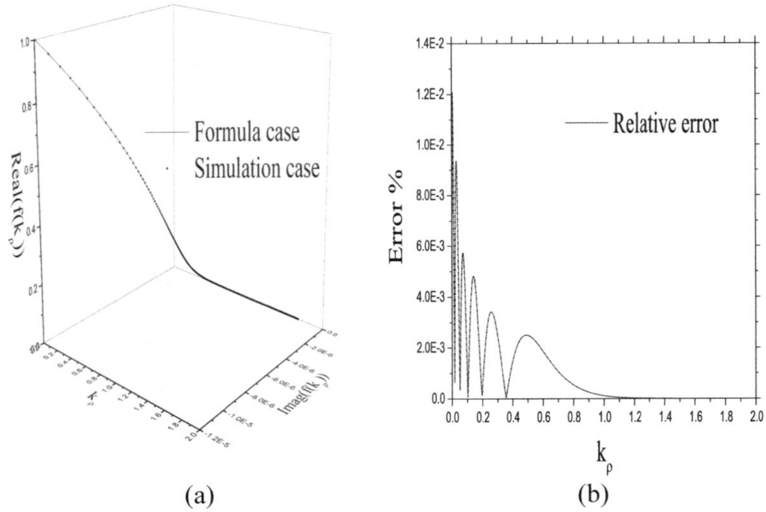

Fig. 3.3 Two spatial curves from function $f(k_\rho)$ and its simulation for 2-layer earth case

Table 3.2 Coefficients of Quasi-static complex images for a two-layer earth model

n	α_n	β_n
1	$0.520\angle - 3.803 \times 10^{-4}$	$10.091\angle + 180$
2	$0.362\angle - 9.018 \times 10^{-4}$	$23.396\angle + 180$
3	$0.116\angle - 1.374 \times 10^{-3}$	$53.181\angle + 180$

For the three-layer earth model, the earth's conductivities and permittivities are $\sigma_1 = 100^{-1}\ S/m$, $\sigma_2 = 200^{-1}\ S/m$, $\sigma_3 = 300^{-1}\ S/m$, $\varepsilon_1 = 10\ \varepsilon_0$, $\varepsilon_2 = 12\ \varepsilon_0$ and $\varepsilon_3 = 10\ \varepsilon_0$, respectively. The first layer earth height is 4 m, and the secondary layer earth height is 0.6 m, and frequency is 50 Hz. With the 1-d MP approach, three terms of quasi-static complex images will arrive at relatively error 0.21%. From Fig. 3.4, we can see that the two spatial curves are superposed. The four quasi-static complex images's coefficients are given in Table 3.3. From the Table 3.3, we know that there is no a pair of conjugate complex number.

To verify whether there are conjugate complex pairs in the three-layer soil model, reverse the order of soil resistivity in the above example. So for the three-layer earth model, the earth's conductivities and permittivities are $\sigma_1 = 300^{-1}\ S/m$, $\sigma_2 = 200^{-1}\ S/m$, $\sigma_3 = 100^{-1}\ S/m$, $\varepsilon_1 = 10\ \varepsilon_0$, $\varepsilon_2 = 12\ \varepsilon_0$ and $\varepsilon_3 = 10\ \varepsilon_0$, respectively. The first layer earth height is 4 m, and the secondary layer earth height is 0.6 m, and frequency is still 50 Hz. With the 1-d MP approach, three terms of quasi-static complex images will arrive at relatively error 0.0092%. From Fig. 3.5, we can see that the two spatial curves are superposed. The three quasi-static complex images's coefficients are given in Table 3.3. From the Table 3.3, we know that there is a pair of conjugate complex number, α_n and β_n ($n = 4, 5$) (Table 3.4).

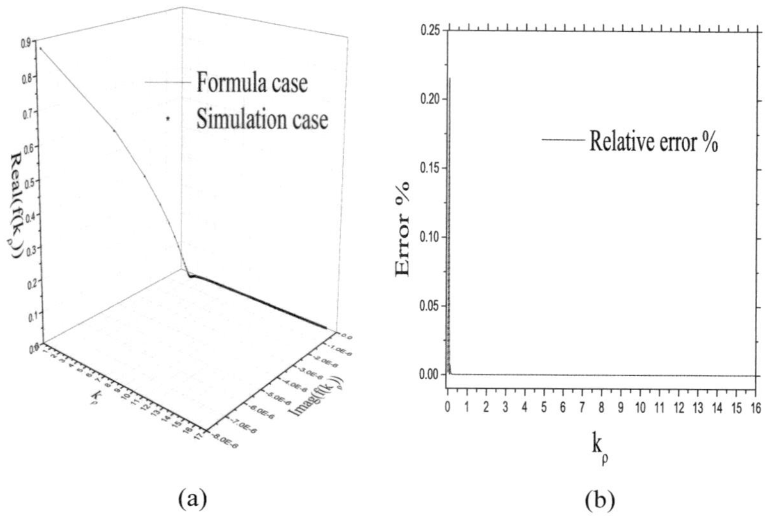

Fig. 3.4 Two spatial curves from $f(k_\rho)$ and its simulation for 3-layer earth case

Table 3.3 Coefficients of Quasi-static complex images for a three-layer earth model

n	α_n	β_n
1	$6.650 \times 10^{-2}\angle + 180$	$1.199\angle + 180$
2	$4.585 \times 10^{-3}\angle - 9.823 \times 10^{-4}$	$2.524\angle - 180$
3	$0.466\angle - 2.583 \times 10^{-4}$	$8.265\angle - 180$
4	$0.295\angle - 6.106 \times 10^{-4}$	$17.460\angle + 180$
5	$0.174\angle - 1.005 \times 10^{-3}$	$35.589\angle + 180$

For four-layer earth model, the earth's conductivities and permittivities are $\sigma_1 = 100^{-1}\ S/m, \sigma_2 = 200^{-1}\ S/m, \sigma_3 = 300^{-1}\ S/m, \sigma_4 = 400^{-1}\ S/m, \varepsilon_1 = 10\,\varepsilon_0, \varepsilon_2 = 12\ \varepsilon_0$, $\varepsilon_3 = 10\ \varepsilon_0$ and $\varepsilon_4 = 10\ \varepsilon_0$, respectively. The first layer earth height is 4. m, secondary layer earth height is 3.4 m and third layer earth height is 9.2 m, frequency is $50Hz$. With the 1-d MP approach, eight terms of quasi-static complex images will arrive at a relative error 0.017%. From Fig. 3.6, we can see that the two spatial curves superpose each other. The coefficients of the six quasi-static complex images are given in Table 3.5.

From Table 3.5, we know that there are no pairs of conjugate complex images.

Similarly, in order to observe whether there is a conjugate complex image coefficient, the order of soil resistivity in the above example is reversed, and the remaining parameters remain unchanged. So for four-layer earth model, the earth's conductivities and permittivities are $\sigma_1 = 100^{-1}\ S/m$, $\sigma_2 = 200^{-1}\ S/m$, $\sigma_3 = 300^{-1}\ S/m$, $\sigma_4 = 400^{-1}\ S/m$. With the 1-d MP approach, eight terms of quasi-static complex images will arrive at a relative error 0.007%. From Fig. 3.7, we can see that the two

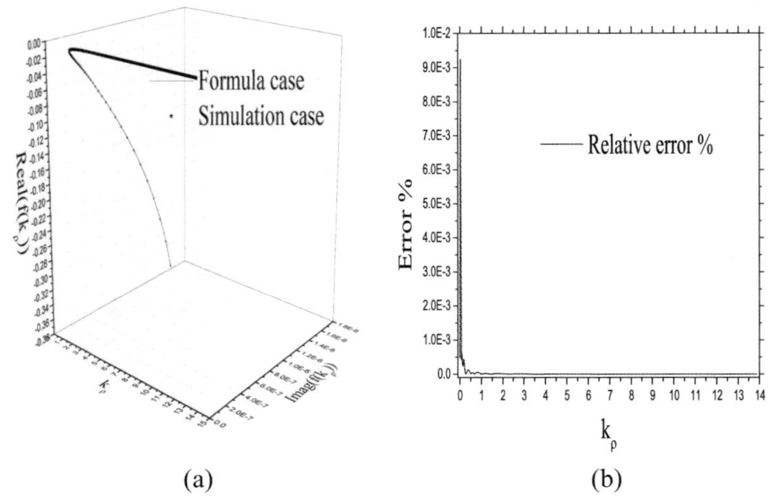

Fig. 3.5 Two spatial curves from $f(k_\rho)$ and its simulation for 3-layer earth case

Table 3.4 Coefficients of Quasi-static complex images for a three-layer earth model

n	α_n	β_n
1	$6.671 \times 10^{-2}\angle + 180$	$1.200\angle - 180$
2	$4.202 \times 10^{-3}\angle - 1.068 \times 10^{-3}$	$2.339\angle + 180$
3	$0.443\angle + 180$	$8.455\angle - 180$
4	$7.298 \times 10^{-2}\angle - 26.338$	$17.677\angle + 165.747$
5	$7.298 \times 10^{-2}\angle + 26.338$	$17.677\angle - 165.747$

Table 3.5 Coefficients of Quasi-static complex images for a four-layer earth model

n	α_n	β_n
1	$6.084 \times 10^{-5}\angle + 180$	$1.980\angle + 180$
2	$0.331\angle - 3.403 \times 10^{-4}$	$8.829\angle + 180$
3	$0.331\angle - 4.570 \times 10^{-4}$	$19.399\angle + 180$
4	$0.404\angle - 8.159 \times 10^{-4}$	$45.194\angle + 180$
5	$0.119\angle - 1.336 \times 10^{-3}$	$104.916\angle + 180$

spatial curves superpose each other. The coefficients of the six quasi-static complex images are given in Table 3.6.

From Table 3.6, we know that there are two pairs of conjugate complex images α_n and β_n (n = 4,and, 5).

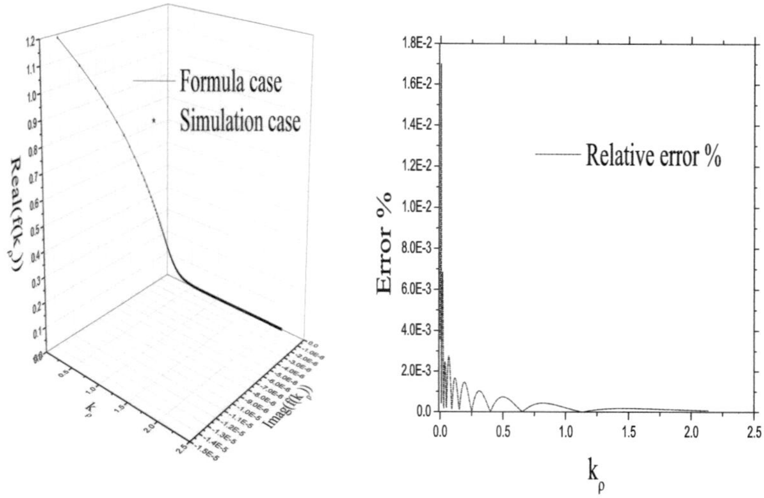

(a) (b)

Fig. 3.6 Two spatial curves from $f(k_\rho)$ and its simulation for 4-layer earth case

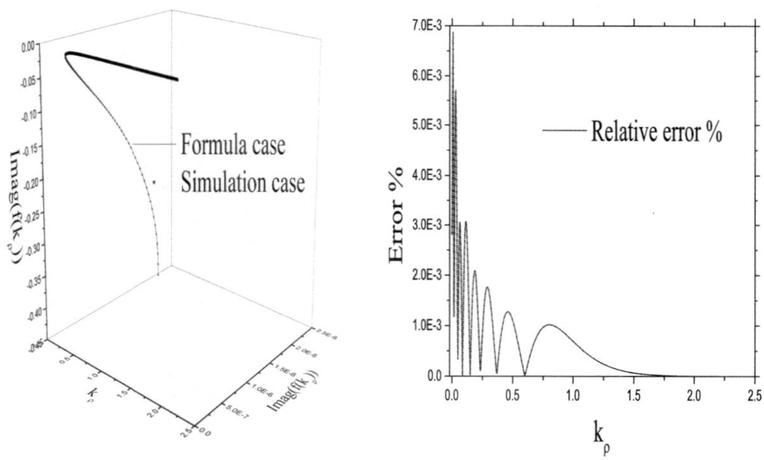

(a) (b)

Fig. 3.7 Two spatial curves from $f(k_\rho)$ and its simulation for 4-layer earth case

Table 3.6 Coefficients of Quasi-static complex images for a four-layer earth model

n	α_n	β_n
1	$0.168\angle + 180$	$7.695\angle - 180$
2	$0.243\angle + 180$	$15.645\angle - 180$
3	$3.837 \times 10^{-2}\angle + 122.297$	$36.263\angle + 153.280$
4	$3.837 \times 10^{-2}\angle - 122.297$	$36.263\angle - 153.280$

Numerical results for a point source in frequency domain

Once the closed form of vector and scalar Green's function of vector dipole and scalar monopole buried in horizontal multilayered earth model have been achieved, the numerical approximation of the electromagnetical field distribution generated by a harmonic current point source can be achieved. The source, in general, is located inside any layer of the horizontally multilayered earth.

Once voltage (reference point at infinite distance) φ and VMP (A) have been obtained for a point source, the electrical field intensity (EFI) (E) and magnetic field intensity (MEI) (B) of the point source can be calculated through following formula.

$$\bar{B} = \nabla \times \bar{A} \tag{3.71}$$

$$\bar{E} = -j\omega\bar{A} - \frac{\partial\varphi}{\partial\tau}\hat{\tau} \tag{3.72}$$

Here $\hat{\tau}$ is unit director vector of the vector dipole.

This chapter adopts a four layer horizontal layered soil model, with four layers of soil conductivity of $100.0^{-1}S/m$, $200.0^{-1}S/m$, $400.0^{-1}S/m$, and $600.0^{-1}S/m$, respectively. The relative dielectric constants of the four layers of soil are 10, 22, 22, and 22, respectively. The thickness of the three layers of soil is 5 meters, 2.4 meters, and 3.4 meters, respectively. The unit point current sources are located 0.5 meters underground, with a frequency of 49.56 kHz. Among them, horizontal dipoles are placed along the x-axis and y-axis respectively, while vertical dipoles are placed along the z-axis.

Figures 3.8, 3.9 and 3.10 show the 3-d distribution of absolute values of voltage, VMP, EFI (E), and MFI (B) on the surface of the earth for a monopole current source and a dipole current source placed along the x-axis, y-axis and z-axis, respectively.

Comparing small figure (a) in the three Figs. 3.8, 3.9 and 3.10, it can be seen that the monopole current has no directional limit, and the three voltage distribution maps are exactly the same.

Comparing the small figures (b), (c) and (d) in the three Figs. 3.8, 3.9 and 3.10, it can be seen that the dipole current has directional limitations, and there are certain differences in the 3-d distribution maps of the three VMP, EFI, and MFI.

3.2.7 Summary

In this chapter, the process how to achieve the closed form of Green's function for a point source (incudes monopole and dipole) within horizontal multilayered earth model by QSCIM has been fully discussed. First, the kernel of the closed form of Green's function for a point source buried in two to four layered horizontal earth model has been introduced. Then, we discussed how the kernel of the Green's function is to be expanded into finite exponential series by 1-d MP method. Lastly, the numerical results about QSCIM for a point source have been discussed.

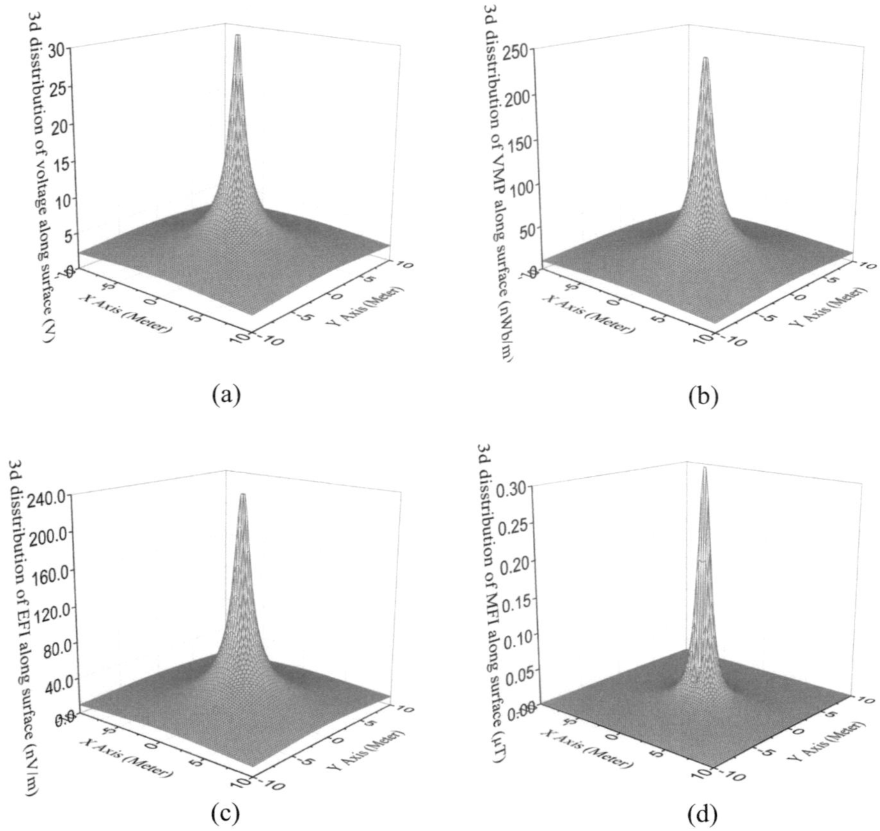

Fig. 3.8 3-d Distribution of $|\varphi|$, $|A|$, $|E|$ and $|B|$ along ground surface above a monopole and an horizontal dipole along x-axis

3.3 QSCIM Based on Matrix Pencil Method in Time Domain and its Closed Form of the Green's Function

3.3.1 Introduction

In recent years, many frequency-domain numerical simulation algorithms based on quasi-static electromagnetic field theory have been developed [16–30], which can be used to calculate the lightning transient impact grounding problem of substation grounding grid. The QSCIM in frequency domain, which can be used to effectively calculate the infinite integral in Green's function, has been developed in these frequency domain algorithms, and is realized through the 1-d MP method. At the same time, the time-domain numerical simulation algorithm based on quasi-static electromagnetic field theory, which can simulate the lightning transient impact grounding

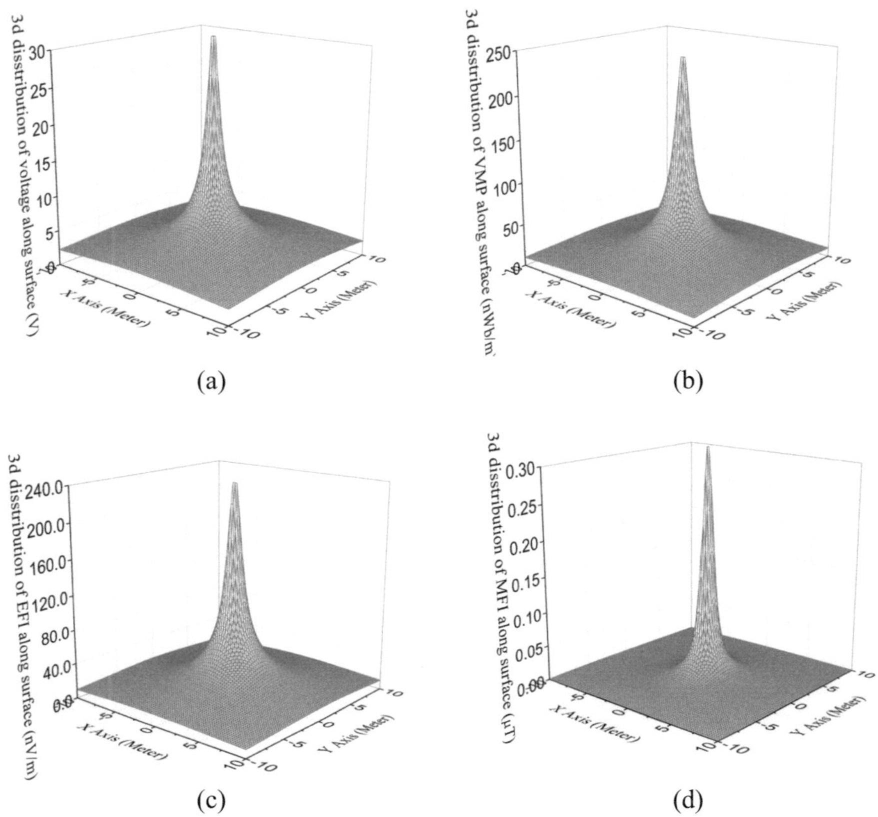

Fig. 3.9 3-d Distribution of $|\varphi|$, $|A|$, $|E|$ and $|B|$ along ground surface above a monopole and an horizontal dipole along y-axis

problem of substation grounding grid, has also been developed [26, 30, 34, 35]. The corresponding infinite integral in the time domain Green's function also needs to be calculated by developing fast and effective algorithms. If a time domain point source is buried in the horizontal multilayered earth, the SEP of a scalar monopole or the VMP of a vector dipole is satisfied with scalar or vector Poisson equation in time domain. After solving the time domain scalar or vector Poisson equation for the monopole or dipole, time domain scalar or vector Green's function for the monopole or dipole can be obtained. The infinite integral for Bessel function of the first kind, of order zero, has also appeared within these time domain Green's functions formulae. Like the frequency domain Green's function, the infinite integral can be transformed into finite exponential series by 1-d MP method [31] and 2-d Matrix Pencil (2-d MP) method [36]. In this way, closed form of scalar or vector Green's function of monopole or dipole can be achieved. Among these derivation procedure we also discuss how to achieve a finite exponential series with 1-d and 2-d MP method, called the time domain QSCIM.

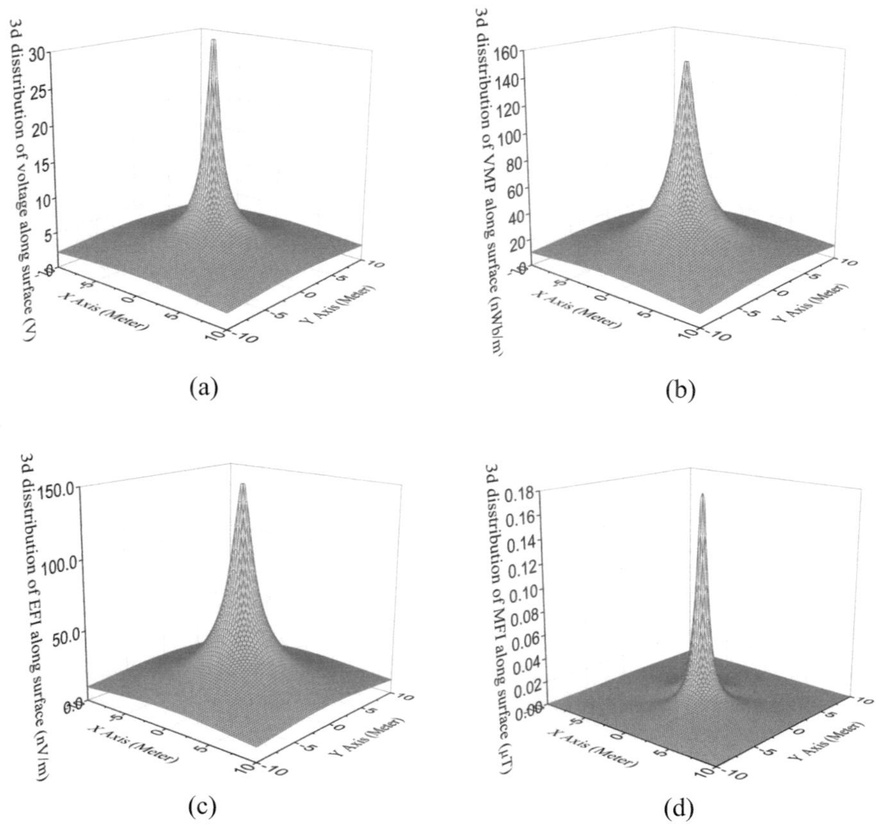

Fig. 3.10 3-d Distribution of $|\varphi|$, $|A|$, $|E|$ and $|B|$ along ground surface above a monopole and a vertical dipole

In this chapter, time domain QSCIM has been fully discussed based on the 1-d and 2-d MP method. First, the procedure for time domain vector or scalar Green's function of vector dipole or scalar monopole buried in horizontal multilayered earth model is compendiously introduced. Then the process to achieve finite exponential series for the kernel of Green's function based on 1-d and 2-d MP method is discussed. Finally, numerical results for the kernel of the closed form of time domain Green's function are carefully studied and discussed.

3.3.2 Theory of QSCIM in Time Domain

The closed form of scalar or vector Green's function of monopole or dipole in time domain

For a time domain point source (scalar monopole or vector dipole) buried in horizontal multilayered earth model, we consider the distributed electromagnetic field around the grounding grid of the substation. Because the size of substation grounding grid is limited, the hysteresis effect of electromagnetic field distributed around it can be ignored. Hence, quasi-static electrical field can be regarded as a better theory for this grounding problem. Further, in this case, the SEP φ of a scalar monopole satisfies the scalar Poisson equation in time domain, and VMP $\bar{A}$ of a vector dipole satisfies the vector Poisson equation in time domain. Here we suppose that the point source current is located at the origin of the co-ordinate system (Fig. 3.11). The horizontal N_e-layer earth model is considered.

$$\nabla^2 \varphi_{\vartheta\theta}(x, y, z; x', y', z', t) = -\frac{I_s(x', y', z')}{\varepsilon_\vartheta} e^{-\frac{\sigma_\vartheta}{\varepsilon_\vartheta} t} u(t) \tag{3.73}$$

where $\vartheta, \theta = 1, \ldots, N_e$, $I_s(x', y', z')$ is scalar unit leakage current for scalar monopole. The (x', y', z') and (x, y, z) are the rectangular coordinate values of source point and field point respectively. σ_ϑ and ε_ϑ are the soil conductivity and dielectric constant of the ϑ-th layer in the layered conductive medium. $u(t)$ is step function.

$$\nabla^2 \bar{A}_{\vartheta\theta}(x, y, z; x', y', z', t) = -\mu \hat{I}_b(x', y', z')\delta(t) \tag{3.74}$$

where $\hat{I}_b(x', y', z')$ is vector unit branch current for vector dipole, and $\hat{I}_b(x', y', z') = I_b(x', y', z') \cos\theta_x \hat{x} + I_b(x', y', z') \cos\theta_y \hat{y} + I_b(x', y', z') \cos\theta_z \hat{z} = I_{b_x}(x', y', z')$ $\hat{x} + I_{b_y}(x', y', z')\hat{y} + I_{b_z}(x', y', z')\hat{z}$. μ is magnetoconductivity.

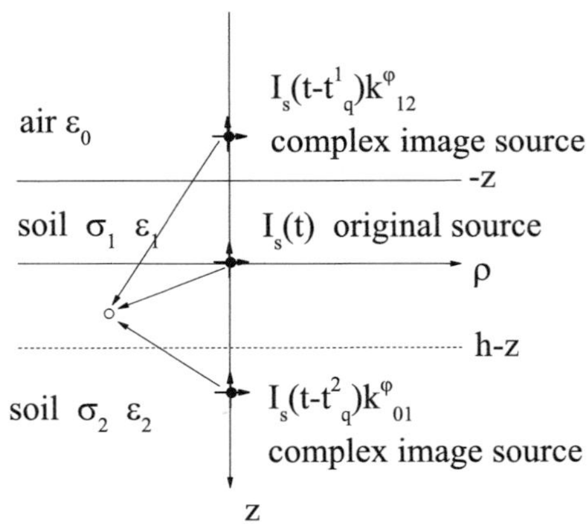

Fig. 3.11 The point source current in time domain

After solving the Poisson equations (3.73) and (3.74), we can get the scalar Green's function for monopole and vector Green's function for vector dipole. However, in practice, the Green's function of point current in time domain is obtained by inverse Fourier transform based on the Green's function of point current in frequency domain. For more detail about the derivation process refer to [26, 34, 35] for time domain scalar and vector Green's function. Like the frequency-domain Green's function, there is a special chapter at the end of this book to discuss the derivation process of the time-domain Green's function in multi-layer layered conductive media in detail. Here, we will simply give, as examples, the formula for scalar Green's function $G_{11}^{\varphi^t}$ of monopole and vector Green's function $G_{11}^{A_{zz}^t}$ or $G_{11}^{A_{xz}^t}$ of dipole in time domain, both of which are buried in horizontal two-layer earth model with source and field point at the same layer.

For monopole case, from [26, 34, 35], we know

$$G_{11}^{\varphi f}(x', y', z'; x, y, z; \omega) = \frac{I_s(x', y', z')}{4\pi\sigma_1} \int_0^{\infty} [f_0^{\varphi}(\omega)e^{-k_\rho|z-z'|} + (f_2^{\varphi}(k_\rho, \omega)e^{-k_\rho(z+z')}$$

$$+ f_3^{\varphi}(k_\rho, \omega)e^{-k_\rho(2h_1-z'-z)} - f_4^{\varphi}(k_\rho, \omega)(e^{-k_\rho(2h_1+z-z')} + e^{-k_\rho(2h_1-z+z')}))]J_0(k_\rho\rho)dk_\rho \tag{3.75}$$

where $f_0^{\varphi}(\omega) = \frac{1}{1+\frac{j\omega\varepsilon_1}{\sigma_1}}$, $f_1^{\varphi}(k_\rho, \omega) = \frac{1}{1+\frac{j\omega\varepsilon_1}{\sigma_1}} \frac{1}{(1+k_{01}^{\varphi}k_{12}^{\varphi}\cdot e^{-2k_\rho h_1})}$, $f_2^{\varphi}(k_\rho, \omega) = \frac{1}{1+\frac{j\omega\varepsilon_1}{\sigma_1}} \frac{k_{01}^{\varphi}}{(1+k_{01}^{\varphi}k_{12}^{\varphi}\cdot e^{-2k_\rho h_1})}$, $f_3^{\varphi}(k_\rho, \omega) = \frac{1}{1+\frac{j\omega\varepsilon_1}{\sigma_1}} \frac{k_{12}^{\varphi}}{(1+k_{01}^{\varphi}k_{12}^{\varphi}\cdot e^{-2k_\rho h_1})}$, $f_4^{\varphi}(k_\rho, \omega) = \frac{1}{1+\frac{j\omega\varepsilon_1}{\sigma_1}} \frac{k_{01}^{\varphi}k_{12}^{\varphi}}{(1+k_{01}^{\varphi}k_{12}^{\varphi}\cdot e^{-2k_\rho h_1})}$, $k_{01}^{\varphi} = \frac{\overline{\sigma_0}-\overline{\sigma_1}}{\overline{\sigma_0}+\overline{\sigma_1}}$, $k_{12}^{\varphi} = \frac{\overline{\sigma_1}-\overline{\sigma_2}}{\overline{\sigma_1}+\overline{\sigma_2}}$, $\overline{\sigma_0} = j\omega\varepsilon_0, \overline{\sigma_1} = \sigma_1 + j\omega\varepsilon_1, \overline{\sigma_2} = \sigma_2 + j\omega\varepsilon_2$, h_1 is thickness of the first layer earth. And $\rho = \sqrt{(x-x')^2 + (y-y')^2}$. It must be pointed out that although the electrical parameters (resistivity and dielectric constant) of horizontally stratified soil given here are constants, if these electrical parameters (resistivity and dielectric constant) of horizontally stratified soil are a function of frequency, or the electrical parameters (resistivity and dielectric constant) of horizontally stratified soil with frequency characteristics, the time-domain QSCIM developed in this chapter can also be applied.

The kernel $f_0^{\varphi}(\omega)$ of scalar Green's function $G_{11}^{\varphi f}$ can be expanded into 1-d finite exponential series by the 1-d MP method, the 1-d MP method has been discussed in detail in the previous chapter, and will not be repeated in this chapter.

$$f_0^{\varphi}(\omega) = \frac{1}{1+\frac{j\omega\varepsilon_1}{\sigma_1}} = 1 + \sum_{i=1}^{M_0} \alpha_i^0 e^{\beta_i^0 \omega} \tag{3.76}$$

where M_0, α_i^0 and β_i^0 are 1-d time domain quasi-static complex image's coefficients.

The kernel $f_1^{\varphi}(k_\rho, \omega)$, $f_2^{\varphi}(k_\rho, \omega)$ and $f_3^{\varphi}(k_\rho, \omega)$ of scalar Green's function $G_{11}^{\varphi f}$ can be expanded into 2-d finite exponential series by the 2-d MP method.

$$f_1^{\varphi}(k_\rho,\omega)=1+\sum_{i=1}^{M_1^{\varphi}}\alpha_i^{\varphi 1}e^{\beta_i^{\varphi 1}k_\rho+\zeta_i^{\varphi 1}\omega} \tag{3.77}$$

$$f_1^{\varphi}(k_\rho,\omega)=\hat{k}_{01}^{\varphi}+\sum_{i=1}^{M_2^{\varphi}}\alpha_i^{\varphi 2}e^{\beta_i^{\varphi 2}k_\rho+\zeta_i^{\varphi 2}\omega} \tag{3.78}$$

$$f_2^{\varphi}(k_\rho,\omega)=\hat{k}_{12}^{\varphi}+\sum_{i=1}^{M_3^{\varphi}}\alpha_i^{\varphi 3}e^{\beta_i^{\varphi 3}k_\rho+\zeta_i^{\varphi 3}\omega} \tag{3.79}$$

$$f_3^{\varphi}(k_\rho,\omega)=\hat{k}_{01}^{\varphi}\hat{k}_{12}^{\varphi}+\sum_{i=1}^{M_4^{\varphi}}\alpha_i^{\varphi 4}e^{\beta_i^{\varphi 4}k_\rho+\zeta_i^{\varphi 4}\omega} \tag{3.80}$$

where $\hat{k}_{01}^{\varphi}=-1$, $\hat{k}_{12}^{\varphi}=\frac{\sigma_1-\sigma_2}{\sigma_1+\sigma_2}$. M_e^{φ}, $\alpha_i^{\varphi_e}$, $\beta_i^{\varphi_e}$ and $\zeta_i^{\varphi_e}$ (e=1,2,3,4) are 2-d time domain quasi-static complex image's coefficients for a monopole.

Replacing $f_0^{\varphi}(\omega)$, $f_1^{\varphi}(k_\rho,\omega)$, $f_2^{\varphi}(k_\rho,\omega)$ and $f_3^{\varphi}(k_\rho,\omega)$ of $G_{11}^{\varphi f}$ with the 1-d and 2-d finite exponential series and employing the Lipschitz integration [23], we have

$$G_{11}^{\varphi f}(x',y',z';x,y,z;\omega)=\frac{I_s(x',y',z')}{4\pi\sigma_1}[\frac{1}{R}+\sum_{i=1}^{M_0}\alpha_i^0 e^{\beta_i^0\omega}\frac{1}{R}-\frac{\hat{k}_{01}^{\varphi}}{R_1}+\frac{\hat{k}_{12}^{\varphi}}{R_2}-\frac{\hat{k}_{01}^{\varphi}\hat{k}_{12}^{\varphi}}{R_3}-\frac{\hat{k}_{01}^{\varphi}\hat{k}_{12}^{\varphi}}{R_4}$$
$$-\sum_{i=1}^{M_2^{\varphi}}\alpha_i^2\frac{e^{\zeta_i^{\varphi 2}\omega}}{R_{i_1}}+\sum_{i=1}^{M_3^{\varphi}}\alpha_i^3\frac{e^{\zeta_i^{\varphi 3}\omega}}{R_{i_2}}-\sum_{i=1}^{M_4^{\varphi}}\alpha_i^4(\frac{e^{\zeta_i^{\varphi 4}\omega}}{R_{i_3}}+\frac{e^{\zeta_i^{\varphi 4}\omega}}{R_{i_4}})] \tag{3.81}$$

where $R=\sqrt{(x-x')^2+(y-y')^2+(z-z')^2}$,
$R_i=\sqrt{(x-x')^2+(y-y')^2+(sign_a z+sign_b z'+z_i^0)^2}$, in which $sign_a=1$ for R_1 and R_3, $sign_b=1$ for R_3 and R_4, $sign_a=sign_b=-1$ for others; for R_{2-4}, $z_i^0=2h$, for R_1, $z_i^0=0$.

$R_{i_{1-4}}=\sqrt{(x-x')^2+(y-y')^2+(sign_a z+sign_b z'+z_i^1)^2}$, in which $sign_a=1$ for R_{i_1} and R_{i_3}, $sign_b=1$ for R_{i_3} and R_{i_4}, $sign_a=sign_b=-1$ for others; for R_{i_1}, $z_i^1=-\beta_i^{\varphi 2}$, for R_{i_2}, $z_i^1=2h-\beta_i^{\varphi 3}$, for $R_{i_{3-4}}$, $z_i^1=2h-\beta_i^{\varphi 4}$.

The time-domain Green's function equation can be obtained through the inverse Fourier transform of the above frequency-domain Green's function equation.

$$G_{11}^{\varphi^t}(x',y',z';x,y,z;t)=\frac{1}{4\pi\sigma_1}[\left(\frac{1}{R}-\frac{\hat{k}_{01}^{\varphi}}{R_1}+\frac{\hat{k}_{12}^{\varphi}}{R_2}-\frac{\hat{k}_{01}^{\varphi}\hat{k}_{12}^{\varphi}}{R_3}-\frac{\hat{k}_{01}^{\varphi}\hat{k}_{12}^{\varphi}}{R_4}\right)I_s(x',y',z';t)$$

$$+\sum_{i=1}^{M_0}\alpha_i^0\frac{I_s(x',y',z';t-t_i^{\varphi_0})}{R}-\sum_{i=1}^{M_2^{\varphi}}\alpha_i^2\frac{I_s(x',y',z';t-t_i^{\varphi_2})}{R_{i_1}}$$

$$+\sum_{i=1}^{M_3^{\varphi}}\alpha_i^3\frac{I_s(x',y',z';t-t_i^{\varphi_3})}{R_{i_2}}-\sum_{i=1}^{M_4^{\varphi}}\alpha_i^4(\frac{1}{R_{i_3}}+\frac{1}{R_{i_4}})I_s(x',y',z';t-t_i^{\varphi_4})] \quad (3.82)$$

where $t_i^{\varphi_0}=j\beta_i^0$, $t_i^{\varphi_2}=j\zeta_i^{\varphi 2}$, $t_i^{\varphi_3}=j\zeta_i^{\varphi 3}$ and $t_i^{\varphi_4}=j\zeta_i^{\varphi 4}$.

For vertical dipole case, from [26, 34, 35], we know

$$G_{11}^{A_{zz}^f}(x',y',z';x,y,z;\omega)=\frac{\mu I_{b_z}(x',y',z')}{4\pi}\int_0^{\infty}[e^{-k_\rho|z-z'|}+(-f_2^{TM}(k_\rho)e^{-k_\rho(z+z')}$$

$$+f_3^{TM}(k_\rho)e^{-k_\rho(2h_1-z'-z)}-f_4^{TM}(k_\rho)(e^{-k_\rho(2h_1+z-z')}+e^{-k_\rho(2h_1-z+z')}))]J_0(k_\rho\rho)dk_\rho \quad (3.83)$$

where $f_1^{TM}(k_\rho)=\frac{1}{(1+k_{01}^{TM}k_{12}^{TM}\cdot e^{-2k_\rho h_1})}$, $f_2^{TM}(k_\rho)=\frac{k_{01}^{TM}}{(1+k_{01}^{TM}k_{12}^{TM}\cdot e^{-2k_\rho h_1})}$, $f_3^{TM}(k_\rho)=\frac{k_{12}^{TM}}{(1+k_{01}^{TM}k_{12}^{TM}\cdot e^{-2k_\rho h_1})}$, and $f_4^{TM}(k_\rho)=\frac{k_{01}^{TM}k_{12}^{TM}}{(1+k_{01}^{TM}k_{12}^{TM}\cdot e^{-2k_\rho h_1})}$, $k_{01}^{TM}=\frac{\overline{\sigma_1}-\overline{\sigma_0}}{\overline{\sigma_1}+\overline{\sigma_0}}$, $k_{12}^{TM}=\frac{\overline{\sigma_2}-\overline{\sigma_1}}{\overline{\sigma_2}+\overline{\sigma_1}}$.

Here we know $k_{01}^{TM}=-k_{01}^{\varphi}$ and $k_{12}^{TM}=-k_{12}^{\varphi}$, and we can observe that the kernel $f_i^{\varphi}(k_\rho)$ and $f_i^{TM}(k_\rho)(i=1,2,3)$ are equal, it implies $f_i^{TM}(k_\rho)=-f_i^{\varphi}(k_\rho)(i=2,3)$ and $f_i^{TM}(k_\rho)=f_i^{\varphi}(k_\rho)(i=1,4)$. Similar to scalar Green's function $G_{11}^{\varphi^f}$, we have

$$f_1^{TM}(k_\rho,\omega)=1+\sum_{i=1}^{M_1^{TM}}\alpha_i^{TM_1}e^{\beta_i^{TM_1}k_\rho+\zeta_i^{TM_1}\omega} \quad (3.84)$$

$$f_2^{TM}(k_\rho,\omega)=\hat{k}_{01}^{TM}+\sum_{i=1}^{M_2^{TM}}\alpha_i^{TM_2}e^{\beta_i^{TM_2}k_\rho+\zeta_i^{TM_2}\omega} \quad (3.85)$$

$$f_3^{TM}(k_\rho,\omega)=\hat{k}_{12}^{TM}+\sum_{i=1}^{M_3^{TM}}\alpha_i^{TM_3}e^{\beta_i^{TM_3}k_\rho+\zeta_i^{TM_3}\omega} \quad (3.86)$$

$$f_4^{TM}(k_\rho,\omega)=\hat{k}_{01}^{TM}\hat{k}_{12}^{TM}+\sum_{i=1}^{M_4^{TM}}\alpha_i^{TM_4}e^{\beta_i^{TM_4}k_\rho+\zeta_i^{TM_4}\omega} \quad (3.87)$$

where M_e^{TM}, $\alpha_i^{TM_e}$, $\beta_i^{TM_e}$ and $\zeta_i^{TM_e}$ (e=1,2,3,4) are 2-d time domain quasi-static complex image's coefficients for a dipole.

Replacing $f_2^{TM}(k_\rho,\omega)$, $f_3^{TM}(k_\rho,\omega)$ and $f_4^{TM}(k_\rho,\omega)$ of $G_{11}^{A^f}$ with the 2-d finite exponential series and employing the Lipschitz integration [23], we have

$$G_{11}^{A_{zz}^f}(x',y',z';x,y,z;\omega)=\frac{\mu I_{b_z}(x',y',z')}{4\pi}[\frac{1}{R}-\frac{\hat{k}_{01}^{TM}}{R_1}+\frac{\hat{k}_{12}^{TM}}{R_2}-\frac{\hat{k}_{01}^{TM}\hat{k}_{12}^{TM}}{R_3}-\frac{\hat{k}_{01}^{TM}\hat{k}_{12}^{TM}}{R_4}$$

$$-\sum_{i=1}^{M_2^{TM}}\alpha_i^{TM_2}\frac{e^{\zeta_i^{TM_2}\omega}}{R_{i_2}}+\sum_{i=1}^{M_3^{TM}}\alpha_i^{TM_3}\frac{e^{\zeta_i^{TM_3}\omega}}{R_{i_3}}-\sum_{i=1}^{M_4^{TM}}\alpha_i^{TM_4}(\frac{e^{\zeta_i^{TM_4}\omega}}{R_{i_4}}+\frac{e^{\zeta_i^{TM_4}\omega}}{R_{i_4}})] \quad (3.88)$$

where R, R_{1-4}, $R_{i_{1-4}}$ are same as in scalar Green's function $G_{11}^{\varphi^f}$ in Eq. (3.81) except TM for φ.

The time-domain Green's function equation can be obtained through the inverse Fourier transform of the above frequency-domain Green's function equation.

$$G_{11}^{A_{zz}^t}(x',y',z';x,y,z;t)=\frac{\mu}{4\pi}[\left(\frac{1}{R}-\frac{\hat{k}_{01}^{TM}}{R_1}+\frac{\hat{k}_{12}^{TM}}{R_2}-\frac{\hat{k}_{01}^{TM}\hat{k}_{12}^{TM}}{R_3}-\frac{\hat{k}_{01}^{TM}\hat{k}_{12}^{TM}}{R_4}\right)I_{b_z}(x',y',z';t)$$

$$-\sum_{i=1}^{M_2^{TM}}\alpha_i^{TM_2}\frac{I_{b_z}(x',y',z';t-t_i^{TM_2})}{R_{i_1}}+\sum_{i=1}^{M_3^{TM}}\alpha_i^{TM_3}\frac{I_{b_z}(x',y',z';t-t_i^{TM_3})}{R_{i_2}}$$

$$-\sum_{i=1}^{M_4^{TM}}\alpha_i^{TM_4}(\frac{1}{R_{i_3}}+\frac{1}{R_{i_4}})I_{b_z}(x',y',z';t-t_i^{TM_4}] \quad (3.89)$$

where $t_i^{TM_2}=j\zeta_i^{TM_2}$, $t_i^{TM_3}=j\zeta_i^{TM_3}$ and $t_i^{TM_4}=j\zeta_i^{TM_4}$.

For horizontal dipole case, from [26, 34, 35], we know

$$G_{11}^{A_{xz}^f}(x',y',z';x,y,z;\omega)=\frac{\mu I_{b_x}(x',y',z')}{4\pi}\frac{\partial}{\partial x}\int_0^\infty(-f_2^{TM}(k_\rho)e^{-k_\rho(z+z')}$$

$$+f_3^{TM}(k_\rho)e^{-k_\rho(2h_1-z'-z)}-f_4^{TM}(k_\rho)(e^{-k_\rho(2h_1+z-z')}+e^{-k_\rho(2h_1-z+z')}))J_0(k_\rho\rho)\frac{dk_\rho}{k_\rho} \quad (3.90)$$

Just as in Green's function $G_{11}^{A_{zz}^f}$, considering the Lipschitz integration's varied form [23], we have

$$G_{11}^{A_{xz}^f}(x',y',z';x,y,z;\omega)=\frac{\mu I_{b_x}(x',y',z')}{4\pi}\frac{\partial}{\partial x}[-k_{01}^{TM}\ln(z_1+R_1)+k_{12}^{TM}\ln(z_2+R_2)$$

$$-k_{01}^{TM}k_{12}^{TM}\ln(z_3+R_3)-k_{01}^{TM}k_{12}^{TM}\ln(z_4+R_4)-\sum_{i=1}^{M_2^{TM}}\alpha_i^{TM_2}\ln(z_{i_1}+R_{i_1})e^{\zeta_i^{TM_2}\omega}$$

$$+\sum_{i=1}^{M_3^{TM}}\alpha_i^{TM_3}\ln(z_{i_2}+R_{i_2})e^{\zeta_i^{TM_3}\omega}-\sum_{i=1}^{M_4^{TM}}\alpha_i^{TM_4}(\ln(z_{i_3}+R_{i_3})+\ln(z_{i_4}+R_{i_4}))e^{\zeta_i^{TM_4}\omega}] \tag{3.91}$$

where R_{1-4}, $R_{i_{1-4}}$ are same as in scalar Green's function $G_{11}^{\varphi^f}$ in Eq. (3.81). Other parameters are $z_{1-4}=sign_a z+sign_b z'+z_i^0, z_{i_{1-4}}=sign_a z+sign_b z'+z_i^1$, in which $sign_a=1$ for z_1, z_3, z_{i_1} and z_{i_3}, $sign_b=1$ for z_3, z_4, z_{i_3} and z_{i_4}, $sign_a=sign_b=-1$ for others, for z_{2-4}, $z_i^0=2h$, for z_1, $z_i^0=0$, for $z_{i_{e=1,2}}$, $z_i^1=2h-\beta_{e+1}^{TM}$, for $z_{i_{e=3,4}}$, $z_i^1=\beta_4^{TM}$.

The time-domain Green's function equation can be obtained through the inverse Fourier transform of the above frequency-domain Green's function equation.

$$G_{11}^{A_{xz}^t}(x',y',z';x,y,z;t)=\frac{\mu}{4\pi}\frac{\partial}{\partial x}[(-k_{01}^{TM}\ln(z_1+R_1)+k_{12}^{TM}\ln(z_2+R_2)$$

$$-k_{01}^{TM}k_{12}^{TM}\ln(z_3+R_3)-k_{01}^{TM}k_{12}^{TM}\ln(z_4+R_4))I_{b_x}(x',y',z';t)$$

$$-\sum_{i=1}^{M_2^{TM}}\alpha_i^{TM_2}\ln(z_{i_1}+R_{i_1})I_{b_x}(x',y',z';t-t_i^{TM_2})+\sum_{i=1}^{M_3^{TM}}\alpha_i^{TM_3}\ln(z_{i_2}+R_{i_2})I_{b_x}(x',y',z';t-t_i^{TM_3})$$

$$-\sum_{i=1}^{M_4^{TM}}\alpha_i^{TM_4}(\ln(z_{i_3}+R_{i_3})+\ln(z_{i_4}+R_{i_4}))I_{b_x}(x',y',z';t-t_i^{TM_4})] \tag{3.92}$$

If a dipole was lying in general orientation, the dipole can be divided into horizontal and vertical dipole components.

The Green's function of horizontal dipole

$$G_{11}^{A_{xx}^t}(x',y',z';x,y,z;t)=\frac{\mu}{4\pi}\frac{I_{b_x}(x',y',z';t)}{R} \tag{3.93}$$

The formula (3.93) will not be discussed since it is as simple formula.

As we observe the kernels of Green's functions for all vertical and horizontal dipole and monopole, we see $f_i^{TM}(k_\rho)=f_i^{\varphi}(k_\rho)(i=1,2,3)$. So for horizontal two layer earth model, only kernel $f_i^{\varphi}(k_\rho)(i=1,2,3)$ can be considered as the unified kernel formula to be expanded into finite exponential series. This conclusion can apply for all other vector or scalar Green's function of vector dipole or scalar monopole buried in horizontal two-layer earth model with different field point locations, such as $G_{10}^{\varphi^t}, G_{10}^{A_{zz}^t}, G_{10}^{A_{xz}^t}, G_{12}^{\varphi^t}, G_{12}^{A_{zz}^t}, G_{12}^{A_{xz}^t}, G_{21}^{\varphi^t}, G_{21}^{A_{zz}^t}, G_{21}^{A_{xz}^t}, G_{22}^{\varphi^t}, G_{22}^{A_{zz}^t}, G_{22}^{A_{xz}^t}$, etc. Further, the same conclusion can be arrived at with vector or scalar Green's function of vector dipole or scalar monopole buried in the other horizontal multilayered earth model, such as 3-layer or 4-layer earth model cases.

The kernel of closed form of Green's function for both scalar monopole and vector dipole buried in horizontal three-layer and four-layer earth model will be introduced next

3-layer earth model case

$$f_1^{\Phi}(k_\rho,\omega) = \frac{1}{f_{sub_3}^{\Phi}(k_\rho,\omega)} \tag{3.94}$$

$$f_2^{\Phi}(k_\rho,\omega) = \frac{k_{01}^{\Phi}}{f_{sub_3}^{\Phi}(k_\rho,\omega)} \tag{3.95}$$

$$f_3^{\Phi}(k_\rho,\omega) = \frac{k_{12}^{\Phi}}{f_{sub_3}^{\Phi}(k_\rho,\omega)} \tag{3.96}$$

$$f_4^{\Phi}(k_\rho,\omega) = \frac{k_{01}^{\Phi}k_{12}^{\Phi}}{f_{sub_3}^{\Phi}(k_\rho,\omega)} \tag{3.97}$$

$$f_5^{\Phi}(k_\rho,\omega) = \frac{k_{23}^{\Phi}}{f_{sub_3}^{\Phi}(k_\rho,\omega)} \tag{3.98}$$

$$f_6^{\Phi}(k_\rho,\omega) = \frac{k_{01}^{\Phi}k_{23}^{\Phi}}{f_{sub_3}^{\Phi}(k_\rho,\omega)} \tag{3.99}$$

$$f_7^{\Phi}(k_\rho,\omega) = \frac{k_{01}^{\Phi}k_{12}^{\Phi}k_{23}^{\Phi}}{f_{sub_3}^{\Phi}(k_\rho,\omega)} \tag{3.100}$$

and

$$f_{sub_3}^{\Phi}(k_\rho,\omega) = 1 + k_{01}^{\Phi}k_{12}^{\Phi}e^{-2k_\rho h_1} + k_{12}^{\Phi}k_{23}^{\Phi}e^{-2k_\rho h_2} + k_{01}^{\Phi}k_{23}^{\Phi}e^{-2k_\rho (h_1+h_2)} \tag{3.101}$$

where $\Phi = TM$ for vector dipole case and $\Phi = \varphi$ for scalar monopole case; $k_{23}^{\varphi} = \frac{\overline{\sigma_2}-\overline{\sigma_3}}{\overline{\sigma_2}+\overline{\sigma_3}}$, $\overline{\sigma_3} = \sigma_3 + j\omega\varepsilon_3$, $k_{23}^{TM} = -k_{23}^{\varphi}$, h_2 is thickness of second layer earth.

4-layer earth model case

$$f_1^{\Phi}(k_\rho,\omega) = \frac{1}{f_{sub_4}^{\Phi}(k_\rho,\omega)} \tag{3.102}$$

$$f_2^{\Phi}(k_\rho,\omega) = \frac{k_{01}^{\Phi}}{f_{sub_4}^{\Phi}(k_\rho,\omega)} \tag{3.103}$$

$$f_3^\Phi(k_\rho,\omega)=\frac{k_{12}^\Phi}{f_{sub_4}^\Phi(k_\rho,\omega)} \tag{3.104}$$

$$f_4^\Phi(k_\rho,\omega)=\frac{k_{01}^\Phi k_{12}^\Phi}{f_{sub_4}^\Phi(k_\rho,\omega)} \tag{3.105}$$

$$f_5^\Phi(k_\rho,\omega)=\frac{k_{23}^\Phi}{f_{sub_4}^\Phi(k_\rho,\omega)} \tag{3.106}$$

$$f_6^\Phi(k_\rho,\omega)=\frac{k_{01}^\Phi k_{23}^\Phi}{f_{sub_4}^\Phi(k_\rho,\omega)} \tag{3.107}$$

$$f_7^\Phi(k_\rho,\omega)=\frac{k_{01}^\Phi k_{12}^\Phi k_{23}^\Phi}{f_{sub_4}^\Phi(k_\rho,\omega)} \tag{3.108}$$

$$f_8^\Phi(k_\rho,\omega)=\frac{k_{34}^\Phi}{f_{sub_4}^\Phi(k_\rho,\omega)} \tag{3.109}$$

$$f_9^\Phi(k_\rho,\omega)=\frac{k_{01}^\Phi k_{34}^\Phi}{f_{sub_4}^\Phi(k_\rho,\omega)} \tag{3.110}$$

$$f_{10}^\Phi(k_\rho,\omega)=\frac{k_{01}^\Phi k_{12}^\Phi k_{34}^\Phi}{f_{sub_4}^\Phi(k_\rho,\omega)} \tag{3.111}$$

$$f_{11}^\Phi(k_\rho,\omega)=\frac{k_{01}^\Phi k_{23}^\Phi k_{34}^\Phi}{f_{sub_4}^\Phi(k_\rho,\omega)} \tag{3.112}$$

$$f_{12}^\Phi(k_\rho,\omega)=\frac{k_{12}^\Phi k_{23}^\Phi k_{34}^\Phi}{f_{sub_4}^\Phi(k_\rho,\omega)} \tag{3.113}$$

$$f_{13}^\Phi(k_\rho,\omega)=\frac{k_{01}^\Phi k_{12}^\Phi k_{23}^\Phi k_{34}^\Phi}{f_{sub_4}^\Phi(k_\rho,\omega)} \tag{3.114}$$

and

$$f_{sub_4}^\Phi(k_\rho,\omega)=1+k_{01}^\Phi k_{12}^\Phi e^{-2k_\rho h_1}+k_{12}^\Phi k_{23}^\Phi e^{-2k_\rho h_2}+k_{23}^\Phi k_{34}^\Phi e^{-2k_\rho h_3}+k_{01}^\Phi k_{23}^\Phi e^{-2k_\rho(h_1+h_2)}$$

$$+k_{12}^\Phi k_{34}^\Phi e^{-2k_\rho(h_2+h_3)}+k_{01}^\Phi k_{12}^\Phi k_{23}^\Phi k_{34}^\Phi e^{-2k_\rho(h_1+h_3)}+k_{01}^\Phi k_{34}^\Phi e^{-2k_\rho(h_1+h_2+h_3)} \tag{3.115}$$

where $\Phi = TM$ for vector dipole case and $\Phi = \varphi$ for scalar monopole case; $k_{34}^{\varphi} = \frac{\overline{\sigma_3}-\overline{\sigma_4}}{\overline{\sigma_3}+\overline{\sigma_4}}$, $\overline{\sigma_4} = \sigma_4 + j\omega\varepsilon_4$, $k_{34}^{TM} = -k_{34}^{\varphi}$, h_3 is thickness of third layer earth.

For simplified description for 2-d MP in this chapter, only scalar monopole case is discussed in below, which means that $f(k_\rho, \omega)$ is used to represent $f^{\varphi}(k_\rho, \omega)$.

Application range for the time domain QSCIM

The maximum frequency of applicability of the method is limited by the quasi-stationary approximation of the electromagnetic fields, which means the propagation effect of the electromagnetic field around the dipole can be neglected, so

$$\delta(t - \frac{R}{c_i}) \approx \delta(t) \tag{3.116}$$

where $c_i = \frac{1}{\sqrt{\mu(\varepsilon_i + \frac{\sigma_i}{j\omega})}}$, $i = 1, \ldots, N_s$.

3.3.3 2-d MP Method

If a function $f(k_\rho, \omega)$ can be expanded into finite exponential series, as below,

$$f(k_\rho, \omega) = f_{k_\rho \to \infty \ and \ \omega \to zero} + \sum_{i=1}^{M_c} \alpha_i e^{\beta_i k_\rho + \zeta_i \omega} \tag{3.117}$$

The three parameters to be decided are M_c, α_i, β_i and ζ_i.

To get these parameters, we first decide the maximum sample points according to the characteristic of function $f(k_\rho, \omega)$, the maximum value for $k_{\rho_{max}}$ and ω_{max} can thus be obtained according minimum $f(k_\rho, \omega)$ decided by set truncation precise. Once $k_{\rho_{max}}$ and ω_{max} is known, we can get uniformly discrete values of $f(k_\rho, \omega)$ within scope $(0 \le k_\rho \le k_{\rho_{max}}, 0 \le \omega \le \omega_{max})$. Therefore, we have uniform discrete function values of $f(k_\rho, \omega) \approx f(m, n)$ with $k_\rho-$ direction: $(f(0, n), f(1, n), f(2, n), \ldots, f(M, n))$ corresponding to value of k_ρ as $(0, k_{\rho_1}, k_{\rho_2}, \ldots, k_{\rho_{max}})$ or $(0, \Delta k_\rho, \ldots, m\Delta k_\rho, \ldots, (M-1)\Delta k_\rho)$. Here Δk_ρ is rate of sampling. And with $\omega-$ direction: $(f(m, 0), f(m, 1), f(m, 2), \ldots, f(m, , N))$ corresponding to value of ω as $(0, \omega_1, \omega_2, \ldots, \omega_{max})$ or $(0, \Delta\omega, \ldots, n\Delta\omega, \ldots, (N-1)\Delta\omega)$. Here $\Delta\omega$ is rate of sampling. we can obtain

$$f(k_\rho, \omega) \approx f(m, n) = f_{k_\rho \to \infty \ and \ \omega \to zero} + \sum_{i=1}^{M_c} \alpha_i x_i^m y_i^n \qquad (m = 0, \ldots, M-1; n = 0, \ldots, N-1) \tag{3.118}$$

here

$$x_i = e^{\beta_i \Delta k_\rho} \qquad (i = 1, \ldots, M_c) \tag{3.119}$$

$$y_i = e^{\zeta_i \Delta\omega} \qquad (i = 1, \ldots, M_c) \tag{3.120}$$

Method to decide number of M_c

Since we have total MN number of uniform discrete function values of $f(m,n)$, we can get the enhanced matrix $[F_e]$ from the sampling data $f(m,n)$ through partition-and-stacking as fellows:

$$[F_x] = \begin{bmatrix} f_y(0,n) & f_y(1,n) & \dots f_y(M-K-1,n) \\ f_y(1,n) & f_y(2,n) & \dots\ \ f_y(M-K,n) \\ \vdots & \vdots & \vdots \\ f_y(K,n) & f_y(K+1,n) & \dots\ \ f_y(M-1,n) \end{bmatrix} \tag{3.121}$$

where

$$[F_y] = \begin{bmatrix} f(m,0) & f(m,1) & \dots f(m,N-L-1) \\ f(m,1) & f(m,2) & \dots\ \ f(m,N-L) \\ \vdots & \vdots & \vdots \\ f(m,L) & f(m,L+1) & \dots\ \ f(m,N-1) \end{bmatrix} \tag{3.122}$$

$$[F_e] = [F_x(F_y)] \tag{3.123}$$

$[F_e]$ is a $(K+1)(L+1) \times (M-K)(N-L)$ Hankel block matrix, $[F_x]$ is a $(K+1) \times (M-K)$ Hankel block matrix, and $[F_y]$ is a $(L+1) \times (N-L)$ Hankel block matrix. Each column of $[F_x]$ is windowed segment of sequence $f(0,n), f(1,n), \dots, f(M-1,n)$ with window length $K+1$. Each column of $[F_y]$ is windowed segment of sequence $f(m,0), f(m,1), \dots, f(m,N-1)$ with window length $L+1$. where K and L are referred as the pencil parameter [36]. The parameter K, L can be chosen between $M/3 \sim M/2$ and $N/3 \sim N/2$.

Next, a singular-value decomposition (SVD) of the matrix $[F_e]$ can be implemented out as

$$[F_e] = [U][\Sigma][V]^H, \tag{3.124}$$

here, $[U]$ and $[V]$ are unitary matrices, comprised of the eigenvectors of $[F_e][F_e]^H$ and $[F_e]^H[F_e]$, respectively, and $[\Sigma]$ is a diagonal matrix including the singular values of $[F_e]$, i. e. where the superscript "H" denotes the conjugate transpose.

$$[U]^H[F_e][V] = [\Sigma], \tag{3.125}$$

The choice of the parameter M_c can be achieved at this stage by studying the ratios of various singular values to the largest one. Typically, the singular values beyond M_c are set equal to zero. The way which M_c is chosen is as follows. Observe the singular value σ_c such that

$$\frac{\sigma_c}{\sigma_{max}} \approx 10^{-p}, \tag{3.126}$$

where p is the number of significant decimal digits in the data. For example, if the sampling data is accurate up to three significant digits, the singular values for which the ratio in Eq. (3.126) is below 10^{-3} are essentially useless singular values, and they should not be used in the reconstruction of the sampling data.

We next introduce the "filtered" matrix, $[U']$, and $[V']$, constructed so that it contains only M_c predominant right-singular vectors of $[U]$ and $[V]$;

$$[U'] = [u_1, u_2, \ldots, u_{M_c}], \tag{3.127}$$

$$[V'] = [v_1, v_2, \ldots, v_{M_c}], \tag{3.128}$$

The right-singular vectors from $M + 1$ to L, corresponding to the small singular values, are omitted. Therefore,

$$[F_e^1] = [U_1'][\Sigma'][V']^H, \tag{3.129}$$

$$[F_e^2] = [U_2'][\Sigma'][V']^H, \tag{3.130}$$

where $[U_1']$ is obtained from $[U']$ with the last $L + 1$ row of $[U']$ omitted, $[U_2']$ is obtained by deleting the first $L + 1$ row of $[U']$; and $[\Sigma']$ is obtained from the M_c columns of $[\Sigma]$ corresponding to the M_c predominant singular values.

Method to decide β_i

To motivate the 2-d MP method, we can use the two $K(L + 1) \times (M - K)(N - L)$ matrices, $[F_e^1]$ and $[F_e^2]$. We can rewrite

$$[F_e^1] = [E_l][A][E_r], \tag{3.131}$$

$$[F_e^2] = [E_l][X_d][A][E_r], \tag{3.132}$$

where

$$[E_l] = \begin{bmatrix} Y_l \\ Y_l X_d \\ Y_l X_d^2 \\ \vdots \\ Y_l X_d^{K-1} \end{bmatrix} \tag{3.133}$$

where $[E_l]$ is $K(L + 1) \times M_c$ matrix.

$$[E_r] = \begin{bmatrix} Y_r & X_d Y_r & X_d^2 Y_r & \ldots & X_d^{M-K-1} Y_r \end{bmatrix} \tag{3.134}$$

where $[E_r]$ is $M_c \times (M-K)(N-L)$ matrix.

$$[Y_l] = \begin{bmatrix} 1 & 1 & \dots & 1 \\ y_1 & y_2 & \dots & y_{M_c} \\ \vdots & \vdots & & \vdots \\ y_1^L & y_2^L & \dots & y_{M_c}^L \end{bmatrix} \tag{3.135}$$

where $[Y_l]$ is $(L+1) \times M_c$ matrix.

$$[Y_r] = \begin{bmatrix} 1 & y_1 & \dots & y_1^{N-L-1} \\ 1 & y_2 & \dots & y_2^{N-L-1} \\ \vdots & \vdots & & \vdots \\ 1 & y_M & \dots & y_{M_c}^{N-L-1} \end{bmatrix} \tag{3.136}$$

where $[Y_r]$ is $M_c \times (N-L)$ matrix.

$$[X_d^m] = diag[x_1^m, x_2^m, \dots, x_{M_c}^m], \tag{3.137}$$

$$[A] = diag[\alpha_1, \alpha_2, \dots, \alpha_{M_c}], \tag{3.138}$$

where $diag[\bullet]$ represents a $M_c \times M_c$ diagonal matrix.

Now we introduce the matrix pencil

$$[F_e^2] - \lambda[F_e^1] = [E_1]\{[X_d] - \lambda[I]\}[A][E_r], \tag{3.139}$$

where $[I]$ is the $M_c \times M_c$ identity matrix. We can demonstrate that, in general, the rank of $\{[F_e^2] - \lambda[F_e^1]\}$ will be M_c, provided $M_c \leq K \leq M - M_c$. However, if $\lambda = x_i, i = 1, 2, \dots, M_c$, the ith row of $\{[X_d] - \lambda[I]\}$ is zero, and the rank of this matrix is $M_c - 1$. Hence, the parameters x_i, may be found as generalized eigenvalues of the matrix pair $\{[F_e^2]; [F_e^1]\}$. Equivalently, the problem of solving for x_i can be transformed into an ordinary eigenvalue problem,

$$\{[F_e^1]^+[F_e^2] - \lambda[I]\}, \tag{3.140}$$

where $[F_e^1]^+$ is Moore-Penrose pseudo-inverse of $[F_e^1]$, and is defined as

$$[F_e^1]^+ = \{[F_e^1]^H[F_e^1]\}^{-1}[F_e^1]^H, \tag{3.141}$$

where the superscript "H" denotes the conjugate transpose.

The eigenvalues of the matrix

$$\{[F_e^2] - \lambda[F_e^1]\} \Longrightarrow \{[F_e^1]^+[F_e^2] - \lambda[I]\} \tag{3.142}$$

are equivalent to the eigenvalues of the matrix

$$\{[U_2'] - \lambda[U_1']\} \Longrightarrow \{[U_1']\}^+\{[U_2']\} - \lambda[I] \tag{3.143}$$

This methodology can be used to solve for x_i.

Lastly, we point out that in this case x_i represent β_i according to Eq. (3.119).

Method to decide ζ_i

we introduce the permutation (shuffling) matrix.

$$[P] = \begin{bmatrix} p^T(1) \\ p^T(1+(L+1)) \\ \vdots \\ p^T(1+K(L+1)) \\ p^T(2) \\ p^T(2+(L+1)) \\ \vdots \\ p^T(2+K(L+1)) \\ \vdots \\ \vdots \\ p^T((L+1)) \\ p^T((L+1)+(L+1)) \\ \vdots \\ p^T((L+1)+K(L+1)) \end{bmatrix} \tag{3.144}$$

where $p(i)$ is the $(K+1)(L+1) \times 1$ vector with one at the ith position and zero everywhere else. The superscript T denotes the transposition.

so

$$[U''] = [P][U'] \tag{3.145}$$

here $[U_1'']$ is obtained from $[U'']$ with the last $(K+1)$ row of $[U'']$ omitted, $[U''2]$ is obtained by deleting the first $(K+1)$ row of $[U'']$.

Then, we can write the matrix pencil $[U_2''] - \lambda[U_1'']$, which are equivalent to the eigenvalues of the following matrix

$$\{[U_2''] - \lambda[U_1'']\} \Longrightarrow \{[U_1'']\}^+\{[U_2'']\} - \lambda[I] \tag{3.146}$$

This methodology can be used to solve for y_i.

Lastly, we point out that in this case y_i represent ζ_i according to Eq. (3.120).

How to pole paring for β_i and ζ_i

After finding the poles β_i and ζ_i, it needs to be paired to get the correct pairs. This is because the two poles are extracted separately by the 2-d MP, and the order of each subset of the two poles is unknown. To obtain the (correct but not necessarily ordered) pairs $\{(\beta_i, \zeta_i); i = 1, \ldots, M_c\}$, we need to pair the two sets $\{\beta_i, i = 1, \ldots, M_c\}$ and $\{\zeta_i, i = 1, \ldots, M_c\}$ together correctly. The pairs can be matched together by maximizing the criterion below. But the difference here again is that the pairing algorithm is real valued which reduce the computation time as well.

$$J_s(i, j) = \sum_{t=1}^{M_c} \|u_t^H e_L(X_i, Y_i)\|^2 \tag{3.147}$$

where $e_L = X_i \bigotimes Y_i$, $\bigotimes$ is Kronecker product. $X_i = [1, x_i, \ldots, x_i^K]^T$, and $Y_i = [1, y_i, \ldots, y_i^L]^T$, $(i = 1, 2, \ldots, M_c)$. Here $\{u_i; i = 1, 2, \ldots, M_c\}$ are the principle eigenvectors.

This optimization method can not guarantee to give a perfectly matched pole order. And even give unreasonable matching order.

Improved method for solving two poles (β_i and ζ_i) **simultaneously**

According to the previous method of solving two poles (β_i and ζ_i), two matrices ($[F_{e_x}]$ and $[F_{e_y}]$) can be obtained.

$$[F_{e_x}] = \{[U_1']\}^+\{[U_2']\} \tag{3.148}$$

$$[F_{e_y}] = \{[U_1'']\}^+\{[U_2'']\} \tag{3.149}$$

Through linear combination of matrix $[F_{e_x}]$ and $[F_{e_y}]$ obtained from Eqs. (3.148) and (3.149), the problem of parameter pairing caused by parameter overlap or approximation is avoided. The eigenvector matrix $[T]$ can be obtained by eigenvalue decomposition of the combined matrix:

$$[F_{e_c}] = \vartheta[F_{e_x}] + (1 - \vartheta)[F_{e_y}] \tag{3.150}$$

$$[F_{e_c}] = [T]^{-1}[\Phi][T] \tag{3.151}$$

where ϑ is a variable pairing parameter, the value range is [0, 1], set ϑ repetitive eigenvalues can be avoided $\vartheta = 0.3$.

Solve the eigenvalue matrices $[\Phi_{e_x}]$ and $[\Phi_{e_y}]$ corresponding to matrices $[F_{e_x}]$ and $[F_{e_y}]$ according to Eqs. (3.148) and (3.149)

$$[\Phi_{e_x}] = [T][F_{e_x}][T]^{-1} \tag{3.152}$$

$$[\Phi_{e_y}] = [T][F_{e_y}][T]^{-1} \tag{3.153}$$

where $[\Phi_{e_x}] = diag[x_1, x_2, \ldots, x_{M_c}]$, $[\Phi_{e_y}] = diag[y_1, y_2, \ldots, y_{M_c}]$.

This methodology can be used to solve for x_i and y_i.

Lastly, we point out that in this case x_i represent β_i and y_i represent ζ_i according to Eqs. (3.119) and (3.120).

How to decide α_i

Once M_c, the β_i and ζ_i are known, α_i, are solved for from the following least-squares problem:

$$\begin{bmatrix} f(0,0) \\ f(0,1) \\ \vdots \\ f(0,N) \\ f(1,0) \\ f(1,1) \\ \vdots \\ f(1,N) \\ \vdots \\ \vdots \\ f(M,0) \\ f(M,1) \\ \vdots \\ f(M,N) \end{bmatrix} = \begin{bmatrix} 1 & 1 & \dots & 1 \\ y_1 & y_2 & \dots & y_{M_c} \\ \vdots & \vdots & & \vdots \\ y_1^N & y_2^N & \dots & y_{M_c}^N \\ x_1 & x_2 & \dots & x_{M_c} \\ x_1 y_1 & x_2 y_2 & \dots & x_{M_c} y_{M_c} \\ \vdots & \vdots & & \vdots \\ x_1 y_1^N & x_2 y_2^N & \dots & x_{M_c} y_{M_c}^N \\ \vdots & \vdots & & \vdots \\ \vdots & \vdots & & \vdots \\ x_1^M & x_2^M & \dots & x_{M_c}^M \\ x_1^M y_1 & x_2^M y_2 & \dots & x_{M_c}^M y_{M_c} \\ \vdots & \vdots & & \vdots \\ x_1^M y_1^N & x_2^M y_2^N & \dots & x_{M_c}^M y_{M_c}^N \end{bmatrix} \begin{bmatrix} \alpha_1 \\ \alpha_2 \\ \vdots \\ \alpha_{M_c} \end{bmatrix} \tag{3.154}$$

3.3.4 *Closed Form of the Green's Function of a Point Source in Horizontal Multilayered Earth Model in Time Domain*

The derivation process of the closed form Green's function of the point current source in a two-layer soil model was introduced earlier. Using the same method, the closed form Green's function of a point current source in any horizontally layered soil can be derived. Here, the closed form Green's function of a point current source in a four layer horizontally layered soil model is directly presented.

The closed form Green's function of a monopole in a four layer horizontally layered soil model in time domain is firstly presented as below.

$$G_{11}^{\varphi^t}(x', y', z'; x, y, z; t) = \frac{1}{4\pi\sigma_1}[(\frac{1}{R} - \frac{\hat{k}_{01}^{\varphi}}{R_1} + \frac{\hat{k}_{12}^{\varphi}}{R_2} - \frac{\hat{k}_{01}^{\varphi}\hat{k}_{12}^{\varphi}}{R_3} - \frac{\hat{k}_{01}^{\varphi}\hat{k}_{12}^{\varphi}}{R_4} + \frac{\hat{k}_{23}^{\varphi}}{R_5} - \frac{\hat{k}_{01}^{\varphi}\hat{k}_{23}^{\varphi}}{R_6}$$

$$-\frac{\hat{k}_{01}^{\varphi}\hat{k}_{23}^{\varphi}}{R_7} - \frac{\hat{k}_{01}^{\varphi}\hat{k}_{12}^{\varphi}\hat{k}_{23}^{\varphi}}{R_8} + \frac{\hat{k}_{34}^{\varphi}}{R_9} - \frac{\hat{k}_{01}^{\varphi}\hat{k}_{34}^{\varphi}}{R_{10}} - \frac{\hat{k}_{01}^{\varphi}\hat{k}_{34}^{\varphi}}{R_{11}} - \frac{\hat{k}_{01}^{\varphi}\hat{k}_{23}^{\varphi}\hat{k}_{34}^{\varphi}}{R_{12}} - \frac{\hat{k}_{01}^{\varphi}\hat{k}_{12}^{\varphi}\hat{k}_{34}^{\varphi}}{R_{13}} + \frac{\hat{k}_{12}^{\varphi}\hat{k}_{23}^{\varphi}\hat{k}_{34}^{\varphi}}{R_{14}}$$

$$-\frac{\hat{k}_{01}^{\varphi}\hat{k}_{12}^{\varphi}\hat{k}_{23}^{\varphi}\hat{k}_{34}^{\varphi}}{R_{15}}-\frac{\hat{k}_{01}^{\varphi}\hat{k}_{12}^{\varphi}\hat{k}_{23}^{\varphi}\hat{k}_{34}^{\varphi}}{R_{16}})I_s(x',y',z';t)+\sum_{i=1}^{M_0}\alpha_i^0\frac{I_s(x',y',z';t-t_i^{\varphi_0})}{R}$$

$$-\sum_{i=1}^{M_2^{\varphi}}\alpha_i^{\varphi_2}\frac{I_s(x',y',z';t-t_i^{\varphi_2})}{R_{i_1}}+\sum_{i=1}^{M_3^{\varphi}}\alpha_i^{\varphi_3}\frac{I_s(x',y',z';t-t_i^{\varphi_3})}{R_{i_2}}-\sum_{i=1}^{M_4^{\varphi}}\alpha_i^{\varphi_4}(\frac{1}{R_{i_3}}+\frac{1}{R_{i_4}})I_s(x',y',z';t-t_i^{\varphi_4})$$

$$+\sum_{i=1}^{M_5^{\varphi}}\alpha_i^5\frac{I_s(x',y',z';t-t_i^{\varphi 5})}{R_{i_5}}-\sum_{i=1}^{M_6^{\varphi}}\alpha_i^{\varphi_6}(\frac{1}{R_{i_6}}+\frac{1}{R_{i_7}})I_s(x',y',z';t-t_i^{\varphi_6})$$

$$-\sum_{i=1}^{M_7^{\varphi}}\alpha_i^{\varphi_7}\frac{I_s(x',y',z';t-t_i^{\varphi_7})}{R_{i_8}}+\sum_{i=1}^{M_8^{\varphi}}\alpha_i^{\varphi_8}\frac{I_s(x',y',z';t-t_i^{\varphi_8})}{R_{i_9}}$$

$$-\sum_{i=1}^{M_9^{\varphi}}\alpha_i^{\varphi_9}(\frac{1}{R_{i_{10}}}+\frac{1}{R_{i_{11}}})I_s(x',y',z';t-t_i^{\varphi_9})-\sum_{i=1}^{M_{10}^{\varphi}}\alpha_i^{\varphi_{10}}\frac{I_s(x',y',z';t-t_i^{\varphi_{10}})}{R_{i_{12}}}$$

$$-\sum_{i=1}^{M_{11}^{\varphi}}\alpha_i^{\varphi_{11}}\frac{I_s(x',y',z';t-t_i^{\varphi_{11}})}{R_{i_{13}}}+\sum_{i=1}^{M_{12}^{\varphi}}\alpha_i^{\varphi_{12}}\frac{I_s(x',y',z';t-t_i^{\varphi_{12}})}{R_{i_{14}}}$$

$$-\sum_{i=1}^{M_{13}^{\varphi}}\alpha_i^{\varphi_{13}}(\frac{1}{R_{i_{15}}}+\frac{1}{R_{i_{16}}})I_s(x',y',z';t-t_i^{\varphi_{13}})] \tag{3.155}$$

where $t_i^{\varphi 0}=j\beta_i^0, t_i^{\varphi 2}=j\zeta_i^{\varphi 2}, t_i^{\varphi 3}=j\zeta_i^{\varphi 3}, t_i^{\varphi 4}=j\zeta_i^{\varphi 4}, t_i^{\varphi 5}=j\zeta_i^{\varphi 5}, t_i^{\varphi 6}=j\zeta_i^{\varphi 6}, t_i^{\varphi 7}=j\zeta_i^{\varphi 7}, t_i^{\varphi 8}=j\zeta_i^{\varphi 8}, t_i^{\varphi 9}=j\zeta_i^{\varphi 9}, t_i^{\varphi 10}=j\zeta_i^{\varphi 10}, t_i^{\varphi 11}=j\zeta_i^{\varphi 11}, t_i^{\varphi 12}=j\zeta_i^{\varphi 12}$, and $t_i^{\varphi 13}=j\zeta_i^{\varphi 13}$. $R=\sqrt{(x-x')^2+(y-y')^2+(z-z')^2}$, $R_i=\sqrt{(x-x')^2+(y-y')^2+(sign_a z+sign_b z'+z_i^0)^2}$, in which $sign_a=1$ for $R_1, R_3, R_6, R_8, R_{10}, R_{12}, R_{13}$ and R_{16}, $sign_b=1$ for $R_1, R_4, R_7, R_8, R_{11}, R_{12}, R_{13}$ and R_{15}, $sign_a=sign_b=-1$ for others; for R_{2-4}, $z_i^0=2h_1$, for R_{5-7}, $z_i^0=2(h_1+h_2)$, for R_{9-11}, $z_i^0=2(h_1+h_2+h_3)$, for R_{14-16}, $z_i^0=2(h_1+h_3)$, for R_8, $z_i^0=2h_2$, for R_{12}, $z_i^0=2(h_2+h_3)$, for R_{13}, $z_i^0=2h_3$, for R_1, $z_i^0=0$.

$R_{i_{1-16}}=\sqrt{(x-x')^2+(y-y')^2+(sign_a z+sign_b z'+z_i^1)^2}$, in which $sign_a=1$ for R_{i_1}, R_{i_3}, R_{i_6}, R_{i_8}, $R_{i_{10}}$, $R_{i_{12}}$, $R_{i_{13}}$ and $R_{i_{16}}$, $sign_b=1$ for R_1, R_4, R_7, R_8, $R_{i_{11}}$, $R_{i_{12}}$, $R_{i_{13}}$ and $R_{i_{15}}$, $sign_a=sign_b=-1$ for others; for R_{i_2}, R_{i_3}, and R_{i_4}, $z_i^1=2h_1-\beta_i^{\varphi_2}$, $z_i^1=2h_1-\beta_i^{\varphi_3}$ and $z_i^1=2h_1-\beta_i^{\varphi_4}$, respectively; for R_{i_5}, R_{i_6} and R_{i_7}, $z_i^1=2(h_1+h_2)-\beta_i^{\varphi_5}$, $z_i^1=2(h_1+h_2)-\beta_i^{\varphi_6}$ and $z_i^1=2(h_1+h_2)-\beta_i^{\varphi_7}$, respectively; for R_{i_9}, $R_{i_{10}}$ and $R_{i_{11}}$, $z_i^1=2(h_1+h_2+h_3)-\beta_i^{\varphi_9}$, $z_i^1=2(h_1+h_2+h_3)-\beta_i^{\varphi_{10}}$ and $z_i^1=2(h_1+h_2+h_3)-\beta_i^{\varphi_{11}}$, respectively;for $R_{i_{14}}$, $R_{i_{15}}$ and $R_{i_{16}}$, $z_i^1=2(h_1+h_3)-\beta_i^{\varphi_{14}}$, $z_i^1=2(h_1+h_3)-\beta_i^{\varphi_{15}}$, $z_i^1=2(h_1+h_3)-\beta_i^{\varphi_{16}}$, respectively; for R_{i_8}, $z_i^1=2h_2-\beta_i^{\varphi_8}$, for $R_{i_{12}}$, $z_i^1=2(h_2+h_3)-\beta_i^{\varphi_{12}}$, for $R_{i_{13}}$, $z_i^1=2h_3-\beta_i^{\varphi_{13}}$, for R_1, $z_i^1=-\beta_i^{\varphi_2}$.

Formula (3.155) can be further written into two parts, namely the part without delay effect and the part with delay effect, as shown in the following equation.

$$G_{11}^{\varphi_1^t}(x', y', z'; x, y, z; t) = G_{11}^{\varphi_s^t}(x', y', z'; x, y, z) + G_{11}^{\varphi_q^t}(x', y', z'; x, y, z) \tag{3.156}$$

$$\begin{aligned} G_{11}^{\varphi_1^t}(x', y', z'; x, y, z; t) = \frac{1}{4\pi\sigma_1}[(\frac{1}{R} - \frac{\hat{k}_{01}^{\varphi}}{R_1} + \frac{\hat{k}_{12}^{\varphi}}{R_2} - \frac{\hat{k}_{01}^{\varphi}\hat{k}_{12}^{\varphi}}{R_3} - \frac{\hat{k}_{01}^{\varphi}\hat{k}_{12}^{\varphi}}{R_4} + \frac{\hat{k}_{23}^{\varphi}}{R_5} - \frac{\hat{k}_{01}^{\varphi}\hat{k}_{23}^{\varphi}}{R_6} - \frac{\hat{k}_{01}^{\varphi}\hat{k}_{23}^{\varphi}}{R_7} \\ - \frac{\hat{k}_{01}^{\varphi}\hat{k}_{12}^{\varphi}\hat{k}_{23}^{\varphi}}{R_8} + \frac{\hat{k}_{34}^{\varphi}}{R_9} - \frac{\hat{k}_{01}^{\varphi}\hat{k}_{34}^{\varphi}}{R_{10}} - \frac{\hat{k}_{01}^{\varphi}\hat{k}_{34}^{\varphi}}{R_{11}} - \frac{\hat{k}_{01}^{\varphi}\hat{k}_{23}^{\varphi}\hat{k}_{34}^{\varphi}}{R_{12}} - \frac{\hat{k}_{01}^{\varphi}\hat{k}_{12}^{\varphi}\hat{k}_{34}^{\varphi}}{R_{13}} + \frac{\hat{k}_{12}^{\varphi}\hat{k}_{23}^{\varphi}\hat{k}_{34}^{\varphi}}{R_{14}} - \frac{\hat{k}_{01}^{\varphi}\hat{k}_{12}^{\varphi}\hat{k}_{23}^{\varphi}\hat{k}_{34}^{\varphi}}{R_{15}} \\ - \frac{\hat{k}_{01}^{\varphi}\hat{k}_{12}^{\varphi}\hat{k}_{23}^{\varphi}\hat{k}_{34}^{\varphi}}{R_{16}} + \sum_{i=1}^{M_0}\alpha_i^0\frac{I_s(x', y', z'; t - t_i^{\varphi 0})}{R} - \sum_{i=1}^{M_2^{\varphi}}\alpha_i^{\varphi 2}\frac{I_s(x', y', z'; t - t_i^{\varphi 2})}{R_{i_1}})I_s(x', y', z'; t) \\ + \sum_{i=1}^{M_3^{\varphi}}\alpha_i^{\varphi_3}\frac{I_s(x', y', z'; t - t_i^{\varphi_3})}{R_{i_2}} - \sum_{i=1}^{M_4^{\varphi}}\alpha_i^{\varphi_4}(\frac{1}{R_{i_3}} + \frac{1}{R_{i_4}})I_s(x', y', z'; t - t_i^{\varphi_4}) \\ + \sum_{i=1}^{M_5^{\varphi}}\alpha_i^5\frac{I_s(x', y', z'; t - j\zeta_i^5)}{R_{i_5}} - \sum_{i=1}^{M_6^{\varphi}}\alpha_i^{\varphi_6}(\frac{1}{R_{i_6}} + \frac{1}{R_{i_7}})I_s(x', y', z'; t - t_i^{\varphi_6}) \\ - \sum_{i=1}^{M_7^{\varphi}}\alpha_i^{\varphi_7}\frac{I_s(x', y', z'; t - t_i^{\varphi_7})}{R_{i_8}} + \sum_{i=1}^{M_8^{\varphi}}\alpha_i^{\varphi_8}\frac{I_s(x', y', z'; t - t_i^{\varphi_8})}{R_{i_9}} \\ - \sum_{i=1}^{M_9^{\varphi}}\alpha_i^{\varphi_9}(\frac{1}{R_{i_{10}}} + \frac{1}{R_{i_{11}}})I_s(x', y', z'; t - t_i^{\varphi_9}) - \sum_{i=1}^{M_{10}^{\varphi}}\alpha_i^{\varphi_{10}}\frac{I_s(x', y', z'; t - t_i^{\varphi_{10}})}{R_{i_{12}}} \\ - \sum_{i=1}^{M_{11}^{\varphi}}\alpha_i^{\varphi_{11}}\frac{I_s(x', y', z'; t - t_i^{\varphi_{11}})}{R_{i_{13}}} + \sum_{i=1}^{M_{12}^{\varphi}}\alpha_i^{\varphi_{12}}\frac{I_s(x', y', z'; t - t_i^{\varphi_{12}})}{R_{i_{14}}} \\ - \sum_{i=1}^{M_{13}^{\varphi}}\alpha_i^{\varphi_{13}}(\frac{1}{R_{i_{15}}} + \frac{1}{R_{i_{16}}})I_s(x', y', z'; t - t_i^{\varphi_{13}})] \end{aligned} \tag{3.157}$$

$$G_{11}^{\varphi_s^t}(x', y', z'; x, y, z) = \frac{1}{4\pi\sigma_1}[(\frac{1}{R} - \frac{\hat{k}_{01}^{\varphi}}{R_1} + \frac{\hat{k}_{12}^{\varphi}}{R_2} - \frac{\hat{k}_{01}^{\varphi}\hat{k}_{12}^{\varphi}}{R_3} - \frac{\hat{k}_{01}^{\varphi}\hat{k}_{12}^{\varphi}}{R_4} + \frac{\hat{k}_{23}^{\varphi}}{R_5} - \frac{\hat{k}_{01}^{\varphi}\hat{k}_{23}^{\varphi}}{R_6} - \frac{\hat{k}_{01}^{\varphi}\hat{k}_{23}^{\varphi}}{R_7}$$

$$-\frac{\hat{k}_{01}^{\varphi}\hat{k}_{12}^{\varphi}\hat{k}_{23}^{\varphi}}{R_8}+\frac{\hat{k}_{34}^{\varphi}}{R_9}-\frac{\hat{k}_{01}^{\varphi}\hat{k}_{34}^{\varphi}}{R_{10}}-\frac{\hat{k}_{01}^{\varphi}\hat{k}_{34}^{\varphi}}{R_{11}}-\frac{\hat{k}_{01}^{\varphi}\hat{k}_{23}^{\varphi}\hat{k}_{34}^{\varphi}}{R_{12}}-\frac{\hat{k}_{01}^{\varphi}\hat{k}_{12}^{\varphi}\hat{k}_{34}^{\varphi}}{R_{13}}+\frac{\hat{k}_{12}^{\varphi}\hat{k}_{23}^{\varphi}\hat{k}_{34}^{\varphi}}{R_{14}}$$
$$-\frac{\hat{k}_{01}^{\varphi}\hat{k}_{12}^{\varphi}\hat{k}_{23}^{\varphi}\hat{k}_{34}^{\varphi}}{R_{15}}-\frac{\hat{k}_{01}^{\varphi}\hat{k}_{12}^{\varphi}\hat{k}_{23}^{\varphi}\hat{k}_{34}^{\varphi}}{R_{16}})I_s(x',y',z';t)] \tag{3.158}$$

$$G_{11}^{\varphi_q^t}(x',y',z';x,y,z)=\frac{1}{4\pi\sigma_1}[\sum_{i=1}^{M_0}\alpha_i^0\frac{I_s(x',y',z';t-t_i^{\varphi_0})}{R}-\sum_{i=1}^{M_2^{\varphi}}\alpha_i^{\varphi_2}\frac{I_s(x',y',z';t-t_i^{\varphi_2})}{R_{i_1}}$$
$$+\sum_{i=1}^{M_3^{\varphi}}\alpha_i^{\varphi_3}\frac{I_s(x',y',z';t-t_i^{\varphi_3})}{R_{i_2}}-\sum_{i=1}^{M_4^{\varphi}}\alpha_i^{\varphi_4}(\frac{1}{R_{i_3}}+\frac{1}{R_{i_4}})I_s(x',y',z';t-t_i^{\varphi_4})$$
$$+\sum_{i=1}^{M_5^{\varphi}}\alpha_i^{5}\frac{I_s(x',y',z';t-j\zeta_i^{\varphi_5})}{R_{i_5}}-\sum_{i=1}^{M_6^{\varphi}}\alpha_i^{\varphi_6}(\frac{1}{R_{i_6}}+\frac{1}{R_{i_7}})I_s(x',y',z';t-t_i^{\varphi_6})-\sum_{i=1}^{M_7^{\varphi}}\alpha_i^{\varphi_7}\frac{I_s(x',y',z';t-t_i^{\varphi_7})}{R_{i_8}}$$
$$+\sum_{i=1}^{M_8^{\varphi}}\alpha_i^{\varphi_8}\frac{I_s(x',y',z';t-t_i^{\varphi_8})}{R_{i_9}}-\sum_{i=1}^{M_9^{\varphi}}\alpha_i^{\varphi_9}(\frac{1}{R_{i_{10}}}+\frac{1}{R_{i_{11}}})I_s(x',y',z';t-t_i^{\varphi_9})$$
$$-\sum_{i=1}^{M_{10}^{\varphi}}\alpha_i^{\varphi_{10}}\frac{I_s(x',y',z';t-t_i^{\varphi_{10}})}{R_{i_{12}}}-\sum_{i=1}^{M_{11}^{\varphi}}\alpha_i^{\varphi_{11}}\frac{I_s(x',y',z';t-t_i^{\varphi_{11}})}{R_{i_{13}}}$$
$$+\sum_{i=1}^{M_{12}^{\varphi}}\alpha_i^{\varphi_{12}}\frac{I_s(x',y',z';t-t_i^{\varphi_{12}})}{R_{i_{14}}}-\sum_{i=1}^{M_{13}^{\varphi}}\alpha_i^{\varphi_{13}}(\frac{1}{R_{i_{15}}}+\frac{1}{R_{i_{16}}})I_s(x',y',z';t-t_i^{\varphi_{13}})] \tag{3.159}$$

Formula (3.156) can be generalized to the scalar Green's function where the scalar monopole is located in the ξ-th layer of multi-layer horizontal layered soil and the field point is located in the ζ-th layer.

$$G_{\xi\zeta}^{\varphi^t}(x',y',z';x,y,z;t)=G_{\xi\zeta}^{\varphi_s^t}(x',y',z';x,y,z)+G_{\xi\zeta}^{\varphi_q^t}(x',y',z';x,y,z) \tag{3.160}$$

Meanwhile, since $\bar{r}_s$ is the position vector of the source point, the position coordinates in the Cartesian coordinate system are (x', y', z'); And $\bar{r}_f$ is the position vector of the field point, and the position coordinates in the Cartesian coordinate system are (x, y, z). Therefore, it is shown in the following equation.

$$G^{\varphi^t}_{\xi\zeta}(\bar{r}_s;\bar{r}_f;t)=G^{\varphi^t_s}_{\xi\zeta}(\bar{r}_s;\bar{r}_f)+G^{\varphi^t_q}_{\xi\zeta}(\bar{r}_s;\bar{r}_f) \tag{3.161}$$

The closed form Green's function of a dipole in a four layer horizontally layered soil model in time domain is then presented as below.

- Green's function of vertical dipole in multilayered stratified soil Model

$$\begin{aligned}
G^{A^t_{zz}}_{11}(x',y',z';x,y,z;t)&=\frac{\mu}{4\pi}[(\frac{1}{R}-\frac{\hat{k}^{TM}_{01}}{R_1}+\frac{\hat{k}^{TM}_{12}}{R_2}-\frac{\hat{k}^{TM}_{01}\hat{k}^{TM}_{12}}{R_3}-\frac{\hat{k}^{TM}_{01}\hat{k}^{TM}_{12}}{R_4}+\frac{\hat{k}^{TM}_{23}}{R_5}\\
&-\frac{\hat{k}^{TM}_{01}\hat{k}^{TM}_{23}}{R_6}-\frac{\hat{k}^{TM}_{01}\hat{k}^{TM}_{23}}{R_7}-\frac{\hat{k}^{TM}_{01}\hat{k}^{TM}_{12}\hat{k}^{TM}_{23}}{R_8}+\frac{\hat{k}^{TM}_{34}}{R_9}-\frac{\hat{k}^{TM}_{01}\hat{k}^{TM}_{34}}{R_{10}}-\frac{\hat{k}^{TM}_{01}\hat{k}^{TM}_{34}}{R_{11}}-\frac{\hat{k}^{TM}_{01}\hat{k}^{TM}_{23}\hat{k}^{TM}_{34}}{R_{12}}\\
&-\frac{\hat{k}^{TM}_{01}\hat{k}^{TM}_{12}\hat{k}^{TM}_{34}}{R_{13}}+\frac{\hat{k}^{TM}_{12}\hat{k}^{TM}_{23}\hat{k}^{TM}_{34}}{R_{14}}-\frac{\hat{k}^{TM}_{01}\hat{k}^{TM}_{12}\hat{k}^{TM}_{23}\hat{k}^{TM}_{34}}{R_{15}}-\frac{\hat{k}^{TM}_{01}\hat{k}^{TM}_{12}\hat{k}^{TM}_{23}\hat{k}^{TM}_{34}}{R_{16}})I_{b_z}(x',y',z';t)\\
&-\sum_{i=1}^{M^{TM}_2}\alpha_i^{TM_2}\frac{I_{b_z}(x',y',z';t-t_i^{TM_2})}{R_{i_1}}+\sum_{i=1}^{M^{TM}_3}\alpha_i^{TM_3}\frac{I_{b_z}(x',y',z';t-t_i^{TM_3})}{R_{i_2}}\\
&-\sum_{i=1}^{M^{TM}_4}\alpha_i^{TM_4}(\frac{1}{R_{i_3}}+\frac{1}{R_{i_4}})I_{b_z}(x',y',z';t-t_i^{TM_4})+\sum_{i=1}^{M^{TM}_5}\alpha_i^{TM_5}\frac{I_{b_z}(x',y',z';t-t_i^{TM_5})}{R_{i_5}}\\
&-\sum_{i=1}^{M^{TM}_6}\alpha_i^{TM_6}(\frac{1}{R_{i_6}}+\frac{1}{R_{i_7}})I_{b_z}(x',y',z';t-t_i^{TM_6})-\sum_{i=1}^{M^{TM}_7}\alpha_i^{TM_7}\frac{I_{b_z}(x',y',z';t-t_i^{TM_7})}{R_{i_8}}\\
&+\sum_{i=1}^{M^{TM}_8}\alpha_i^{TM_8}\frac{I_{b_z}(x',y',z';t-t_i^{TM_8})}{R_{i_9}}-\sum_{i=1}^{M^{TM}_9}\alpha_i^{TM_9}(\frac{1}{R_{i_{10}}}+\frac{1}{R_{i_{11}}})I_{b_z}(x',y',z';t-t_i^{TM_9})\\
&-\sum_{i=1}^{M^{TM}_{10}}\alpha_i^{TM_{10}}\frac{I_{b_z}(x',y',z';t-t_i^{TM_{10}})}{R_{i_{12}}}-\sum_{i=1}^{M^{TM}_{11}}\alpha_i^{TM_{11}}\frac{I_{b_z}(x',y',z';t-t_i^{TM_{11}})}{R_{i_{13}}}\\
&+\sum_{i=1}^{M^{TM}_{12}}\alpha_i^{TM_{12}}\frac{I_{b_z}(x',y',z';t-t_i^{TM_{12}})}{R_{i_{14}}}-\sum_{i=1}^{M^{TM}_{13}}\alpha_i^{TM_{13}}(\frac{1}{R_{i_{15}}}+\frac{1}{R_{i_{16}}})I_{b_z}(x',y',z';t-t_i^{TM_{13}})]
\end{aligned} \tag{3.162}$$

R, R_i and $R_{i_{1-16}}$ are completely consistent with Eq. (3.155), except that TM replaces φ.

Formula (3.162) can be further written into two parts, namely the part without delay effect and the part with delay effect, as shown in the following equation.

$$G_{11}^{A^t_{zz}}(x',y',z';x,y,z;t)=G_{11}^{A^t_{szz}}(x',y',z';x,y,z)+G_{11}^{A^t_{qzz}}(x',y',z';x,y,z) \tag{3.163}$$

so

$$G_{11}^{A^t_{szz}}(x',y',z';x,y,z)=\frac{\mu}{4\pi}[\frac{1}{R}-\frac{\hat{k}_{01}^{TM}}{R_1}+\frac{\hat{k}_{12}^{TM}}{R_2}-\frac{\hat{k}_{01}^{TM}\hat{k}_{12}^{TM}}{R_3}-\frac{\hat{k}_{01}^{TM}\hat{k}_{12}^{TM}}{R_4}+\frac{\hat{k}_{23}^{TM}}{R_5}$$

$$-\frac{\hat{k}_{01}^{TM}\hat{k}_{23}^{TM}}{R_6}-\frac{\hat{k}_{01}^{TM}\hat{k}_{23}^{TM}}{R_7}-\frac{\hat{k}_{01}^{TM}\hat{k}_{12}^{TM}\hat{k}_{23}^{TM}}{R_8}+\frac{\hat{k}_{34}^{TM}}{R_9}-\frac{\hat{k}_{01}^{TM}\hat{k}_{34}^{TM}}{R_{10}}-\frac{\hat{k}_{01}^{TM}\hat{k}_{34}^{TM}}{R_{11}}-\frac{\hat{k}_{01}^{TM}\hat{k}_{23}^{TM}\hat{k}_{34}^{TM}}{R_{12}}$$

$$-\frac{\hat{k}_{01}^{TM}\hat{k}_{12}^{TM}\hat{k}_{34}^{TM}}{R_{13}}+\frac{\hat{k}_{12}^{TM}\hat{k}_{23}^{TM}\hat{k}_{34}^{TM}}{R_{14}}-\frac{\hat{k}_{01}^{TM}\hat{k}_{12}^{TM}\hat{k}_{23}^{TM}\hat{k}_{34}^{TM}}{R_{15}}-\frac{\hat{k}_{01}^{TM}\hat{k}_{12}^{TM}\hat{k}_{23}^{TM}\hat{k}_{34}^{TM}}{R_{16}}]I_{b_z}(x',y',z';t) \tag{3.164}$$

$$G_{11}^{A^t_{qzz}}(x',y',z';x,y,z;t)=\frac{\mu}{4\pi}[-\sum_{i=1}^{M_2^{TM}}\alpha_i^{TM_2}\frac{I_{b_z}(x',y',z';t-t_i^{TM_2})}{R_{i_1}}$$

$$+\sum_{i=1}^{M_3^{TM}}\alpha_i^{TM_3}\frac{I_{b_z}(x',y',z';t-t_i^{TM_3})}{R_{i_2}}-\sum_{i=1}^{M_4^{TM}}\alpha_i^{TM_4}(\frac{1}{R_{i_3}}+\frac{1}{R_{i_4}})I_{b_z}(x',y',z';t-t_i^{TM_4})$$

$$+\sum_{i=1}^{M_5^{TM}}\alpha_i^{TM_5}\frac{I_{b_z}(x',y',z';t-t_i^{TM_5})}{R_{i_5}}-\sum_{i=1}^{M_6^{TM}}\alpha_i^{TM_6}(\frac{1}{R_{i_6}}+\frac{1}{R_{i_7}})I_{b_z}(x',y',z';t-t_i^{TM_6})$$

$$-\sum_{i=1}^{M_7^{TM}}\alpha_i^{TM_7}\frac{I_{b_z}(x',y',z';t-t_i^{TM_7})}{R_{i_8}}+\sum_{i=1}^{M_8^{TM}}\alpha_i^{TM_8}\frac{I_{b_z}(x',y',z';t-t_i^{TM_8})}{R_{i_9}}$$

$$-\sum_{i=1}^{M_9^{TM}}\alpha_i^{TM_9}(\frac{1}{R_{i_{10}}}+\frac{1}{R_{i_{11}}})I_{b_z}(x',y',z';t-t_i^{TM_9})-\sum_{i=1}^{M_{10}^{TM}}\alpha_i^{TM_{10}}\frac{I_{b_z}(x',y',z';t-t_i^{TM_{10}})}{R_{i_{12}}}$$

$$-\sum_{i=1}^{M_{11}^{TM}}\alpha_i^{TM_{11}}\frac{I_{b_z}(x',y',z';t-t_i^{TM_{11}})}{R_{i_{13}}}+\sum_{i=1}^{M_{12}^{TM}}\alpha_i^{TM_{12}}\frac{I_{b_z}(x',y',z';t-t_i^{TM_{12}})}{R_{i_{14}}}$$

$$-\sum_{i=1}^{M_{13}^{TM}}\alpha_i^{TM_{13}}(\frac{1}{R_{i_{15}}}+\frac{1}{R_{i_{16}}})I_{b_z}(x',y',z';t-t_i^{TM_{13}})] \tag{3.165}$$

- Green's function of horizontal dipole in multilayered stratified soil Model
  - Green's function of the vertical component of horizontal dipole in multilayered stratified soil Model

$$G_{11}^{A_{xz}^t}(x', y', z'; x, y, z; t) = \frac{\mu}{4\pi}\frac{\partial}{\partial x}[(-\hat{k}_{01}^{TM}\ln(z_1 + R_1) + \hat{k}_{12}^{TM}\ln(z_2 + R_2)$$

$$-\hat{k}_{01}^{TM}\hat{k}_{12}^{TM}\ln(z_3 + R_3) - \hat{k}_{01}^{TM}\hat{k}_{12}^{TM}\ln(z_4 + R_4) + \hat{k}_{23}^{TM}\ln(z_5 + R_5) - \hat{k}_{01}^{TM}\hat{k}_{23}^{TM}\ln(z_6 + R_6)$$

$$-\hat{k}_{01}^{TM}\hat{k}_{23}^{TM}\ln(z_7 + R_7) - \hat{k}_{01}^{TM}\hat{k}_{12}^{TM}\hat{k}_{23}^{TM}\ln(z_8 + R_8) + \hat{k}_{34}^{TM}\ln(z_9 + R_9)$$

$$-\hat{k}_{01}^{TM}\hat{k}_{34}^{TM}\ln(z_{10} + R_{10}) - \hat{k}_{01}^{TM}\hat{k}_{34}^{TM}\ln(z_{11} + R_{11}) - \hat{k}_{01}^{TM}\hat{k}_{23}^{TM}\hat{k}_{34}^{TM}\ln(z_{12} + R_{12})$$

$$-\hat{k}_{01}^{TM}\hat{k}_{12}^{TM}\hat{k}_{34}^{TM}\ln(z_{13} + R_{13}) + \hat{k}_{12}^{TM}\hat{k}_{23}^{TM}\hat{k}_{34}^{TM}\ln(z_{14} + R_{14})$$

$$-\hat{k}_{01}^{TM}\hat{k}_{12}^{TM}\hat{k}_{23}^{TM}\hat{k}_{34}^{TM}\ln(z_{15} + R_{15}) - \hat{k}_{01}^{TM}\hat{k}_{12}^{TM}\hat{k}_{23}^{TM}\hat{k}_{34}^{TM}\ln(z_{16} + R_{16}))I_{b_x}(x', y', z'; t)$$

$$-\sum_{i=1}^{M_2^{TM}}\alpha_i^{TM_2}\ln(z_{i_1} + R_{i_1})I_{b_x}(x', y', z'; t - t_i^{TM_2})$$

$$+\sum_{i=1}^{M_3^{TM}}\alpha_i^{TM_3}\ln(z_{i_2} + R_{i_2})I_{b_x}(x', y', z'; t - t_i^{TM_3})$$

$$-\sum_{i=1}^{M_4^{TM}}\alpha_i^{TM_4}(\ln(z_{i_3} + R_{i_3}) + \ln(z_{i_4} + R_{i_4}))I_{b_x}(x', y', z'; t - t_i^{TM_4})$$

$$+\sum_{i=1}^{M_5^{TM}}\alpha_i^{TM_5}\ln(z_{i_5} + R_{i_5})I_{b_x}(x', y', z'; t - t_i^{TM_5})$$

$$-\sum_{i=1}^{M_6^{TM}}\alpha_i^{TM_6}(\ln(z_{i_6} + R_{i_6}) + \ln(z_{i_7} + R_{i_7}))I_{b_x}(x', y', z'; t - t_i^{TM_6})$$

$$-\sum_{i=1}^{M_7^{TM}}\alpha_i^{TM_7}\ln(z_{i_8} + R_{i_8})I_{b_x}(x', y', z'; t - t_i^{TM_7})$$

$$+\sum_{i=1}^{M_8^{TM}} \alpha_i^{TM_8} \ln(z_{i_9}+R_{i_9}) I_{b_x}(x',y',z';t-t_i^{TM_8})$$

$$-\sum_{i=1}^{M_9^{TM}} \alpha_i^{TM_9} (\ln(z_{i_{10}}+R_{i_{10}})+\ln(z_{i_{11}}+R_{i_{11}})) I_{b_x}(x',y',z';t-t_i^{TM_9})$$

$$-\sum_{i=1}^{M_{10}^{TM}} \alpha_i^{TM_{10}} \ln(z_{i_{12}}+R_{i_{12}}) I_{b_x}(x',y',z';t-t_i^{TM_{10}})$$

$$-\sum_{i=1}^{M_{11}^{TM}} \alpha_i^{TM_{11}} \ln(z_{i_{13}}+R_{i_{13}}) I_{b_x}(x',y',z';t-t_i^{TM_{11}})$$

$$+\sum_{i=1}^{M_{12}^{TM}} \alpha_i^{TM_{12}} \ln(z_{i_{14}}+R_{i_{14}}) I_{b_x}(x',y',z';t-t_i^{TM_{12}})$$

$$-\sum_{i=1}^{M_{13}^{TM}} \alpha_i^{TM_{13}} (\ln(z_{i_{15}}+R_{i_{15}})+\ln(z_{i_{16}}+R_{i_{16}})) I_{b_x}(x',y',z';t-t_i^{TM_{13}})] \tag{3.166}$$

Formula (3.166) can also be theoretically divided into two parts: the part without delay effects and the part with delay effects, observing formula (3.166), there are formula as below.

$$G_{11}^{A^t_{xz}}(x',y',z';x,y,z;t) = G_{11}^{A^t_{sxz}}(x',y',z';x,y,z) + G_{11}^{A^t_{qxz}}(x',y',z';x,y,z) \tag{3.167}$$

$$G_{11}^{A^t_{sxz}}(x',y',z';x,y,z) = \frac{\mu}{4\pi}\frac{\partial}{\partial x}[(-\hat{k}_{01}^{TM}\ln(z_1+R_1)+\hat{k}_{12}^{TM}\ln(z_2+R_2)$$

$$-\hat{k}_{01}^{TM}\hat{k}_{12}^{TM}\ln(z_3+R_3)-\hat{k}_{01}^{TM}\hat{k}_{12}^{TM}\ln(z_4+R_4)+\hat{k}_{23}^{TM}\ln(z_5+R_5)-\hat{k}_{01}^{TM}\hat{k}_{23}^{TM}\ln(z_6+R_6)$$

$$-\hat{k}_{01}^{TM}\hat{k}_{23}^{TM}\ln(z_7+R_7)-\hat{k}_{01}^{TM}\hat{k}_{12}^{TM}\hat{k}_{23}^{TM}\ln(z_8+R_8)+\hat{k}_{34}^{TM}\ln(z_9+R_9)$$

$$-\hat{k}_{01}^{TM}\hat{k}_{34}^{TM}\ln(z_{10}+R_{10})-\hat{k}_{01}^{TM}\hat{k}_{34}^{TM}\ln(z_{11}+R_{11})-\hat{k}_{01}^{TM}\hat{k}_{23}^{TM}\hat{k}_{34}^{TM}\ln(z_{12}+R_{12})$$

$$-\hat{k}_{01}^{TM}\hat{k}_{12}^{TM}\hat{k}_{34}^{TM}\ln(z_{13}+R_{13})+\hat{k}_{12}^{TM}\hat{k}_{23}^{TM}\hat{k}_{34}^{TM}\ln(z_{14}+R_{14})$$

$$
- \hat{k}_{01}^{TM}\hat{k}_{12}^{TM}\hat{k}_{23}^{TM}\hat{k}_{34}^{TM}\ln(z_{15}+R_{15}) - \hat{k}_{01}^{TM}\hat{k}_{12}^{TM}\hat{k}_{23}^{TM}\hat{k}_{34}^{TM}\ln(z_{16}+R_{16}))I_{b_x}(x',y',z';t)] \tag{3.168}
$$

$$
\begin{aligned}
G_{11}^{A^t_{q\,xz}}(x',y',z';x,y,z) = \frac{\mu}{4\pi}\frac{\partial}{\partial x}[&-\sum_{i=1}^{M_2^{TM}} \alpha_i^{TM_2}\ln(z_{i_1}+R_{i_1})I_{b_x}(x',y',z';t-t_i^{TM_2}) \\
&+\sum_{i=1}^{M_3^{TM}} \alpha_i^{TM_3}\ln(z_{i_2}+R_{i_2})I_{b_x}(x',y',z';t-t_i^{TM_3}) \\
&-\sum_{i=1}^{M_4^{TM}} \alpha_i^{TM_4}(\ln(z_{i_3}+R_{i_3})+\ln(z_{i_4}+R_{i_4}))I_{b_x}(x',y',z';t-t_i^{TM_4}) \\
&+\sum_{i=1}^{M_5^{TM}} \alpha_i^{TM_5}\ln(z_{i_5}+R_{i_5})I_{b_x}(x',y',z';t-t_i^{TM_5}) \\
&-\sum_{i=1}^{M_6^{TM}} \alpha_i^{TM_6}(\ln(z_{i_6}+R_{i_6})+\ln(z_{i_7}+R_{i_7}))I_{b_x}(x',y',z';t-t_i^{TM_6}) \\
&-\sum_{i=1}^{M_7^{TM}} \alpha_i^{TM_7}\ln(z_{i_8}+R_{i_8})I_{b_x}(x',y',z';t-t_i^{TM_7}) \\
&+\sum_{i=1}^{M_8^{TM}} \alpha_i^{TM_8}\ln(z_{i_9}+R_{i_9})I_{b_x}(x',y',z';t-t_i^{TM_8}) \\
&-\sum_{i=1}^{M_9^{TM}} \alpha_i^{TM_9}(\ln(z_{i_{10}}+R_{i_{10}})+\ln(z_{i_{11}}+R_{i_{11}}))I_{b_x}(x',y',z';t-t_i^{TM_9}) \\
&-\sum_{i=1}^{M_{10}^{TM}} \alpha_i^{TM_{10}}\ln(z_{i_{12}}+R_{i_{12}})I_{b_x}(x',y',z';t-t_i^{TM_{10}}) \\
&-\sum_{i=1}^{M_{11}^{TM}} \alpha_i^{TM_{11}}\ln(z_{i_{13}}+R_{i_{13}})I_{b_x}(x',y',z';t-t_i^{TM_{11}})
\end{aligned}
$$

$$+\sum_{i=1}^{M_{12}^{TM}} \alpha_i^{TM_{12}} \ln(z_{i_{14}} + R_{i_{14}}) I_{b_x}(x', y', z'; t - t_i^{TM_{12}})$$

$$-\sum_{i=1}^{M_{13}^{TM}} \alpha_i^{TM_{13}} (\ln(z_{i_{15}} + R_{i_{15}}) + \ln(z_{i_{16}} + R_{i_{16}})) I_{b_x}(x', y', z'; t - t_i^{TM_{13}})] \tag{3.169}$$

– Green's function of the horizontal component of horizontal dipole in multilayered stratified soil Model

$$G_{11}^{A_{xx}^t}(x', y', z'; x, y, z; t) = \frac{\mu}{4\pi} \frac{I_{b_x}(x', y', z'; t)}{R} \tag{3.170}$$

Formula (3.170) can also be theoretically divided into two parts: the part without delay effects and the part with delay effects, observing formula (3.170), there are formula as below.

$$G_{11}^{A_{xx}^t}(x', y', z'; x, y, z; t) = G_{11}^{A_{sxx}^t}(x', y', z'; x, y, z) + G_{11}^{A_{qxx}^t}(x', y', z'; x, y, z) \tag{3.171}$$

$$G_{11}^{A_{sxx}^t}(x', y', z'; x, y, z) = \frac{\mu}{4\pi} \frac{I_{b_x}(x', y', z'; t)}{R} \tag{3.172}$$

$$G_{11}^{A_{qxx}^t}(x', y', z'; x, y, z; t) = 0 \tag{3.173}$$

Obviously, for the three-layer horizontally layered soil model, the $\hat{k}_{34}^{\varphi} = 0, \hat{k}_{34}^{TM} = 0$, $M_e^{\varphi} = 0$ and $M_e^{TM} = 0$, $e = 8 \sim 16$ in Eqs. (3.155), (3.162), and (3.166) will degenerate into the closed form Green's function of the point current source in the three-layer horizontally layered soil model. For the two-layer horizontally layered soil model, the $\hat{k}_{23}^{\varphi} = 0$, $\hat{k}_{34}^{\varphi} = 0$, $\hat{k}_{23}^{TM} = 0$, $\hat{k}_{34}^{TM} = 0$, $M_e^{\varphi} = 0$ and $M_e^{TM} = 0$, $e = 5 \sim 16$ in Eqs. (3.155), (3.162), and (3.166) will degenerate into the closed form Green function of the point current source in the two-layer horizontally layered soil model.

The dyadic Green's function for arbitrarily placed vector dipoles in the first layer of soil in the multilayer horizontally stratified soil model can be simply written as follows.

$$\bar{\bar{G}}_{11}^{A^t}(x', y', z'; x, y, z; t) =$$

$$\begin{bmatrix} G_{11}^{A_{xx}^t}(x', y', z'; x, y, z; t) & 0 & G_{11}^{A_{xz}^t}(x', y', z'; x, y, z; t) \\ 0 & G_{11}^{A_{yy}^t}(x', y', z'; x, y, z; t) & G_{11}^{A_{yz}^t}(x', y', z'; x, y, z; t) \\ 0 & 0 & G_{11}^{A_{zz}^t}(x', y', z'; x, y, z; t) \end{bmatrix} \tag{3.174}$$

$$\bar{\bar{G}}_{11}^{A^t}(x', y', z'; x, y, z; t) = \bar{\bar{G}}_{11}^{A_s^t}(x', y', z'; x, y, z) + \bar{\bar{G}}_{11}^{A_q^t}(x', y', z'; x, y, z) \tag{3.175}$$

$$\bar{\bar{G}}_{11}^{A_s^t}(x',y',z';x,y,z)=\begin{bmatrix} G_{11}^{A_s^t xx}(x',y',z';x,y,z) & 0 & G_{11}^{A_s^t xz}(x',y',z';x,y,z) \\ 0 & G_{11}^{A_s^t yy}(x',y',z';x,y,z) & G_{11}^{A_s^t yz}(x',y',z';x,y,z) \\ 0 & 0 & G_{11}^{A_s^t zz}(x',y',z';x,y,z) \end{bmatrix} \quad (3.176)$$

$$\bar{\bar{G}}_{11}^{A_q^t}(x',y',z';x,y,z;t)=\begin{bmatrix} 0 & 0 & G_{11}^{A_q^t xz}(x',y',z';x,y,z) \\ 0 & 0 & G_{11}^{A_q^t yz}(x',y',z';x,y,z) \\ 0 & 0 & G_{11}^{A_q^t zz}(x',y',z';x,y,z) \end{bmatrix} \quad (3.177)$$

Formula (3.175) can be generalized to the scalar Green's function where the scalar monopole is located in the ξ-th layer of multi-layer horizontal layered soil and the field point is located in the ζ-th layer.

$$\bar{\bar{G}}_{\xi\zeta}^{A^t}(x',y',z';x,y,z;t)=\bar{\bar{G}}_{\xi\zeta}^{A_s^t}(x',y',z';x,y,z)+\bar{\bar{G}}_{\xi\zeta}^{A_q^t}(x',y',z';x,y,z) \quad (3.178)$$

In addition, there is a vector position expression as follows

$$\bar{\bar{G}}_{\xi\zeta}^{A^t}(\bar{r}_s;\bar{r}_f;t)=\bar{\bar{G}}_{\xi\zeta}^{A_s^t}(\bar{r}_s;\bar{r}_f)+\bar{\bar{G}}_{\xi\zeta}^{A_q^t}(\bar{r}_s;\bar{r}_f) \quad (3.179)$$

3.3.5 Simulation Result Analysis in Time Domain

For the 2-d time-domain QSCIM, as it is generated by transforming from the frequency domain to the time domain through Fourier transform, it must be explained in conjunction with lightning waveform. The following examples all use double exponential function lightning wave waveform $i(t)=1.29(e^{-0.019010t}-e^{-0.292288t})kA$, the maximum frequency used is 0.8 MHz.

Numerical results for QSCIM in time domain

For two-layer earth model, we give conductivities 300^{-1} S/m, 100^{-1} S/m and permittivities 10 ε_0, 12 ε_0 to the first and second layer of the earth, respectively. The thickness of first layer earth is 5 m. Applying the 2-d MP approach, The core function of the time domain Green's function needs only six quasi-static complex image coefficients to be fitted with a set relative error of 1% to stop the calculation. Figure 3.12a, b show these monotone function and superposition situation of two spatial surfaces from functions $f_1(k_\rho,\omega)$, $f_2(k_\rho,\omega)$, $f_3(k_\rho,\omega)$, and $f_4(k_\rho,\omega)$ and their simulations for real and imagine parts. The figure (c) reflect the relative error distribution of the four fitting function cores. The coefficients of quasi-static complex images for the four core functions are given in Table 3.7.

From these graphs, it can be seen that for core functions $f_1(k_\rho,\omega)$, $f_2(k_\rho,\omega)$, $f_3(k_\rho,\omega)$, and $f_4(k_\rho,\omega)$, both real and imaginary surfaces can be fitted numerically well. The maximum relative errors of their surface fitting are 0.90%, 0.79%, 0.85%,

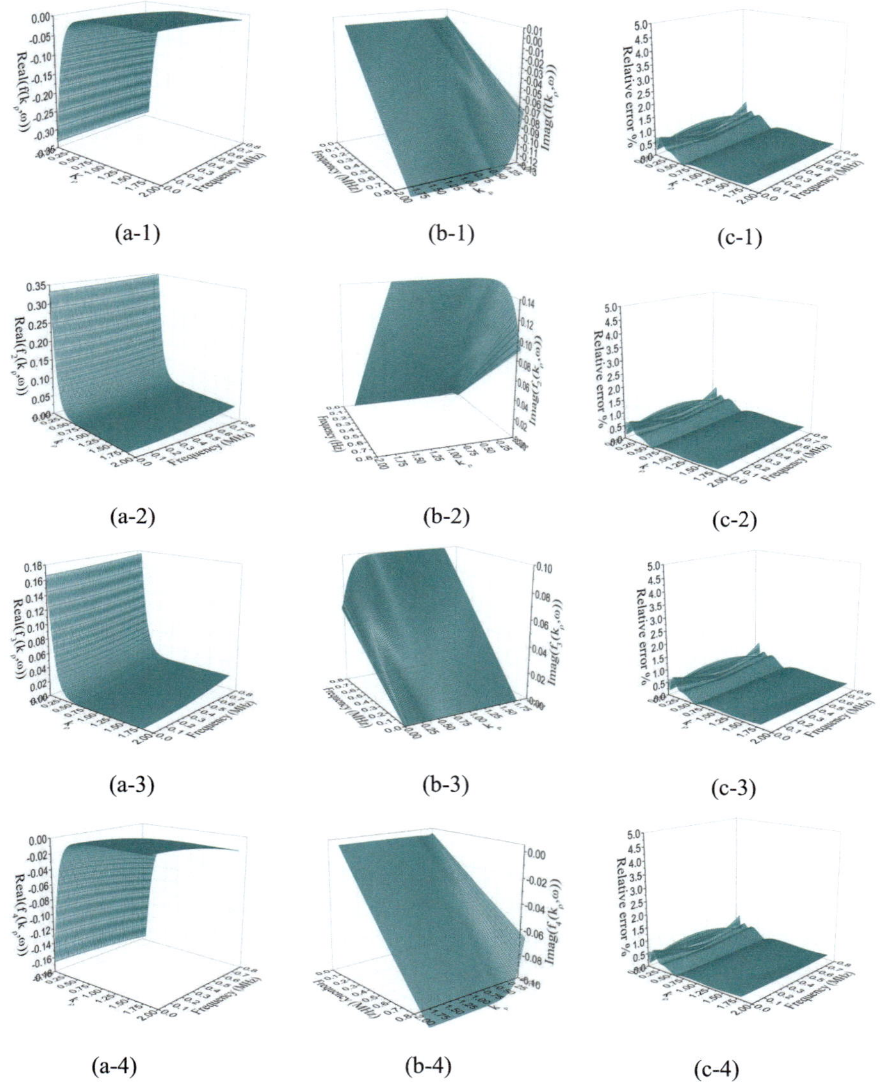

Fig. 3.12 Two spatial surfaces from function $f_1(k_\rho)$, $f_2(k_\rho)$, $f_3(k_\rho)$ and $f_4(k_\rho)$, and theirs simulation for 2-layer earth case with $(\sigma_1 < \sigma_2)$

Table 3.7 Coefficients of Quasi-static complex images for a two-layer earth model ($\sigma_1 < \sigma_2$)

$f(k_\rho, \omega)$	n	α_n	ζ_n	β_n
$f_1(k_\rho, \omega)$	1	$0.289\angle-177.211$	$1.935\times10^{-8}\angle+83.516$	$1.624\times10^{-3}\angle-18.758$
	2	$0.291\angle+2.790$	$7.241\times10^{-8}\angle-91.738$	$1.670\times10^{-3}\angle+175.661$
	3	$0.114\angle+48.789$	$5.886\times10^{-8}\angle-103.096$	$20.616\angle+170.834$
	4	$4.471\times10^{-2}\angle-115.487$	$6.368\times10^{-8}\angle-87.039$	$20.971\angle-170.221$
	5	$0.285\angle-171.623$	$1.805\times10^{-8}\angle-59.209$	$9.883\angle-176.407$
	6	$0.107\angle-178.745$	$1.198\times10^{-7}\angle-94.046$	$7.452\angle+176.861$
$f_2(k_\rho, \omega)$	1	$0.297\angle+3.174$	$2.227\times10^{-8}\angle+82.502$	$1.349\times10^{-3}\angle-20.143$
	2	$0.298\angle-176.821$	$8.506\times10^{-8}\angle-91.971$	$1.398\times10^{-3}\angle+176.761$
	3	$0.106\angle-134.330$	$6.417\times10^{-8}\angle-102.573$	$20.603\angle+170.833$
	4	$3.781\times10^{-2}\angle+67.790$	$6.916\times10^{-8}\angle-87.844$	$20.962\angle-170.258$
	5	$0.290\angle+6.553$	$1.953\times10^{-8}\angle-69.371$	$9.770\angle-176.645$
	6	$0.104\angle+3.463$	$1.326\times10^{-7}\angle-92.876$	$7.693\angle+177.126$
$f_3(k_\rho, \omega)$	1	$0.195\angle+2.682$	$1.833\times10^{-8}\angle+82.932$	$1.171\times10^{-3}\angle-20.206$
	2	$0.195\angle-177.314$	$8.044\times10^{-8}\angle-91.616$	$1.206\times10^{-3}\angle+175.340$
	3	$5.516\times10^{-2}\angle-126.690$	$7.048\times10^{-8}\angle-101.260$	$20.609\angle+170.856$
	4	$2.501\times10^{-2}\angle+72.226$	$7.564\times10^{-8}\angle-87.990$	$20.956\angle-170.229$
	5	$0.135\angle+6.998$	$2.408\times10^{-8}\angle-69.109$	$9.782\angle-176.005$
	6	$5.794\times10^{-2}\angle+3.052$	$1.328\times10^{-7}\angle-93.319$	$7.658\angle+176.462$
$f_4(k_\rho, \omega)$	1	$0.193\angle-176.925$	$2.112\times10^{-8}\angle+81.926$	$1.029\times10^{-3}\angle-21.628$
	2	$0.194\angle+3.075$	$9.211\times10^{-8}\angle-91.860$	$1.069\times10^{-3}\angle+176.216$
	3	$5.231\times10^{-2}\angle+49.832$	$7.580\times10^{-8}\angle-100.956$	$20.596\angle+170.857$
	4	$2.193\times10^{-2}\angle-106.375$	$8.115\times10^{-8}\angle-88.609$	$20.947\angle-170.268$
	5	$0.138\angle-174.299$	$2.692\times10^{-8}\angle-76.240$	$9.692\angle-176.414$
	6	$5.631\times10^{-2}\angle-175.556$	$1.448\times10^{-7}\angle-92.417$	$7.843\angle+176.916$

and 0.75%, respectively. At the same time, it can be seen from these tables that none of the six time domain complex image coefficients required for the numerical fitting of the four core functions has a conjugate complex pair phenomenon, which is inconsistent with the frequency domain complex image.

In order to further compare with the frequency domain quasi static complex image method, the order of the two layer earth resistivity in the above example is reversed ($\sigma_1 > \sigma_2$), and the remaining parameters remain unchanged. Through the 2d M-P, it is known that the core functions ($f_1(k_\rho, \omega)$, $f_2(k_\rho, \omega)$, $f_3(k_\rho, \omega)$, and $f_4(k_\rho, \omega)$) of the Green's function corresponding to such a two-layer soil model can be simulated through five time domain quasi static complex image coefficients. Figure 3.13a, b shows a comparison diagram of the real and imaginary parts of the simulated spatial surface and the analytical spatial surface for the four time domain Green's function core functions. Figure 3.13c shows the relative error distribution of two types of spatial surfaces.

As can be seen from Fig. 3.13, the two types of spatial surfaces are basically consistent, with the maximum relative errors of the four core functions being 1.22%, 1.07%, 0.71%, and 0.68%, respectively. Compared to the above example, the actu-

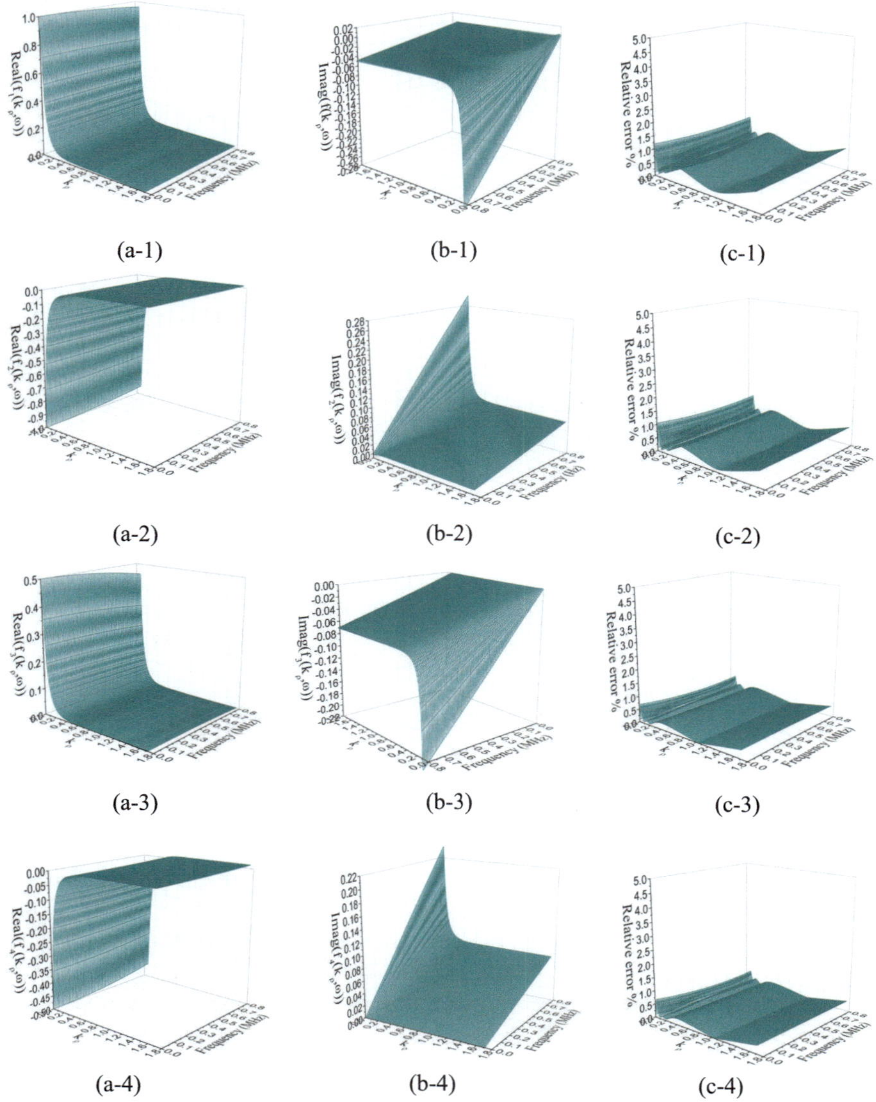

Fig. 3.13 Two spatial surfaces from function $f_1(k_\rho)$, $f_2(k_\rho)$, $f_3(k_\rho)$ and $f_4(k_\rho)$, and theirs simulation for 2-layer earth case with $(\sigma_1 > \sigma_2)$

arial accuracy is relatively reduced. At the same time, it can be seen from Table 3.8 that there is no conjugate complex number phenomenon in the fitting time domain quasi static complex image coefficients of the four core functions.

From the above two numerical examples, it can be seen that there is no conjugate complex pair phenomenon in the time domain complex image coefficient. Therefore,

Table 3.8 Coefficients of Quasi-static complex images for a two-layer earth model ($\sigma_1 > \sigma_2$)

$f(k_\rho, \omega)$	n	α_n	ζ_n	β_n
$f_1(k_\rho, \omega)$	1	$0.255 \times 10^{-2}\angle - 178.562$	$6.620 \times 10^{-9}\angle + 87.718$	$2.004 \times 10^{-2}\angle - 7.542$
	2	$0.266 \times 10^{-2}\angle + 0.760$	$2.423 \times 10^{-8}\angle - 90.599$	$2.046 \times 10^{-2}\angle + 177.000$
	3	$0.429\angle + 0.383$	$2.723 \times 10^{-8}\angle - 92.441$	$10.010\angle - 179.958$
	4	$7.335 \times 10^{-2}\angle - 7.384$	$1.152 \times 10^{-7}\angle - 86.951$	$46.476\angle + 177.048$
	5	$0.382\angle + 1.514$	$4.686 \times 10^{-8}\angle - 86.488$	$22.670\angle - 178.892$
	6	$0.105\angle - 0.111$	$9.584 \times 10^{-8}\angle - 100.644$	$20.250\angle + 176.530$
$f_2(k_\rho, \omega)$	1	$0.278\angle + 1.429$	$7.580 \times 10^{-9}\angle + 87.412$	$2.004 \times 10^{-2}\angle - 7.542$
	2	$0.287\angle - 178.993$	$2.844 \times 10^{-8}\angle - 90.671$	$2.046 \times 10^{-2}\angle + 177.000$
	3	$0.443\angle - 179.614$	$2.890 \times 10^{-8}\angle - 92.642$	$10.010\angle - 179.958$
	4	$8.542 \times 10^{-2}\angle + 178.125$	$1.163 \times 10^{-7}\angle - 87.293$	$46.476\angle + 177.048$
	5	$0.391\angle - 177.585$	$5.060 \times 10^{-8}\angle - 88.675$	$22.670\angle - 178.892$
	6	$7.353 \times 10^{-2}\angle + 168.544$	$9.444 \times 10^{-8}\angle - 99.310$	$20.250\angle + 176.530$
$f_3(k_\rho, \omega)$	1	$0.228\angle - 178.436$	$9.835 \times 10^{-9}\angle + 85.847$	$5.125 \times 10^{-3}\angle - 13.080$
	2	$0.231\angle + 1.460$	$4.677 \times 10^{-8}\angle - 90.867$	$5.268 \times 10^{-3}\angle + 175.312$
	3	$0.233\angle + 0.128$	$4.376 \times 10^{-8}\angle - 93.119$	$9.996\angle - 179.870$
	4	$6.046 \times 10^{-2}\angle + 4.759$	$1.298 \times 10^{-7}\angle - 88.500$	$45.927\angle + 177.568$
	5	$0.188\angle - 0.614$	$6.496 \times 10^{-8}\angle - 92.663$	$21.866\angle + 179.253$
	6	$1.554 \times 10^{-2}\angle - 11.904$	$1.079 \times 10^{-7}\angle - 94.488$	$19.214\angle + 178.200$
$f_4(k_\rho, \omega)$	1	$0.234\angle + 1.568$	$1.002 \times 10^{-8}\angle + 85.711$	$4.769 \times 10^{-3}\angle - 13.417$
	2	$0.236\angle - 178.526$	$4.916 \times 10^{-8}\angle - 90.868$	$4.905 \times 10^{-3}\angle + 175.354$
	3	$0.234\angle - 179.877$	$4.542 \times 10^{-8}\angle - 93.266$	$9.993\angle - 179.854$
	4	$6.196 \times 10^{-2}\angle - 175.132$	$1.314 \times 10^{-7}\angle - 88.630$	$45.879\angle + 177.607$
	5	$0.186\angle + 178.963$	$6.620 \times 10^{-8}\angle - 92.753$	$21.810\angle + 179.281$
	6	$1.606 \times 10^{-2}\angle + 172.466$	$1.097 \times 10^{-7}\angle - 94.143$	$19.081\angle + 178.164$

the order of resistivity is no longer considered in the later calculation examples of multi-layer soil models.

For the three-layer earth model, the earth's conductivities and permittivities are $\sigma_1 = 85.0^{-1}\ S/m, \sigma_2 = 66.0^{-1}\ S/m, \sigma_3 = 108.3^{-1}\ S/m, \varepsilon_1 = 15\,\varepsilon_0, \varepsilon_2 = 12\,\varepsilon_0$ and $\varepsilon_3 = 15\,\varepsilon_0$, respectively. The first layer earth height is 3.8 m, and the secondary layer earth height is 3.4 m. Using the 2-d MP method, only five 2-d time-domain complex image coefficients are required to achieve the specified stop operation relative error of 1%. From Figs. 3.14 and 3.15, we can see that the two spatial surfaces are superposed. The 2-d six quasi-static complex images' coefficients are given in Table 3.9. From the Table 3.9, it can be seen that there is no a pair of conjugate complex number.

As can be seen from Figure 2, the spatial surface fitting accuracy of the eight core functions is basically acceptable, with the maximum relative error of fitting being 0.23%, 0.29%, 0.18%, 0.22%, 0.32%, 0.36%, 0.23%, and 0.27%, respectively. Are less than the accuracy required by conventional numerical calculations, which is 5%.

For four-layer earth model, the earth's conductivities and permittivities are $\sigma_1 = 75.0^{-1}\ S/m,\ \sigma_2 = 96^{-1}\ S/m,\ \sigma_3 = 118^{-1}\ S/m,\ \sigma_4 = 169.31^{-1}\ S/m,\ \varepsilon_1 = 15\ \varepsilon_0$, $\varepsilon_2 = 12\ \varepsilon_0,\ \varepsilon_3 = 13\ \varepsilon_0$ and $\varepsilon_4 = 12\ \varepsilon_0$, respectively. The first layer earth height is

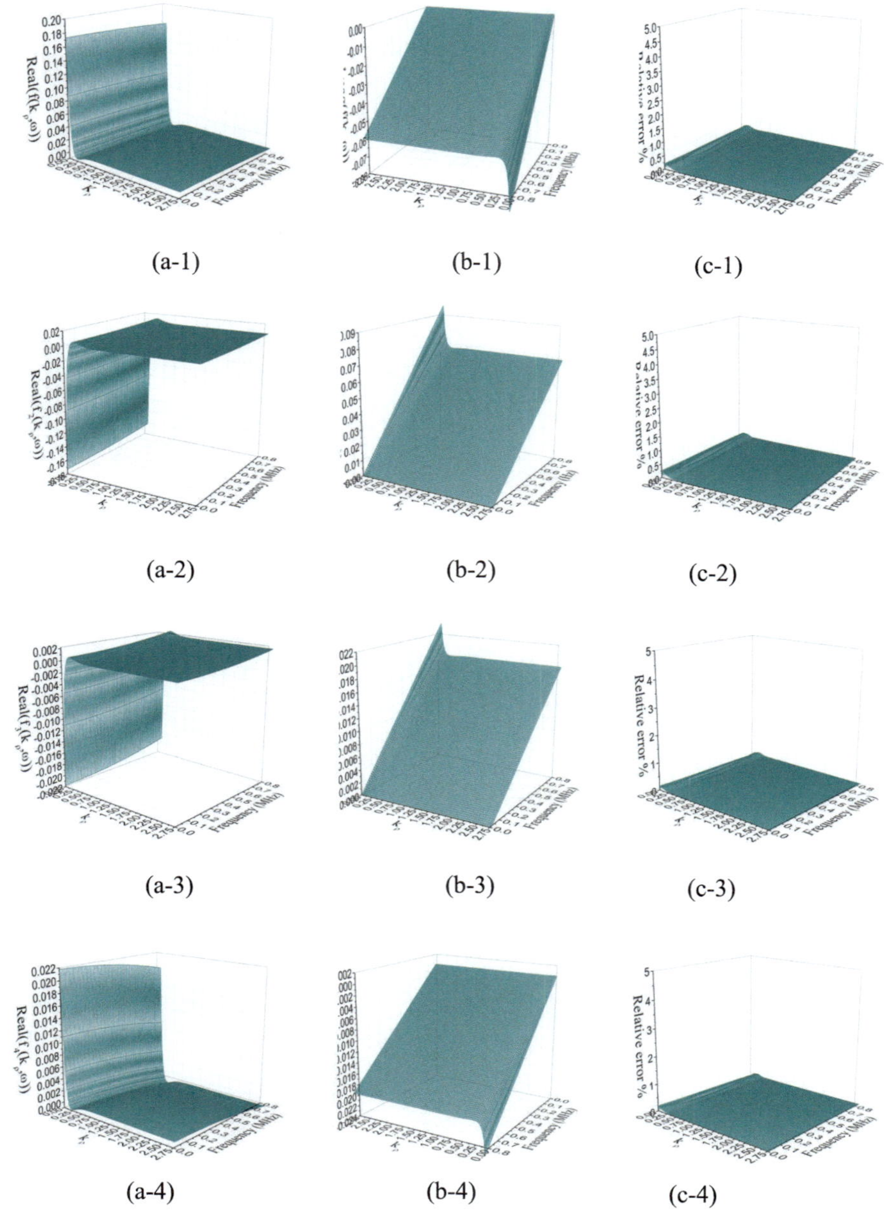

Fig. 3.14 Two curves from $f_1(k_\rho, \omega)$, $f_2(k_\rho, \omega)$, $f_3(k_\rho, \omega)$ and $f_4(k_\rho, \omega)$, and its simulation for 3-layer earth case

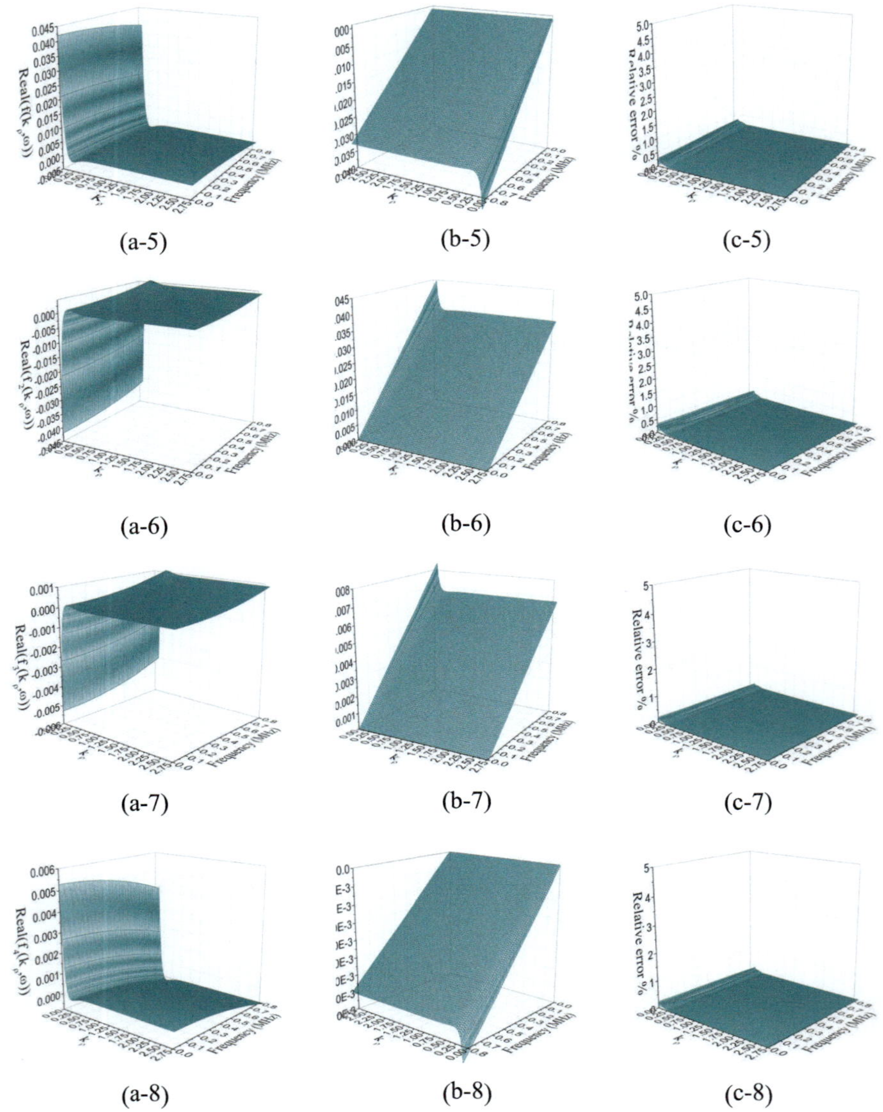

Fig. 3.15 Two curves from $f_5(k_\rho, \omega)$, $f_6(k_\rho, \omega)$, $f_7(k_\rho, \omega)$ and $f_8(k_\rho, \omega)$, and its simulation for 3-layer earth case

Table 3.9 Coefficients of Quasi-static complex images for a three-layer earth model

$f(k_\rho, \omega)$	n	α_n	ζ_n	β_n
$f_1(k_\rho, \omega)$	1	$0.288\angle -178.808$	$8.244 \times 10^{-9}\angle +87.234$	$1.166 \times 10^{-5}\angle -14.267$
	2	$0.288\angle +1.187$	$3.079 \times 10^{-8}\angle -90.741$	$7.659 \times 10^{-5}\angle +175.730$
	3	$0.116\angle +177.481$	$2.068 \times 10^{-8}\angle -89.218$	$8.220\angle +179.735$
	4	$3.990 \times 10^{-2}\angle -6.351$	$4.303 \times 10^{-8}\angle -86.932$	$33.568\angle -179.911$
	5	$0.255\angle -2.376$	$2.490 \times 10^{-8}\angle -94.791$	$13.637\angle -179.697$
	6	$1.118 \times 10^{-2}\angle +118.632$	$6.314 \times 10^{-8}\angle -92.462$	$18.723\angle +179.092$
$f_2(k_\rho, \omega)$	1	$0.292\angle +1.311$	$9.149 \times 10^{-9}\angle +86.927$	$5.034 \times 10^{-5}\angle +172.152$
	2	$0.292\angle -178.691$	$3.452 \times 10^{-8}\angle -90.815$	$7.959 \times 10^{-6}\angle +164.747$
	3	$0.115\angle -2.861$	$2.228 \times 10^{-8}\angle -89.288$	$8.219\angle +179.732$
	4	$0.251\angle +176.908$	$2.590 \times 10^{-8}\angle -95.923$	$13.635\angle -179.637$
	5	$3.892 \times 10^{-2}\angle +170.795$	$4.440 \times 10^{-8}\angle -86.837$	$33.559\angle -179.862$
	6	$1.396 \times 10^{-2}\angle -86.368$	$6.596 \times 10^{-8}\angle -92.143$	$19.110\angle +179.031$
$f_3(k_\rho, \omega)$	1	$7.316 \times 10^{-2}\angle +1.129$	$7.485 \times 10^{-9}\angle +86.403$	$5.689 \times 10^{-6}\angle +134.589$
	2	$7.317 \times 10^{-2}\angle -178.871$	$4.062 \times 10^{-8}\angle -90.663$	$2.249 \times 10^{-5}\angle -176.434$
	3	$1.458 \times 10^{-2}\angle -2.381$	$3.743 \times 10^{-8}\angle -89.596$	$8.219\angle +179.734$
	4	$3.180 \times 10^{-2}\angle +177.926$	$4.134 \times 10^{-8}\angle -92.806$	$13.647\angle -179.676$
	5	$4.928 \times 10^{-3}\angle +173.943$	$5.967 \times 10^{-8}\angle -88.100$	$33.475\angle -179.864$
	5	$1.110 \times 10^{-3}\angle -71.884$	$8.504 \times 10^{-8}\angle -91.414$	$18.941\angle +178.920$
$f_4(k_\rho, \omega)$	1	$7.150 \times 10^{-2}\angle -178.769$	$8.176 \times 10^{-9}\angle +86.130$	$2.611 \times 10^{-5}\angle +162.519$
	2	$7.149 \times 10^{-2}\angle +1.230$	$4.370 \times 10^{-8}\angle -90.724$	$2.488 \times 10^{-6}\angle -93.746$
	3	$1.444 \times 10^{-2}\angle +177.284$	$3.903 \times 10^{-8}\angle -89.602$	$8.219\angle +179.730$
	4	$3.135 \times 10^{-2}\angle -2.722$	$4.229 \times 10^{-8}\angle -93.531$	$13.644\angle -179.612$
	5	$4.786 \times 10^{-3}\angle -8.698$	$6.104 \times 10^{-8}\angle -88.009$	$33.468\angle -179.819$
	6	$1.526 \times 10^{-3}\angle +84.175$	$8.747 \times 10^{-8}\angle -91.244$	$19.294\angle +178.875$
$f_5(k_\rho, \omega)$	1	$0.125\angle -178.823$	$8.003 \times 10^{-9}\angle +86.408$	$5.120 \times 10^{-5}\angle +171.006$
	2	$0.125\angle +1.176$	$4.149 \times 10^{-8}\angle -90.693$	$2.126 \times 10^{-5}\angle +1.880$
	3	$2.770 \times 10^{-2}\angle +176.575$	$3.516 \times 10^{-8}\angle -89.489$	$8.220\angle +179.727$
	4	$5.991 \times 10^{-2}\angle -4.197$	$3.802 \times 10^{-8}\angle -95.362$	$13.625\angle -179.554$
	5	$9.101 \times 10^{-3}\angle -14.014$	$5.704 \times 10^{-8}\angle -87.239$	$33.579\angle -179.826$
	5	$5.031 \times 10^{-3}\angle +77.629$	$7.901 \times 10^{-8}\angle -91.488$	$19.511\angle +178.985$
$f_6(k_\rho, \omega)$	1	$0.123\angle +1.269$	$8.640 \times 10^{-9}\angle +86.152$	$6.943 \times 10^{-5}\angle +170.217$
	2	$0.123\angle -178.730$	$4.444 \times 10^{-8}\angle -90.748$	$4.086 \times 10^{-5}\angle -0.463$
	3	$2.750 \times 10^{-2}\angle -3.788$	$3.674 \times 10^{-8}\angle -89.437$	$8.221\angle +179.722$
	4	$5.920 \times 10^{-2}\angle +175.150$	$3.906 \times 10^{-8}\angle -96.168$	$13.617\angle -179.488$
	5	$8.841 \times 10^{-3}\angle +162.412$	$5.842 \times 10^{-8}\angle -87.169$	$33.578\angle -179.778$
	6	$6.177 \times 10^{-3}\angle -109.109$	$8.166 \times 10^{-8}\angle -91.264$	$19.787\angle +178.907$
$f_7(k_\rho, \omega)$	1	$2.118 \times 10^{-2}\angle +1.101$	$6.733 \times 10^{-9}\angle +85.061$	$3.085\angle +164.822$
	2	$2.117 \times 10^{-2}\angle -178.898$	$5.418 \times 10^{-8}\angle -90.614$	$1.225\angle -8.443$
	3	$3.478 \times 10^{-3}\angle -3.243$	$5.191 \times 10^{-8}\angle -89.649$	$8.220\angle +179.724$
	4	$7.501 \times 10^{-3}\angle +176.289$	$5.437 \times 10^{-8}\angle -93.652$	$13.635\angle -179.523$
	5	$1.113 \times 10^{-3}\angle +166.859$	$7.366 \times 10^{-8}\angle -88.125$	$33.488\angle -179.778$
	6	$5.705 \times 10^{-4}\angle -109.239$	$1.001 \times 10^{-7}\angle -90.889$	$19.662\angle +178.837$
$f_8(k_\rho, \omega)$	1	$2.085 \times 10^{-2}\angle -178.821$	$7.258 \times 10^{-9}\angle +84.847$	$4.084 \times 10^{-5}\angle +165.271$
	2	$2.085 \times 10^{-2}\angle +1.178$	$5.681 \times 10^{-8}\angle -90.659$	$2.270 \times 10^{-5}\angle -5.854$
	3	$3.457 \times 10^{-3}\angle +176.403$	$5.349 \times 10^{-8}\angle -89.602$	$8.221\angle +179.718$
	4	$7.426 \times 10^{-3}\angle -4.325$	$5.537 \times 10^{-8}\angle -94.239$	$13.628\angle -179.454$
	5	$1.078 \times 10^{-3}\angle -16.537$	$7.504 \times 10^{-8}\angle -88.055$	$33.488\angle -179.734$
	5	$7.116 \times 10^{-4}\angle +65.343$	$1.025 \times 10^{-7}\angle -90.749$	$19.917\angle +178.770$

4.867 m, secondary layer earth height is 5.432 m and third layer earth height is 6.34 m. Using the 2-d MP method, six 2-d time domain complex image coefficients can be used to simulate the spatial surface of the core function when the stop operation error is set to 1%. From Figs. 3.16 and 3.17, we can see that the two spatial surfaces superpose each other. The coefficients of the six quasi-static complex images are given in Tables 3.10 and 3.11.

From Figs. 3.16, 3.17, 3.18, and 3.19a, it can be seen that the real part of a 3-d space surface is basically consistent, and from Figs. 3.16, 3.17, 3.18, and 3.19b, it can be seen that the imaginary part of a 3-d space surface is also basically consistent. Although there are significant differences, it also meets the 5% principle for numerical calculation accuracy requirements. As can be seen from Figs. 3.16, 3.17, 3.18, and 3.19c, the maximum relative error for the 16 core functions is 2.95%, 4.08%, 1.53%, 1.17%, 1.36%, 1.19%, 2.39%, 1.69%, 2.93%, 2.66%, 2.71%, 2.78%, 1.44%, 1.41%, 2.25%, and 2.22%, respectively.

Numerical results for a point source in time domain

By using the time-domain QSCIM, the closed form Green's function of the point current source in multi-layer horizontally layered soil can be obtained. In this way, the electromagnetic field distribution of the buried point current source on the surface of the earth above it can be calculated.

Firstly, calculate the voltage (reference point at infinite distance) and VMP on the surface of the earth above the point current source buried in multi-layer horizontal layered soil. Once voltage φ and VMP (A) have been obtained for a point source, the EFI (E) and MFI (B) of the point source can be calculated through following formula.

$$\bar{B} = \nabla \times \bar{A} \tag{3.180}$$

$$\bar{E} = -\frac{d\bar{A}}{dt} - \frac{\partial \varphi}{\partial \tau}\hat{\tau} \tag{3.181}$$

Here $\hat{\tau}$ is unit director vector of the vector dipole.

This chapter uses a four layer horizontally layered soil model as an example to calculate the distribution of electromagnetic field on the surface of the earth above the buried current source. The resistivity of the four horizontally layered soil layers is 600Ωm, 400Ωm, 200Ωm, and 100Ω m, respectively, and the relative dielectric constants are 10, 22, 22, and 22, respectively. The thickness of the three layers of soil is $5m$, $2.4m$, and $3.4m$, respectively.

Figures 3.20, 3.21, and 3.22 respectively show the distribution of electromagnetic fields on the surface of the earth above a monopole point current source and a dipole point current source in three directions. Observing the three figures, it can be seen that due to the lack of directionality of the monopole point current source, the voltage distribution is basically identical. Due to the directionality of dipole point current source, the VMP (A), EFI (E) and MFI (B) distribution are slightly different.

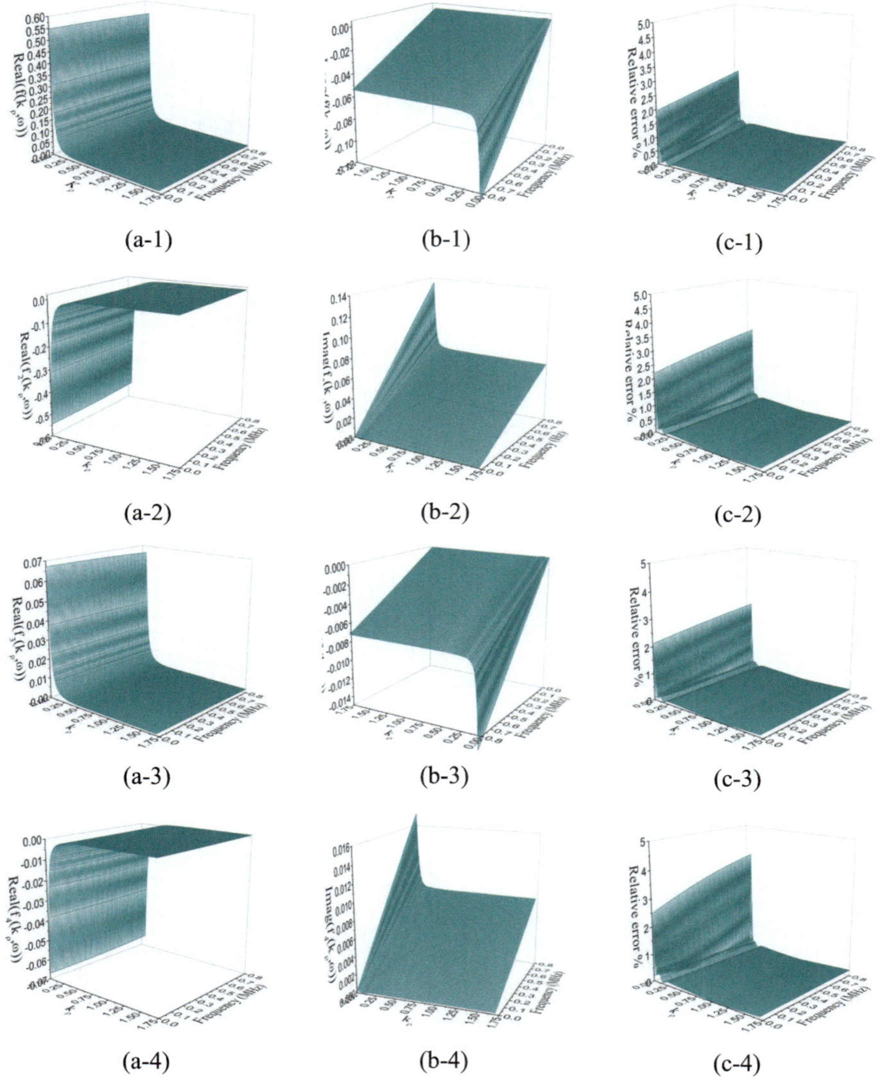

Fig. 3.16 Two curves from $f_1(k_\rho, \omega)$, $f_2(k_\rho, \omega)$, $f_3(k_\rho, \omega)$ and $f_4(k_\rho, \omega)$, and its simulation for 3-layer earth case

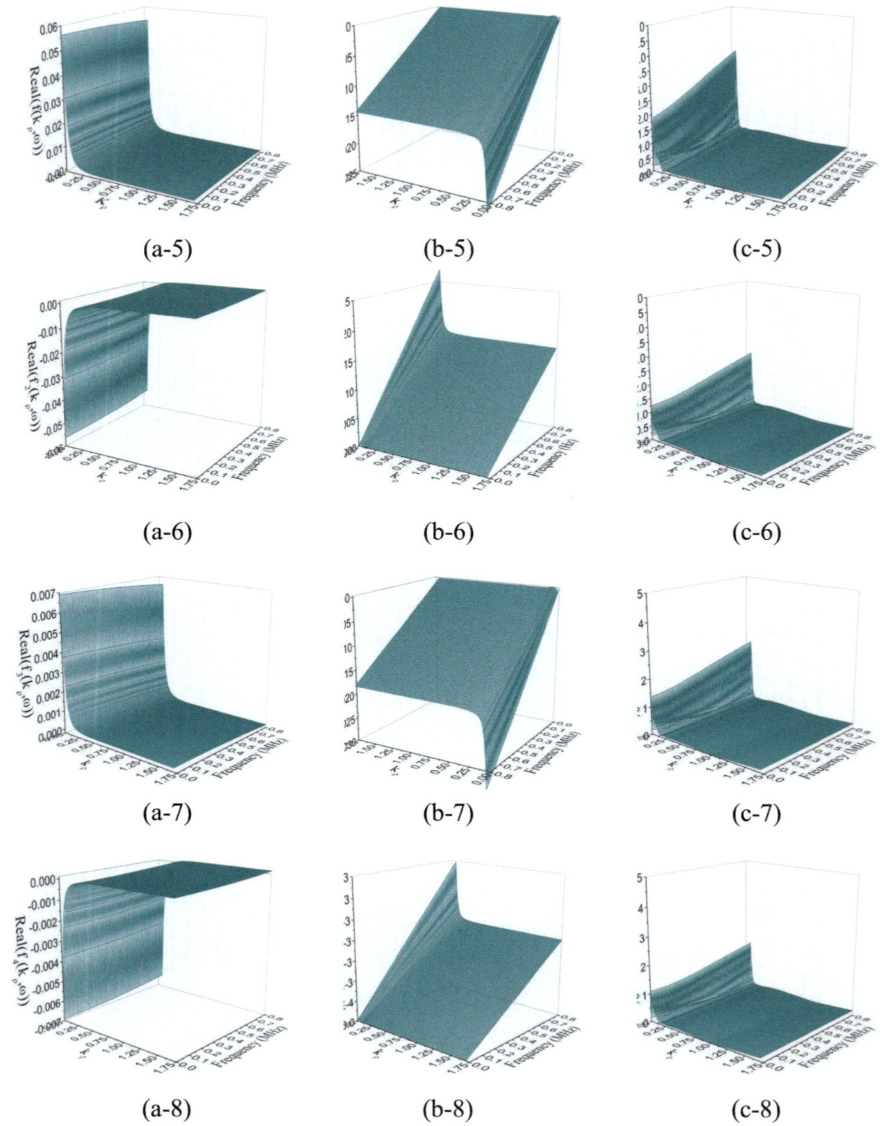

Fig. 3.17 Two curves from $f_5(k_\rho, \omega)$, $f_6(k_\rho, \omega)$, $f_7(k_\rho, \omega)$ and $f_8(k_\rho, \omega)$, and its simulation for 3-layer earth case

Table 3.10 Coefficients of Quasi-static complex images for a three-layer earth model

$f(k_\rho,\omega)$	n	α_n	ζ_n	β_n
$f_1(k_\rho,\omega)$	1	$0.288\angle-179.001$	$7.280\times10^{-9}\angle+87.561$	$1.136\times10^{-2}\angle+0.808$
	2	$0.288\angle+1.001$	$2.716\times10^{-8}\angle-90.653$	$1.342\times10^{-3}\angle-175.435$
	3	$0.104\angle-0.800$	$9.776\times10^{-9}\angle-90.627$	$9.608\angle-179.939$
	4	$0.173\angle+7.636$	$2.605\times10^{-8}\angle-93.436$	$25.341\angle+179.921$
	5	$0.204\angle+6.534$	$1.418\times10^{-8}\angle-101.285$	$43.847\angle-173.935$
	6	$6.802\times10^{-2}\angle+6.534$	$8.097\times10^{-8}\angle-88.868$	$45.991\angle+171.085$
$f_2(k_\rho,\omega)$	1	$0.291\angle+1.061$	$8.081\times10^{-9}\angle+87.289$	$1.254\times10^{-3}\angle-1.512$
	2	$0.292\angle-178.939$	$3.046\times10^{-8}\angle-90.719$	$1.438\times10^{-3}\angle-176.642$
	3	$0.104\angle+178.846$	$1.109\times10^{-8}\angle-90.765$	$9.608\angle-179.939$
	4	$0.169\angle-168.554$	$2.739\times10^{-8}\angle-93.936$	$25.330\angle+179.944$
	5	$0.195\angle+168.927$	$7.368\times10^{-9}\angle-108.644$	$42.718\angle-169.450$
	6	$8.241\times10^{-2}\angle-175.916$	$9.044\times10^{-8}\angle-89.031$	$43.450\angle+167.011$
$f_3(k_\rho,\omega)$	1	$3.583\times10^{-2}\angle-178.958$	$7.816\times10^{-9}\angle+87.379$	$1.225\times10^{-3}\angle-0.919$
	2	$3.593\times10^{-2}\angle+1.041$	$2.947\times10^{-8}\angle-90.695$	$1.414\times10^{-3}\angle-176.349$
	3	$1.283\times10^{-2}\angle-1.013$	$1.072\times10^{-8}\angle-90.722$	$9.608\angle-179.939$
	4	$2.096\times10^{-2}\angle+10.184$	$2.702\times10^{-8}\angle-93.776$	$25.334\angle+179.936$
	5	$2.443\times10^{-2}\angle-10.218$	$1.012\times10^{-8}\angle-106.000$	$43.017\angle-171.083$
	5	$9.542\times10^{-3}\angle+5.035$	$8.693\times10^{-8}\angle-88.785$	$44.377\angle+168.399$
$f_4(k_\rho,\omega)$	1	$3.663\times10^{-2}\angle+1.029$	$8.453\times10^{-9}\angle+87.151$	$1.302\times10^{-3}\angle-2.496$
	2	$3.674\times10^{-2}\angle-178.974$	$3.242\times10^{-8}\angle-90.742$	$1.472\times10^{-3}\angle-177.097$
	3	$1.301\times10^{-2}\angle+178.201$	$1.204\times10^{-8}\angle-90.852$	$9.608\angle-179.939$
	4	$2.068\times10^{-2}\angle-164.956$	$2.834\times10^{-8}\angle-94.275$	$25.322\angle+179.962$
	5	$2.279\times10^{-2}\angle+166.536$	$3.149\times10^{-9}\angle+74.888$	$42.098\angle-164.171$
	6	$1.175\times10^{-2}\angle-178.231$	$1.023\times10^{-7}\angle-90.717$	$40.567\angle+163.076$
$f_5(k_\rho,\omega)$	1	$5.281\times10^{-2}\angle-178.720$	$8.255\times10^{-9}\angle+86.303$	$7.967\times10^{-4}\angle-6.271$
	2	$5.290\times10^{-2}\angle+1.271$	$4.266\times10^{-8}\angle-90.715$	$8.779\times10^{-4}\angle-179.066$
	3	$9.960\times10^{-3}\angle-2.904$	$2.598\times10^{-8}\angle-90.694$	$9.608\angle-179.937$
	4	$2.128\times10^{-2}\angle+19.204$	$4.216\times10^{-8}\angle-94.229$	$25.286\angle-179.949$
	5	$1.523\times10^{-2}\angle+3.258$	$6.003\times10^{-8}\angle-50.502$	$71.793\angle-164.804$
	6	$1.319\times10^{-2}\angle-33.629$	$8.963\times10^{-8}\angle-115.441$	$31.212\angle+175.481$
$f_6(k_\rho,\omega)$	1	$5.277\times10^{-2}\angle+1.312$	$8.630\times10^{-9}\angle+86.137$	$8.167\times10^{-4}\angle-6.790$
	2	$5.285\times10^{-2}\angle-178.694$	$4.492\times10^{-8}\angle-90.742$	$8.946\times10^{-4}\angle-179.146$
	3	$1.001\times10^{-2}\angle+177.429$	$2.729\times10^{-8}\angle-90.775$	$9.608\angle-179.937$
	4	$2.149\times10^{-2}\angle-159.638$	$4.339\times10^{-8}\angle-94.634$	$25.272\angle-179.909$
	5	$1.634\times10^{-2}\angle-173.969$	$6.035\times10^{-8}\angle-58.483$	$68.901\angle-172.145$
	6	$1.312\times10^{-2}\angle+138.404$	$8.453\times10^{-8}\angle-111.873$	$31.567\angle+176.517$
$f_7(k_\rho,\omega)$	1	$6.491\times10^{-3}\angle-178.701$	$8.501\times10^{-9}\angle+86.190$	$8.106\times10^{-4}\angle-6.631$
	2	$6.501\times10^{-3}\angle+1.291$	$4.425\times10^{-8}\angle-90.731$	$8.893\times10^{-4}\angle-179.120$
	3	$1.229\times10^{-3}\angle-2.629$	$2.692\times10^{-8}\angle-90.750$	$9.608\angle-179.937$
	4	$2.624\times10^{-3}\angle+20.050$	$4.306\times10^{-8}\angle-94.507$	$25.276\angle-179.922$
	5	$1.967\times10^{-3}\angle+5.474$	$6.034\times10^{-8}\angle-56.586$	$70.106\angle-170.656$
	6	$1.620\times10^{-3}\angle-39.316$	$8.548\times10^{-8}\angle-112.916$	$31.502\angle+176.316$
$f_8(k_\rho,\omega)$	1	$6.496\times10^{-3}\angle+1.331$	$8.844\times10^{-9}\angle+86.031$	$8.260\times10^{-4}\angle-7.099$
	2	$6.507\times10^{-3}\angle-178.674$	$4.644\times10^{-8}\angle-90.755$	$9.017\times10^{-4}\angle-179.173$
	3	$1.230\times10^{-3}\angle+177.527$	$2.824\times10^{-8}\angle-90.829$	$9.607\angle-179.936$
	4	$2.679\times10^{-3}\angle-158.998$	$4.427\times10^{-8}\angle-94.911$	$25.262\angle-179.879$
	5	$2.083\times10^{-3}\angle-172.971$	$6.093\times10^{-8}\angle-61.938$	$66.811\angle-174.464$
	6	$1.602\times10^{-3}\angle+133.656$	$8.302\times10^{-8}\angle-109.966$	$31.602\angle+176.893$
$f_9(k_\rho,\omega)$	1	$6.655\times10^{-2}\angle-178.378$	$1.097\times10^{-8}\angle+85.869$	$1.259\times10^{-3}\angle-6.824$
	2	$6.672\times10^{-2}\angle+1.613$	$4.762\times10^{-8}\angle-90.952$	$1.360\times10^{-3}\angle-178.341$
	3	$1.724\times10^{-2}\angle-2.216$	$2.160\times10^{-8}\angle-91.459$	$9.607\angle-179.934$
	4	$4.059\times10^{-2}\angle+20.632$	$3.724\times10^{-8}\angle-97.338$	$25.235\angle-179.744$
	5	$3.245\times10^{-2}\angle+6.543$	$5.313\times10^{-8}\angle-64.547$	$62.424\angle-176.628$
	6	$2.007\times10^{-2}\angle-59.614$	$7.527\times10^{-8}\angle-106.876$	$31.329\angle+177.187$

Table 3.11 Coefficients of Quasi-static complex images for a three-layer earth model (continue)

$f(k_\rho, \omega)$	n	α_n	ζ_n	β_n
$f_{10}(k_\rho, \omega)$	1	$6.832 \times 10^{-2}\angle + 1.629$	$1.113 \times 10^{-8}\angle + 85.762$	$1.244 \times 10^{-3}\angle - 7.142$
	2	$6.850 \times 10^{-2}\angle - 178.378$	$4.940 \times 10^{-8}\angle - 90.955$	$1.341 \times 10^{-3}\angle - 178.356$
	3	$1.719 \times 10^{-2}\angle + 177.919$	$2.292 \times 10^{-8}\angle - 91.530$	$9.606\angle - 179.933$
	4	$3.324 \times 10^{-2}\angle - 173.615$	$5.432 \times 10^{-8}\angle - 66.619$	$61.679\angle - 177.269$
	5	$4.151 \times 10^{-2}\angle - 159.589$	$3.836 \times 10^{-8}\angle - 97.759$	$25.223\angle - 179.685$
	6	$1.948 \times 10^{-2}\angle + 115.832$	$7.572 \times 10^{-8}\angle - 105.484$	$31.129\angle + 177.310$
$f_{11}(k_\rho, \omega)$	1	$8.341 \times 10^{-3}\angle - 178.375$	$1.107 \times 10^{-8}\angle + 85.796$	$1.248 \times 10^{-3}\angle - 7.042$
	2	$8.363 \times 10^{-3}\angle + 1.616$	$4.886 \times 10^{-8}\angle - 90.953$	$1.345 \times 10^{-3}\angle - 178.351$
	3	$2.113 \times 10^{-3}\angle - 2.125$	$2.255 \times 10^{-8}\angle - 91.507$	$9.606\angle - 179.933$
	4	$4.054 \times 10^{-3}\angle + 6.490$	$5.402 \times 10^{-8}\angle - 66.058$	$61.897\angle - 177.118$
	5	$5.065 \times 10^{-3}\angle + 20.532$	$3.806 \times 10^{-8}\angle - 97.627$	$25.226\angle - 179.704$
	6	$2.422 \times 10^{-3}\angle - 62.838$	$7.551 \times 10^{-8}\angle - 105.900$	$31.190\angle + 177.285$
$f_{12}(k_\rho, \omega)$	1	$8.553 \times 10^{-3}\angle + 1.634$	$1.123 \times 10^{-8}\angle + 85.688$	$1.232 \times 10^{-3}\angle - 7.347$
	2	$8.575 \times 10^{-3}\angle - 178.373$	$5.062 \times 10^{-8}\angle - 90.957$	$1.325 \times 10^{-3}\angle - 178.360$
	3	$2.106 \times 10^{-3}\angle + 178.020$	$2.387 \times 10^{-8}\angle - 91.574$	$9.606\angle - 179.932$
	4	$4.140 \times 10^{-3}\angle - 173.803$	$5.522 \times 10^{-8}\angle - 67.818$	$61.281\angle - 177.613$
	5	$5.172 \times 10^{-3}\angle - 159.832$	$3.916 \times 10^{-8}\angle - 98.039$	$25.214\angle - 179.642$
	6	$2.341 \times 10^{-3}\angle + 112.932$	$7.618 \times 10^{-8}\angle - 104.638$	$30.988\angle + 177.370$
$f_{13}(k_\rho, \omega)$	1	$9.889 \times 10^{-3}\angle - 178.432$	$1.054 \times 10^{-8}\angle + 84.902$	$8.192 \times 10^{-4}\angle - 9.566$
	2	$9.905 \times 10^{-3}\angle + 1.560$	$6.019 \times 10^{-8}\angle - 90.893$	$8.786 \times 10^{-4}\angle - 179.260$
	3	$1.762 \times 10^{-3}\angle - 1.495$	$3.780 \times 10^{-8}\angle - 91.260$	$9.605\angle - 179.929$
	4	$3.613 \times 10^{-3}\angle + 5.044$	$6.857 \times 10^{-8}\angle - 73.977$	$60.249\angle - 178.480$
	5	$4.495 \times 10^{-3}\angle + 18.685$	$5.225 \times 10^{-8}\angle - 97.287$	$25.188\angle - 179.448$
	6	$1.762 \times 10^{-3}\angle - 76.257$	$8.903 \times 10^{-8}\angle - 100.561$	$30.456\angle + 177.451$
$f_{14}(k_\rho, \omega)$	1	$9.961 \times 10^{-3}\angle + 1.583$	$1.070 \times 10^{-9}\angle + 84.786$	$8.157 \times 10^{-4}\angle - 9.821$
	2	$9.978 \times 10^{-3}\angle - 178.423$	$6.196 \times 10^{-8}\angle - 90.901$	$8.735 \times 10^{-4}\angle - 179.228$
	3	$1.759 \times 10^{-3}\angle + 178.700$	$3.911 \times 10^{-8}\angle - 91.315$	$9.604\angle - 179.927$
	4	$3.645 \times 10^{-3}\angle - 175.537$	$6.987 \times 10^{-8}\angle - 74.690$	$59.998\angle - 178.654$
	5	$4.539 \times 10^{-3}\angle - 162.139$	$5.322 \times 10^{-8}\angle - 97.599$	$25.181\angle - 179.367$
	6	$1.667 \times 10^{-3}\angle + 101.197$	$9.038 \times 10^{-8}\angle - 99.873$	$30.273\angle + 177.429$
$f_{15}(k_\rho, \omega)$	1	$1.221 \times 10^{-3}\angle - 178.422$	$1.064 \times 10^{-8}\angle + 84.821$	$8.161 \times 10^{-4}\angle - 9.742$
	2	$1.223 \times 10^{-3}\angle + 1.570$	$6.144 \times 10^{-8}\angle - 90.898$	$8.743 \times 10^{-4}\angle - 179.236$
	3	$2.162 \times 10^{-4}\angle - 1.362$	$3.875 \times 10^{-8}\angle - 91.297$	$9.604\angle - 179.928$
	4	$4.466 \times 10^{-4}\angle + 4.662$	$6.952 \times 10^{-8}\angle - 74.493$	$60.072\angle - 178.615$
	5	$5.561 \times 10^{-4}\angle + 18.141$	$5.297 \times 10^{-8}\angle - 97.505$	$25.182\angle - 179.393$
	6	$2.087 \times 10^{-4}\angle - 78.123$	$8.994 \times 10^{-8}\angle - 100.076$	$30.328\angle + 177.441$
$f_{16}(k_\rho, \omega)$	1	$1.230 \times 10^{-3}\angle + 1.592$	$1.080 \times 10^{-8}\angle + 84.706$	$8.121 \times 10^{-4}\angle - 9.991$
	2	$1.232 \times 10^{-3}\angle - 178.414$	$6.319 \times 10^{-8}\angle - 90.905$	$8.687 \times 10^{-4}\angle - 179.203$
	3	$2.159 \times 10^{-4}\angle + 178.837$	$4.006 \times 10^{-8}\angle - 91.350$	$9.604\angle - 179.926$
	4	$4.498 \times 10^{-4}\angle - 175.928$	$7.082 \times 10^{-8}\angle - 75.134$	$59.856\angle - 178.764$
	5	$5.604 \times 10^{-4}\angle - 162.717$	$5.391 \times 10^{-8}\angle - 97.805$	$25.177\angle - 179.309$
	6	$1.968 \times 10^{-4}\angle + 99.676$	$9.137 \times 10^{-8}\angle - 99.439$	$30.150\angle + 177.407$

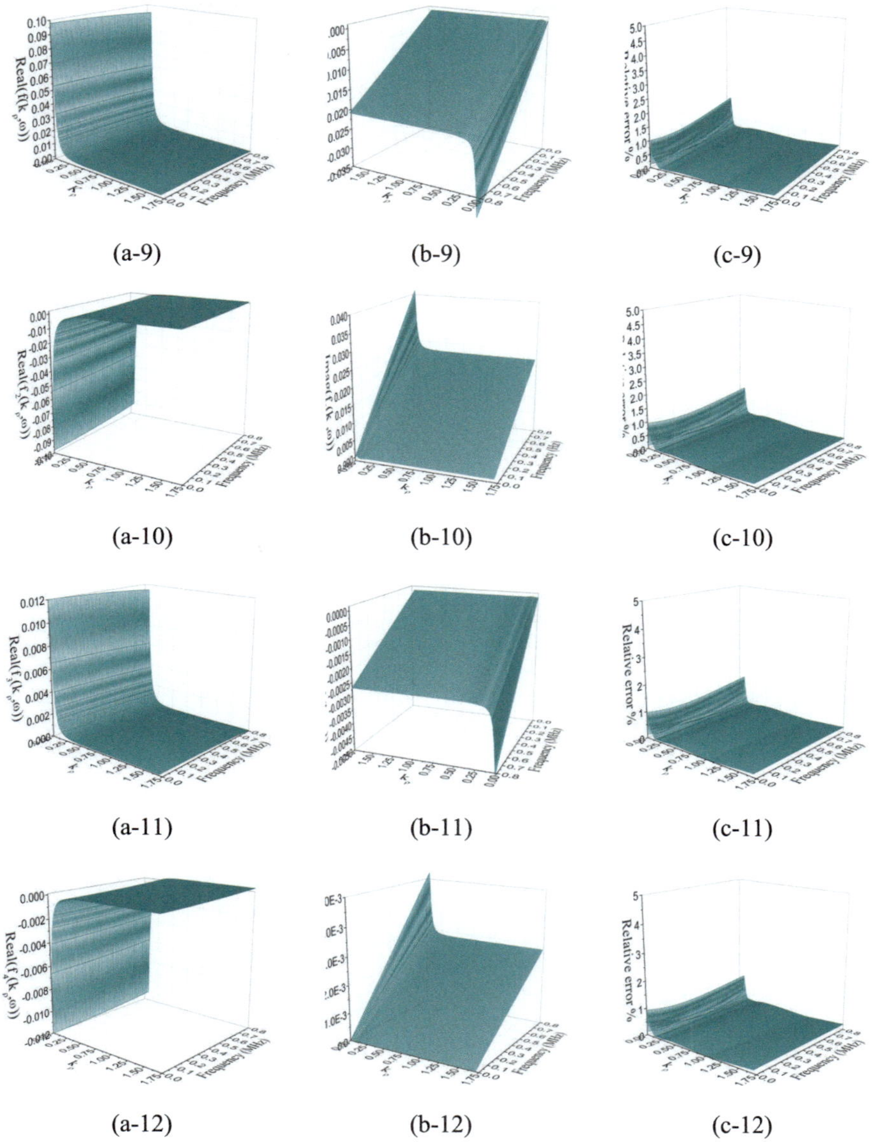

Fig. 3.18 Two curves from $f_9(k_\rho, \omega)$, $f_{10}(k_\rho, \omega)$, $f_{11}(k_\rho, \omega)$ and $f_{12}(k_\rho, \omega)$, and its simulation for 3-layer earth case

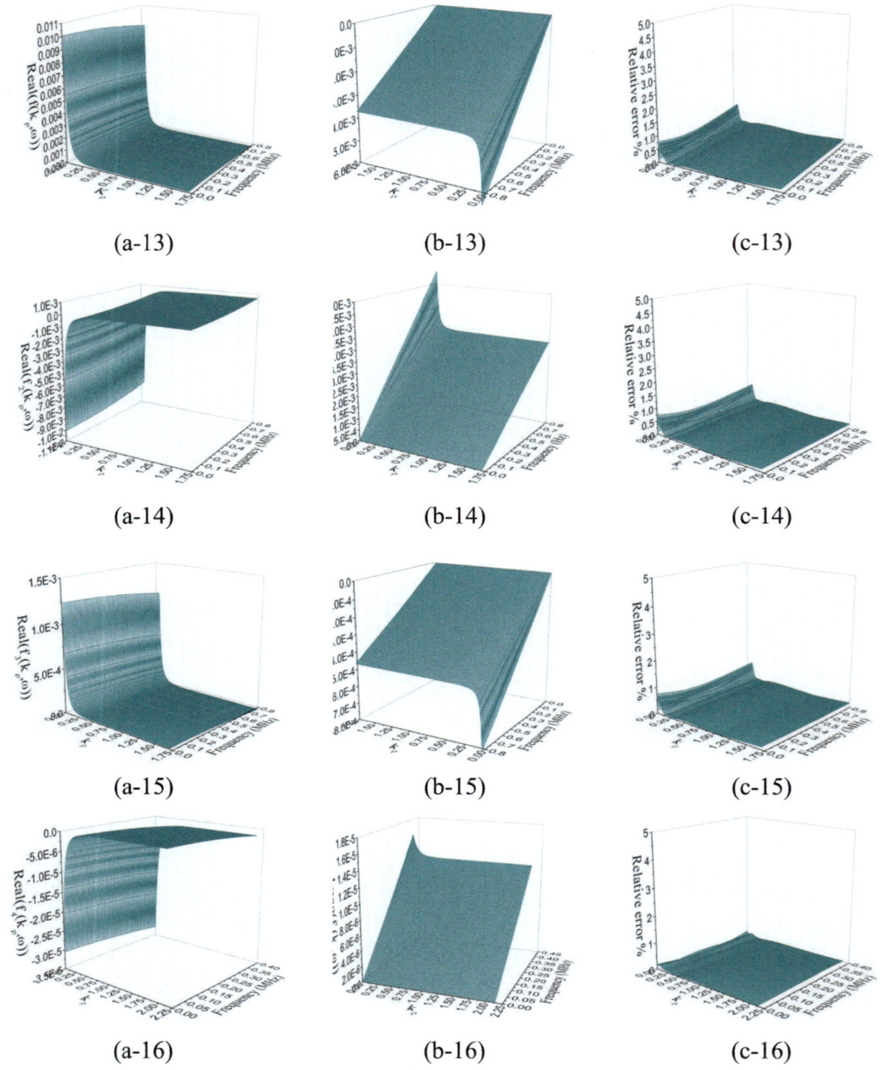

Fig. 3.19 Two curves from $f_{13}(k_\rho, \omega)$, $f_{14}(k_\rho, \omega)$, $f_{15}(k_\rho, \omega)$ and $f_{16}(k_\rho, \omega)$, and its simulation for 3-layer earth case

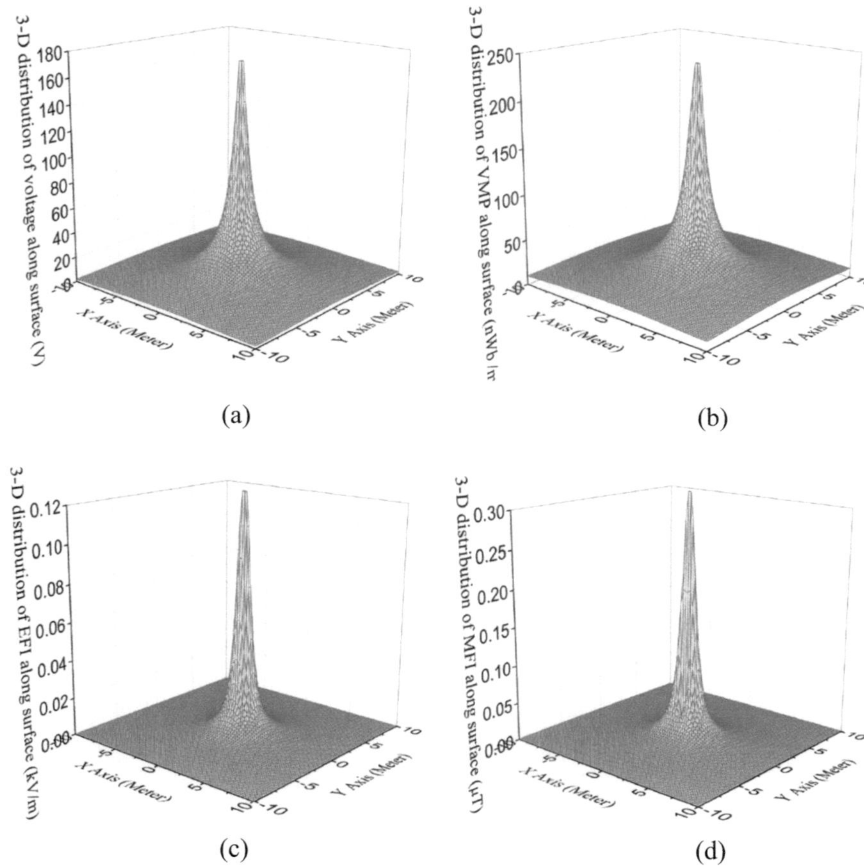

Fig. 3.20 3-d Distribution of $|\varphi|$, $|A|$, $|E|$ and $|B|$ along ground surface above a monopole and an horizontal dipole along x-axis

3.3.6 Summary

This chapter mainly introduces the 2-d time-domain QSCIM, which first introduces the time-domain Green's function and further elaborates on the core function of the spatial surface that needs to be fitted using the 2-d time-domain QSCIM. And mainly introduced the core function of the time-domain Green's function of the 2-4 layer horizontal layered soil model. Next, the 2-d MP method was introduced, which uses the 2-d time-domain QSCIM to fit the core function of the spatial surface. Finally, the core functions of the time-domain Green's function of soil models based on different layers are presented, respectively

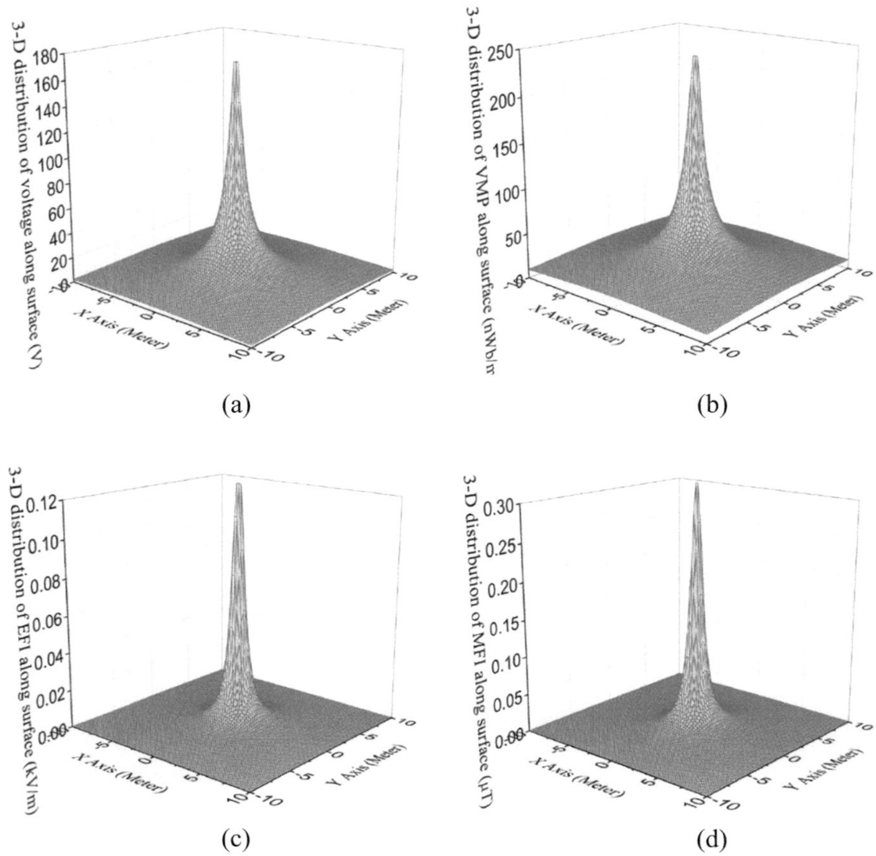

Fig. 3.21 3-d Distribution of $|\varphi|$, $|A|$, $|E|$ and $|B|$ along ground surface above a monopole and an horizontal dipole along y-axis

3.4 Summary

The frequency-domain and time-domain QSCIM introduced in this chapter overcome the lengthy calculation of infinite integration and greatly improve the computational efficiency of the Green's function of point current sources in multi-layer horizontal layered soil conductive media. It is one of the fundamental foundations of the electromagnetic field numerical methods introduced later in this book.

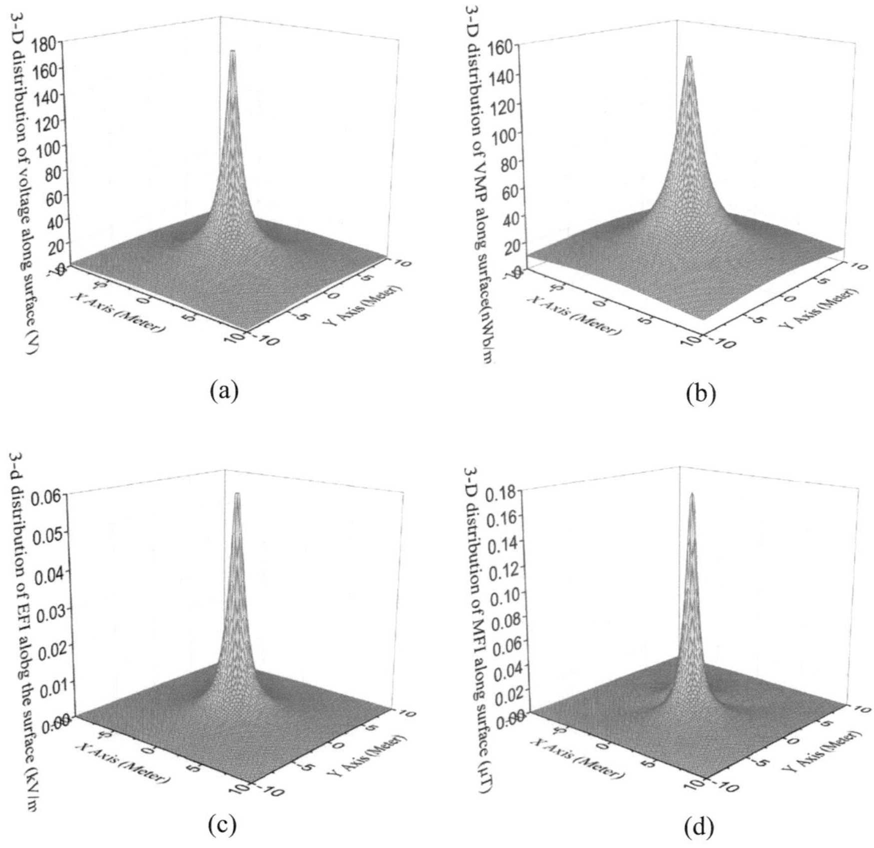

Fig. 3.22 3-d Distribution of $|\varphi|$, $|A|$, $|E|$ and $|B|$ along ground surface above a monopole and a vertical dipole

References

1. A.P. Sakis Meliopoulos, F. Xia, G.J. Cokkinides, An advanced computer model for grounding system analysis. IEEE Trans. Power Delivery **8**(1), 13–23 (1993)
2. F. Navarrina, I. Colominas, M. Casteleiro, Surface voltages and resistance of grounding systems of grid and rods in two-layer earth by rapid galerkin's moment method. IEEE Trans. Power Delivery **12**(1), 179–185 (1997)
3. S. Vujevic, Numerical analysis of potential distribution for a point current source in horizontally stratified multilayer earth. Eng. Model. **8**, 21–29 (1995)
4. Z.-X. Li, W. Chen, J.-B. Fan, L. Jiayu, Numerical analysis of earthing grids buried in horizontal stratified multilayer earth. Int. J. Numer. Methods Eng. **41**, 1297–1319 (1998)
5. L.P. Zhang, J.S. Yuan, Z.X. Li, The complex image method and its application in numerical simulation of substation grounding grids. Commun. Numer. Methods Eng. **15**, 835–839 (1999)
6. I. Colominas, F. Navarrina, M. Casteleiro, A numerical formula for grounding system analysis in stratified soils. IEEE Trans. Power Delivery **17**(2), 587–595 (2002)

7. J.S. Yuan, H.N. Yang, L.P. Zhang, X. Cui, X.S. Ma, Simulation of substation grounding grids with unequal-potential. IEEE Trans. Magn. **36**(4), 1468–1471 (2000)
8. L. Huang, D.G. Kasten, Model of ground grid and metallic conductor currents in high voltage a.c. substations for the computation of electromagnetic fields. Electr. Power Syst. Res. **59**, 31–37 (2001)
9. Z.X. Li, G.F. Li, J.B. Fan, C.X. Zhang, Numerical calculation of grounding system buried in vertical earth model in low frequency domain based on the boundary element method. Europ. Trans. Electr. Power **19**(8), 1177–1190 (2009)
10. P.D. Rancic, L.V. Stefanovic, D.R. Djordjevic, A new model of the vertical ground rod in two-layer earth. IEEE Trans. Magn. **28**(2), 1497–1500 (1992)
11. P.D. Rancic, L.V. Stefanovic, D.R. Djordjevic, An improved linear grounding system analysis in two-layer earth. IEEE Trans. Magn. **32**(5), 5179–5187 (1996)
12. Z.X. Li, J.B. Fan, W.J. Chen, Numerical simulation of substation grounding grids buried in both horizontal and vertical multilayer earth model. Int. J. Numer. Methods En. **69**(11), 2359–2380 (2007)
13. S. Vujevic, P. Sarajcev, Potential distribution for a harmonic current point source in horizontally stratified multilayer medium. COMPEL Int. J. Comput. Mathem. Electr. Electron. Eng. **27**, 624–637 (2008)
14. Z.X. Li, G.F. Li, J.B. Fan, Numerical simulation of substation grounding system in low-frequency domain based on the boundary element method with linear basis function. Commun. Numer. Methods Eng. **26**(12), 1819–1914 (2010)
15. Z.X. Li, G.F. Li, F. JB, Y. Yin, Numerical simulation of substation grounding grids buried in vertical earth model based on the thin-wire approximation with linear basis functions. Appl. Comput. Electromagn. Soc. J. **26**(4), 353–366 (2011)
16. Z.X. Li, G.F. Li, J.B. Fan, C.X. Zhang, Numerical calculation of ac substation grounding systems buried in a horizontal multilayered earth by higher order basis function. Int. J. Power Energy Syst. **32**(1), 21–30 (2012)
17. Z.X. Li, G.F. Li, J.B. Fan, Y. Yin, Numerical calculation of ac substation grounding systems buried in a vertical multilayered earth model by the boundary element method with higher order basis function. Int. J. Numer. Modelling Electron. Network Dev. Fields **25**(3), 258–274 (2012)
18. P. Sarajcev, S. Vujevic, D. Lovric, Time-harmonic current distribution on conductor grid in horizontally stratified multilayer medium. Progress Electromagn. Res. B **31**, 67–87 (2011)
19. A.F. Otero, J. Cidras, J.L. Alamo, Frequency-dependent grounding system calculation by means of a conventional nodal analysis technology. IEEE Trans. Power Delivery **14**(3), 873–877 (1999)
20. F. Dawalibi, Electromagnegtic fields generated by overhead and buried short conductors, part 1—single conductor. IEEE Trans. Power Delivery **1**(4), 105–111 (1986)
21. F. Dawalibi, R.D. Southey, Analysis of electrical interference from power lines to gas pipelines, part 1 - computational methods. IEEE Trans. Power Delivery **4**(3), 1840–1846 (1989)
22. F. Dawalibi, Electromagnegtic fields generated by overhead and buried short conductors, part 2 - ground networks. IEEE Trans. Power Delivery **1**(4), 112–119 (1986)
23. Z.X. Li, W.J. Chen, J.B. Fan, J.Y. Lu, A novel mathematical modeling of grounding system buried in multilayer earth. IEEE Trans. Power Delivery **21**(3), 1267–1272 (2006)
24. Z.X. Li, W.J. Chen, Numerical simulation grounding system buried within horizontal multilayer earth in frequency domain. Commun. Numer. Methods Eng. **32**(1), 11–17 (2007)
25. Z.X. Li, J.B. Fan, Numerical calculation of grounding system in low-frequency domain based on the boundary element method. Int. J. Numer. Methods Eng. **73**(5), 685–705 (2008)
26. Z.X. Li, G.F. Li, J.B. F, Y. Yin, A novel mathematical model for the lightning response of a grounding system buried in multilayered earth based on the quasi-static complex image method. Europ. Trans. Electr. Power **22**(6), 831–845 (2012)
27. Z.X. Li, K.-L. Gao, Y. Yin, D. Ge, Transient lightning responses of grounding systems buried in horizontal multilayered earth with a hybrid method. J. Electrost. **72**(5), 381–386 (2014)

28. Z.X. Li, K.-L. Gao, Y. Yu, S-W. Rao, Lightning response of a grounding system buried in multiple layers in earth using the peec method based on the quasi-static complex image method. Electr. Power Components Syst. **48**(3), 291–303 (2020)
29. Z.X. Li, Y. Yin, K.-C. Wang, Lightning response of a grounding system buried in multilayered earth based on the galerkin's boundary element and quasi-static complex image methods. IEEE Trans. Electromagn. Compat. **63**(4), 1118–1127 (2021)
30. Z.X. Li, W.-J. Chen, K.-C. Wang, Frequency domain and time domain analysis of the transient behavior of buried grounding grids in horizontal multilayered earth model. Electr. Eng. **104**(4), 2515–2529 (2022)
31. T.K. Sarkar, O. Pereira, Using the matrix pencil method to estimate the parameters of a sum of complex expontentials. IEEE Antenna Propagation Mag. **37**(1), 48–55 (1995)
32. Y.L. Chow, J.J. Yang, K.D. Srivastava, Complex images of a ground electrode in layered soils. J. Appl. Phys. **71**(2), 569–574 (1992)
33. Y. Hua, T.K. Sarkar, Generalized pencil-of-function method for extracting poles of an em system from its transient response. IEEE Trans. Antenna Propagation **37**(2), 229–234 (1989)
34. Z.-X. Li, S.W. Rao, Transient lightning response of grounding grids in a horizontal, multi-layered soil model that considers soil ionization effects with time-domain quasi-static complex images. Int. J. Numer. Modelling Electron. Network Dev. Fields **32**(1), e2463 (2019)
35. Z.-X. Li, P. Li, K.-C. Wang, Transient lightning response to grounding grids buried in horizontal multilayered earth model considering time domain quasi-static complex image method and soil ionization effect. COMPEL Int. J. Comput. Mathem. Electr. Electron. Eng. **40**(3), 516–534 (2021)
36. Y.B. Hua, Estimating two-dimension frequencies by matrix enhancement and matrix pencil. IEEE Trans. Signal Process. **40**(9), 2267–2281 (1992)

Chapter 4
Fundamentals of Numerical Calculation Techniques III

Mathematical Formulas For Different Types of Lightning Currents.

4.1 Preface

This chapter will provide a detailed introduction to three mathematical formulas used to generate the external excitation sources required for the lightning impulse response problem described in later chapters, namely the double exponential function, Heidler function, and composite Heidler function. These three mathematical formulas that can generate lightning currents have their own characteristics. Below, we will introduce in detail the relevant formulas and their parameter inversion methods.

4.2 Introduction to Three Different Lightning Waveforms

This small chapter will specifically introduce different mathematical forms of lightning current expressions.

An accurate assessment of lightning effects on power systems depends, among other factors, on an appropriate representation of the lightning current. In the field of electromagnetic analysis of lightning phenomena a number of mathematical functions are available for modelling the channel-base lightning current [1]. The simplest of these functions is the double-exponential function [2], which, despite its numerous drawbacks, continues to be in use mainly due to its simplicity.

In recent years, a number of experiments have been conducted with the purpose of directly measuring and recording the current waveform produced by a lightning strike. The most important data were accumulated and analysed by Berger, who recorded on top of a telecommunications tower situated on the mountain San Salvatore in Switzerland [3, 4]. Based on this data, a new analytical function has been

Z.-X. Li, *Numerical Simulation Algorithm of Electromagnetic Field for Grounding Problems in Power System Substation Grounding Grids*,
https://doi.org/10.1007/978-981-96-0904-8_4

proposed by Heidler in Reference [5]. Avariant of this new analytical function, for a current steepness factor $n = 10$, has been incorporated in IEC 623051 as the power function recommended for lightning current modelling in both lightning research and engineering applications [6]. The advantage of the Heidler function in relation to some of the previously used functions such as the double-exponential function is that it much more realistically approximates the lightning return stroke. For example, the double-exponential function's first-time derivation for $t = 0$ does not equal zero, which is not physically valid because in such a way the function has a discontinuity at the start. In addition, the double-exponential function deviated from the measured lightning current waveshapes [4]. Unlike the double-exponential function, the Heidler function does not have a discontinuity at the start. In addition, it permits a good separation of the characteristic lightning current quantities.

In general, two such Heidler functions are sufficient to represent the subsequent stroke current [7]. However, the use of a couple of Heidler functions fails to reproduce the pronounced concavity existing at the front first stroke currents. To cope with this problem, the International Council on Large Electric Systems ($CIGR\acute{E}$) brochure recommends the use of two independent expressions, one leading to a gradually increasing steepness up to 90% of the current peak and the other describing the tail of the desired waveform [8].

Although satisfactory in terms of creating a pronounced concave profile in the rising part of the obtained current waveform and also for allowing the control of its front time, peak value, and duration, $CIGR\acute{E}'s$ recommended procedure has the disadvantage of leading to a discontinuity at the transition point between the expressions describing the front and the tail of the synthesized curve, if the adopted parameters are not carefully chosen. Furthermore, the occurrence of subsidiary peaks in first stroke currents is not considered in $CIGR\acute{E}'s$ representation. As indicated by Visacro [9], an appropriate description of typical first stroke currents should include multiple peaks.

Let's first take a look at the lightning current waveform, as shown in Fig. 4.1 below.

The lightning current approximation of the first return stroke is depicted in Fig. 4.1, where I_0 is the current peak value, t_0 is the virtual starting time, t_1 is the time to 10% of peak value, t_2 is the time to 90% of peak value, t_h is the total time to half value of the peak value, t_{max} is the time to the peak value, T_1 is the front duration and T_2 is the time to half value [6].

A discussion of how to model lightning current waveforms using mathematical representations will begin below.

4.2.1 The First Method—Double Exponential Function

The standard double exponential function of lightning current is expressed as follows

$$i(t) = \frac{I}{\eta}(e^{-\alpha t} - e^{-\beta t}) \tag{4.1}$$

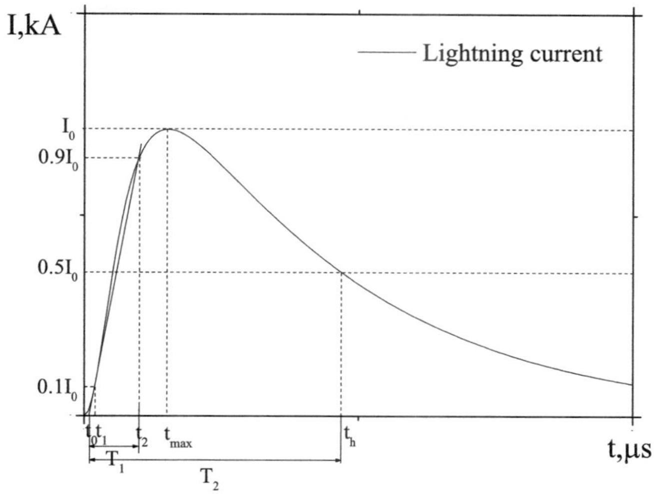

Fig. 4.1 The waveform of lighting current

where η is the correction coefficient of the current peak value, α and β are the parameters of the double-exponential function.

4.2.2 The Second Method—Heidler Function

The current function most frequently used in the lightning research is Heidler function, which can be expressed by the following equation [6, 10].

$$i(t) = \frac{I_0}{\eta} \frac{(\frac{t}{\tau_1})^n}{1 + (\frac{t}{\tau_1})^n} e^{-\frac{t}{\tau_2}} \tag{4.2}$$

where I_0 is the current peak value, η is the correction coefficient of the current peak value, τ_1 is the time constant determining current rise-time, τ_2 is the time constant determining current decay-time, and n is the current steepness factor.

$$\eta = e^{-\frac{\tau_1}{\tau_2}(n\frac{\tau_2}{\tau_1})^{\frac{1}{n}}} \tag{4.3}$$

4.2.3 The Third Method—Composite Heidler Function

In order to satisfy the real lightning waveform in reality, the composite Heidler function is used to describe the lightning waveform. The method employed here to synthesize a typical lightning current waveform consists of using (4.4) and (4.5) to sum over m Heidler functions as shown below.

$$i(t) = \sum_{k=1}^{m} \frac{I_{0k}}{\eta_k} \frac{(\frac{t}{\tau_{1k}})^{n_k}}{1 + (\frac{t}{\tau_{1k}})^{n_k}} e^{-\frac{t}{\tau_{2k}}} \tag{4.4}$$

$$\eta_k = e^{-\frac{\tau_{1k}}{\tau_{2k}} (n \frac{\tau_{2k}}{\tau_{1k}})^{\frac{1}{n_k}}} \tag{4.5}$$

where I_{0k} controls the amplitude, τ_{1k} is the front-time constant, τ_{2k} is the decay-time constant, η_k is the amplitude correction factor, and n_k is an exponent controlling the steepness of each current waveform k applied to build i(t) [11].

4.3 Estimation of Typical Common Mathematical Formulae Parameters for Numerical Simulation of Electromagnetic Fields for Lightning Waves

For commonly used mathematical function formulas describing lightning waveforms, such as the Heidler function and the double exponential function, how to estimate the parameters of the formula is a fundamental problem in the numerical simulation of electromagnetic fields to simulate the lightning impact grounding problem.

4.3.1 Estimation of the Parameters of a Typical Commonly Used Double-Exponential Mathematical Function Equation

The text of this subchapter describes an algorithm for estimating the parameters of a double exponential function. The algorithm can simultaneously solve systems consisting of two nonlinear equations, depending on the available input data: peak current, front duration, and half-value time.

Lightning Return Stroke Current—Double Exponential Function

The double exponential function can be used to approximate the simulated lightning return current, the expression of which is given in Eq. (4.1) and is further rewritten as follows.

$$i(t) = \frac{I_0}{\eta}(e^{-\alpha t} - e^{-\beta t}) \tag{4.6}$$

Introduction to systems of nonlinear equations for double exponential function case

Observing Fig. 4.1, the two basic requirements for estimating the parameters η, α and β of the double exponential function can be written as:

$$i(t) = 0.9I_0 \quad for \quad t = t_2 \tag{4.7}$$

$$i(t) = 0.5I_0 \quad for \quad t = t_h \tag{4.8}$$

According to Eqs. (4.7–4.8), the following four normalized nonlinear equations can be obtained:

$$R_1 = \frac{1}{0.9\eta}\left(e^{-\alpha t_2} - e^{-\beta t_2}\right) - 1 \tag{4.9}$$

$$R_2 = \frac{1}{0.5\eta}\left(e^{-\alpha t_h} - e^{-\beta t_h}\right) - 1 \tag{4.10}$$

Least square estimates of η, α and β for double exponential function case

Figure 4.2 is a flowchart depicting the estimation procedure of the lightning current parameters η, α and β introduced by Eqs. (4.6). The parameters α and β are computed from the corresponding nonlinear system of equations by Marquardt's least squares method depending on the available input requirements, while the parameter η is computed from the linear equations at the beginning of the iterative procedure and at the end of each iteration.

Here the peak current I_0, the leading edge duration T_1 and the half-value time T_2 are the lightning currents specified in the regulations. The estimation of the double exponential function parameters α and β is based on the Marquardt least squares method (also known as Levenberg-Marquardt method) [12, 13].

The current $i(t)$ attains its peak value I_0 at the time t_{max}. After substituting $I = I_0$ and $t = t_{max}$ in Eq. (4.6), the following equation is obtained:

$$I_0 = \frac{I_0}{\eta}(e^{-\alpha t_{max}} - e^{-\beta t_{max}}) \tag{4.11}$$

After dividing both sides of Eq. (4.11) by $\frac{I_0}{\eta}$, the equation for computing correction coefficient η is obtained:

$$\eta = e^{-\alpha t_{max}} - e^{-\beta t_{max}} \tag{4.12}$$

At each j-th iteration, the parameters t_{max}, η, and t_1 must be calculated, where the auxiliary parameters t_{max} and t_1 are computed by solving the corresponding nonlinear equations and the coefficients η can be computed by using Eq. (4.12).

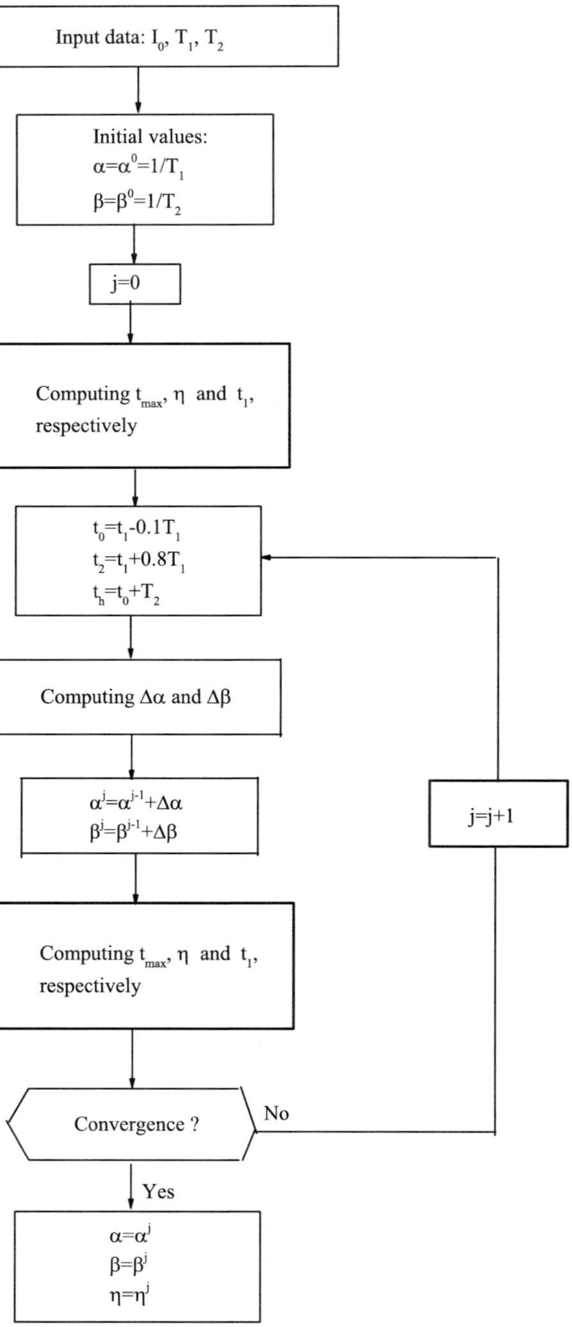

Fig. 4.2 Least squares estimation of double-exponential function parameters

The lightning current parameters α and β can be estimated from the available input requirements. The unknown parameters α and β can both be estimated by solving a system of m nonlinear equations using the least squares method:

$$S = \sum_{i=1}^{m} E_i^2 = minimum \tag{4.13}$$

where $m = 2$, $E_1 = R_1$ and $E_2 = R_2$.

The system of nonlinear Eqs. (4.9) and (4.10) is solved simultaneously using Marquardt's least squares method:

$$[\mathbf{A} + \lambda \cdot \mathbf{D}] \cdot \begin{bmatrix} \Delta\alpha \\ \Delta\beta \end{bmatrix} = [\mathbf{A} + 3 \cdot \mathbf{D}] \cdot \begin{bmatrix} \Delta\alpha \\ \Delta\beta \end{bmatrix} = [\mathbf{B}] \tag{4.14}$$

where λ is an adjustable positive parameter which is used to control the iteration.

Observe that on the left hand side, the matrix $[\mathbf{A}]$ becomes trivial with respect to the principal elements of the matrix $[\lambda \cdot \mathbf{D}]$ as λ goes to infinity. Thus, the solution of the current iteration is only affected by the elements of the matrix $[\lambda \cdot \mathbf{D}]$, which reduces the Marquardt least squares method to the steepest slope descent method. On the other hand, if the parameter λ tends to 0, the solution of the current iteration is only affected by the elements of the matrix $[\mathbf{A}]$, This reduces Marquardt's least squares to Gauss-Newton or generalized least squares [13]. It is well known that the steepest descent method converges quickly but slowly, while the Gauss-Newton method converges quickly but less reliably.

The main advantage of Marquardt's least squares method [12] is that the parameter λ can be chosen to follow the Gauss-Newton method as closely as possible, ensuring fast convergence while retaining some of the properties of the steepest descent method to prevent divergence. After several numerical experiments, the parameter λ was chosen to be 3 for fast and reliable convergence.

Matrix $[\mathbf{D}]$ is a diagonal matrix whose diagonal elements are identical to those of matrix $[\mathbf{A}]$ and is defined by the following formula:

$$[\mathbf{A}] = [\mathbf{J}]^T \cdot [\mathbf{J}] \tag{4.15}$$

where the Jacobian matrix $[\mathbf{J}]$ can be approximated:

$$[\mathbf{J}] = \begin{bmatrix} \frac{\partial E_1}{\partial \alpha} & \frac{\partial E_1}{\partial \beta} \\ \frac{\partial E_2}{\partial \alpha} & \frac{\partial E_2}{\partial \beta} \end{bmatrix} \tag{4.16}$$

According to Marquardt's least squares method, the matrix on the right-hand side of Eq. (4.14) is

$$[\mathbf{B}] = -[\mathbf{J}]^T \cdot \begin{bmatrix} E_1 \\ E_2 \end{bmatrix} \tag{4.17}$$

The iterations will continue until the parameters α and β reach the convergence condition. The condition for the convergence of the values of α and β is

$$S = \sum_{i=1}^{m} E_i^2 < 10^{-12} \tag{4.18}$$

When this condition is reached, the iterative process ends and the final values of the parameters α, β and η are obtained.

Calculation of the parameter t_{max} for double exponential function case

As shown in Fig. 4.3, the input data for calculating the unknown auxiliary parameter t_{max} are the time constants α^j and β^j obtained in the current j-th iteration.

The time derivative of the double-exponential function described by Eq. (4.6) is equal to zero at $t = t_{max}$ when the lightning current peaks:

$$\frac{di(t)}{dt}|_{t=t_{max}} = \frac{I_0}{\eta}(-\alpha e^{-\alpha t} + \beta e^{-\beta t})|_{t=t_{max}} = 0 \tag{4.19}$$

the following equation for $t = t_{max}$ can be obtained:

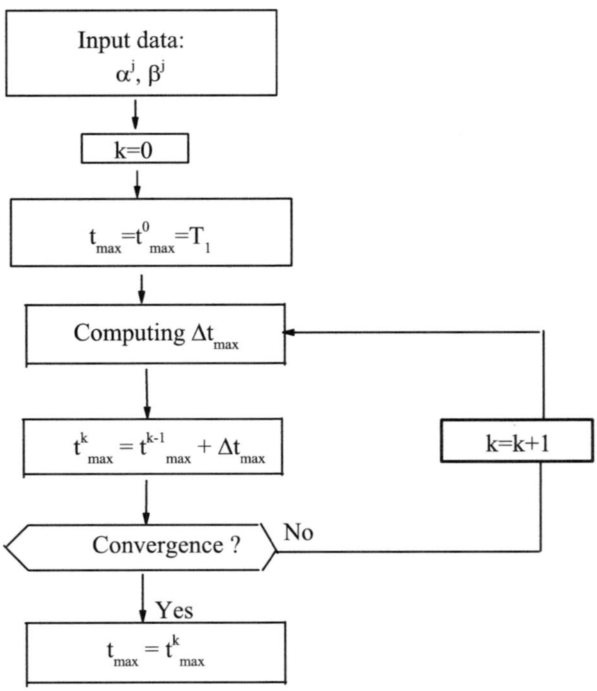

Fig. 4.3 Computation of t_{max} for double exponential function case

$$-\alpha e^{-\alpha t_{max}} + \beta e^{-\beta t_{max}} = 0 \tag{4.20}$$

which can be transformed into the next nonlinear equation for computing t_{max}:

$$F_{max} = -\alpha e^{-\alpha t_{max}} + \beta e^{-\beta t_{max}} = 0 \tag{4.21}$$

The Jacobian J_{max} of F_{max} is defined by:

$$J_{max} = \frac{\partial F_{max}}{\partial t_{max}} \tag{4.22}$$

so

$$J_{max} = \frac{\partial F_{max}}{\partial t_{max}} = \alpha^2 e^{-\alpha t_{max}} - \beta^2 e^{-\beta t_{max}} \tag{4.23}$$

In each j-th and k-th iteration, the parameter t_{max} can be computed using nonlinear Eq. (4.21).

The nonlinear Eq. (4.21) is solved using a special case of Marquardt's least squares method for a single nonlinear equation. Similarly to Eq. (4.14), it can be expressed as

$$J_{max}^2 \cdot (1 + \lambda) \cdot \Delta t_{max} = B \tag{4.24}$$

where

$$B = -J \cdot F_{max} \tag{4.25}$$

In each j-th and k-th iteration, the Jacobian J_{max} can be computed using Eq. (4.23).

Then the growth of the parameter Δt_{max} can be computed by:

$$\Delta t_{max} = -\frac{F_{max}}{J_{max} \cdot (1 + \lambda)} = -\frac{F_{max}}{4 \cdot J_{max}} \tag{4.26}$$

where $\lambda = 3$.

The value of t_{max} for the k-th iteration is computed from the following expression:

$$t_{max}^{k} = t_{max}^{k-1} + \Delta t_{max} \tag{4.27}$$

As stated in Eq. (4.27), the value of parameter t_{max} will grow in every iteration by the value of Δt_{max}. The iterations will continue until the condition for the convergence of the parameter t_{max} is achieved:

$$S = F_{max}^2 < 10^{-12} \tag{4.28}$$

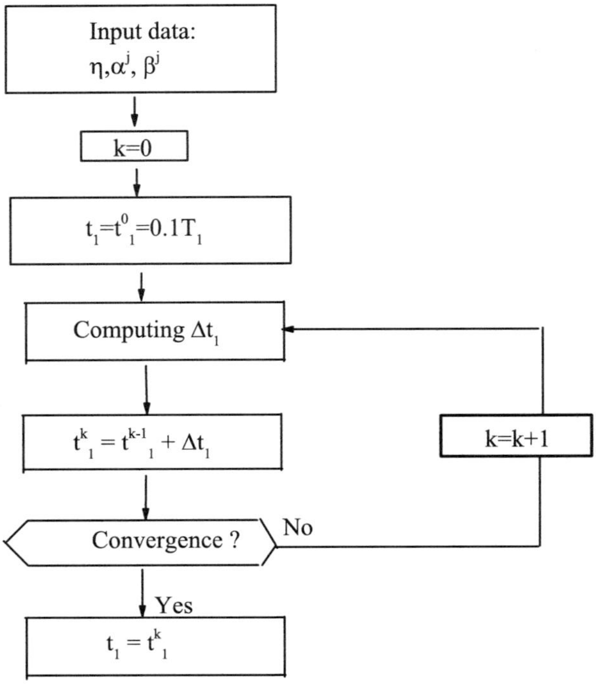

Fig. 4.4 Computation of t_1 for double exponential function case

Calculation of the parameter t_1 for double exponential function case

The input data for the calculation of the unknown auxiliary parameter t_1 are the coefficients η and the time constants α and β obtained in the j-th iteration. the estimation of the parameter t_1 is shown in Fig. 4.4.

The current $i(t)$ reaches 10% of the peak current I_0 at t_1. Substituting $I = 0.1I_0$ and $t = t_1$ into Eq. (4.6), the following equation is obtained:

$$0.1I_0 = \frac{I_0}{\eta}(e^{-\alpha t_1} - e^{-\beta t_1}) \tag{4.29}$$

After dividing both sides of Eq. (4.30) by $\frac{I_0}{\eta}$, follows the next equation:

$$1 = \frac{1}{0.1\eta}(e^{-\alpha t_1} - e^{-\beta t_1}) \tag{4.30}$$

Rearranging the previous expression the next nonlinear equation for computing t_1 is obtained:

$$F_1 = \frac{1}{0.1\eta}(e^{-\alpha t_1} - e^{-\beta t_1}) - 1 \tag{4.31}$$

The Jacobian J_1 of F_1 is defined by:

$$J_1 = \frac{\partial F_1}{\partial t_1} \tag{4.32}$$

Substitution of Eq. (4.31) into Eq. (4.32) and partial differentiation with respect to t_1 yields:

$$\frac{\partial F_1}{\partial t_1} = \frac{1}{0.1\eta}(-\alpha e^{-\alpha t_1} + \beta e^{-\beta t_1}) \tag{4.33}$$

In each j-th and k-th iteration, the parameter t_1 can be computed using nonlinear Eq. (4.31).

The nonlinear Eq. (4.31) can be solved in the same way as described for the parameter t_{max}. Again, in the first iteration ($k = 0$), the first step is to assume the value of the variable t_1. Analogously, as in Eqs. (4.24)–(4.26), the growth of the iteration for the parameter t_1 can be found from:

$$\Delta t_1 = -\frac{F_1}{J_1 \cdot (1 + \lambda)} = -\frac{F_1}{4 \cdot J_1} \tag{4.34}$$

where $\lambda = 3$.

In each j-th and k-th iteration, the Jacobian J_1 can be computed using Eq. (4.33).

The value of t_1 for the k-th iteration is then computed from the following expression:

$$t_1^k = t_1^{k-1} + \Delta t_1 \tag{4.35}$$

As presented in Eq. (4.35), the value of parameter t_1 will grow in each iteration by the value of Δt_1. The iterations will continue until the condition for the convergence of parameter t_1 is achieved:

$$S = F_1^2 < 10^{-12} \tag{4.36}$$

Numerical examples for double exponential function case

In this subchapter, several examples of solving arithmetic for lightning wave parameters will be presented.

The first calculation example first gives the head time and half-wave time of the lightning wave and the maximum lightning current amplitude as follows: $I_0 = 1$ kA, $T_1/T_2 = 5/20$ μs. Double-exponential function parameters are estimated using a computer program that implements the method described in the previous section. The following parameters have been obtained: $\eta = 0.09806$, $\alpha = 0.11828$ μs and $\beta = 0.15453$ μs. Double-exponential function with these parameters is depicted in Fig. 4.5.

Observing Fig. 4.5, it can be seen that: although the T_1 and T_2 parameters have been specified as $T_1/T_2 = 5/20$, the T_1 and T_2 parameters obtained by the program generated by the algorithm are $T_1/T_2 = 4.7/20.1$, with a certain error, which is thankfully acceptable.

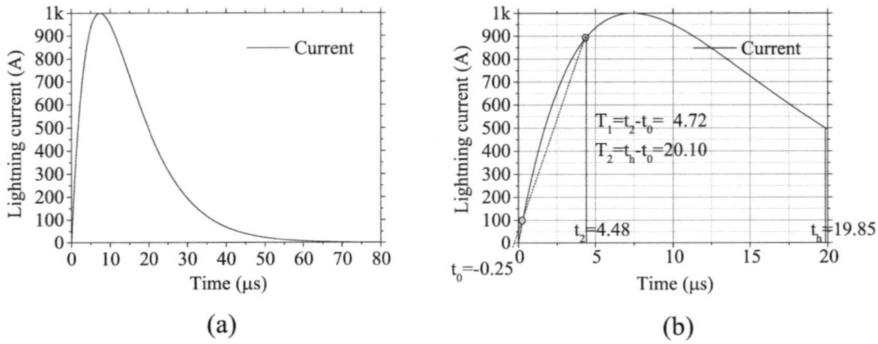

Fig. 4.5 Double exponential function

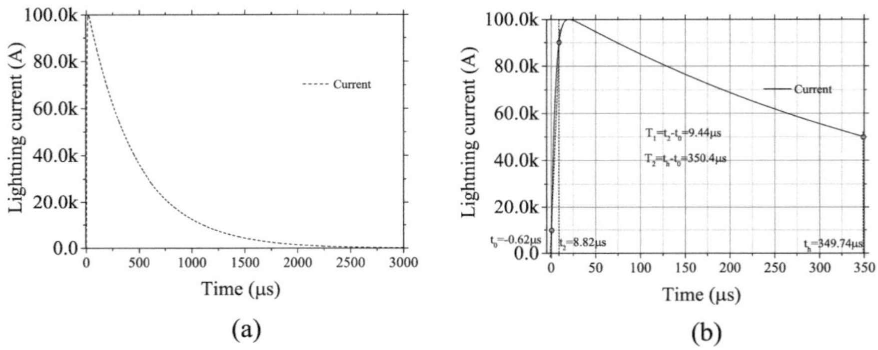

Fig. 4.6 Double exponential function

The second calculation example first gives the head time and half-wave time of the lightning wave and the maximum lightning current amplitude as follows: $I_0 = 100\,\text{kA}$, $T_1/T_2 = 10/350\,\mu\text{s}$. Double-exponential function parameters are estimated using a computer program that implements the method described in the previous section. The following parameters have been obtained: $\eta = 0.94887$, $\alpha = 0.00213\,\mu\text{s}$ and $\beta = 0.23334\,\mu\text{s}$. Double-exponential function with these parameters is depicted in Fig. 4.6.

The second computational example starts with the head time and half-wave time of the lightning wave, as well as the maximum lightning current amplitude as follows: $I_0 = 100$ kA, and $T_1/T_2 = 10/350\,\mu\text{s}$. The parameters of the biexponential function are estimated by means of a computer program, which implements the method described in the previous section. The following parameters were obtained: $\eta = 0.94887, \alpha = 0.00213\,\mu\text{s}$ and $\beta = 0.23334\,\mu\text{s}$. The biexponential function with these parameters is shown in Fig. 4.6.

Observing Fig. 4.6, it can be noticed that although the T_1 and T_2 parameters are specified as $T_1/T_2 = 10/350$, the program generated by the above algorithm obtains

Table 4.1 Double-exponential function parameters

T_1/T_2 (appointed)	T_1/T_2 (computed)	$\alpha(\mu s^{-1})$	$\beta(\mu s^{-1})$	η
0.5/5 μs	0.47/5.01 μs	0.179082	3.778495	0.818514
4/16 μs	3.78/16.14 μs	0.141898	0.200206	0.126015
1.2/50 μs	1.13/50.04 μs	0.014760	1.973811	0.956578

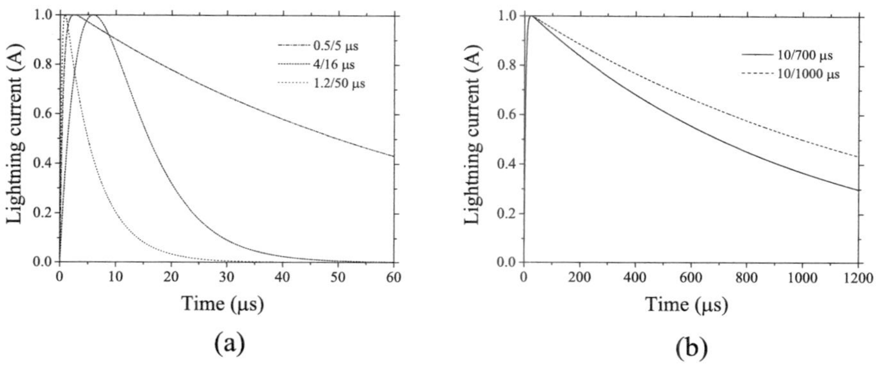

Fig. 4.7 Double-exponential function approximation of fast-decaying lightning current waveshapes

T_1 and T_2 parameters as $T_1/T_2 = 9.4/350.4$, which is an error, but good enough to be acceptable.

The third arithmetic example is described below: Other often used lightning current waveshapes sometimes used for designing low-voltage power lines within structures are the $T_1/T_2 = 0.2/5\,\mu s$ waveshape, the $T_1/T_2 = 4/16\,\mu s$ waveshape and $T_1/T_2 = 1.2/50\,\mu s$ waveshape [1, 14]. Since these waveshapes are only defined by T_1 and T_2 values, only two nonlinear equations are simultaneously solved. Results of the estimation for these three waveshapes are presented in Table 4.1. Corresponding double-exponential functions approximating these waveshapes are depicted in Fig. 4.7a. All functions have a current peak value of $I_0 = 1A$ for plotting clarity purposes. Peak values can be changed accordingly since the parameters in Table 4.1 are independent of the peak current value.

The fourth example concerns lightning waves on communication lines. In the case of communication lines which have a different exposure to lightning than power lines, different waveshapes are used for designing lightning protection system. These waveshapes are characterized by a relatively sharp rise followed by a very slow current decay: the $T_1/T_2 = 10/700\,\mu s$ waveshape and the $T_1/T_2 = 10/1000\,\mu s$ waveshape [14]. Again, these waveshapes are only defined by T_1 and T_2 values, so only two nonlinear equations are simultaneously solved.

Table 4.2 Double-exponential function parameters

T_1/T_2 (appointed)	T_1/T_2 (computed)	$\alpha(\mu s^{-1})$	$\beta(\mu s^{-1})$	η
10/700 μs	9.44/700.34 μs	0.001030	0.245028	0.973067
10/1000 μs	9.45/1000.33 μs	0.000713	0.249078	0.980514

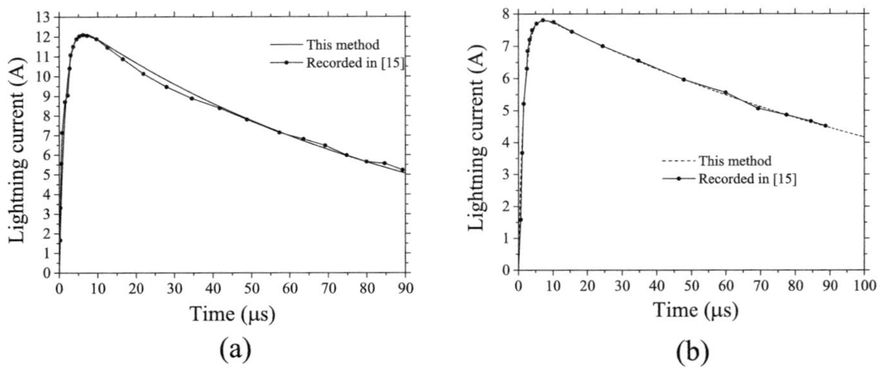

Fig. 4.8 Recorded impulse current and its double-exponential function approximation

Table 4.3 Double-exponential function parameters

T_1/T_2 (appointed)	T_1/T_2 (computed)	$\alpha(\mu s^{-1})$	$\beta(\mu s^{-1})$	η
3.37/73 μs	3.19/73.12 μs	0.0106652	0.6557783	0.918980
3.70/110 μs	3.50/110.13 μs	0.006867	0.620579	0.940333

Results of the estimation for these two waveshapes are presented in Table 4.2. Double-exponential functions approximating these waveshapes are depicted in Fig. 4.7b.

The fifth example is described as follows: In addition to normalizing the lightning current, the least squares method presented can easily be used to approximate the waveforms recorded for various pulse stroke currents. Figure 4.8 shows the recorded waveform from [15] along with the double exponential function approximation. The input data for the recorded waveform current from [15] were $I_0 = 12.15A$, $T_1 = 3.37\,\mu s$, $T_2 = 73\,\mu s$, and $I_0 = 7.79A$, $T_1 = 3.7\,\mu s$, $T_2 = 110\,\mu s$. The resulting double exponential function parameters are given in Table 4.3.

The sixth example is described as follows: In much the same way as the previous example, the least squares method described earlier in this paper is used to approximate the waveforms recorded for various pulse stroke curves. Figure 4.9 shows the waveform recorded in [16] and the double exponential function approximation. The input data for the recorded waveform current from [16] were $I_0 = 12.1A$,

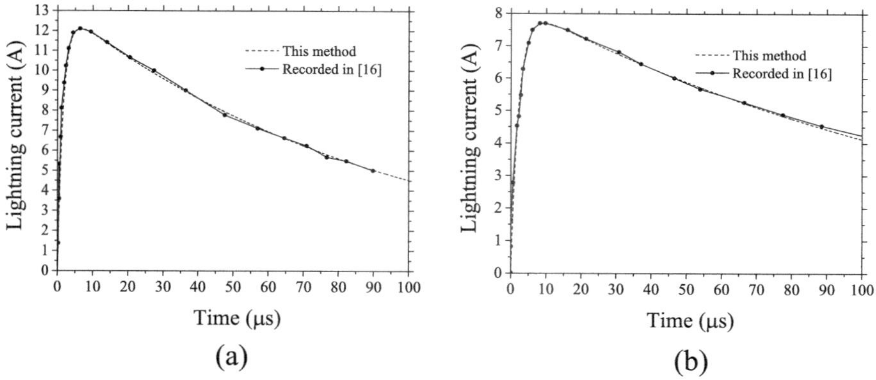

Fig. 4.9 Recorded impulse current and its double-exponential function approximation

Table 4.4 Double-exponential function parameters

T_1/T_2 (appointed)	T_1/T_2 (computed)	$\alpha(\mu s^{-1})$	$\beta(\mu s^{-1})$	η
3.5/73 μs	3.31/73.12 μs	0.0107115	0.6280774	0.915906
5/100 μs	4.73/110.18 μs	0.007066	0.442934	0.920202

$T_1 = 3.5\,\mu s$, $T_2 = 73\,\mu s$, and $I_0 = 7.7A$, $T_1 = 5.0\,\mu s$, $T_2 = 100\,\mu s$. The resulting double exponential function parameters are given in Table 4.4.

4.3.2 *Estimation of the Parameters of Typical Common Heidler Function Formulas*

The Heidler function can be used to approximate the simulated lightning return current, the expression of which is given in Eq. (4.2) and is further rewritten as follows.

$$i(t) = \frac{I_0}{\eta} \frac{(\frac{t}{\tau_1})^n}{1 + (\frac{t}{\tau_1})^n} e^{-\frac{t}{\tau_2}} \tag{4.37}$$

Introduction to systems of nonlinear equations for Heidler function case

Observing Fig. 4.1, just like double exponential function , the two basic requirements for estimating the parameters η, τ_1 and τ_2 of the Heidler function can be written as:

$$i(t) = 0.9 I_0 \qquad for \qquad t = t_2 \tag{4.38}$$

$$i(t) = 0.5I_0 \qquad for \qquad t = t_h \tag{4.39}$$

According to Eqs. (4.38–4.39), the following four normalized nonlinear equations can be obtained:

$$R_1 = 0.9\eta + (0.9\eta - e^{-\frac{t_2}{\tau_2}})(\frac{t_2}{\tau_1})^n \tag{4.40}$$

$$R_2 = 0.5\eta + (0.5\eta - e^{-\frac{t_h}{\tau_2}})(\frac{t_h}{\tau_1})^n \tag{4.41}$$

Least square estimates of η, τ_1 and τ_2 for Heidler function case

Fig. 4.10 is a flowchart depicting the estimation procedure of the lightning current parameters η, τ_1 and τ_2 introduced by Eqs. (4.37). The parameters τ_1 and τ_2 are computed from the corresponding nonlinear system of equations by Marquardt's least squares method depending on the available input requirements, while the parameter η is computed from the linear equations at the beginning of the iterative procedure and at the end of each iteration.

Here the peak current I_0, the leading edge duration T_1 and the half-value time T_2 are the lightning currents specified in the regulations. Just like double exponential function, the estimation of the Heidler function parameters τ_1 and τ_2 is also based on the Marquardt least squares method [12, 13].

The current $i(t)$ attains its peak value I_0 at the time t_{max}. After substituting $I = I_0$ and $t = t_{max}$ in Eq. (4.37), the following equation is obtained:

$$I_0 = \frac{I_0}{\eta}\frac{(\frac{t_{max}}{\tau_1})^n}{1 + (\frac{t_{max}}{\tau_1})^n}e^{-\frac{t_{max}}{\tau_2}} \tag{4.42}$$

After dividing both sides of Eq. (4.42) by $\frac{I_0}{\eta}$, the equation for computing correction coefficient η is obtained:

$$\eta = \frac{(\frac{t_{max}}{\tau_1})^n}{1 + (\frac{t_{max}}{\tau_1})^n}e^{-\frac{t_{max}}{\tau_2}} \tag{4.43}$$

At each j-th iteration, the parameters t_{max}, η, and t_1 must be calculated, where the auxiliary parameters t_{max} and t_1 are computed by solving the corresponding nonlinear equations and the coefficients η can be computed by using Eq. (4.43).

The lightning current parameters τ_1 and τ_2 can be estimated from the available input requirements. The unknown parameters τ_1 and τ_2 can both be estimated by solving a system of m nonlinear equations using the least squares method.

$$S = \sum_{i=1}^{m} E_i^2 = minimum \tag{4.44}$$

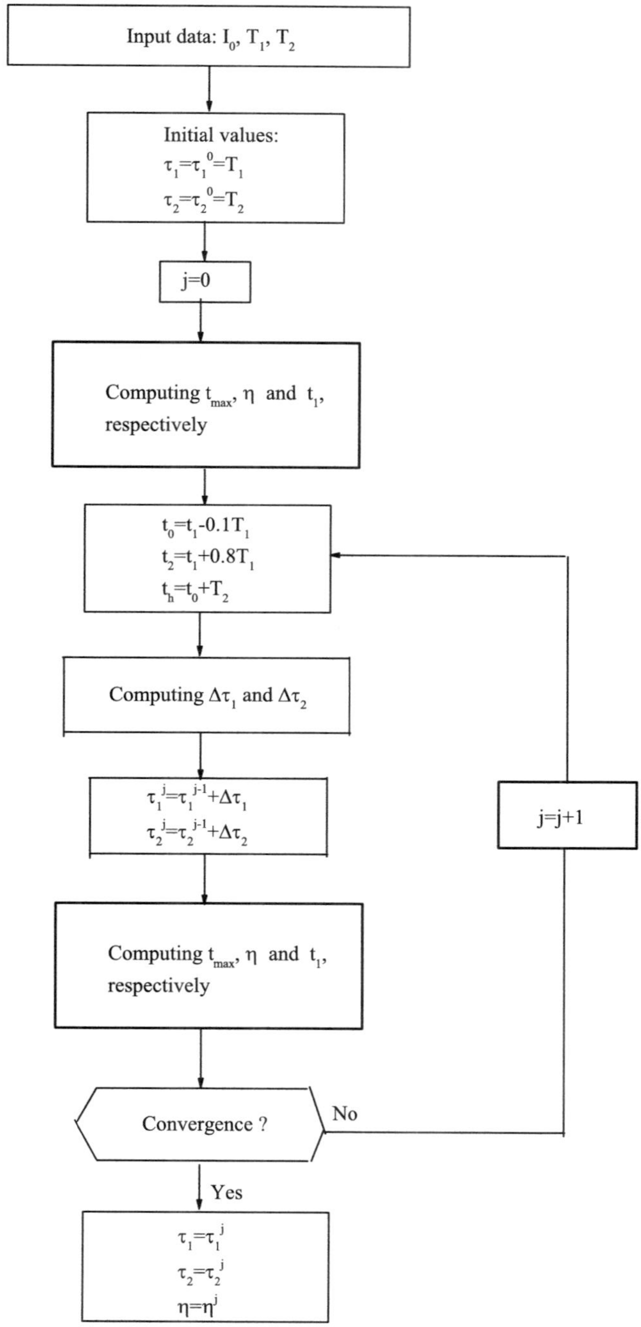

Fig. 4.10 Least squares estimation of Heidler function parameters

where $m = 2$, $E_1 = R_1$ and $E_2 = R_2$.

The system of nonlinear Eqs. (4.40) and (4.41) is solved simultaneously using Marquardt's least squares method:

$$[\mathbf{A} + \lambda \cdot \mathbf{D}] \cdot \begin{bmatrix} \Delta\tau_1 \\ \Delta\tau_2 \end{bmatrix} = [\mathbf{A} + 3 \cdot \mathbf{D}] \cdot \begin{bmatrix} \Delta\tau_1 \\ \Delta\tau_2 \end{bmatrix} = [\mathbf{B}] \tag{4.45}$$

where λ is an adjustable positive parameter which is used to control the iteration.

It is clear that Eqs. (4.14) and (4.45) are formally identical, but the connotations of the elements of the two are very different. Note that on the left-hand side of the equation, as λ tends to infinity, the matrix [**A**] becomes too small to play any role relative to the principal elements of the matrix $[\lambda \cdot \mathbf{D}]$. Thus, the solution of the current iteration is only affected by the elements of the matrix $[\lambda \cdot \mathbf{D}]$, which reduces the Marquardt least squares method to the steepest slope descent method. On the other hand, if the parameter λ tends to 0, the solution of the current iteration is only affected by the elements of the matrix [**A**], This reduces Marquardt's least squares to Gauss-Newton or generalized least squares [13]. It is well known that the steepest descent method converges quickly but slowly, while the Gauss-Newton method converges quickly but less reliably.

The main advantage of Marquardt's least squares method [12] is that the parameter λ can be chosen to follow the Gauss-Newton method as closely as possible, ensuring fast convergence while retaining some of the properties of the steepest descent method to prevent divergence. After several numerical experiments, the parameter λ was chosen to be 3 for fast and reliable convergence.

Matrix [**D**] is a diagonal matrix whose diagonal elements are identical to those of matrix [**A**] and is defined by the following formula:

$$[\mathbf{A}] = [\mathbf{J}]^T \cdot [\mathbf{J}] \tag{4.46}$$

where the Jacobian matrix [**J**] can be approximated:

$$[\mathbf{J}] = \begin{bmatrix} \frac{\partial E_1}{\partial \tau_1} & \frac{\partial E_1}{\partial \tau_2} \\ \frac{\partial E_2}{\partial \tau_1} & \frac{\partial E_2}{\partial \tau_2} \end{bmatrix} \tag{4.47}$$

According to Marquardt's least squares method, the matrix on the right-hand side of Eq. (4.45) is

$$[\mathbf{B}] = -[\mathbf{J}]^T \cdot \begin{bmatrix} E_1 \\ E_2 \end{bmatrix} \tag{4.48}$$

The iterations will continue until the parameters τ_1 and τ_2 reach the convergence condition. The condition for the convergence of the values of τ_1 and τ_2 is

$$S = \sum_{i=1}^{m} E_i^2 < 10^{-12} \tag{4.49}$$

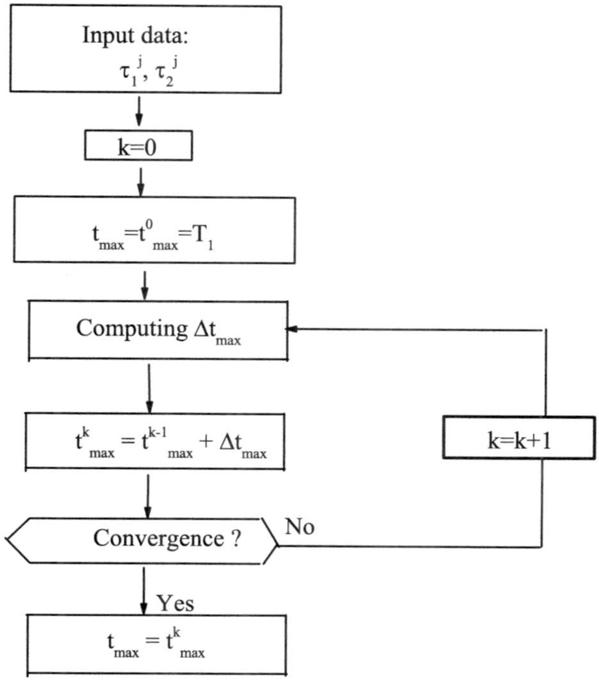

Fig. 4.11 Computation of t_{max} for Heidler function case

When this condition is reached, the iterative process ends and the final values of the parameters τ_1 and τ_2 and η are obtained.

Calculation of the parameter t_{max} for Heidler function case

As shown in Fig. 4.11, the input data for calculating the unknown auxiliary parameter t_{max} are the time constants τ_1^j and τ_2^j obtained in the current j-th iteration.

The time derivative of the double-exponential function described by Eq. (4.37) is equal to zero at $t = t_{max}$ when the lightning current peaks:

$$\frac{di(t)}{dt}|_{t=t_{max}} = \frac{I_0}{\eta} e^{-\frac{t_{max}}{\tau_2}} (\frac{t_{max}}{\tau_1})^{n-1} \left[\frac{\frac{n}{\tau_1} - \frac{1}{\tau_2}(\frac{t_{max}}{\tau_1}) - \frac{1}{\tau_2}(\frac{t_{max}}{\tau_1})^{n+1}}{(1 + (\frac{t_{max}}{\tau_1})^n)^2} \right] = 0 \qquad (4.50)$$

the following equation for $t = t_{max}$ can be obtained:

$$\frac{n\tau_2}{\tau_1} - (\frac{t_{max}}{\tau_1}) - (\frac{t_{max}}{\tau_1})^{n+1} = 0 \qquad (4.51)$$

which can be transformed into the next nonlinear equation for computing t_{max}:

$$F_{max} = (\frac{t_{max}}{\tau_1})^{n+1} + (\frac{t_{max}}{\tau_1}) - \frac{n\tau_2}{\tau_1} = 0 \tag{4.52}$$

The Jacobian J_{max} of F_{max} is defined by:

$$J_{max} = \frac{\partial F_{max}}{\partial t_{max}} \tag{4.53}$$

so

$$J_{max} = \frac{\partial F_{max}}{\partial t_{max}} = \frac{1}{\tau_1^k}\left[1 + (n+1)(\frac{t_{max}}{\tau_1^k})^n\right] \tag{4.54}$$

In each j-th and k-th iteration, the parameter t_{max} can be computed using nonlinear Eq. (4.52).

The nonlinear Eq. (4.52) is solved using a special case of Marquardt's least squares method for a single nonlinear equation. Similarly to Eq. (4.45), it can be expressed as

$$J_{max}^2 \cdot (1 + \lambda) \cdot \Delta t_{max} = B \tag{4.55}$$

where

$$B = -J \cdot F_{max} \tag{4.56}$$

In each j-th and k-th iteration, the Jacobian J_{max} can be computed using Eq. (4.54).

Then the growth of the parameter Δt_{max} can be computed by:

$$\Delta t_{max} = -\frac{F_{max}}{J_{max} \cdot (1 + \lambda)} = -\frac{F_{max}}{4 \cdot J_{max}} \tag{4.57}$$

where $\lambda = 3$.

The value of t_{max} for the k-th iteration is computed from the following expression:

$$t_{max}^k = t_{max}^{k-1} + \Delta t_{max} \tag{4.58}$$

As stated in Eq. (4.58), the value of parameter t_{max} will grow in every iteration by the value of Δt_{max}. The iterations will continue until the condition for the convergence of the parameter t_{max} is achieved:

$$S = F_{max}^2 < 10^{-12} \tag{4.59}$$

Calculation of the parameter t_1 for Heidler function case

The input data for the calculation of the unknown auxiliary parameter t_1 are the coefficients η and the time constants τ_1 and τ_2 obtained in the j-th iteration. the estimation of the parameter t_1 is shown in Fig. 4.13.

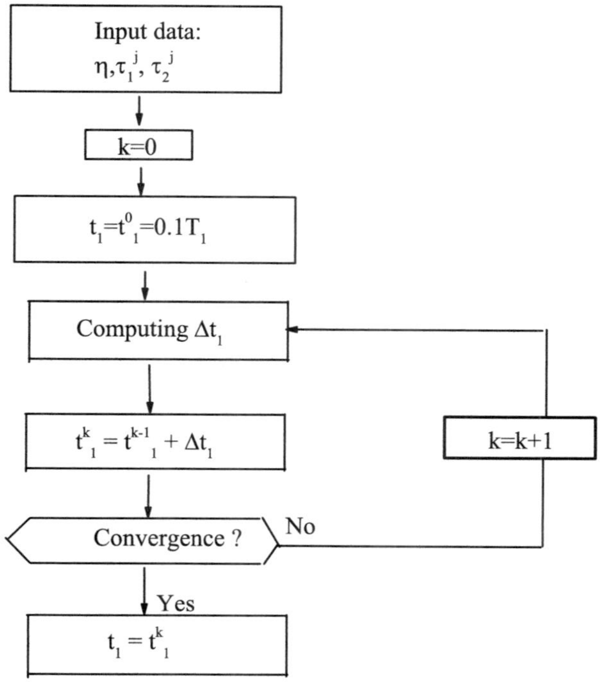

Fig. 4.12 Computation of t_1 for Heidler function case

The current $i(t)$ reaches 10% of the peak current I_0 at t_1. Substituting $I = 0.1I_0$ and $t = t_1$ into Eq. (4.37), the following equation is obtained (Fig. 4.12).

$$0.1I_0 = 0.1I_0 = \frac{I_0}{\eta}\frac{(\frac{t_1}{\tau_1})^n}{1+(\frac{t_1}{\tau_1})^n}e^{-\frac{t_1}{\tau_2}} \tag{4.60}$$

After dividing both sides of Eq. (4.30) by $\frac{I_0}{\eta}$, and after some more manipulation, the following formula can be obtained.

$$0.1\eta + (\frac{t_1}{\tau_1})^n\left[0.1\eta - e^{-\frac{t_1}{\tau_2}}\right] = 0 \tag{4.61}$$

Rearranging the previous expression the next nonlinear equation for computing t_1 is obtained:

$$F_1 = 0.1\eta + (\frac{t_1}{\tau_1})^n(0.1\eta - e^{-\frac{t_1}{\tau_2}}) \tag{4.62}$$

The Jacobian J_1 of F_1 is defined by:

$$J_1 = \frac{\partial F_1}{\partial t_1} \tag{4.63}$$

Substitution of Eq. (4.62) into Eq. (4.63) and partial differentiation with respect to t_1 yields:

$$\frac{\partial F_1}{\partial t_1} = \frac{n}{\tau_1^k}(\frac{t_1}{\tau_1^k})^{n-1}(0.1\eta - e^{-\frac{t_1}{\tau_2^k}}) + \frac{1}{\tau_2^k}(\frac{t_1}{\tau_1^k})^n e^{-\frac{t_1}{\tau_2^k}} \tag{4.64}$$

In each j-th and k-th iteration, the parameter t_1 can be computed using nonlinear Eq. (4.62).

The nonlinear Eq. (4.62) can be solved in the same way as described for the parameter t_{max}. Again, in the first iteration ($k = 0$), the first step is to assume the value of the variable t_1. Analogously, as in Eqs. (4.55)–(4.57), the growth of the iteration for the parameter t_1 can be found from:

$$\Delta t_1 = -\frac{F_1}{J_1 \cdot (1 + \lambda)} = -\frac{F_1}{4 \cdot J_1} \tag{4.65}$$

where $\lambda = 3$.

In each j-th and k-th iteration, the Jacobian J_1 can be computed using Eq. (4.64).

The value of t_1 for the k-th iteration is then computed from the following expression:

$$t_1^k = t_1^{k-1} + \Delta t_1 \tag{4.66}$$

As presented in Eq. (4.35), the value of parameter t_1 will grow in each iteration by the value of Δt_1. The iterations will continue until the condition for the convergence of parameter t_1 is achieved:

$$S = F_1^2 < 10^{-12} \tag{4.67}$$

Numerical examples for Heidler function case

In this first example, the unknown parameters of Heidler function are computed by simultaneously solving the nonlinear equations E_1 and E_2. The input data are taken from Reference [17], and they represent the maximum values of lightning current quantities for the subsequent short stroke for Lightning Protection Level I: $I_0 = 50$ kA, $T1/T2 = 0.25/100$ μs. Parameters τ_1, τ_2, and η are computed using the Marquardt least squares method, are presented in Table 4.5 for the current parameters, while the Heidler lightning function is shown in Fig. 4.13.

The second arithmetic example is described as follows: the unknown parameters of Heidler function are computed by simultaneously solving the nonlinear equations E_1 and E_2. The input data are taken from Reference [4], and they represent the maximum

Table 4.5 Computed parameters of Heidler function of the subsequent stroke 0.25/100 μs for $I_0 = 50$ kA

n	$\tau_1(\mu s)$	$\tau_2(\mu s)$	η	$T_1/T_2(set)$	$T_1/T_2(calculated)$
2	−0.0889279	141.4801616	0.9862388	$T1/T2 =$ 0.25/100 μs	$T1/T2 =$ 0.25/100.03 μs

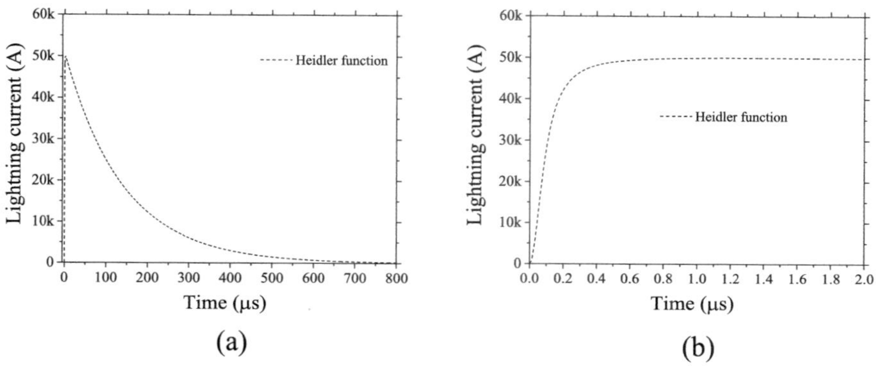

Fig. 4.13 Heidler function case

Table 4.6 Computed parameters of Heidler function of the subsequent stroke 10/350 μs for $I_0 =$ 200 kA

n	$\tau_1(\mu s)$	$\tau_2(\mu s)$	η	$T_1/T_2(set)$	$T_1/T_2(calculated)$
2	−4.4166943	450.9185294	0.9175015	$T1/T2 =$ 10/350 μs	$T1/T2 =$ 10.02/351.01 μs

values of lightning current quantities for the first short stroke for Lightning Protection Level I: $I_0 = 200$ kA, $T_1/T_2 = 10/350\,\mu$s. Parameters τ_1, τ_2, and η are computed using a computer program written in FORTRAN that implements the Marquardt least squares method described previously. The results of the computations are shown in Table 4.6 for the current parameters, while the Heidler lightning function is shown in Fig. 4.14.

The third arithmetic example is described as follows: For the first negative impulse, the input data provided by IEC 623051 Ed. 2 consist of I_0, T_1 and T_2. This implies that the lightning function parameters τ_1, τ_2 and η will be estimated by solving a set of two nonlinear equations.

The input data are taken from Reference [6], and they represent the maximum values of lightning current quantities for the first negative impulse: LPL I : $I_0 = 100$ kA; $T_1/T_2 = 1/200\,\mu$s; Parameters τ_1, τ_2, and η are computed using a computer program written in FORTRAN that implements the Marquardt least squares method described previously. The results of the computations are shown in Table 4.7 for the current parameters, while the Heidler lightning function is shown in Fig. 4.15.

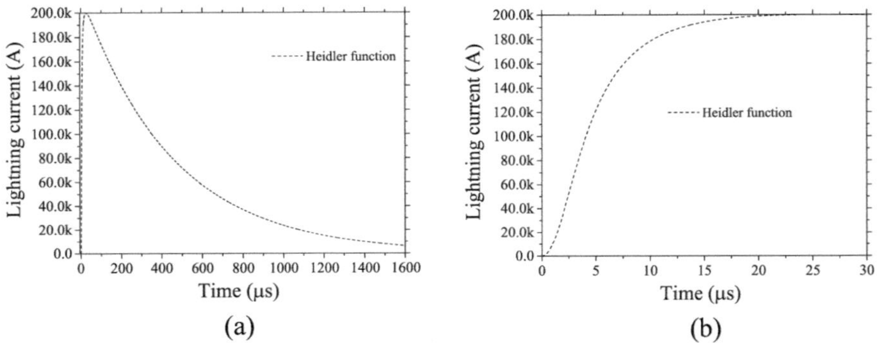

Fig. 4.14 Heidler function case

Table 4.7 Computed parameters of Heidler function of the subsequent stroke 1/200 μs for $I_0 =$ 100 kA

n	$\tau_1(\mu s)$	$\tau_2(\mu s)$	η	$T_1/T_2(set)$	$T_1/T_2(calculated)$
2	−0.3680589	279.5562237	0.9775816	$T1/T2 =$ 1/200 μs	$T1/T2 =$ 1.0/200.10 μs

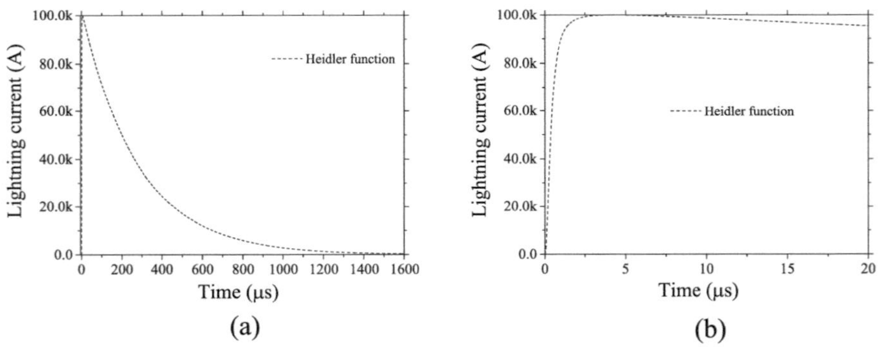

Fig. 4.15 Heidler function case

The fourth arithmetic example is described as follows: The most common waveshapes used for designing low-voltage power lines within structures are the $T_1/T_2 = 8/20\,\mu s$ waveshape [14, 18], the lightning waveshapes can be approximated by Heidler function. In Fig. 4.16, approximations of the 8/20 μs waveshape using the Heidler function with different values of the current steepness factor n are presented in order to illustrate the effect of n on the resulting waveshape. For all other waveshapes, $n = 10$ is used although any n can be used with the proposed method. As can be observed, n directly affects the time offset or the delay of the start of the transient.

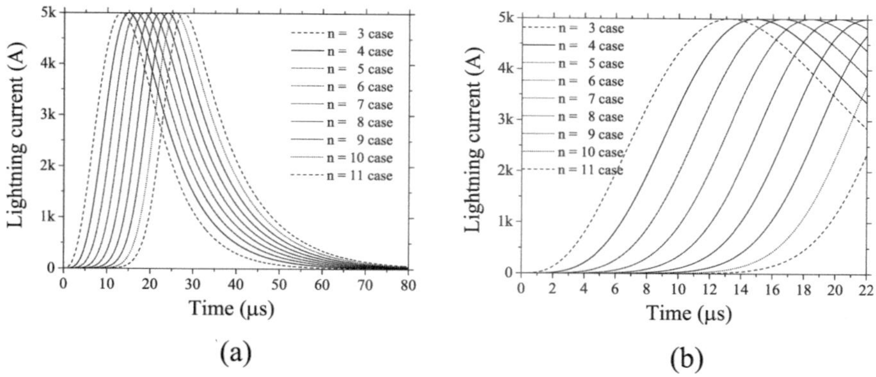

Fig. 4.16 Heidler function case

Table 4.8 Computed parameters of Heidler function of the subsequent stroke 8/20 μs for $I_0 = 50$ kA

n	$\tau_1 (\mu s)$	$\tau_2 (\mu s)$	η	$T_1/T_2(set)$	$T_1/T_2(calculated)$
3	15.4028121	7.22332784	0.06212568	$T1/T2 = 8/20$ μs	$T1/T2 = 8.01/20.81$ μs
4	13.8142165	8.78063493	0.10538416	$T1/T2 = 8/20$ μs	$T1/T2 = 8.00/20.81$ μs
5	14.6905244	9.54413614	0.11385093	$T1/T2 = 8/20$ μs	$T1/T2 = 8.01/20.81$ μs
6	16.0788471	10.03350792	0.11056371	$T1/T2 = 8/20$ μs	$T1/T2 = 8.01/20.81$ μs
7	17.6691267	10.37856103	0.10260371	$T1/T2 = 8/20$ μs	$T1/T2 = 8.01/20.81$ μs
8	19.3605117	10.63596716	0.09295446	$T1/T2 = 8/20$ μs	$T1/T2 = 8.01/20.81$ μs
9	21.1098733	10.83563595	0.08301466	$T1/T2 = 8/20$ μs	$T1/T2 = 8.01/20.80$ μs
10	22.8955944	10.99511258	0.07354406	$T1/T2 = 8/20$ μs	$T1/T2 = 8.01/20.81$ μs
11	24.7056252	11.12544199	0.06458439	$T1/T2 = 8/20$ μs	$T1/T2 = 8.01/20.81$ μs

Regardless of the chosen n, one can easily obtain Heidler function parameters for mentioned waveshapes using the Marquardt least squares method [12] because only T_1 and T_2 are known. Complete set of Heidler function parameters for all waveshapes is given in Table 4.8. Note that, unlike in all other cases, Heidler function approximation of the 8/20 μs waveshape features a time constant τ_1 that is greater than τ_2. This is because the 8/20 μs waveshape, unlike other waveshapes, has a sharp rise, which is followed closely by a sharp decay.

The last arithmetic example is described below: Heidler function can also be used to approximate actual recorded data of a lightning waveshape. This approximation can have certain limitations because actual lightning waveshapes can significantly differ from the standardized lightning waveshapes. Nevertheless, Marquardt least squares method [12] can also be used to estimate the parameters of Heidler function to approximate the recorded data. Two measured lightning strokes [19] are approximated with the Heidler function to illustrate this.

The first lightning return-stroke current taken from [19] has a peak current of $I_0 = 9127.7A$, front duration $T_1 = 3.33$ μs and time to half value $T_2 = 36.83$ μs

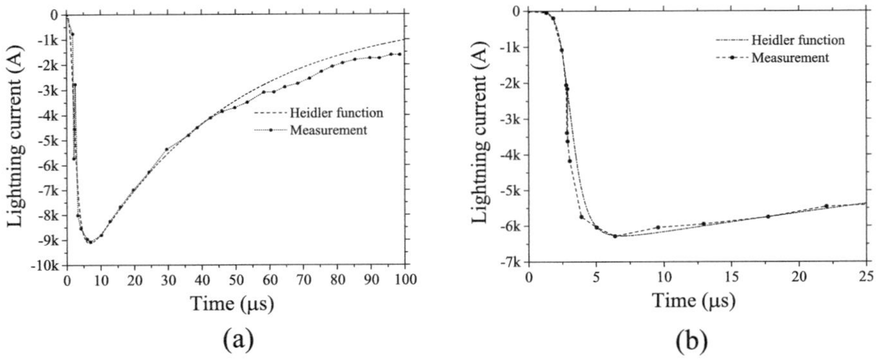

Fig. 4.17 Heidler function case

Table 4.9 Computed parameters of Heidler function of the subsequent stroke 8/20 µs for $I_0 = 50$ kA

n	$\tau_1(\mu s)$	$\tau_2(\mu s)$	η	$T_1/T_2(set)$	$T_1/T_2(calculated)$
3	2.6087	41.6807	0.8041	$T1/T2 =$ 3.33/36.83 µs	$T1/T2 =$ 3.33/37.17 µs
6	−3.23529	109.200375	0.9289	$T1/T2 =$ 2.5/81.55 µs	$T1/T2 =$ 2.50/81.80 µs

(Fig. 4.17a). Heidler function approximation of the measured lightning waveshape is depicted also in Fig. 4.4 with the following parameters obtained using the Marquardt least squares method [12]: $n = 3$, $\eta = 0.8041$, $\tau_1 = 2.6087$ µs and $\tau_2 = 41.6807$ µs, which can be seen in Table 4.9. Note the good agreement in the rise time part of the waveshape unlike in the decay part.

The second lightning return-stroke current also taken from [19] has a peak current of $I_0 = 6278A$, front duration $T_1 = 2.5$ µs and time to half value $T_2 = 81.55$ µs (Fig. 4.17b). Heidler function approximation of the measured lightning waveshape is depicted in Fig. 4.5 with the following parameters: $n = 6$, $\eta = 0.9289$, $\tau_1 = -3.23529$ µs and $\tau_2 = 109.200375$ µs, which can be seen in Table 4.9. Heidler function approximation accurately models the recorded data, except at the start of the measurement [19].

4.4 Summary

This chapter provides a detailed and in-depth study of three mathematical formulas for generating external excitation sources lightning currents, namely the double exponential function, Heidler function, and composite Heidler function. The inversion of these mathematical formulas and their parameters is one of the foundations for the content of the relevant chapters that follow.

References

1. V. Cooray, *Lightning protection* (The Institution of Engineering and Technology, Stevenage, 2007)
2. Chas.E.R. Bruce, R.H. Golde, The lightning discharge. J. Inst. Electr. Eng. Part II Power Eng. **88**(6), 487–505 (1941)
3. K. Berger, R.B. Anderson, H. Kroninger, Parameters of lightning flashes. Electra **41**, 23–37 (1975)
4. R.B. Anderson, A.J. Erikson, Lightning parameters for engineering application. Electra **69**, 65–101 (1980)
5. L.P. Zhang, J.S. Yuan, Z.X. Li, A class of analytical functions to study the lightning effects associated with the current front. Europ. Trans. Electric Power **12**(2), 141–150 (2002)
6. V.P. Larionov, *Protection against lightning, part 1 : General principles*. Iec Edition (2010)
7. C.A. Nucci, F. Rachidi, M.V. Ianoz, C. Mazzetti, Lightning-induced voltages on overhead lines. IEEE Trans. Electromagn. Compat. **35**(1), 75–86 (1993)
8. A. Eriksson, L. Dellera, J. Elovaara, A. Hileman, K. . Weck, Guide to procedures for estimating the lightning performance of transmission lines. Cigre Brochure (1991)
9. S. Visacro, A representative curve for lightning current waveshape of first negative stroke. Geophys. Res. Lett. **31**, L07112 (2004)
10. F. Heidler, J. Cvetic, B.V. Stanic, Calculation of lightning current parameters. IEEE Trans. Power Delivery **14**(2), 399–404 (1999)
11. P.D. Rancic, L.V. Stefanovic, D.R. Djordjevic, Analytische blitzstromfunktion zur lemp-berechnung, in *Proceedings 18th International Conference Lightning Protection*, pp. 63–66, Munich, Germany (1985)
12. D.W. Marquardt, An algorithm for least-squares estimation of nonlinear parameters. J. Soc. Indus. Appl. Mathem. **11**(2), 431–441 (1963)
13. S.A. Lill, *Optimization in action* (Academic Press, London, 1976)
14. M.A. Uman, *The Art and Science of Lightning Protection* (Cambridge University Press, New York, 2008)
15. Z. Stojkovic, M.S. Savic, J.M. Nahman, D. Salamon, B. Bukorovic, Experimental investigation of grounding grid impulse characteristics. Europ. Trans. Electr. Power **8**(6), 417–421 (1998)
16. Z. Stojkovic, M.S. Savic, J.M. Nahman, D. Salamon, B. Bukorovic, Sensitivity analysis of experimentally determined grounding grid impulse characteristics. IEEE Trans. Power Delivery **13**(4), 1136–1142 (1998)
17. V.P. Larionov, *Protection Against Lightning, Part 1 : General Principles*. Iec Edition (2006)
18. V. Cooray, *Lightning Protection* (The Institution of Engineering and Technology, London, United Kingdom, 2010)
19. N. Theethayi, V.A. Rakov, R. Thottappillil, Responses of airport runway lighting system to direct lightning strikes: comparisons of tlm predictions with experimental data. IEEE Trans. Electromagn. Compat. **50**(3), 660–668 (2008)

Chapter 5
Numerical Methods for Grounding Problem at Steady-State I

Various frequency-domain numerical methods for steady-state grounding problems based on equipotential electromagnetic field modeling.

5.1 Preface

For the equipotential model simulating the steady-state problem of substation grounding grid, the scalar field numerical calculation method is mainly used. The main work of this chapter is to develop and consider the spatial 1-d scalar BEM and spatial 1-d scalar MM.

5.2 Spatial 1-d Scalar MM and BEM

This chapter is based on the basic work of the electromagnetic field numerical calculation methods introduced earlier, and will comprehensively introduce the electromagnetic field numerical methods of the equipotential model, namely the spatial 1-d Galerkin MM and BEM.

5.2.1 Introduction

The performance of an a.c. substation grounding system under low 50 or 60 Hz frequency are always analyzed based on analogy electrostatic field (or direct current electrical field) theory [1–13] in recent years. There have been many papers studying the performance of AC substation grounding systems in the low-frequency domain (about 50 to 60 Hz) [14–16]. All these papers are based on the equipotential math-

Z.-X. Li, *Numerical Simulation Algorithm of Electromagnetic Field for Grounding Problems in Power System Substation Grounding Grids*,
https://doi.org/10.1007/978-981-96-0904-8_5

ematical model, with some variations: direct current electric field theory is used in [1–13], while quasi-static electric field theory is used in [14–16].

The spatial 1-d scalar BEM for solving the grounding system problem has been pioneer researched in reference [8–13], and the spatial 1-d scalar MM for solving the grounding system problem has been pioneered in [1–6, 14–16]. However, the application of the spatial 1-d scalar BEM or MM for grounding system is also based on analogy electrostatic field (or direct current electrical field) theory.

In this chapter, based on quasi-static electrical field, the QSCIM and the closed form of Green's function for horizontal multilayered earth model are introduced. Fast convergent spatial 1-d scalar BEM or MM combined with closed form of Green's function of a point current source in horizontal multilayered earth model is developed, so it can be used to calculate the currents distribution along both the grounding system and the floating metallic conductors buried in horizontal multilayered earth model, meanwhile, not only the conductive effect of current leaking into the multilayered earth model, but also capacitive effects from the multilayered earth model, have been considered in the method. It should be noted here that the authors of this book published articles [15, 16] on numerical algorithms of equipotential electromagnetic fields in the frequency domain for numerical simulation of steady-state grounding problems of substation grounding grids in multilayer horizontally stratified soils.

5.2.2 Spatial 1-d Scalar Galerkin MM

In the numerical simulation of the grounding system including floating electrode buried in horizontal N_e-layer earth model, the grounding system including floating electrode is divided into small cylindrical segments. First, it can be supposed that the system consists of total N_b segments of cylindrical bars and total N_p discreted nodes along the grounding grid, and consider the currents density $J_{l_i}(i = 1, \ldots, N_b)$ emanating from the surface of each segment as unknowns. Next an equation system can be constructed, and suppose the grounding systems are equal-potential. In this way, the currents can be obtained by solving the equation system. Finally the voltage (zero potential reference point at infinity) at any points can be calculated by the contributions of all the currents.

According to the previous chapter on spatial 1-d SGWRM, it can be seen that

$$\int_{l_\theta} [\bar{W}_{l_\theta}^m(\bar{r}_\theta)]dr_\theta[\varphi_{l_\theta}(\bar{r}_\theta)] = \sum_{\vartheta=1}^{N_b} \int_{l_\theta} \int_{l_\vartheta} G_{\xi\zeta}^{\varphi f}(\bar{r}_\theta, \bar{r}_\vartheta)[\bar{W}_{l_\theta}^m(\bar{r}_\theta)][\bar{N}_{l_\vartheta}^n(\bar{r}_\vartheta)]dr_\vartheta dr_\theta[\bar{J}_{l_\vartheta}^n(\bar{r}_\vartheta)] \tag{5.1}$$

where $\xi = 0, 1, \ldots, N_e$; $\zeta = 0, 1, \ldots, N_e.\theta = 1, \ldots, N_b$. $\bar{r}_\theta$ and $\bar{r}_\vartheta$ are the position vectors of the field point and source point, respectively. l_θ and l_ϑ are the length of θ-th and ϑ-th conductor segments, respectively. $[\bar{W}_{l_\theta}^m(\bar{r}_\theta)]$ and $[\bar{N}_{l_\vartheta}^n(\bar{r}_\vartheta)]$ are vectors for test functions and basic functions, respectively. $[\varphi_{l_\theta}(\bar{r}_\theta)]$ is the voltage of the θ-th conductor segment. $[\bar{J}_{l_\vartheta}^n(\bar{r}_\vartheta)]$ is a vector for leakage line current density along the ϑ-th conductor segment. $G_{\xi\zeta}^{\varphi^f}(\bar{r}_\theta, \bar{r}_\vartheta)$ is frequency domain monopole Green's function. r_ϑ and r_θ are parameter variables for the distribution of the ϑ-th and θ-th thin conductor segments, respectively.

To get Galerkin type of the spatial 1-d scalar Galaerkin MM, Galerkin type weighting approaches (test functions are identical to basic functions) must be used, $[\bar{W}_{l_\theta}^m(\bar{r}_\theta)] \Rightarrow [\bar{N}_{l_\vartheta}^n(\bar{r}_\vartheta)]^T$ (Here T is the matrix transpose symbol), and suppose line current density $[\bar{J}_{l_\vartheta}(\bar{r}_\vartheta)]$ ($\vartheta = 1, \ldots, N_b$) are uniformly emanating from the surface of each segment, hence $m = n = 0$, $[\bar{N}_{l_\vartheta}^0(\bar{r}_\vartheta)] = \frac{1}{l_\vartheta}$, and $[\bar{N}_{l_\vartheta}^0(\bar{r}_\vartheta)][\bar{J}_{l_\vartheta}^0(\bar{r}_\vartheta)] = \frac{I_{l_\vartheta}(\bar{r}_\vartheta)}{l_\vartheta}$, $[W_{l_\theta}^0(\bar{r}_\theta)] \Rightarrow [N_{l_\vartheta}^0(\bar{r}_\vartheta)]^T = \frac{1}{l_\vartheta}$, ($\vartheta = 1, \ldots, N_b$). Then it can be obtained as below

$$\int_{l_\theta} \frac{1}{l_\theta} \varphi_{l_\theta}(\bar{r}_\theta) dr_\theta = \sum_{\vartheta=1}^{N_b} \int_{l_\theta} \frac{1}{l_\theta} \int_{l_\vartheta} G_{\xi\zeta}^{\varphi^f}(\bar{r}_\theta, \bar{r}_\vartheta) \frac{I_{l_\vartheta}(\bar{r}_\vartheta)}{l_\vartheta} dr_i dr_\theta \tag{5.2}$$

$$\sum_{\vartheta=1}^{N_b} \frac{1}{l_\theta} \int_{l_\theta} \frac{1}{l_\vartheta} \int_{l_\vartheta} G_{\xi\zeta}^{\varphi^f}(\bar{r}_\theta, \bar{r}_\vartheta) dr_\vartheta dr_\theta I_{l_\vartheta}(\bar{r}_\vartheta) = \varphi_{l_\theta}(\bar{r}_\theta) \tag{5.3}$$

where $\theta = 1, \ldots, N_b$.

So matrix equation can be obtained as below

$$[\bar{\bar{Z}}_{b_s}][\bar{I}_l^b] = [\bar{\varphi}_l^b] \tag{5.4}$$

where $[\bar{\bar{Z}}_{b_s}]$ is $N_b \times N_b$ matrix for mutual impedance coefficients of grounding system. $[\bar{I}_l^b]$ is $N_b \times 1$ vector for leakage line currents. $[\bar{\varphi}_l^b]$ is $1 \times N_b$ vector for voltage of the grounding system. It must be pointed out that the coefficients matrix in linear system Eq. (5.4) is symmetric and positive definite [17].

Once the leakage line current is solved, the voltage at any point can be calculated by all leakage line currents, as

$$\varphi(\bar{r}) = \sum_{\vartheta=1}^{N_b} \int_{l_\vartheta} G_{\xi\zeta}^{\varphi^f}(\bar{r}, \bar{r}_\vartheta) \frac{1}{l_\vartheta} I_{l_\vartheta}(\bar{r}_\vartheta) dr_\vartheta \tag{5.5}$$

Since this algorithm considers that the potential of the entire grounding system is the same, it can be assumed that the voltage of the grounding system is constant 1, so the Eq. (5.4) can be solved easies, which means that all leakage line current can be obtained.

Let the total leakage current from grounding system be I_0, and the total current into and out of floating electrode be zero, so

$$I_{all} = \sum_{\vartheta=1}^{N_b} I_{l_\vartheta}(\bar{r}_\vartheta) \tag{5.6}$$

The voltage of the grounding grid is equal to the ratio of the known incident current to the calculated leakage line current $\frac{I_0}{I_{all}}$. All calculated leakage line current values need to be multiplied by this ratio. If there is a floating electrode present, its floating voltage can be calculated from all the currents entering and exiting the floating electrode according to formula (5.5).

The grounding impedance and admittance are given by the relation

$$\frac{\varphi_1}{I_0} = Z_g = \frac{1}{Y_g} = R_g + jX_g \tag{5.7}$$

The element Z_{ij} in Eq. (5.4) is defined as:

$$Z_{\vartheta\theta} = \frac{1}{l_\vartheta l_\theta} \int_{l_\vartheta} \int_{l_\theta} G_{\xi\zeta}^{\varphi f}(\bar{r}_\theta, \bar{r}_\vartheta) dr_\vartheta dr_\theta \tag{5.8}$$

For half infinite homogenous earth model, we have

$$Z_{\vartheta\theta} = \frac{1}{4\pi\bar{\sigma}_1 l_\vartheta l_\theta} \int_{l_\vartheta} \int_{l_\theta} (\frac{1}{\bar{R}} - \frac{k_{01}}{\bar{R}'}) dr_\vartheta dr_\theta \tag{5.9}$$

5.2.3 *Spatial 1-d Scalar Galerkin BEM*

Just like Spatial 1-d scalar Galerkin MM case, in the numerical simulation of the grounding system including floating electrode buried in horizontal N_s-layer earth model, the grounding system including floating electrode is divided into small cylindrical segments. First, the system consists of total N_b segments of cylindrical bars and total N_p discreted nodes along the grounding grid was supposed, and consider the currents density $J_{l_i} (i = 1, \ldots, N_b)$ emanating from the surface of each segment as unknowns. Next an equation system can be constructed, and suppose the grounding systems are equal-potential. In this way, the currents can be obtained by solving the equation system. Finally the voltage (zero potential reference point at infinity) at any points can be calculated by the contributions of all the currents.

An approximate spatial 1-d SGWRM derived from the spatial 3-d SGWRM has been studied in the previous chapter. The spatial 1-d scalar BEM can also be briefly introduced below.

From the previous chapter on spatial 1-d SGWRM, the linear system of equations can be obtained as below

$$\int_{l_\theta} [\bar{W}_{l_\theta}^m(\bar{r}_\theta)] dr_\theta [\bar{\varphi}_{l_\theta}(\bar{r}_\theta)] = \sum_{\vartheta=1}^{N_b} \int_{l_\theta} \int_{l_\vartheta} G_{\xi\zeta}^{\varphi f}(\bar{r}_\theta, \bar{r}_\vartheta) [\bar{W}_{l_\theta}^m(\bar{r}_\theta)][\bar{N}_{l_\vartheta}^n(\bar{r}_\vartheta)] dr_i dr_\theta [\bar{J}_{l_\vartheta}^n(\bar{r}_\vartheta)] \tag{5.10}$$

where $\theta = 1, \ldots, N_b$.

To apply Galerkin type of the spatial 1-d scalar BEM, Galerkin type weighting approaches (test functions are identical to trial functions) must be used, $\left[\bar{W}_{l_\theta}^m(\bar{r}_\theta)\right] \equiv \left[\bar{N}_{l_\vartheta}^n(\bar{r}_\vartheta)\right]^T$, and suppose current density $[J_{l_\vartheta}^{n,k}(\bar{r}_\vartheta)]$ ($\vartheta = 1, \ldots, N_b; k = 1, \ldots, N_p$) is a vector for leakage current of the k-th node on the ϑ-th discrete conductor segment, hence $m = n = 0$, $\left[\bar{N}_{l_\vartheta}^0(\bar{r}_\vartheta)\right] = \left[\frac{1}{l_\vartheta}\ -\frac{1}{l_\vartheta}\right]$, and $[\bar{J}_{l_\vartheta}^{0,k}(\bar{r}_\vartheta)] = \begin{bmatrix} I_{l_\vartheta}^{k_1}(\bar{r}_\vartheta) \\ I_{l_\vartheta}^{k_2}(\bar{r}_\vartheta) \end{bmatrix}$ ($k_1, k_2 = 1, 2, \ldots, N_p$), $[W_{l_\theta}^0(\bar{r}_\theta)] \equiv [N_{l_\theta}^0(\bar{r}_\theta)]^T = \begin{bmatrix} \frac{1}{l_\theta} \\ -\frac{1}{l_\theta} \end{bmatrix}$. Then we have

$$\int_{l_\theta} \begin{bmatrix} \frac{1}{l_\theta} \\ -\frac{1}{l_\theta} \end{bmatrix} \varphi_{l_\theta}(\bar{r}_\theta) dr_\theta = \sum_{\vartheta=1}^{N_b} \int_{l_\theta} \int_{l_\vartheta} G_{\xi\zeta}^{\varphi f}(\bar{r}_\theta, \bar{r}_\vartheta) \begin{bmatrix} \frac{1}{l_\theta} \\ -\frac{1}{l_\theta} \end{bmatrix} \left[\frac{1}{l_\vartheta}\ -\frac{1}{l_\vartheta}\right] dr_\vartheta dr_\theta \begin{bmatrix} I_{l_\theta}^{k_1}(\bar{r}_\vartheta) \\ I_{l_\theta}^{k_2}(\bar{r}_\vartheta) \end{bmatrix} \tag{5.11}$$

$$\sum_{i=1}^{N_b} \int_{l_\theta} \int_{l_\vartheta} G_{\xi\zeta}^{\varphi f}(\bar{r}_\theta, \bar{r}_\vartheta) \begin{bmatrix} \frac{1}{l_\vartheta}\frac{1}{l_\theta} & -\frac{1}{l_\vartheta}\frac{1}{l_\theta} \\ -\frac{1}{l_\vartheta}\frac{1}{l_\theta} & \frac{1}{l_\vartheta}\frac{1}{l_\theta} \end{bmatrix} dr_\vartheta dr_\theta \begin{bmatrix} I_{l_\theta}^{k_1}(\bar{r}_\vartheta) \\ I_{l_\theta}^{k_2}(\bar{r}_\vartheta) \end{bmatrix} = \begin{bmatrix} 1 \\ -1 \end{bmatrix} \varphi_{l_\theta}(\bar{r}_\theta) \tag{5.12}$$

where $\theta = 1, \ldots, N_b$. k_1 and k_2 are the numbers of the two nodes in the ϑ-th conductor segment, respectively, $k_1, k_2 = 1, \ldots, N_p$.

By assembling the contributions from each element, Eq. (5.12) can be expressed in matrix form as

$$[\bar{\bar{Z}}_{p_s}][\bar{I}_l^P] = [\bar{\varphi}_l^P] \tag{5.13}$$

where $[\bar{\bar{Z}}_{p_s}]$ is the $N_p \times N_p$ matrix about mutual impedance coefficient between a pair of segments in grounding system, $[\bar{I}_l^P]$ is a $N_p \times 1$ vector for leakage currents emanating from grounding system's discrete conductor's node, $[\bar{\varphi}_l^P]$ is a $N_p \times 1$ vector for voltage of grounding system's discrete conductor's node. It must be pointed out that the coefficients matrix in linear system Eq. (5.13) is symmetric and positive definite [17].

Since this algorithm considers that the potential of the entire grounding system is the same, it can be assumed that the voltage of the grounding system is constant 1, so the Eq. (5.13) can be solved easies, which means that all leakage line current can be obtained.

Let the total leakage current from grounding system be I_0, and the total current into and out of floating electrode be zero, so

$$I_{all} = \sum_{\vartheta=1}^{N_b} (I_{l_\vartheta}^{k_1}(\bar{r}_\vartheta) - I_{l_\vartheta}^{k_2}(\bar{r}_\vartheta)) \tag{5.14}$$

where k_1 and k_2 are the numbers of the two nodes in the ϑ-th conductor segment, respectively, $k_1, k_2 = 1, \ldots, N_p$.

The voltage of the grounding grid is equal to the ratio of the known incident current to the calculated leakage line current $\frac{I_0}{I_{all}}$. All calculated leakage line current values need to be multiplied by this ratio. If there is a floating electrode present, its floating voltage can be calculated from all the currents entering and exiting the floating electrode according to formula (5.5).

The grounding impedance and admittance are given by the relation

$$\frac{\varphi_1}{I_0} = Z_g = \frac{1}{Y_g} = R_g + jX_g \tag{5.15}$$

The element Z_{ij} in Eq. (5.14) is defined as:

$$Z_{\vartheta\theta} = \frac{1}{4\pi\bar{\sigma}_1 l_\vartheta l_\theta} \int_{l_\vartheta} \int_{l_\theta} G_{\xi\zeta}^{\varphi f}(\bar{r}_\theta, \bar{r}_\vartheta) dr_\vartheta dr_\theta \tag{5.16}$$

For half infinite homogenous earth model, It can be seen as below

$$Z_{\vartheta\theta} = \frac{1}{4\pi\bar{\sigma}_1 l_\vartheta l_\theta} \int_{l_\vartheta} \int_{l_\theta} (\frac{1}{\bar{R}_{\vartheta\theta}} - \frac{k_{01}}{\bar{R'}_{\vartheta\theta}}) dr_\vartheta dr_\theta \tag{5.17}$$

The voltage at any point can be calculated by all leakage currents, as

$$\varphi(\bar{r}) = \sum_{\vartheta=1}^{N_b} \int_{l_\vartheta} G_{\xi\zeta}^{\varphi f}(\bar{r}, \bar{r}_\vartheta)[\frac{I_{l_\vartheta}^{k_1}}{l_\vartheta} - \frac{I_{l_\vartheta}^{k_2}}{l_\vartheta}] dr_\vartheta \tag{5.18}$$

where $k_1, k_2 = 1, 2, \ldots, N_p$.

5.3 The Closed Form of Green's Functions

The main task of using the spatial 1-d scalar MM and the BEM equipotential model to simulate the grounding system is to calculate the mutual impedance coefficient shown in Eqs. (5.8) and (5.16). If the earth is half infinite homogenous conductive medium, Eqs. (5.9) and (5.17) will be calculated easily. However, as the earth is sometime regarded as horizontal multilayered conductive medium in practice, a infinite integral about Bessel function in the Green's function must be calculated, and some fast calculation techniques must adopted to avoid the integral. As the result, the corresponding closed form of Green's function of a point source in horizontal multilayered earth model have been defined and fully discussed in Chap. 3.

Once the closed form of Green's function of a point source in horizontal multilayered earth model has been derived, the mutual impedance coefficient can be fastly calculated through Eqs. (5.8) and (5.16). It should be pointed out that the integral of

Eqs. (5.9) and (5.17) can be analytically calculated with the closed form of Green's function, as described in reference [1]. It must be pointed out that the analytical formula in the literature [1] only applies to the mutual impedance matrix element coefficients between vertical grounding rods and horizontal grid grounding grids. The mutual impedance matrix element coefficients between grounding grids of any shape can also be analytically calculated, which will be discussed in the next section.

5.4 General Analytical Formulas of Mutual Impedance

The element $Z_{i\theta}^{nm}$ in Eqs. (5.8) and (5.16) is the mutual impedance coefficient between a pair of segments with leakage currents $I_{l_i}^{n}$ or $I_{l_\theta}^{m}$ in the grounding system. It has been defined in (5.9) and (5.17) (for semi-homogenous earth model); for a general case, we have

$$Z_{\vartheta\theta} = \int_{l_\vartheta}\int_{l_\theta} \frac{1}{\bar{R}_{\vartheta\theta}} dr_\vartheta dr_\theta, \tag{5.19}$$

Here $\vartheta = 1, \ldots, N_b, \theta = 1, \ldots, N_b$, and $\bar{R}_{\vartheta\theta} = |\bar{r}_\theta - \bar{r}_\vartheta|$.

The mutual impedance matrix element coefficients between this arbitrarily shaped grounding grid can also be analytically calculated, which will have been discussed in depth in Chap. 2 and will not be further discussed here.

5.5 Simulation Result and Analysis

According to the approach introduced in this chapter, a fortran language program has been implemented. The program can simulate a grounding system buried in horizontal up to four-layer earth model. After the QSCIM was introduced, the closed form of Green's function of horizontal multilayered earth model have been derived, the QSCIM can avoid the infinite Maclaurin's series by just a sum of only 3~5 complex exponential terms. Meanwhile, analytical formula about mutual impedance coefficient Eqs. (5.8) and (5.16) [1] was adopted in the program, so no numerical integral calculation is required. On the other hand, the fast convergent scalar Galerkin 1-d MM and BEM with zero order basic functions has been used.

5.5.1 Verification of the Approach for Spatial 1-d Scalar Galerkin MM and BEM

To verify the result of the method proposed in this work, some cases solved by other authors are studied.

The first case, from [14], is a horizontal grounding electrode of 20 m length and 0.16 m radius, buried at 1 m depth in the two-layer earth, with $p = \frac{\sigma_2}{\sigma_1}$, $\rho_1 = 100.0\,\Omega\text{m}$, $\varepsilon_1 = 81\varepsilon_0$, $\varepsilon_2 = 5\varepsilon_0$, $h = 2$ m, $f = 50$ Hz, and $I_e = 300A$. The result is given in Table 5.1.

The second example comes from measurement data from public literature [18]. Four multi-layer soil models are presented in this study, as shown in Table 5.2. The grounding grids modelled in the study are the following: (1) S4–20 m×20 m square four-mesh grid. (2) S16–20 m×20 m square sixteen-mesh grid. The radius of all grounding grid conductors is set to 9.8 mm. The calculated grounding resistance with various algorithms and data from [18] can be seen in Tables 5.3 and 5.4. By referring to these tables, it can be seen that the calculation results of various algorithms are basically consistent with the data in the literature [18].

The third case, from [19], two types of grounding grids with different sizes are used to measure grounding impedance. Because of overcoming the influence of

Table 5.1 Comparison with publication paper results

	Reference [14]	MM	BEM
$p = \frac{\sigma_2}{\sigma_1}$	$Z_g[\Omega]$	$Z_g[\Omega]$	$Z_g[\Omega]$
10	$(2.526, -0.046 \times 10^{-3})$	$(2.502, -0.698 \times 10^{-9})$	$(2.502, 0.479 \times 10^{-4})$
5	$(2.894, -0.049 \times 10^{-3})$	$(2.873, -0.488 \times 10^{-9})$	$(2.873, 0.0)$
2	$(3.870, -0.055 \times 10^{-3})$	$(3.869, -0.540 \times 10^{-4})$	$(3.869, -0.540 \times 10^{-4})$
1	$(5.210, -0.067 \times 10^{-3})$	$(5.250, -0.119 \times 10^{-3})$	$(5.247, -0.119 \times 10^{-3})$
0.5	$(7.266, -0.096 \times 10^{-3})$	$(7.370, -0.959 \times 10^{-4})$	$(7.370, -0.959 \times 10^{-4})$
0.2	$(11.252, -0.180 \times 10^{-3})$	$(11.466, -0.184 \times 10^{-3})$	$(11.466, -0.184 \times 10^{-3})$
0.1	$(15.137, -0.305 \times 10^{-3})$	$(15.461, -0.312 \times 10^{-3})$	$(15.461, -0.312 \times 10^{-3})$

Table 5.2 Four multi-layer soil models

Model	Layer	Resistivity (Ωm)	Thickness (m)
(a)	1	2000	3.0
	2	100	3.0
	3	1000	∞
(b)	1	50	3.0
	2	1000	3.0
	3	100	∞

Table 5.3 Soil model (a) and grid type : S4 case

Grid type	S4 case			S16 case		
Depth (m)	$R_e(\Omega)$ [18]	MM(eq)	BEM(eq)	$R_e(\Omega)$ [18]	MM(eq)	BEM(eq)
1.5	24.46	24.45	24.45	19.40	19.43	19.45
4.5	7.99	8.02	8.02	7.86	8.02	8.02
7.5	14.63	14.71	14.71	12.23	12.31	12.32
15	16.16	16.24	16.24	13.86	13.95	13.97
50	16.32	16.40	16.40	14.05	14.13	14.16

Table 5.4 Soil model (b) and grid type : S4 case

Grid type	S4 case			S16 case		
Depth (m)	$R_e(\Omega)$ [18]	MM(eq)	BEM(eq)	$R_e(\Omega)$ [18]	MM(eq)	BEM(eq)
1.5	2.57	2.59	2.59	2.51	2.52	2.53
4.5	8.67	8.75	8.75	6.15	6.23	6.24
7.5	2.13	2.14	2.14	1.91	1.92	1.93
15	1.85	1.86	1.86	1.63	1.64	1.64
50	1.66	1.67	1.67	1.44	1.44	1.44

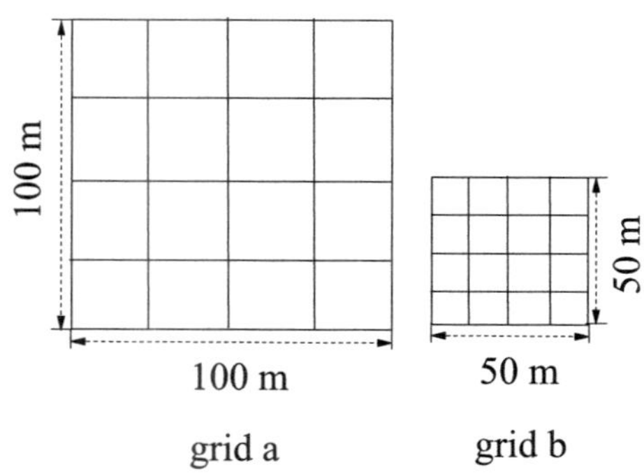

Fig. 5.1 The grounding grid's sketch map

mutual inductance of measuring equipment, the grounding impedance measured in this paper is very accurate and scientific. Two frequencies (80 and 500 hertz) are used to measure the grounding impedance of the grounding grid buried in uniform, two-layer and three-layer horizontally stratified earth. The sizes of the two types of grounding grids are 100 m multiplied by 100 m (grid a) and 50 m multiplied by 50 m (grid b) respectively, which can be seen in Fig. 5.1. They are all composed of 16-mesh grids, and the radius of the grid conductors is 0.5 cm, and buried at a depth of 0.5 m. In this simulation calculation, the relative dielectric constant of all the earth is 10. The measured and numerical simulated results were given in Tables 5.5 and 5.7 and 5.6.

Table 5.5 Comparison calculated $Z_g[\Omega]$ with publication paper measured result $Z_g[\Omega]$ (half uniform earth model case)

Earth model	$\rho = 100\,\Omega\text{m}$, $f = 80$ Hz	$\rho = 10\,\Omega\text{m}$, $f = 80$ Hz	$\rho = 100\,\Omega\text{m}$, $f = 500$ Hz	$\rho = 100\,\Omega\text{m}, f =$ 500 Hz	$\rho = 10\,\Omega\text{m}, f =$ 500 Hz
Grid type	grid a	grid a	grid a	grid b	grid b
Measurement	0.523	0.052	0.523	1.025	0.100
MM	$(0.532, -2.56 \times 10^{-5})$	$(0.053, -2.56 \times 10^{-7})$	$(0.532, -1.60 \times 10^{-5})$	$(1.016, -3.05 \times 10^{-5})$	$(0.102, -3.05 \times 10^{-7})$
BEM	$(0.537, -2.59 \times 10^{-5})$	$(0.054, -2.59 \times 10^{-7})$	$(0.539, -1.62 \times 10^{-5})$	$(1.029, -3.09 \times 10^{-5})$	$(0.103, -3.09 \times 10^{-7})$

Table 5.6 Comparison calculated $Z_g[\Omega]$ with publication paper measured result $Z_g[\Omega]$ (mutilayered earth model case, $f = 80$ Hz)

Earth model	$\frac{\rho_1}{\rho_2} = \frac{25\,\Omega m}{120\,\Omega m}, h = 5$ m	$\frac{\rho_1}{\rho_2} = \frac{120\,\Omega m}{25\,\Omega m}, h = 5$ m	$\frac{\rho_1}{\frac{\rho_2}{\rho_3}} = \frac{120\,\Omega m}{\frac{25\,\Omega m}{250\,\Omega m}}, \frac{h_1}{h_2} = \frac{5m}{9m}$	$\frac{\rho_1}{\frac{\rho_2}{\rho_3}} = \frac{25\,\Omega m}{\frac{120\,\Omega m}{10\,\Omega m}}, \frac{h_1}{h_2} = \frac{5m}{9m}$,	$\frac{\rho_1}{\frac{\rho_2}{\rho_3}} = \frac{25\,\Omega m}{\frac{120\,\Omega m}{10\,\Omega m}}, \frac{h_1}{h_2} = \frac{5m}{100m}$
Grid type	grid a	grid a	grid a	grid a	grid a
Measurement	0.390	0.276	0.658	0.130	0.300
MM	$(0.403, -1.83 \times 10^{-6})$	$(0.297, -1.26 \times 10^{-6})$	$(0.667, -4.63 \times 10^{-6})$	$(0.133, 3.41 \times 10^{-7})$	$(0.300, -1.17 \times 10^{-6})$
BEM	$(0.413, -1.88 \times 10^{-6})$	$(0.298, -1.26 \times 10^{-6})$	$(0.672, -4.66 \times 10^{-6})$	$(0.135, 3.51 \times 10^{-7})$	$(0.309, -1.21 \times 10^{-6})$

The fourth case is a simulative testing result from the paper [5]. First-layer resistivity in the two-layer earth model was 100.0 Ωm with 5 m thickness, the resistivity of second layer earth will be determined by the refractive factor K of the resistivity of the two-layer earth. The structure and size of the grounding grid for testing is $20 \times 20\,\text{m}^2$ grid of 2×2 meshes with radius of 0.75 cm, sunken at 0.5 m, various test grounding grids vary with the number of vertical grounding rods at different locations, which can be seen Fig. 5.2 The numerical results from published and numerical simulated results from this method were given in Table 5.7.

5.5.2 *The Computational Cost and Accuracy of the Approach*

First, by considering some reasonable simplification, the approach calculates the grounding system based on the spatial 1-d scalar Galerkin type of MM and BEM. Second, the main task of simulating grounding system is to calculate the mutual impedance coefficients as shown in Eqs. (5.8) and (5.16). Although multilayered earth model must be considered in real practice problem, the closed form of Green's function of a point source in the multilayered earth model can be taken to avoid verbose calculation about infinite integral or Maclaurin's series by applying the

Table 5.7 Comparison with publication paper results (two-layer earth model case)

Grid type	K	$Z_g[\Omega]$([5])	$Z_g[\Omega]$(MM)	$Z_g[\Omega]$(BEM)	Grid type	K	$Z_g[\Omega]$([5])	$Z_g[\Omega]$(MM)	$Z_g[\Omega]$(BEM)
grid a	0.9	7.78	8.013	8.045	grid d	0.9	7.99	8.128	8.161
	0.5	3.38	3.507	3.522		0.5	3.6	3.692	3.705
	0.0	1.77	1.780	1.800		0.0	2.01	2.037	2.042
	−0.5	0.81	0.804	0.805		-0.5	1.05	1.047	1.048
	−0.9	0.16	0.158	0.158		−0.9	0.25	0.255	0.256
grid b	0.9	7.8	8.019	8.051	grid e	0.9	8.28	8.355	8.408
	0.5	3.4	3.524	3.537		0.5	3.89	3.946	3.976
	0.0	1.8	1.822	1.825		0.0	2.3	2.322	2.337
	−0.5	0.84	0.830	0.831		−0.5	1.35	1.351	1.357
	−0.9	0.17	0.167	0.167		−0.9	0.42	0.426	0.426
grid c	0.9	7.96	8.119	8.156	grid f	0.9	8.44	8.454	8.517
	0.5	3.56	3.667	3.683		0.5	4.08	4.099	4.134
	0.0	1.97	1.990	1.996		0.0	2.55	2.572	2.591
	−0.5	0.99	0.985	0.987		−0.5	1.79	1.810	1.819
	−0.9	0.22	0.223	0.223		−0.9	1.39	1.413	1.418

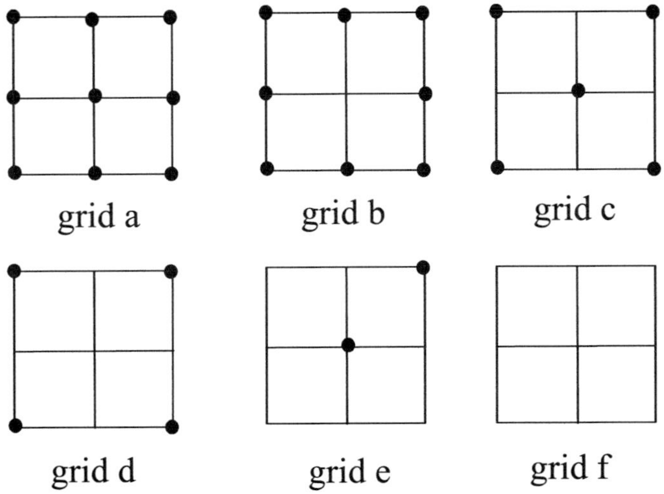

Fig. 5.2 The grounding grid's sketch map

QSCIM. Third, the analytical formula about mutual impedance is adopted to avoid numerical integral in the approach. Therefore the approach is very fast.

Verification of the approach has been shown in above section, by comparing our results with those from different method in other reference, especially with measurement data. We can see that the accuracy of our approach is good.

5.5.3 Numerical Results Analysis

A general grounding system can be seen in Fig. 5.3, which consists of 6 × 6 square grids and the burial depth of the grounding grid is 0.5 meters. The earth is modelled as horizontal four layers conductive media, whose resistivities and permittivities are $\rho_1 = 125\ \Omega\text{m}$ and $\rho_2 = 220\ \Omega\text{m}$, $\rho_3 = 100\ \Omega\text{m}$, $\rho_4 = 30\ \Omega\text{m}$, $\varepsilon_1 = 10\varepsilon_0, \varepsilon_2 = 10\varepsilon_0$, $\varepsilon_3 = 10\varepsilon_0$ and $\varepsilon_4 = 10\varepsilon_0$, respectively. The first, second and third layer thickness are $h_1 = 5$ m, $h_2 = 3$ m and $h_3 = 9$ m. All the conductor radius of grids are 5 mm. The external excited a.c. current $300A$ with power frequency 0 Hz, 500 Hz and 50000 Hz is injected from the corner of the grounding system.

The calculated numerical results can be seen in Table 5.8. Note: Z_g is the grounding impedance; GPR_g is the grounding potential rise (GPR) of the grounding grid; I_0 is the total leakage currents of the grounding system.

It can be seen that the total leakage current of the grounding system is close to the external injected current for all frequencies.

The 3-d distribution diagram of leakage current and node voltage on the grounding grid at different frequencies obtained from the two algorithms can be seen in Figs. 5.4 and 5.5.

Table 5.8 Numerical results

Frequencies	f=0 Hz case		f=500 Hz case		f=50000 Hz case	
Methods	BEM case	MM case	BEM case	MM case	BEM case	MM case
$Z_g(\Omega)$	$(0.826, -1.40e-14)$	$(0.823, -1.41e-14)$	$(0.826, -2.86e-5)$	$(0.823, -2.84e-5)$	$(0.286, -2.83e-3)$	$(0.283, -2.84e-3)$
GPR_g(V)	$(247.90, -4.21e-12)$	$(246.93, -4.23e-12)$	$(247.93, -8.57e-3)$	$(246.93, -8.53e-3)$	$(247.91, -0.848)$	$(246.93, -0.853)$
I_0(A)	$(300.0, 1.64e-14)$	$(300.0, 8.21e-19)$	$(300.0, -3.23e-9)$	$(300.0, -3.17e-9)$	$(300.0, -3.60e-7)$	$(300.0, 1.24e-6)$

Fig. 5.3 Grounding system

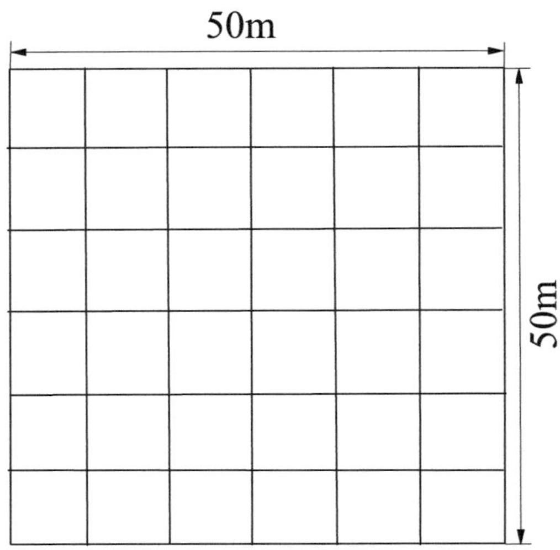

Comparing (a) and (b), (c) and (d), as well as (e) and (f) in Fig. 5.4, it can be seen that the results calculated by the two algorithms are basically consistent; At the same time, it can be seen that due to the constant amplitude of the incident current, only the frequency changes. Therefore, at different frequencies, the distribution of leakage current is basically consistent.

Basically consistent with the distribution pattern of leakage current, comparing (a) and (b), (c) and (d), as well as (e) and (f) in Fig. 5.5, it can be seen that the results calculated by the two algorithms are basically consistent; At the same time, it can be seen that due to the constant amplitude of the incident current, only the frequency changes. Therefore, at different frequencies, the distribution of voltage is basically consistent.

The 3-d distribution of voltage absolute value along the horizontal multilayered earth surface is given in Fig. 5.6 for $f = 0$ Hz, $f = 500$ Hz and $f = 50000$ Hz case, and the 3-d distribution of step voltage's absolute value along the horizontal multilayered earth surface is given in Fig. 5.7 for $f = 0$ Hz, $f = 500$ Hz and $f = 50000$ Hz case. and the 3-d distribution of E absolute value along the horizontal multilayered earth surface is given in Fig. 5.8 for $f = 0$ Hz, $f = 500$ Hz and $f = 50000$ Hz case.

It can be seen from Fig. 5.6a, b for $f = 0$ Hz case, and Fig. 5.6c, d for $f = 500$ Hz case, and Fig. 5.6e, f for $f = 50k$ Hz case, (1) the calculation results for voltage of the two algorithms are basically consistent, (2) because the amplitude of the incident current remains unchanged, only the frequency changes. Therefore, the calculation results for voltage for frequencies of zero and 50 Hz do not change much, while for frequencies of 50 kHz, the amplitude changes increase, and the distribution trend remains basically consistent.

It can be seen from Fig. 5.6a, b for $f = 0$ Hz case, and Fig. 5.6c, d for $f = 500$ Hz case, and Fig. 5.6e, f for $f = 50k$ Hz case, (1) the calculation results for voltage of

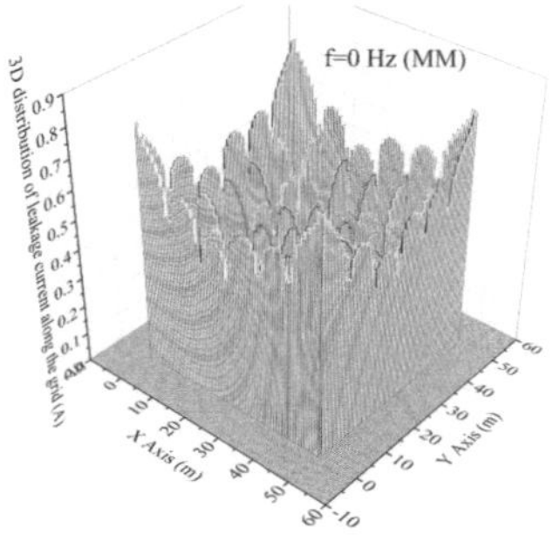

(a) f=0 Hz (MM case)

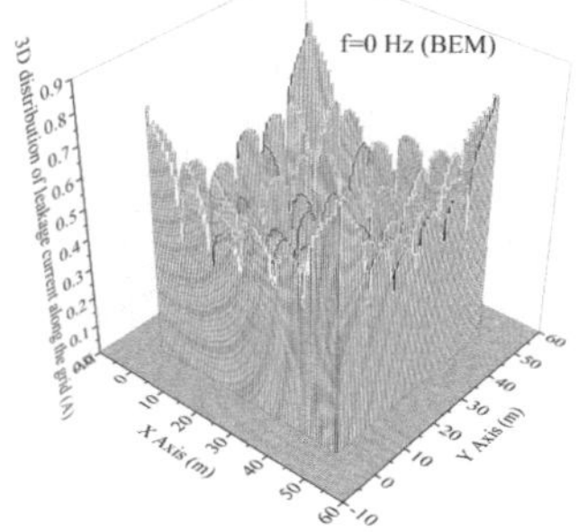

(b) f=0 Hz (BEM case)

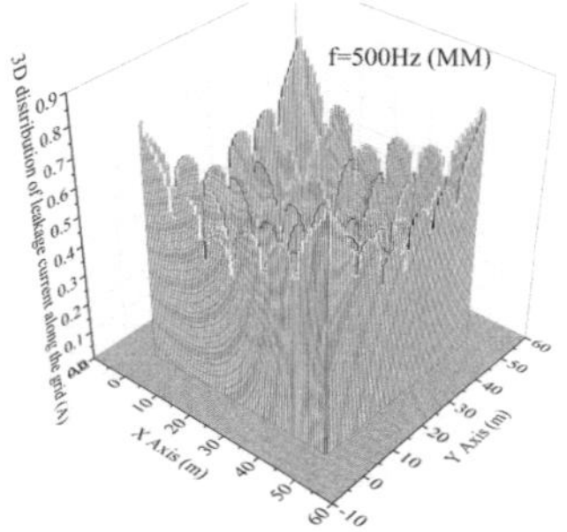

(c) f=500 Hz (MM case)

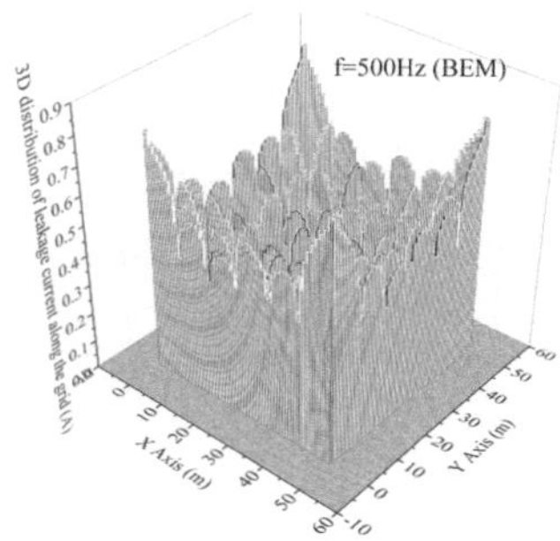

(d) f=500 Hz (BEM case)

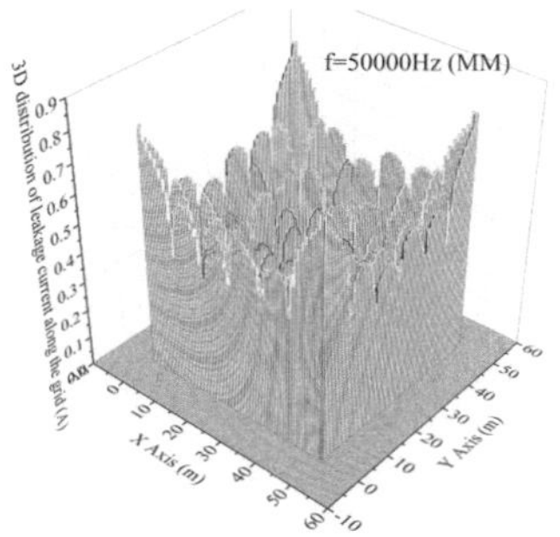

(e) f=50000 Hz (MM case)

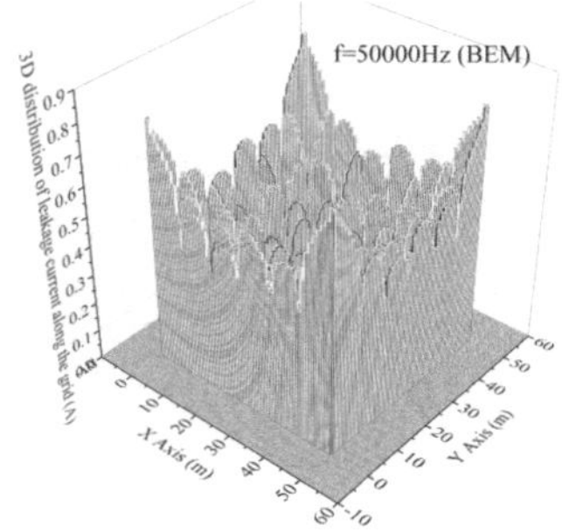

(f) f=50000 Hz (BEM case)

Fig. 5.4 The 3-d distribution of the leakage current's absolute value along the grounding grid about $f = 0$ Hz, $f = 500$ Hz and $f = 5000$ Hz case

the two algorithms are basically consistent, (2) because the amplitude of the incident current remains unchanged, only the frequency changes. Therefore, the calculation results for voltage for frequencies of zero and 50 Hz do not change much, while for

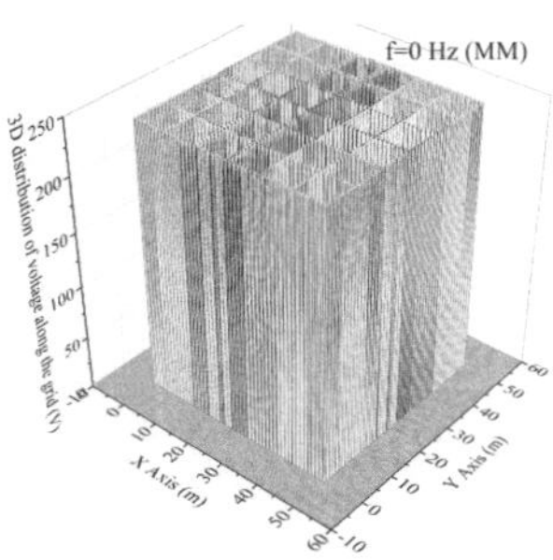

(a) f=0 Hz (MM case)

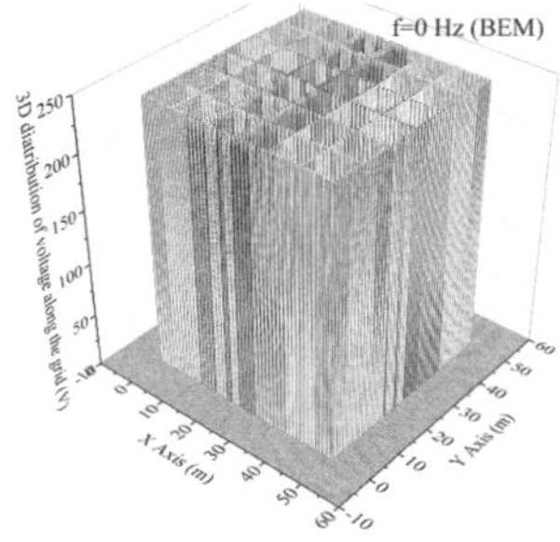

(b) f=0 Hz (BEM case)

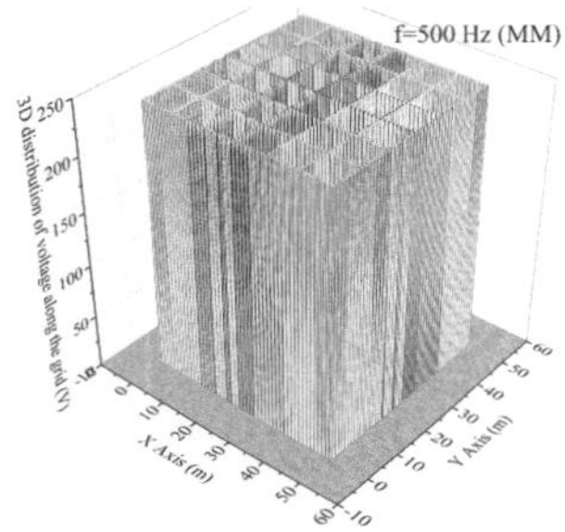

(c) f=500 Hz (MM case)

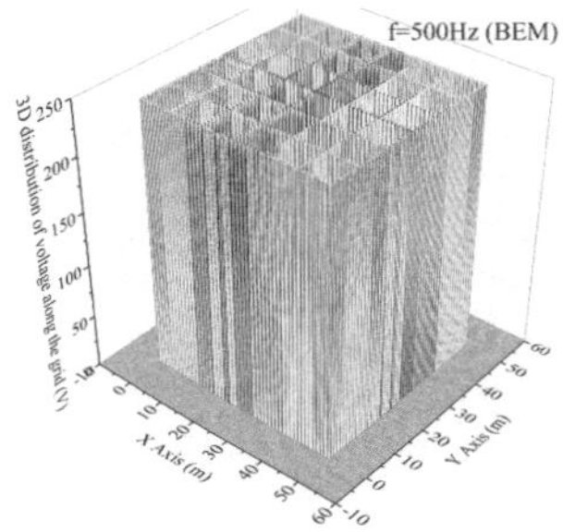

(d) f=500 Hz (BEM case)

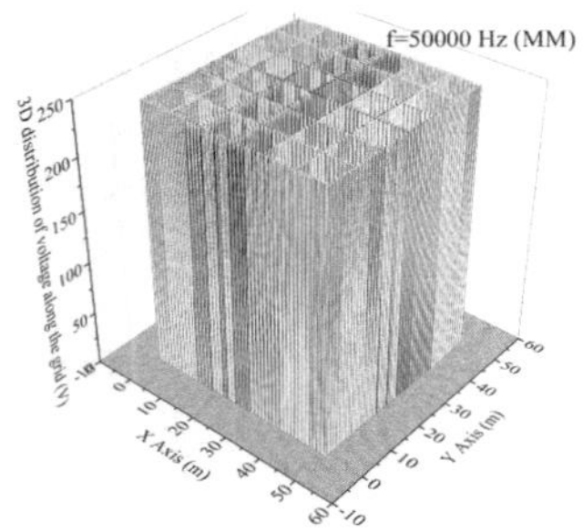

(e) f=50000 Hz (MM case)

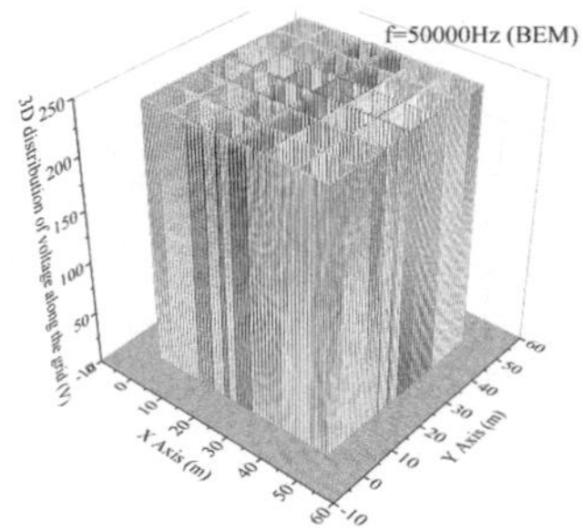

(f) f=50000 Hz (BEM case)

Fig. 5.5 The 3-d distribution of the voltage's absolute value along the grounding grid about $f = 0$ Hz, $f = 500$ Hz and $f = 5000$ Hz case

frequencies of 50 kHz, the amplitude changes increase, and the distribution trend remains basically consistent.

It can also be seen from Fig. 5.8a, b for $f = 0$ Hz case, and Fig. 5.8c, d for $f = 500$ Hz case, and Fig. 5.8e, f for $f = 50k$ Hz case, (1) the calculation results for electric field intensity E of the two algorithms are basically consistent, (2) because the amplitude of the incident current remains unchanged, only the frequency changes.

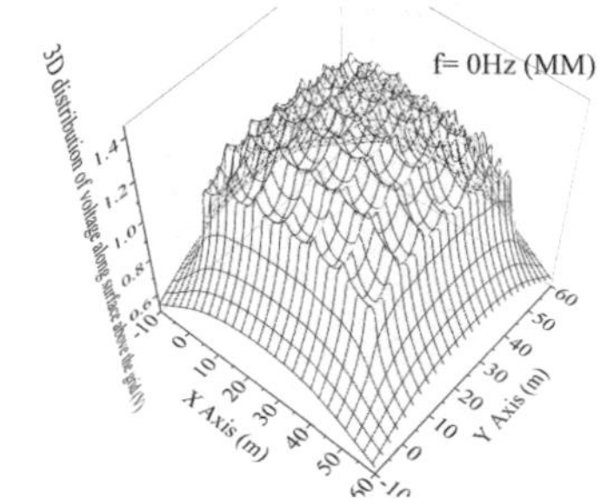

(a) f=0 Hz (MM case)

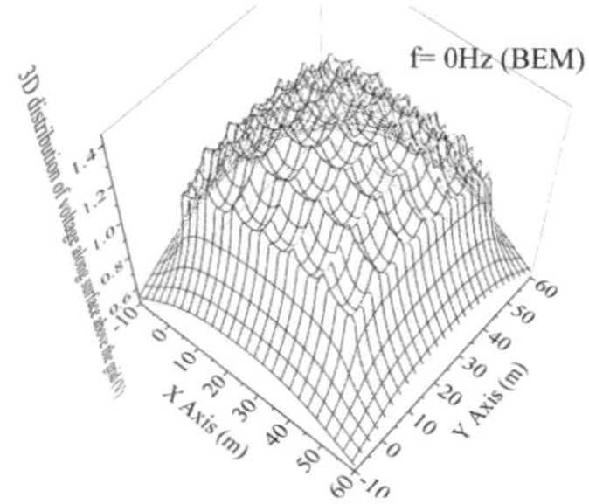

(b) f=0 Hz (BEM case)

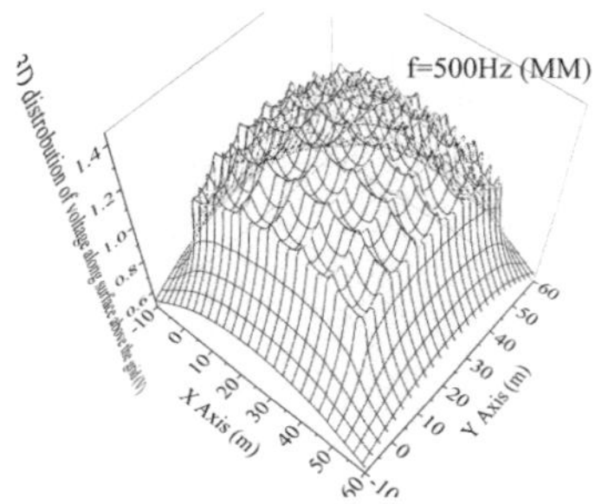

(c) f=500 Hz (MM case)

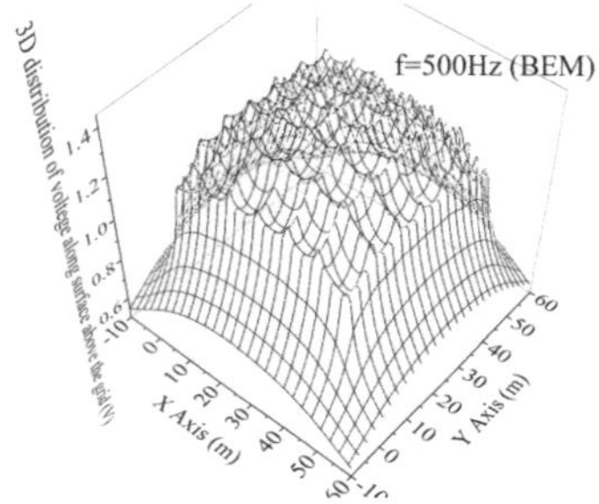

(d) f=500 Hz (BEM case)

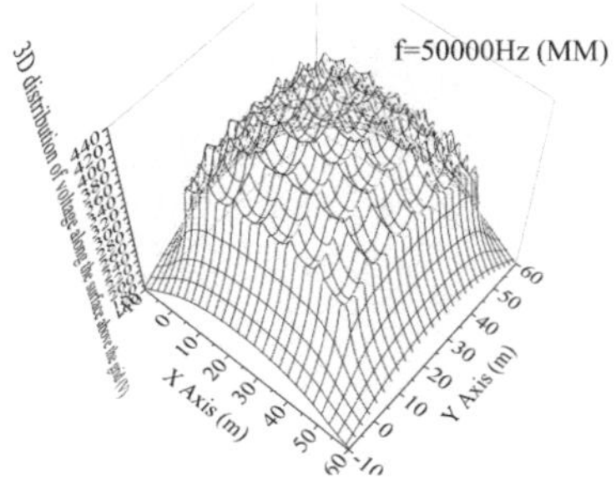

(e) f=50000 Hz (MM case)

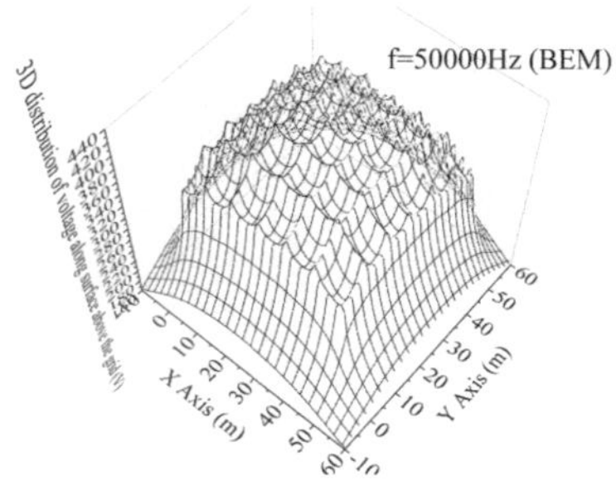

(f) f=50000 Hz (BEM case)

Fig. 5.6 The 3-d distribution of the voltage's absolute value along surface above the grounding grid about $f = 0$ Hz, $f = 500$ Hz and $f = 5000$ Hz case

Therefore, the calculation results for electric field intensity E for frequencies of zero and 50 Hz do not change much, while for frequencies of 50 kHz, the amplitude changes increase, and the distribution trend remains basically consistent.

Similarly, by observing Figs. 5.4, 5.5, 5.6, 5.7, and 5.8, it can be seen that since the two calculation methods are based on the equipotential mathematical model, the distribution of leakage current and voltage on the grounding grid, voltage distribution

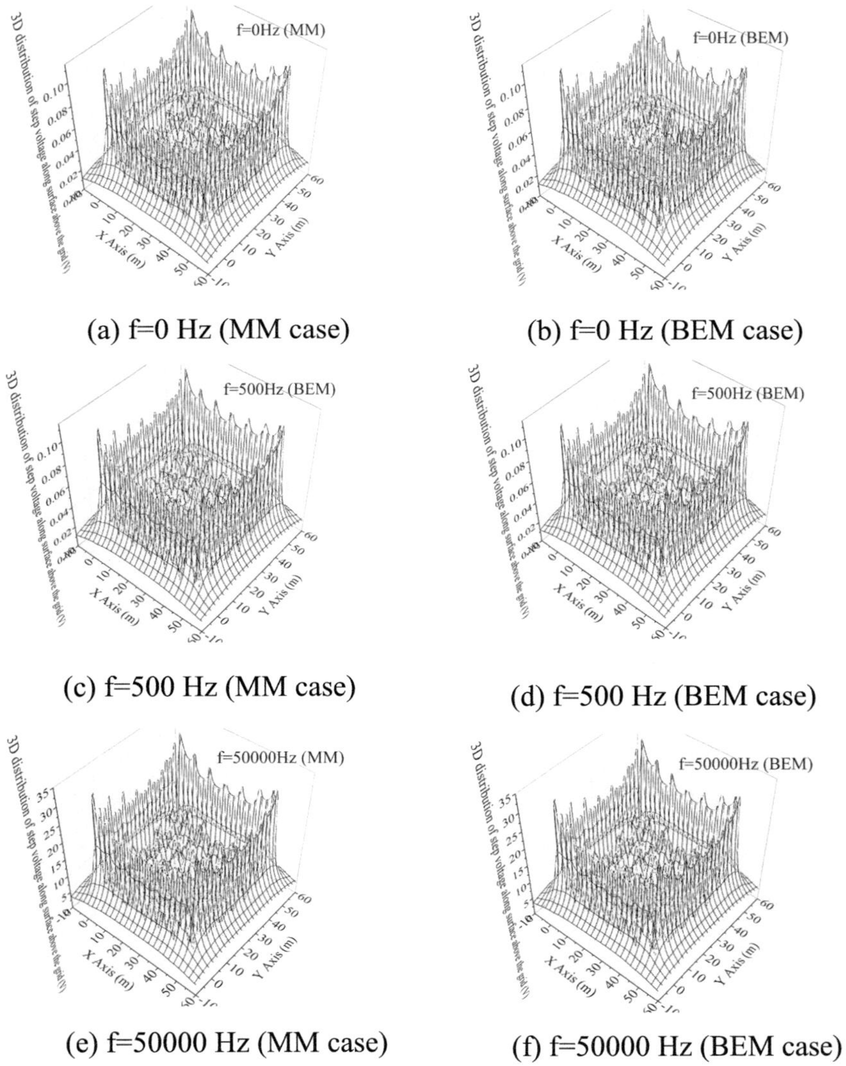

Fig. 5.7 The 3-d distribution of the step voltage's absolute value along surface above the grounding grid about $f = 0$ Hz, $f = 500$ Hz and $f = 5000$ Hz case

on the ground surface, step voltage distribution and electric field intensity distribution are basically evenly distributed around, and there is no phenomenon that whichever is higher is lower.

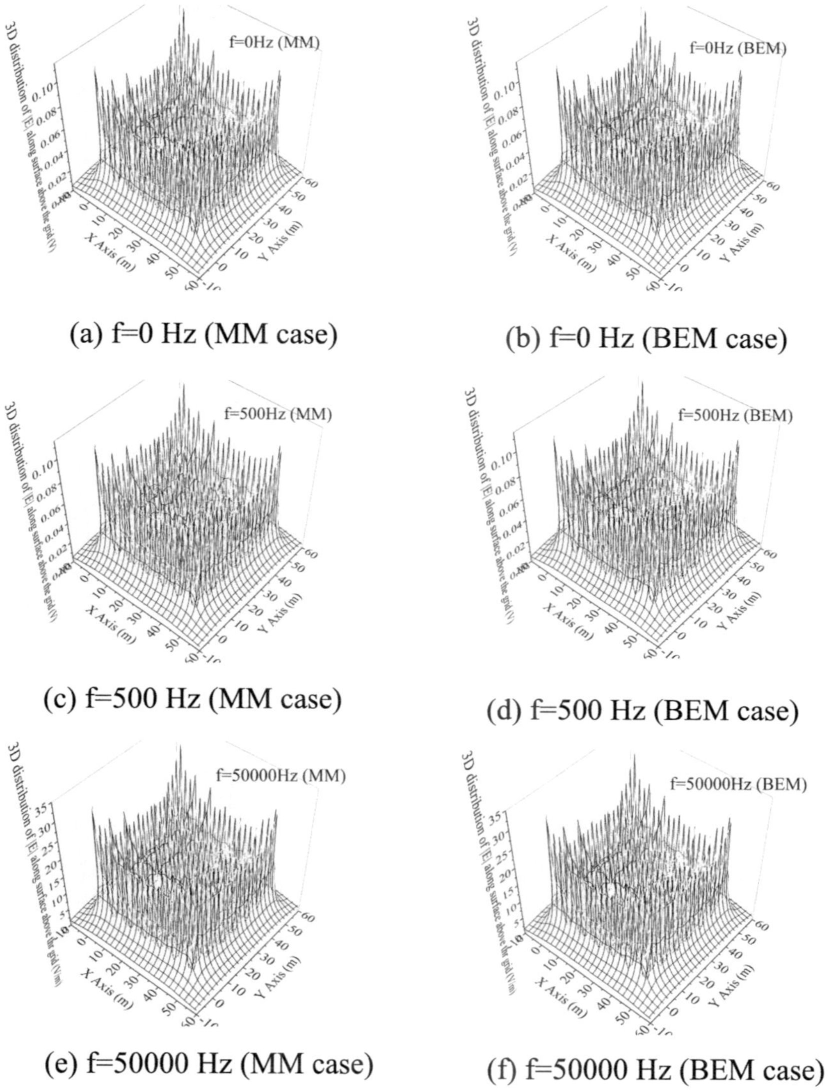

(a) f=0 Hz (MM case) (b) f=0 Hz (BEM case)

(c) f=500 Hz (MM case) (d) f=500 Hz (BEM case)

(e) f=50000 Hz (MM case) (f) f=50000 Hz (BEM case)

Fig. 5.8 The 3-d distribution of the E absolute value along surface above the grounding grid about $f = 0$ Hz, $f = 500$ Hz and $f = 5000$ Hz case

5.6 Summary

- Through the analysis of numerical calculation results, it is found that the calculation results of the spatial 1-d scalar MM and BEM are basically consistent, and the calculation accuracy of the two methods is also basically the same.

- Because they are based on the equipotential mathematical model, the 3-d distribution of the numerical results (leakage current and voltage on the grounding grid, potential, step voltage and electric field strength on the ground surface) calculated by the space 1-d scalar MM and BEM is uniform.

References

1. R.J. Heppe, Computation of potential at surface above an energized grid or other electrode allowing for non-uniform current distribution. IEEE Trans. Power Apparatus Syst. **98**(6), 1978–1989 (1979)
2. Y.L. Chow, J.J. Yang, K.D. Srivastavaa, Complex images of a ground electrode in layered soils. J. Appl. Phys. **71**(2), 569–574 (1992)
3. A.P. Sakis Meliopoulos, F Xia, G.J. Cokkinides, An advanced computer model for grounding system analysis. IEEE Trans. Power Delivery **8**(1), 13–23 (1993)
4. Y.L. Chow, M.M. Elsherbiny, M.M.A. Salama, Efficient computation of rodbed grounding resistance in a homogenous earth by galerkin's moment method. IEE Proc. Gener. Transm. Distrib. **142**(6), 653–660 (1995)
5. Y.L. Chow, M.M. Elsherbiny, M.M.A. Salama, Surface voltages and resistance of grounding system of grid and rods in two-layer earth by the rapid galerkin's moment method. IEEE Trans. Power Delivery **12**(1), 179–185 (1997)
6. L. Zhang, J. Yuan, Z.X. Li, The complex image method and its application in numerical simulation of substation grounding grids. Commun. Numer. Methods Eng. **15**, 835–839 (1999)
7. V. Slavko, K. Mate, Numerical analysis of earthing grids buried in horizontal stratified multilayered earth. Int. J. Numer. Methods Eng. **41**, 1297–1319 (1998)
8. I. Colominas, F. Navarrina, M. Casteleiro, A boundary element numerical approach for earthing grid computation. Comput. Methods Appl. Mech. Eng. **174**, 73–90 (1999)
9. I. Colominas, F. Navarrina, M. Casteleiro, A numerical formula for grounding system analysis in stratified soils. IEEE Trans. Power Delivery **17**(2), 587–595 (2002)
10. F. Navarrina, I. Colominas, M. Casteleiro, Why do computer methods for grounding analysis produce anomalous results? IEEE Trans. Power Delivery **18**(4), 1192–1202 (2003)
11. U. Adriano, 0. Bottauscio, M. Zucca, Boundary element approach for the analysis and design of grounding systems in presence of non-homogeneousness. IEE Proc. Gener. Distrib. Distrib. **150**(3), 360–366 (1985)
12. I. Colominas, F. Navarrina, M. Casteleiro, Analysis of transferred earth potentials in grounding systems: A bem numerical approach. IEEE Trans. Power Delivery **20**(1), 339–345 (2005)
13. I. Colominas, F. Navarrina, M. Casteleiro, Numerical simulation of transferred potentials in earthing grids considering layered soil models. IEEE Trans. Power Delivery **22**(3), 1514–1522 (2007)
14. P.D. Rancic, L.V. Stefanovic, D.R. Djordjevic, An improved linear grounding system analysis in two-layer earth. IEEE Trans. Magn. **32**, 5179–5187 (1996)
15. Z.X. Li, J.B. Fan, W.J. Chen, Numerical simulation of substation grounding grids buried in both horizontal and vertical multilayered earth model. Int. J. Numer. Methods Eng. **69**(11), 2359–2380 (2007)
16. Z.X. Li, G.F. Li, J.B. Fan, Numerical simulation of substation grounding system in low frequency domain based on the boundary element method with linear basis function. Commun. Numer. Methods Eng. **26**(12), 1819–1914 (2010)

17. C. Johnson, *Numerical Solution of Partial Differential Equations by the Finite Element Method* (Cambridge Univ. Press, Cambridge, U.K., 1987)
18. F.P. Dawalibi, J. Ma, R.D. Southey, Behaviour of grounding systems in multilayer soil: a parameteric analysis. IEEE Trans. Power Delivery **9**(1), 334–342 (1994)
19. J. Ma, F.P. Dawalibi, Influence of inductive coupling between leads on ground impedance measurements using the fall-of-potential method. IEEE Trans. Power Delivery **16**(4), 739–743 (2001)

Chapter 6
Numerical Methods for Grounding Problem at Steady-State II

Various frequency-domain numerical methods for steady-state grounding problems based on unequal potential electromagnetic field modeling.

6.1 Preface

The previous chapter mainly introduced numerical methods based on equipotential electromagnetic fields, namely spatial 1-d scalar Galerkin BEM and MM. This chapter will further introduce numerical methods based on unequal potential electromagnetic fields, namely the spatial 1-d vector Galerkin BEM and MM, as well as the spatial 1-d hybrid algorithm based on various basis functions.

6.2 Introduction

This chapter mainly introduces the grounding problem of the substation grounding grid under steady-state conditions of the power system. Therefore, the main introduction is the numerical calculation method of electromagnetic fields at a single frequency. Because even in the steady state of the power system, there are certain harmonics present, the electromagnetic field numerical calculation methods for simulating grounding problems in this chapter include zero frequency DC grounding problems, power frequency (50 or 60 Hz), and high-frequency AC grounding problems.

Among the numerous numerical calculation methods for electromagnetic fields obtained by simulating grounding problems in grounding grids buried in horizontal N_e-layer earth model, this chapter only introduces the spatial 1-d vector Galerkin

Z.-X. Li, *Numerical Simulation Algorithm of Electromagnetic Field for Grounding Problems in Power System Substation Grounding Grids*,
https://doi.org/10.1007/978-981-96-0904-8_6

BEM, vector Galerkin MM, and their variant vector Galerkin hybrid algorithms based on various basis functions.

It should be said that the numerical calculation methods based on electromagnetic field theory are very scientific and accurate. In these numerical methods, the literatures [1–10] are based on the BEM, the literatures [11–16] are based on the MM, vector Galerkin hybrid methods [17–42]. Among these hybrid algorithms, references [17–34] are based on PI basis functions, references [35–37] are based on PEEC basis functions, references [38–41] are based on T basis functions, and references [42] are based on TPEEC basis functions. It should be pointed out that the articles [10, 23–25, 29–31, 37, 39, 40], published by the author of this book are about the frequency domain electromagnetic field numerical algorithm for simulating the steady-state grounding problem of substation grounding grid in multi-layer horizontal layered soil. Among them, references [23–25, 29–31, 37, 39, 40] are hybrid algorithms based on various basis functions in the frequency domain electromagnetic field numerical algorithm, while reference [10] is the Galerkin BEM algorithm in the frequency domain electromagnetic field numerical algorithm.

6.3 The Spatial 1-d Vector Galerkin BEM and MM & Hybrid Methods Based on Various Basis Functions

The following chapters will provide a detailed and systematic introduction to vector Galerkin BEM, vector Galerkin MM, and vector Galerkin hybrid algorithm based on various basis functions.

The grounding system is assumed to be a network made up of set of interconnected cylindrical thin conductors placed in any position or orientation. The grounding grid's conductors are assumed to be completely buried in a conductive N_e-layer media (earth) with conductivity σ_i and permittivity $\varepsilon_i = \varepsilon_r \cdot \varepsilon_0$ (here $i = 1, \ldots, N_e$). The methodology proposed is based on the study of all the inductive, capacitive and conductive couplings between the different grounding system conductors. First, the electrode is divided in N_b pieces of segments that can be studied as elemental unit, where the discrete grounding system has N_p nodes. A higher segmented rate of the electrode can enhance the model's accuracy but increase its computational time. Therefore, it is necessary to achieve a compromise solution between the two determinants. From the thin wire approximation, currents are assumed to be distributed along the wire axis.

6.3.1 Mathematical Model of the Grounding System for the Spatial 1-d Vector Galerkin BEM

Completely buried in a conductive N_e-layer media (earth) with conductivity σ_e and permittivity $\varepsilon_e = \varepsilon_{r_e} \cdot \varepsilon_0$ (here $e = 1, \cdots\cdots, N_e$). The grounding grid is divided into N_b pieces of segments that can be studied as elemental units, where the discrete grounding system has N_p nodes.

According to the content of Chap. 2 on the spatial 1-d VGWRM, it can be known that

$$Z_{l_\theta} \int_{l_\theta} [\bar{W}^n_{l_\theta}(\bar{r}_\theta)][\bar{N}^n_{l_\theta}(\bar{r}_\theta)]dr_\theta[\bar{I}^n_{l_\theta}(\bar{r}_\theta)]$$

$$+\sum_{\vartheta=1}^{N_b} \int_{l_\theta}\int_{l_\vartheta} \hat{r}_\theta \cdot j\omega \bar{\bar{G}}^{Af}_{\xi\zeta}(\bar{r}_\theta, \bar{r}_\vartheta) \cdot \hat{r}^T_\vartheta [\bar{W}^n_{l_\theta}(\bar{r}_\theta)][\bar{N}^n_{l_\vartheta}(\bar{r}_\vartheta)]dr_\vartheta dr_\theta[\bar{I}^n_{l_\vartheta}(\bar{r}_\vartheta)]$$

$$+\sum_{\vartheta=1}^{N_b} \int_{l_\theta}\int_{l_\vartheta} G^{\varphi f}_{\xi\zeta}(\bar{r}_\theta, \bar{r}_\vartheta)\frac{\partial[\bar{W}^n_{l_\theta}(\bar{r}_\theta)]}{\partial r_\theta}\frac{d[\bar{N}^n_{l_\vartheta}(\bar{r}_\vartheta)]}{dr_\vartheta}dr_\vartheta dr_\theta[\bar{I}^n_{l_\vartheta}(\bar{r}_\vartheta)] = 0 \tag{6.1}$$

where $\xi = 0, 1, \ldots, N_e$; $\zeta = 0, 1, \ldots, N_e$. $\theta = 1, \ldots, N_b$. $\bar{r}_\theta$ and $\bar{r}_\vartheta$ is location vector along θ-th and ϑ-th conductors segment, respectively; r_θ and r_ϑ, respectively, parameter variable along the θ-th and ϑ-th thin conductor axis. Z_{l_θ} is the internal impedance of the θ-th thin conductor segment, $[\bar{N}^n_{l_\theta}(\bar{r}_\theta)]$, $[\bar{W}^n_{l_\theta}(\bar{r}_\theta)]$ and $[\bar{N}^n_{l_\vartheta}(\bar{r}_\vartheta)]$, $[\bar{W}^n_{l_\vartheta}(\bar{r}_\vartheta)]$ $(n = 1, \ldots, M)$ are, respectively, $M - 1$ order basic function and test function defined on θ-th and ϑ-th thin conductor segment with length l_θ and l_ϑ. $[\bar{I}^n_{l_\theta}(\bar{r}_\theta)]$ and $[\bar{I}^n_{l_\vartheta}(\bar{r}_\vartheta)]$ are, respectively, the branch current along θ-th and ϑ-th thin conductor segment axis. $\hat{r}_\theta$ and $\hat{r}_\vartheta$ are, respectively, the 1×3 dimensional unit director vector along the axis of the θ-th and ϑ-th cylindrical conductor segments. The symbol T represents matrix transposition. j is a imaginary unit, ω is angular frequency. $G^{\varphi f}_{\xi\zeta}(\bar{r}_\theta, \bar{r}_\vartheta)$ and $\bar{\bar{G}}^{Af}_{\xi\zeta}(\bar{r}_\theta, \bar{r}_\vartheta)$ are, respectively, scalar monopole Green's function and vector dipole dyadic Green's function within horizontal multilayered earth model, which had been fully discussed in front Chap. 3.

The unknown current $I^n_{l_\theta}(\bar{r}_\theta)$ along the conductor segment l_θ is expressed by the sum of a finite number of independent $M - 1$ order basis functions $p_m(\bar{r}_\theta)$, with unknown complex coefficients $I^\chi_\theta(\bar{r}_\theta)$:

$$[\bar{N}^n_{l_\theta}(\bar{r}_\theta)][\bar{I}^n_{l_\theta}(\bar{r}_\theta)] = \sum_{m=1}^{M} p_m(\bar{r}_\theta)I^\chi_\theta(\bar{r}_\theta) = [\bar{p}(\bar{r}_\theta)][\bar{I}^\chi_\theta(\bar{r}_\theta)] \tag{6.2}$$

where M is maximum value of independent $M - 1$ order basis functions; χ is nodal number belong to the θ-th conductor segment, which means $1 \le \chi \le N_p$. $[\bar{p}(\bar{r}_\theta)]$

is $1 \times M$ vector of linearly independent $M-1$ order basis functions, $[\bar{I}_{\theta}^{\chi}(\bar{r}_{\theta})]$ is a $M \times 1$ vector of unknown currents of θ-th conductor segment.

Since spatial 1-d vector Garlerkin's BEM will be used in the model, so $M-1$ order test function $[\bar{q}(\bar{r}_{\theta})]$ can obtained as

$$[\bar{q}(\bar{r}_{\theta})] = [\bar{p}(\bar{r}_{\theta})]^{T} \tag{6.3}$$

where $[\bar{q}(\bar{r}_{\theta})]$ is $M \times 1$ vector. The symbol T represents a matrix transpose operation.

Assembling the contributions from each element of the Eq. (6.1) is transferred into the following matrix equation:

$$\sum_{\vartheta=1}^{N_b}[\bar{\bar{Z}}]_{\theta\vartheta}[\bar{I}_{\vartheta}^{\chi}] = 0 \tag{6.4}$$

where $\theta = 1, 2, \ldots, N_b$; $\chi = 1, 2, \ldots, N_p$; $[\bar{\bar{Z}}]_{\theta\vartheta}$ the mutual impedance and induction matrix representing the inter-action of the ϑ-th source with the θ-th observation segment, respectively:

$$[\bar{\bar{Z}}]_{\theta\vartheta} = Z_{l_\theta}\int_{l_\theta}[\bar{q}(\bar{r}_{\theta})][\bar{p}(\bar{r}_{\theta})]dr_{\theta} + \sum_{\vartheta=1}^{N_b}\int_{l_\theta}\int_{l_\vartheta}\hat{r}_{\theta}\cdot j\omega\bar{\bar{G}}_{\xi\zeta}^{Af}(\bar{r}_{\theta},\bar{r}_{\vartheta})\cdot\hat{r}_{\vartheta}^{T}[\bar{q}(\bar{r}_{\theta})][\bar{p}(\bar{r}_{\vartheta})]dr_{\vartheta}dr_{\theta}$$

$$+\sum_{\vartheta=1}^{N_b}\int_{l_\theta}\int_{l_\vartheta}G_{\xi\zeta}^{\varphi f}(\bar{r}_{\theta},\bar{r}_{\vartheta})\frac{\partial[\bar{q}(\bar{r}_{\theta})]}{\partial r_{\theta}}\frac{d[\bar{p}(\bar{r}_{\vartheta})]}{dr_{\vartheta}}dr_{\vartheta}dr_{\theta} \tag{6.5}$$

A linear approximation over a boundary element is used in this work, i.e. So $M = 2$, and the 1 order basic functions are of the form:

$$p_1(\bar{r}_{\theta}) = \frac{l_{\theta} - r_{\theta}}{l_{\theta}} \qquad p_2(\bar{r}_{\theta}) = \frac{r_{\theta}}{l_{\theta}} \tag{6.6}$$

Resulting matrix system (6.5) has N_p number of equations with N_p number of unknowns, the matrix can be obtained as below

$$[\bar{\bar{\mathbf{Z}}}_p][\bar{\mathbf{I}}_p] = 0 \tag{6.7}$$

where $[\bar{\bar{\mathbf{Z}}}_p]$ is $N_p \times N_p$ matrix, who includes the mutual impedance and induction interaction betweens a parts of conductor segments currents. $[\bar{\mathbf{I}}_p]$ is $N_p \times 1$ vector of unknown currents.

The excitation is incorporated into the formulation through the boundary condition:

$$I_{p_1} = I_g \tag{6.8}$$

where I_g denotes the current generator and I_{p_1} current in the injection point.

At a junction consisting of two or more segments the continuity properties of the electric field must be satisfied, which is ensured by applying the Kirchhoff current law

$$\sum_{k=1}^{n} I_{p_k} = 0 \tag{6.9}$$

where n is the total number of conductor segments connecting the node.

And continuity equation

$$[\frac{\partial I_{p_1}}{\partial r_1}]_{at\ junction} = [\frac{\partial I_{p_2}}{\partial r_2}]_{at\ junction} = \cdots = [\frac{\partial I_{p_n}}{\partial r_n}]_{at\ junction} \tag{6.10}$$

The condition (6.10) ensures the discontinuities in charge per unit length to be ruled out in passing from one conductor to another across the junction.

At the conductor free ends, the total current vanishes,i.e.is forced to be zero.

Once Eq. (6.7) is solved, the unknown currents were known, the branch and leakage currents can be obtained easily. The voltage (zero potential reference point at infinity), the EFI (E) and the MFI (B) at any point can be calculated by [29–39].

The study of the performance of the grounding system for spatial 1-d Galerkin BEM has been reduced to the computation of $[\bar{\bar{\mathbf{Z}}}]$ matrix.

Computation of $[\bar{\bar{\mathbf{Z}}}]$ matrix

From [20–39], it can be seen that each segment owns internal impedance. Mutual inductances or impedances between segments' currents are also included in the model; To discuss the calculation technique of computation of $[\bar{\bar{\mathbf{Z}}}]$ matrix, the Eq. (6.5) has been used as reference.

1. The first term of the Eq. (6.5):

$$Z_{l_\theta} \int_{l_\theta} [\bar{q}(\bar{r}_\theta)][\bar{p}(\bar{r}_\theta)]dr_\theta = Z_{l_\theta} \int_{l_\theta} \begin{bmatrix} \frac{r_\theta^2 - 2l_\theta r_\theta + l_\theta^2}{l_\theta^2} & \frac{-r_\theta^2 + l_\theta r_\theta}{l_\theta} \\ \frac{-r_\theta^2 + l_\theta r_\theta}{l_\theta^2} & \frac{r_\theta^2}{l_\theta^2} \end{bmatrix} dr_\theta \tag{6.11}$$

   where Z_{l_θ} consists of internal impedance, the formula for internal impedance can be found in [20–41].

2. The core of second term of the Eq. (6.5): The core is formula of mutual induction interaction betweens a pars of currents along conductor segments.

$$M_{\theta,\vartheta} = \int_{l_\theta} \int_{l_\vartheta} \hat{r}_\theta \cdot j\omega \bar{\bar{G}}_{\xi\zeta}^{A^f}(\bar{r}_\theta, \bar{r}_\vartheta) \cdot \hat{r}_\vartheta^T [\bar{p}(\bar{r}_\theta)][\bar{q}(\bar{r}_\vartheta)]dr_\vartheta dr_\theta \tag{6.12}$$

$$= \int_{l_\theta}\int_{l_\vartheta} \hat{r}_\theta \cdot j\omega \bar{\bar{G}}^{Af}_{\xi\zeta}(\bar{r}_\theta, \bar{r}_\vartheta) \cdot \hat{r}^T_\vartheta \begin{bmatrix} \frac{r_\theta r_\vartheta - l_\vartheta r_\theta - l_\theta r_\vartheta + l_\theta l_\vartheta}{l_\theta l_\vartheta} & \frac{-r_\theta r_\vartheta + l_\theta r_\vartheta}{l_\theta r_\vartheta} \\ \frac{-r_\theta r_\vartheta + l_\vartheta r_\theta}{l_\theta l_\vartheta} & \frac{r_\theta r_\vartheta}{l_\theta l_\vartheta} \end{bmatrix} dr_\vartheta dr_\theta \quad (6.13)$$

For an infinite homogeneous conductivity medium, one has $\bar{\bar{G}}^{Af}_{11}(\bar{r}_\vartheta, \bar{r}_\theta) = \frac{\mu}{4\pi}\frac{1}{R_{\theta\vartheta}}\bar{\bar{I}}^0$, where $\bar{\bar{I}}^0$ is the diagonal unit matrix, $R_{\theta\vartheta} = |\bar{r}_\theta - \bar{r}_\vartheta|$.

$$M_{\theta,\vartheta} = \frac{j\omega\mu}{4\pi}\int_{l_\theta}\int_{l_\vartheta} \frac{\hat{r}_\theta \cdot \hat{r}^T_\vartheta}{R_{\theta\vartheta}} \begin{bmatrix} \frac{r_\theta r_\vartheta - l_\vartheta r_\theta - l_\theta r_\vartheta + l_\theta l_\vartheta}{l_\theta l_\vartheta} & \frac{-r_\theta r_\vartheta + l_\theta r_\vartheta}{l_\theta l_\vartheta} \\ \frac{-r_\theta r_\vartheta + l_\vartheta r_\theta}{l_\theta l_\vartheta} & \frac{r_\theta r_\vartheta}{l_\theta l_\vartheta} \end{bmatrix} dr_\vartheta dr_\theta \quad (6.14)$$

3. The core of third term of the Eq. (6.5): The core is formula of mutual impedance interaction betweens a pars of currents along conductor segments.

$$Z_{\theta,\vartheta} = \int_{l_\theta}\int_{l_\vartheta} G^{\varphi f}_{\xi\zeta}(\bar{r}_\theta, \bar{r}_\vartheta)\frac{d[\bar{q}(\bar{r}_\theta)]}{dr_\theta}\frac{d[\bar{p}(\bar{r}_\vartheta)]}{dr_\vartheta} dr_\vartheta dr_\theta \quad (6.15)$$

$$Z_{\theta,\vartheta} = \int_{l_\theta}\int_{l_\vartheta} G^{\varphi f}_{\xi\zeta}(\bar{r}_\theta, \bar{r}_\vartheta)\begin{bmatrix} \frac{1}{l_\theta l_\vartheta} & -\frac{1}{l_\theta l_\vartheta} \\ -\frac{1}{l_\theta l_\vartheta} & \frac{1}{l_\theta l_\vartheta} \end{bmatrix} dr_\vartheta dr_\theta \quad (6.16)$$

For an infinite homogeneous conductivity medium, one has $G^{\varphi f}_{11}(\bar{r}_\vartheta, \bar{r}_\theta) = \frac{1}{4\pi\bar{\sigma}_1}\frac{1}{R_{\theta\vartheta}}$, where $\bar{\sigma}_1 = \sigma_1 + j\omega\varepsilon_1$, so

$$Z_{\theta,\vartheta} = \frac{1}{4\pi\bar{\sigma}_1 l_\theta l_\vartheta}\int_{l_\theta}\int_{l_\vartheta}\frac{1}{R_{\theta\vartheta}}\begin{bmatrix} 1 & -1 \\ -1 & 1 \end{bmatrix} dr_\theta dr_\vartheta \quad (6.17)$$

Note the medium surrounding the point current source here was considered as homogeneous and infinite. However, in any practical case, the earth is represented via a multilayered earth model. The QSCIM can be used to dealt with the multilayered earth model, this will be discussed next.

6.3.2 Mathematical Model of the Grounding System for the Spatial 1-d Vector Galerkin MM

Completely buried in a conductive N_e-layer media (earth) with conductivity σ_e and permittivity $\varepsilon_e = \varepsilon_{r_e} \cdot \varepsilon_0$ (here $e = 1, \cdots\cdots, N_e$). The grounding grid is divided into N_b pieces of segments that can be studied as elemental units, where the discrete grounding system has N_p nodes.

According to the content of Chap. 2 on the spatial 1-d VGWRM, it can be known that

$$
\begin{aligned}
&Z_{l_\theta}\int_{l_\theta}[\bar{W}^n_{l_\theta}(\bar{r}_\theta)][\bar{N}^n_{l_\theta}(\bar{r}_\theta)]dr_\theta[\bar{I}^n_{l_\theta}(\bar{r}_\theta)]\\
&+\sum_{\vartheta=1}^{N_b}\int_{l_\theta}\int_{l_\vartheta}\hat{r}_\theta\cdot j\omega\bar{\bar{G}}^{Af}_{\xi\zeta}(\bar{r}_\theta,\bar{r}_\vartheta)\cdot\hat{r}^T_\vartheta[\bar{W}^n_{l_\theta}(\bar{r}_\theta)][\bar{N}^n_{l_\vartheta}(\bar{r}_\vartheta)]dr_\vartheta dr_\theta[\bar{I}^n_{l_\vartheta}(\bar{r}_\vartheta)]\\
&+\sum_{\vartheta=1}^{N_b}\int_{l_\theta}\int_{l_\vartheta}G^{\varphi f}_{\xi\zeta}(\bar{r}_\theta,\bar{r}_\vartheta)\frac{\partial[\bar{W}^n_{l_\theta}(\bar{r}_\theta)]}{\partial r_\theta}\frac{d[\bar{N}^n_{l_\vartheta}(\bar{r}_\vartheta)]}{dr_\vartheta}dr_\vartheta dr_\theta[\bar{I}^n_{l_\vartheta}(\bar{r}_\vartheta)]=0
\end{aligned}
\tag{6.18}
$$

where $\xi=0,1,\ldots,N_e$; $\zeta=0,1,\ldots,N_e$. $\theta=1,\ldots,N_b$. $\bar{r}_\theta$ and $\bar{r}_\vartheta$ is location vector along θ-th and ϑ-th conductors segment, respectively; r_θ and r_ϑ, respectively, parameter variable along the θ-th and ϑ-th thin conductor axis. Z_{l_θ} is the internal impedance of the θ-th thin conductor segment, $[\bar{N}^n_{l_\theta}(\bar{r}_\theta)]$, $[\bar{W}^n_{l_\theta}(\bar{r}_\theta)]$ and $[\bar{N}^n_{l_\vartheta}(\bar{r}_\vartheta)]$, $[\bar{W}^n_{l_\vartheta}(\bar{r}_\vartheta)]$ $(n=1,\ldots,M)$ are, respectively, $M-1$ order basic function and test function defined on θ-th and ϑ-th thin conductor segment with length l_θ and l_ϑ. $[\bar{I}^n_{l_\theta}(\bar{r}_\theta)]$ and $[\bar{I}^n_{l_\vartheta}(\bar{r}_\vartheta)]$ are, respectively, the branch current along θ-th and ϑ-th thin conductor segment axis. $\hat{r}_\theta$ and $\hat{r}_\vartheta$ are, respectively, the 1×3 dimensional unit director vector along the axis of the θ-th and ϑ-th cylindrical conductor segments. j is a imaginary unit, ω is angular frequency. $G^{\varphi f}_{\xi\zeta}(\bar{r}_\theta,\bar{r}_\vartheta)$ and $\bar{\bar{G}}^{Af}_{\xi\zeta}(\bar{r}_\theta,\bar{r}_\vartheta)$ are, respectively, scalar monopole Green's function and vector dipole dyadic Green's function within horizontal multilayered earth model, which had been fully discussed in front Chap. 3.

The unknown current $I^n_{l_\theta}(\bar{r}_\theta)$ along the electrode segment l_θ is expressed by the sum of a finite number of independent basis functions $p^\chi_m(\bar{r}_\theta)$, with unknown complex coefficients $I^\chi_\theta(\bar{r}_\theta)$:

$$
[\bar{I}^n_\theta(\bar{r}_\theta)][\bar{N}^n_{l_\theta}(\bar{r}_\theta)]=\sum_{m=1}^{M}p^{m-1}_m(\bar{r}_\theta)I^{m-1}_\theta=[\bar{p}(\bar{r}_\theta)][\bar{I}^\chi_\theta(\bar{r}_\theta)]
\tag{6.19}
$$

where M is maximum value of independent $M-1$ order basis functions; $\chi=0,1,\ldots,M-1$. $[\bar{p}(\bar{r}_\theta)]$ is $1\times M$ vector of linearly independent $M-1$ basis functions, $[\bar{I}^\chi_\theta(\bar{r}_\theta)]$ is a $M\times 1$ vector of unknown currents of θ-th conductor segment.

Spatial 1-d vector Garlerkin's MM will be used in the model, so $M-1$ order test function $[\bar{q}(\bar{r}_\theta)]$ can obtained as

$$
[\bar{q}(\bar{r}_\theta)]=[\bar{p}(\bar{r}_\theta)]^T
\tag{6.20}
$$

where $[\bar{q}(\bar{r}_\theta)]$ is $M\times 1$ vector.

Assembling the contributions from each element of the Eq. (6.18) is transferred into the following matrix equation:

$$\sum_{\vartheta=1}^{N_b}[\bar{\bar{Z}}]_{\theta\vartheta}[\bar{I}_{\vartheta}^{\chi}] = 0 \tag{6.21}$$

where $\theta = 1, 2, \ldots, N_b$; $\chi = 0, 1, \ldots, M - 1$; $[\bar{\bar{Z}}]_{\theta\vartheta}^{\varsigma\upsilon}$ the mutual impedance and induction impedance matrix representing the inter-action of the ϑ-th source with the θ-th observation segment, respectively:

$$[Z]_{\theta\vartheta} = Z_{l_\theta}\int_{l_\theta}[\bar{q}(\bar{r}_\theta)][\bar{p}(\bar{r}_\theta)]dr_\theta + \sum_{\vartheta=1}^{N_b}\int_{l_\theta}\int_{l_\vartheta}\hat{r}_\theta \cdot j\omega\bar{\bar{G}}_{\xi\zeta}^{Af}(\bar{r}_\theta, \bar{r}_\vartheta)\cdot\hat{r}_\vartheta^T[\bar{q}(\bar{r}_\theta)][\bar{p}(\bar{r}_\vartheta)]dr_\vartheta dr_\theta$$

$$+\sum_{\vartheta=1}^{N_b}\int_{l_\theta}\int_{l_\vartheta}G_{\xi\zeta}^{\varphi f}(\bar{r}_\theta, \bar{r}_\vartheta)\frac{\partial[\bar{q}(\bar{r}_\theta)]}{\partial r_\theta}\frac{d[\bar{p}(\bar{r}_\vartheta)]}{dr_\vartheta}dr_\vartheta dr_\theta \tag{6.22}$$

A linear approximation over an element is used in this work, i.e. So $M = 2$, and the 1 order basic functions are of the form:

$$p_1^0(\bar{r}_\theta) = 1 \qquad p_2^1(\bar{r}_\theta) = -\frac{r_\theta}{l_\theta} \tag{6.23}$$

Resulting matrix system (6.21) has $2N_b$ number of equations with $2N_b$ number of unknowns, the matrix can be obtained as below

$$\begin{bmatrix}\bar{\bar{\mathbf{Z}}}_b^{00} & \bar{\bar{\mathbf{Z}}}_b^{01}\\ \bar{\bar{\mathbf{Z}}}_b^{10} & \bar{\bar{\mathbf{Z}}}_b^{11}\end{bmatrix}\begin{bmatrix}\bar{\mathbf{I}}_b^{0}\\ \bar{\mathbf{I}}_b^{1}\end{bmatrix} = 0 \tag{6.24}$$

where the $[\bar{\mathbf{I}}_b^0]$ is $N_b \times 1$ vector of unknown branch currents and the $[\bar{\mathbf{I}}_b^1]$ is $N_b \times 1$ vector of unknown leakage currents; $[\bar{\bar{\mathbf{Z}}}_b^{00}]$ is $N_b \times N_b$ matrix, who presents the mutual induction interaction betweens a parts of conductor segments branch currents, $[\bar{\bar{\mathbf{Z}}}_b^{01}]$ is $N_b \times N_b$ matrix, who presents the mutual impedance and induction interaction betweens a parts of conductor segments branch and leakage currents, $[\bar{\bar{\mathbf{Z}}}_b^{10}]$ is $N_b \times N_b$ matrix, who presents the mutual impedance and induction interaction betweens a parts of conductor segments branch and leakage currents, $[\bar{\bar{\mathbf{Z}}}_b^{11}]$ is $N_b \times N_b$ matrix, who presents the mutual impedance and induction interaction betweens a parts of conductor segments leakage currents.

The excitation is incorporated into the formulation through the boundary condition:

$$I_{b_1}^0 = I_g \tag{6.25}$$

where I_g denotes the current generator and $I_{b_1}^0$ current in the injection point.

At a junction consisting of two or more segments the continuity properties of the electric field must be satisfied, which is ensured by applying the Kirchhoff current law

$$\sum_{k=1}^{n} I_{b_k}^0 = 0 \tag{6.26}$$

and continuity equation

$$[\frac{\partial I_{b_1}^0}{\partial r_1}]_{at\ junction} = [\frac{\partial I_{b_2}^0}{\partial r_2}]_{at\ junction} = \cdots = [\frac{\partial I_{b_n}^0}{\partial r_n}]_{at\ junction} \tag{6.27}$$

The condition (6.27) ensures the discontinuities in charge per unit length to be ruled out in passing from one conductor to another across the junction.

At the conductor free ends, the total current vanishes,i.e.is forced to be zero.

Once Eq. (6.24) is solved, the unknown currents were known, the branch and leakage currents can be obtained easily. The voltage, the EFI (E) and the MFI (B) at any point can be calculated by [29–39].

The study of the performance of the grounding system for spatial 1-d Galerkin MM has been reduced to the computation of $[\bar{\bar{\mathbf{Z}}}]$ matrix.

Computation of $[\bar{\bar{\mathbf{Z}}}]$ matrix

From [20–39], it can be seen that each segment owns internal impedance. Mutual inductances or impedances between segments' currents are also included in the model; To discuss the calculation technique of computation of $[\bar{\bar{\mathbf{Z}}}]$ matrix, the Eq. (6.22) has been used as reference.

1. The first term of the Eq. (6.22):

$$Z_{l_\theta} \int_{l_\theta} [\bar{q}(\bar{\tau}_\theta)][\bar{p}(\bar{\tau}_\theta)] dr_\theta = Z_{l_\theta} \int_{l_\theta} \begin{bmatrix} 1 & \frac{-r_\theta}{l_\theta} \\ \frac{-r_\theta}{l_\theta} & \frac{r_\theta^2}{l_\theta^2} \end{bmatrix} dr_\theta \tag{6.28}$$

   where Z_{l_θ} consists of internal impedance, the formula for internal impedance can be found in [20–39].

2. The core of second term of the Eq. (6.22): The core is formula of mutual induction interaction betweens a pars of currents along conductor segments.

$$M_{\theta,\vartheta} = \int_{l_\theta} \int_{l_\vartheta} \hat{r}_\theta \cdot j\omega \bar{\bar{G}}_{\xi\zeta}^{Af}(\bar{r}_\theta, \bar{r}_\vartheta) \cdot \hat{r}_\vartheta^T [\bar{q}(\bar{r}_\theta)][\bar{p}(\bar{r}_\vartheta)] dr_\vartheta dr_\theta \tag{6.29}$$

$$= \int_{l_\theta}\int_{l_\vartheta} \hat{r}_\theta \cdot j\omega \bar{\bar{G}}^{A^f}_{\xi\zeta}(\bar{r}_\theta, \bar{r}_\vartheta) \cdot \hat{r}_\vartheta^T \begin{bmatrix} 1 & \frac{-r_\vartheta}{l_\vartheta} \\ \frac{-r_\theta}{l_\theta} & \frac{r_\theta r_\vartheta}{l_\theta l_\vartheta} \end{bmatrix} dr_\vartheta dr_\theta \tag{6.30}$$

For an infinite homogeneous conductivity medium, one has $\bar{\bar{G}}^{A^f}_{11}(\bar{r}_\vartheta, \bar{r}_\theta) = \frac{\mu}{4\pi}\frac{1}{R_{ij}}\bar{\bar{I}}^0$, where $\bar{\bar{I}}^0$ is the diagonal unit matrix, $R_{\theta\vartheta} = |\bar{r}_\theta - \bar{r}_\vartheta|$.

$$M_{\theta,\vartheta} = \frac{j\omega\mu}{4\pi}\int_{l_\theta}\int_{l_\vartheta} \frac{\hat{r}_\theta \cdot \hat{r}_\vartheta^T}{R_{\theta\vartheta}} \begin{bmatrix} 1 & \frac{-r_\vartheta}{l_\vartheta} \\ \frac{-r_\theta}{l_\theta} & \frac{r_\theta r_\vartheta}{l_\theta l_\vartheta} \end{bmatrix} dr_\vartheta dr_\theta \tag{6.31}$$

3. The core of third term of the Eq. (6.22): The core is formula of mutual impedance interaction betweens a pars of currents along conductor segments.

$$Z_{\theta,\vartheta} = \int_{l_\theta}\int_{l_\vartheta} G^{\varphi^f}_{\xi\zeta}(\bar{r}_\theta, \bar{r}_\vartheta)\frac{\partial[\bar{q}(\bar{r}_\theta)]}{\partial r_\theta}\frac{\partial[\bar{q}(\bar{r}_\vartheta)]}{\partial r_\vartheta} dr_\vartheta dr_\theta \tag{6.32}$$

$$Z_{\theta,\vartheta} = \int_{l_\theta}\int_{l_\vartheta} G^{\varphi^f}_{\xi\zeta}(\bar{r}_\theta, \bar{r}_\vartheta)\begin{bmatrix} 0 & 0 \\ 0 & \frac{1}{l_\theta l_\vartheta} \end{bmatrix} dr_\vartheta dr_\theta \tag{6.33}$$

For an infinite homogeneous conductivity medium, one has $G^{\varphi^f}_{11}(\bar{r}_\vartheta, \bar{r}_\theta) = \frac{1}{4\pi\bar{\sigma}_1}\frac{1}{R_{\theta\vartheta}}$, where $\bar{\sigma}_1 = \sigma_1 + j\omega\varepsilon_1$, so

$$Z_{\theta,\vartheta} = \frac{1}{4\pi\bar{\sigma}_1 l_\theta l_\vartheta}\int_{l_\theta}\int_{l_\vartheta} \frac{1}{R_{\theta\vartheta}}\begin{bmatrix} 0 & 0 \\ 0 & 1 \end{bmatrix} dr_\theta dr_\vartheta \tag{6.34}$$

Note the medium surrounding the point current source here was considered as homogeneous and infinite. However, in any practical case, the earth is represented via a multilayered earth model. The QSCIM can be used to dealt with the multilayered earth model, this will be discussed next.

6.3.3 Mathematical Model of the Grounding System for the Spatial 1-D Vector Hybrid Method Based on PI Basic Function

According to Chap. 2, the Eq. (2.26) can be written as below

$$Z_{l_\theta} I_b(\bar{r}_\theta) + \hat{r}_\theta \cdot j\omega\bar{A}(\bar{r}_\theta) + \frac{\partial\varphi(\bar{r}_\theta)}{\partial r_\theta} = 0 \tag{6.35}$$

where $\theta = 1, \ldots, N_b$. $I_b(\bar{r}_\theta)$ is the unknown current along θ-th thin conductor segment with length l_θ. $\bar{A}(\bar{r}_\theta)$ is directional vector magnetic potential located on the surface of the θ-th thin wire conductor; $\varphi(\bar{r}_\theta)$ is voltage on the surface of the θ-th thin wire conductor (relative to infinity). $\bar{r}_\theta$ is location vector along θ-th thin conductors segment; $\hat{r}_\theta$ are direction unit vectors along the θ-th thin conductor segment, r_θ is parameter variable along the θ-th thin conductor segment on surface of the thin conductor respectively, $\hat{r}$ is direction unit vector along the grid thin conductor, $\bar{r}$ is location vector along the grid thin conductor, $\bar{I}(\bar{r})$ is vector of unknown current along the grid thin conductor. j is imaginary unit, ω is angular frequency.

Generally (6.35) can be solved only using a numerical Galerkin MM approach. For the segment θ, the corresponding equation can be obtained as Eq. (6.36).

$$\int_0^{l_\theta} Z_{l_\theta} I_b(\bar{r}_\theta) dr_\theta + \int_0^{l_\theta} \hat{r}_\theta \cdot j\omega \bar{A}(\bar{r}_\theta) dr_\theta + \int_0^{l_\theta} \frac{\partial \varphi(\bar{r}_\theta)}{\partial r_\theta} dr_\theta = 0 \tag{6.36}$$

The Eq. (6.36) can be further written as Eq. (6.37)

$$l_\theta Z_{l_\theta} I_b(\bar{r}_\theta) + \sum_{\vartheta=1}^{N_b} \int_0^{l_\theta} \int_0^{l_\vartheta} \hat{r}_\theta \cdot j\omega \bar{\bar{G}}_{\xi\zeta}^{A^f}(\bar{r}_\theta, \bar{r}_\vartheta) \cdot \hat{r}_\vartheta^T dr_\vartheta dr_\theta I_b(\bar{r}_\vartheta) + \varphi(\bar{r}_\theta)|_0^{l_\theta} = 0 \tag{6.37}$$

where $\theta = 1, 2, \ldots, N_b$, $\xi = 0, 1, \ldots, N_e$; $\zeta = 0, 1, \ldots, N_e$. $\bar{\bar{G}}_{\xi\zeta}^{A^f}(\bar{r}_\theta, \bar{r}_\vartheta)$ is the dyadic Green's function of a vector dipole buried in the ζ-th layer of the horizontally layered N_e-layer soil model, with the field point located in the ξ-th layer of the soil.

The Eq. (6.37) can be written as matrix form as Eq. (6.38)

$$[\bar{\bar{\mathbf{Z}}}_b + j\omega \bar{\bar{\mathbf{M}}}_b][\bar{\mathbf{I}}_b] + [\bar{\mathbf{U}}_b] = 0 \tag{6.38}$$

Where in Eq. (6.38), $[\bar{\bar{\mathbf{Z}}}_b]$ was $N_b \times N_b$ diagonal matrix of internal impedance; $[\bar{\bar{\mathbf{M}}}_b]$ was $N_b \times N_b$ mutual induction matrix; $[\bar{\mathbf{I}}_b]$ was $N_b \times 1$ vector of branch currents; $[\bar{\mathbf{U}}_b]$ was $N_b \times 1$ vector of branch voltage of the discrete conductor segments, which can be replaced with $[\bar{\mathbf{U}}_b] = [\bar{\bar{\mathbf{A}}}]^T [\bar{\mathbf{V}}_p]$ [23, 24], so

$$[\bar{\bar{\mathbf{Z}}}_b + j\omega \bar{\bar{\mathbf{M}}}_b][\bar{\mathbf{I}}_b] + [\bar{\bar{\mathbf{A}}}]^T [\bar{\mathbf{V}}_p] = 0 \tag{6.39}$$

where $[\bar{\bar{\mathbf{A}}}]$ was rectangular incidence matrix of order $N_b \times N_p$, which are used to relate to branches and nodes, whose elements can refer to reference [20–25]. The symbol T represents matrix transpose.

- The mutual influence between branch voltage and branch leakage current case

For the leakage current, to each θ-th conductor segment, it can be obtained Eq. (6.40) according spatial 1-d scalar Garlerkin's MM as below.

$$\frac{1}{l_\theta}\int_0^{l_\theta}\varphi_\theta(\bar{r}_\theta) = \sum_{\vartheta=1}^{N_b}\frac{1}{l_\theta}\int_0^{l_\theta}\int_0^{l_\vartheta} G_{\xi\zeta}^{\varphi f}(\bar{r}_\theta,\bar{r}_\vartheta)\frac{I_s(\bar{r}_\vartheta)}{l_\vartheta}dr_\vartheta dr_\theta \tag{6.40}$$

where in Eq. (6.40), $\theta = 1, 2, \ldots, N_b$.
The Eq. (6.40) can be written as matrix form as Eq. (6.41)

$$[\bar{\mathbf{U}}_a] = [\bar{\bar{\mathbf{Z}}}_{s_b}][\bar{\mathbf{I}}_{s_b}] \tag{6.41}$$

where in Eq. (6.41), $[\bar{\bar{\mathbf{Z}}}_{s_b}]$ was $N_b \times N_b$ mutual impedance matrix; $[\bar{\mathbf{I}}_{s_b}]$ was $N_b \times 1$ vector of leakage currents along discrete conductor segments; $[\bar{\mathbf{U}}_a]$ is $N_b \times 1$ vector of average voltage of discrete conductor segments, which can be replaced with $[\bar{\mathbf{U}}_a] = [\bar{\bar{\mathbf{K}}}][\bar{\mathbf{V}}_p]$, so

$$[\bar{\bar{\mathbf{Z}}}_{s_b}][\bar{\mathbf{I}}_{s_b}] = [\bar{\bar{\mathbf{K}}}][\bar{\mathbf{V}}_p] \tag{6.42}$$

here $[\bar{\bar{\mathbf{K}}}]$ was rectangular matrix of order $N_b \times N_p$, whose elements are $K_{\theta,\vartheta} = 0.5$ if θ-th branch is connected to ϑ-th node, otherwise, $K_{\theta,\vartheta} = 0$.

- The interaction between node voltage and leakage current of the branch connected to the node

The potential of the node numbered ς on the θ-th thin wire conductor segment can be calculated by the leakage current on all thin wire conductor segments connected to that node.

$$\varphi_\varsigma(\bar{r}_\varsigma) = \frac{1}{N_\vartheta^\nu}\sum_{\vartheta=1}^{N_\theta^\nu}\int_0^{l_\vartheta^\nu} G_{\xi\zeta}^{\varphi f}(\bar{r}_\varsigma,\bar{r}_\nu)\frac{I_\vartheta^\nu(\bar{r}_\nu)}{l_\vartheta^\nu}dr_\nu \tag{6.43}$$

according spatial 1-d scalar Garlerkin's MM, it can be obtained Eq. (6.44) as below.

$$\frac{1}{N_\theta^\varsigma}\sum_{\theta=1}^{N_\theta^\varsigma}\frac{1}{l_\theta^\varsigma}\int_0^{l_\theta^\varsigma}\varphi_\varsigma(\bar{r}_\varsigma)dr_\varsigma = \frac{1}{N_\theta^\varsigma}\sum_{\theta=1}^{N_\theta^\varsigma}\frac{1}{l_\theta^\varsigma}\int_0^{l_\theta^\varsigma}\frac{1}{N_\theta^\nu}\sum_{\theta=1}^{N_\theta^\nu}\int_0^{l_\theta^\nu} G_{\xi\zeta}^{\varphi f}(\bar{r}_\varsigma,\bar{r}_\nu)\frac{I_\theta^\nu(\bar{r}_\nu)}{l_\theta^\nu}dr_\nu dr_\varsigma \tag{6.44}$$

Equation (6.44) can be written in matrix form as below

$$[\bar{\bar{\mathbf{Z}}}_{s_p}][\bar{\mathbf{I}}_{s_p}] = [\bar{\mathbf{V}}_p] \tag{6.45}$$

where in Eq. (6.45), $[\bar{\bar{\mathbf{Z}}}_{s_p}]$ was $N_p \times N_p$ mutual impedance matrix; $[\bar{\mathbf{I}}_{s_p}]$ was $N_p \times 1$ vector of leakage currents along discrete conductor segments.

Mathematical model based on the mutual influence between branch voltage and branch leakage current case

Refer to Fig. 6.1a, Eq. (6.46) can be obtained according to kirchhoff's current law (KCL).

$$[\bar{\mathbf{I}}_s^e] = [\bar{\bar{\mathbf{A}}}][\bar{\mathbf{I}}_b] + [\bar{\mathbf{F}}] \tag{6.46}$$

where in Eq. (6.46), $[\bar{\mathbf{I}}_s^e]$ was leakage current arranged at the two ends of a discrete conductor according to $[\bar{\mathbf{I}}_s^e] = [\bar{\bar{\mathbf{K}}}]^T[\bar{\mathbf{I}}_{s_b}]$. $[\bar{\mathbf{F}}]$ is a $N_p \times 1$ vector of external current sources;

Combined Eq. (6.39) with Eq. (6.46), Eq. (6.47) can be obtained as below

$$[\bar{\bar{\mathbf{K}}}]^T[\bar{\bar{\mathbf{Z}}}_{s_b}]^{-1}[\bar{\bar{\mathbf{K}}}][\bar{\mathbf{V}}_p] = [\bar{\bar{\mathbf{A}}}]^T[\bar{\mathbf{I}}_b] + [\bar{\mathbf{F}}] \tag{6.47}$$

Equations (6.39) and (6.47) can be solved consecutively as follows

$$[\bar{\bar{\mathbf{Z}}}_b + j\omega\bar{\bar{\mathbf{M}}}_b][\bar{\mathbf{I}}_b] + [\bar{\bar{\mathbf{A}}}]^T[\bar{\mathbf{V}}_p] = 0 \tag{6.48}$$

$$-[\bar{\bar{\mathbf{A}}}][\bar{\mathbf{I}}_b] + [\bar{\bar{\mathbf{K}}}]^T[\bar{\bar{\mathbf{Z}}}_{s_b}]^{-1}[\bar{\bar{\mathbf{K}}}][\bar{\mathbf{V}}_p] = [\bar{\mathbf{F}}] \tag{6.49}$$

- Further Simplify Method a
  Cancel $[\bar{\mathbf{I}}_b]$ items, combined Eq. (6.48) with Eq. (6.49), equation about hybrid method based on PI basic function about voltage unknown can obtained as below

$$\left[[\bar{\bar{\mathbf{K}}}]^T[\bar{\bar{\mathbf{Z}}}_{s_b}]^{-1}[\bar{\bar{\mathbf{K}}}] + [\bar{\bar{\mathbf{A}}}]\left[[\bar{\bar{\mathbf{Z}}}_b] + j\omega[\bar{\bar{\mathbf{M}}}_b]\right]^{-1}[\bar{\bar{\mathbf{A}}}]^T\right][\bar{\mathbf{V}}_p] = [\bar{\mathbf{F}}] \tag{6.50}$$

  For the convenience of explaining various numerical algorithms in the future, this formula is named PI 1 U case.
- Further Simplify Method b
  Cancel $[\bar{\mathbf{V}}_p]$ items, combined Eq. (6.48) with Eq. (6.49), equation about hybrid method based on PI basic function about current unknown can obtained as below

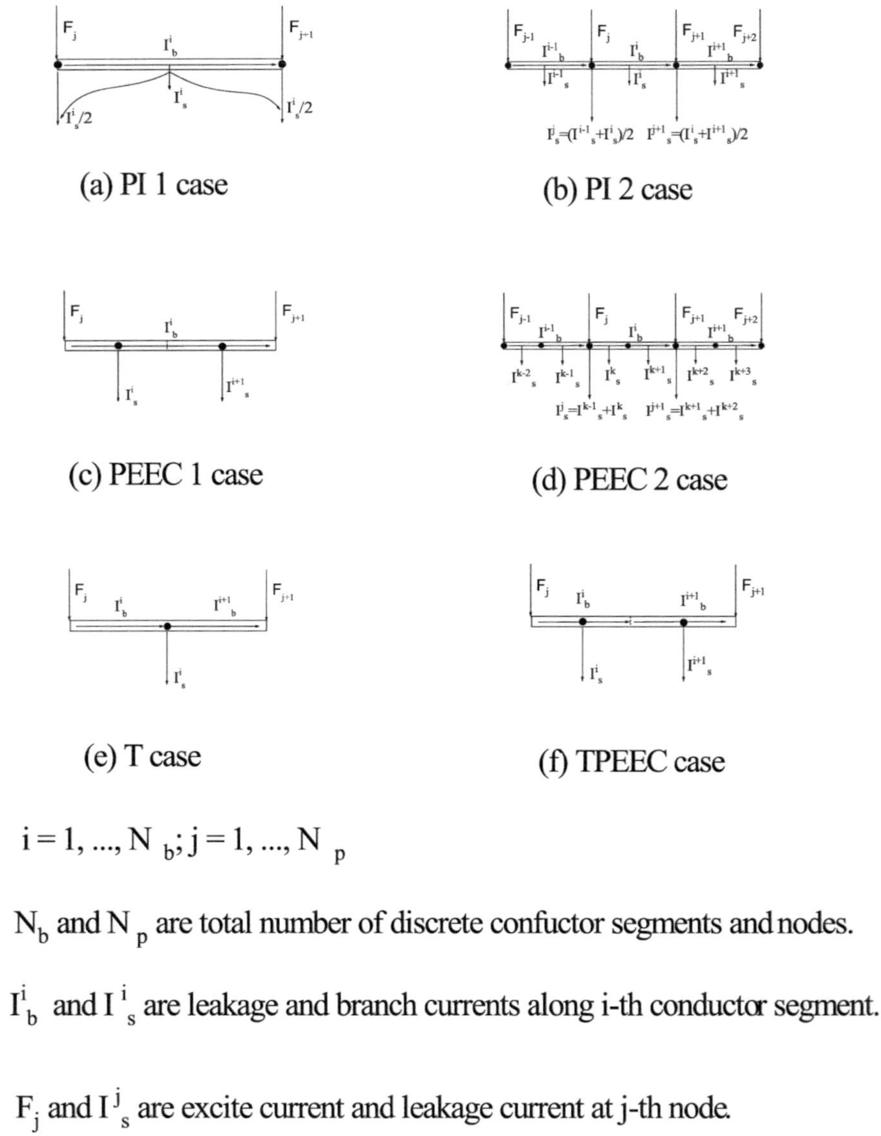

Fig. 6.1 Different basic functions

$$\left[[\bar{\bar{\mathbf{Z}}}_b + j\omega\bar{\bar{\mathbf{M}}}_b] + [\bar{\bar{\mathbf{A}}}]^T\left[[\bar{\bar{\mathbf{K}}}]^T[\bar{\bar{\mathbf{Z}}}_{s_b}]^{-1}[\bar{\bar{\mathbf{K}}}]\right]^{-1}[\bar{\bar{\mathbf{A}}}]\right][\bar{\mathbf{I}}_b] = -[\bar{\bar{\mathbf{A}}}]^T\left[[\bar{\bar{\mathbf{K}}}]^T[\bar{\bar{\mathbf{Z}}}_{s_b}]^{-1}[\bar{\bar{\mathbf{K}}}]\right]^{-1}[\bar{\mathbf{F}}] \tag{6.51}$$

Similarly, for the convenience of explaining various numerical algorithms in the future, this formula is named PI 1 I case.

It must be pointed out that due to the limitations of computer calculation accuracy, the double inversion of the mutual impedance matrix $[\bar{\bar{\mathbf{Z}}}_{s_b}]$ will result in unpredictable calculation errors. Therefore, Eq. (6.51) cannot be used for calculation.

The vector of nodal voltage $[\bar{\mathbf{V}}_p]$ may be calculated through solving the Eq. (6.50). The average voltage $[\bar{\mathbf{U}}_a]$ and leakage currents $[\bar{\mathbf{I}}_s]$, the branch voltage $[\bar{\mathbf{U}}_b]$, branch current $[\bar{\mathbf{I}}_b]$ can also be calculated [20–25].

Mathematical model based on the interaction between node voltage and leakage current of the branch connected to the node

Refer to Fig. 6.1b, Eq. (6.52) can be obtained according to KCL.

$$[\bar{\mathbf{I}}_s^e] = [\bar{\bar{\mathbf{A}}}][\bar{\mathbf{I}}_b] + [\bar{\mathbf{F}}] \tag{6.52}$$

where in Eq. (6.52), $[\bar{\mathbf{I}}_s^e]$ was leakage current arranged at the two ends of a discrete conductor according to $[\bar{\mathbf{I}}_s^e] = [\bar{\mathbf{I}}_{s_p}]$. $[\bar{\mathbf{F}}]$ is a $N_p \times 1$ vector of external current sources;

Combined Eq. (6.45) with Eqs. (6.52), (6.53) can be obtained as below

$$[\bar{\bar{\mathbf{Z}}}_{s_p}]^{-1}[\bar{\mathbf{V}}_p] = [\bar{\bar{\mathbf{A}}}]^T[\bar{\mathbf{I}}_b] + [\bar{\mathbf{F}}] \tag{6.53}$$

Equations (6.39) and (6.53) can be solved consecutively as follows

$$[\bar{\bar{\mathbf{Z}}}_b + j\omega\bar{\bar{\mathbf{M}}}_b][\bar{\mathbf{I}}_b] + [\bar{\bar{\mathbf{A}}}]^T[\bar{\mathbf{V}}_p] = 0 \tag{6.54}$$

$$-[\bar{\bar{\mathbf{A}}}][\bar{\mathbf{I}}_b] + [\bar{\bar{\mathbf{Z}}}_{s_p}]^{-1}[\bar{\mathbf{V}}_p] = [\bar{\mathbf{F}}] \tag{6.55}$$

- Further Simplify Method a
  Cancel $[\bar{\mathbf{I}}_b]$ items, combined Eq. (6.54) with Eq. (6.55), equation about hybrid method based on PI basic function about voltage unknown can obtained as below

$$[\bar{\bar{\mathbf{Z}}}_{s_p}]^{-1}[\bar{\mathbf{V}}_p] + [\bar{\bar{\mathbf{A}}}]\left[[\bar{\bar{\mathbf{Z}}}_b] + j\omega[\bar{\bar{\mathbf{M}}}_b]\right]^{-1}[\bar{\bar{\mathbf{A}}}]^T[\bar{\mathbf{V}}_p] = [\bar{\mathbf{F}}] \tag{6.56}$$

For the convenience of distinguishing various numerical algorithms in the future, this formula is named PI 2 U case.

- Further Simplify Method b
  Cancel $[\bar{\mathbf{V}}_p]$ items, combined Eq. (6.54) with Eq. (6.55), equation about hybrid method based on PI basic function about current unknown can obtained as below

$$\left[[\bar{\bar{\mathbf{Z}}}_b + j\omega\bar{\bar{\mathbf{M}}}_b] + [\bar{\bar{\mathbf{A}}}]^T[\bar{\bar{\mathbf{Z}}}_{s_p}][\bar{\bar{\mathbf{A}}}]\right][\bar{\mathbf{I}}_b] = -[\bar{\bar{\mathbf{A}}}]^T[\bar{\bar{\mathbf{Z}}}_{s_p}][\bar{\mathbf{F}}] \tag{6.57}$$

  Similarly, for the convenience of distinguishing various numerical algorithms in the future, this formula is named PI 2 I case

The vector of nodal voltage $[\bar{\mathbf{V}}_p]$ may be calculated through solving the Eq. (6.56). The average voltage $[\bar{\mathbf{U}}_a]$ and leakage currents $[\bar{\mathbf{I}}_s]$, the branch voltage $[\bar{\mathbf{U}}_b]$, branch current $[\bar{\mathbf{I}}_b]$ can also be calculated [20–25].

Once branch currents and leakage currents have known, the voltage, VMP (A), EFI (E) and MFI (B) at any point can be calculated [20–25].

The study of the grounding systems performance for spatial 1-d hybrid method based on Pi basic function has been reduced to the computation of $[\bar{\bar{\mathbf{Z}}}_b]$, $[\bar{\bar{\mathbf{M}}}_b]$, $[\bar{\bar{\mathbf{Z}}}_{s_b}]$ or $[\bar{\bar{\mathbf{Z}}}_{s_p}]$ matrices.

Internal impedance matrix $[\bar{\bar{\mathbf{Z}}}_b]$

$[\bar{\bar{\mathbf{Z}}}_b]$ is a diagonal matrix, whose elements are the product of the internal impedance and the length of the conductor segment, the formula for internal impedance can be found in [20–39].

Mutual induction matrix $[\bar{\bar{\mathbf{M}}}_b]$

Mutual induction matrix $[\bar{\bar{\mathbf{M}}}_b]$ is matrix with zero diagonal, whose elements about mutual inductance coefficient between two thin conductors, separated by a distance much bigger than their radius, may be obtained with the formula as below

$$M_{\theta,\vartheta} = \int_{l_\theta}\int_{l_\vartheta} \hat{r}_\theta \cdot j\omega\bar{\bar{G}}^{Af}_{\xi\zeta}(\bar{r}_\theta, \bar{r}_\vartheta)\cdot \hat{r}^T_\vartheta d\bar{r}_\theta d\bar{r}_\vartheta \tag{6.58}$$

For an infinite homogeneous conductivity medium, one has $\bar{\bar{G}}^{Af}_{11}(\bar{r}_\vartheta, \bar{r}_\theta) = \frac{\mu}{4\pi}\frac{1}{R_{ij}}\bar{\bar{I}}^0$, where $\bar{\bar{I}}^0$ is the diagonal unit matrix, $R_{\theta\vartheta} = |\bar{r}_\theta - \bar{r}_\vartheta|$.

$$M_{\theta,\vartheta} = \frac{j\omega\mu}{4\pi}\int_{l_\theta}\int_{l_\vartheta} \frac{\hat{r}_\theta \cdot \hat{r}^T_\vartheta}{R_{\theta\vartheta}} dr_\theta dr_\vartheta \tag{6.59}$$

Mutual impedance matrix $[\bar{\bar{\mathbf{Z}}}_{s_b}]$

Mutual impedance matrix $[\bar{\bar{\mathbf{Z}}}_{s_b}]$, whose core is formula of mutual impedance interaction betweens a pars of currents along conductor segments.

$$Z_{s_b}^{\theta,\vartheta} = \frac{1}{l_\theta}\int_{l_\theta}\frac{1}{l_\vartheta}\int_{l_\vartheta} G_{\xi\zeta}^{\varphi^f}(\bar{r}_\theta,\bar{r}_\vartheta)dr_\vartheta dr_\theta \tag{6.60}$$

For an infinite homogeneous conductivity medium, one has $G_{11}^{\varphi^f}(\bar{r}_\vartheta,\bar{r}_\theta) = \frac{1}{4\pi\bar{\sigma}_1}\frac{1}{R_{\theta\vartheta}}$, where $\bar{\sigma}_1 = \sigma_1 + j\omega\varepsilon_1$, so

$$Z_{s_b}^{\theta,\vartheta} = \frac{1}{4\pi\bar{\sigma}_1 l_\theta l_\vartheta}\int_{l_\theta}\int_{l_\vartheta}\frac{1}{R_{\theta\vartheta}}dr_\theta dr_\vartheta \tag{6.61}$$

Mutual impedance matrix $[\bar{\bar{\mathbf{Z}}}_{s_p}]$

Mutual impedance matrix $[\bar{\bar{\mathbf{Z}}}_{s_p}]$, The core of this matrix element formula is the interaction between leakage currents on two sets of conductor segments connected to two nodes.

$$Z_{s_p}^{\varsigma,\upsilon} = \frac{1}{N_\theta^\varsigma}\sum_{\theta=1}^{N_\theta^\varsigma}\frac{1}{l_\theta^\varsigma}\int_{l_\theta^\varsigma}\frac{1}{N_\vartheta^\upsilon}\sum_{\vartheta=1}^{N_\vartheta^\upsilon}\frac{1}{l_\vartheta^\upsilon}\int_{l_\vartheta^\upsilon} G_{\xi\zeta}^{\varphi^f}(\bar{r}_\varsigma,\bar{r}_\upsilon)dr_\varsigma dr_\upsilon \tag{6.62}$$

where ς, υ are the node numbers of discrete conductor segments in the grounding grid. N_θ^ς and N_ϑ^υ are the total number of conductor segments connecting nodes ς and υ. l_θ^ς and l_ϑ^υ are the lengths of the θ-th and ϑ-th conductor segments connecting nodes ς and υ, respectively.

For an infinite homogeneous conductivity medium, one has $G_{11}^{\varphi^f}(\bar{r}_\varsigma,\bar{r}_\upsilon) = \frac{1}{4\pi\bar{\sigma}_1}\frac{1}{R_{\varsigma\upsilon}}$, where $\bar{\sigma}_1 = \sigma_1 + j\omega\varepsilon_1$, $R_{\varsigma\upsilon} = |\bar{r}_\varsigma - \bar{r}_\upsilon|$, so

$$Z_{s_p}^{\varsigma,\upsilon} = \frac{1}{4\pi\bar{\sigma}_1}\frac{1}{N_\theta^\varsigma}\sum_{\theta=1}^{N_\theta^\varsigma}\frac{1}{l_\theta^\varsigma}\int_{l_\theta^\varsigma}\frac{1}{N_\vartheta^\upsilon}\sum_{\vartheta=1}^{N_\vartheta^\upsilon}\frac{1}{l_\vartheta^\upsilon}\int_{l_\vartheta^\upsilon}\frac{1}{R_{\varsigma\upsilon}}dr_\varsigma dr_\upsilon \tag{6.63}$$

Note the medium surrounding the point current here was considered as homogeneous and infinite. However, in any practical case, the earth is represented as a multilayer earth model , at least as two layer media (one is air, another one is earth). The QSCIM can be used to dealt with the multilayered earth model, this will be discussed next.

6.3.4 Mathematical Model of the Grounding System for the Spatial 1-d Vector Hybrid Method Based on PEEC Basic Function

For the derivation process of spatial 1-d hybrid algorithms based on PEEC basis functions please refer to the relevant content in the previous section on spatial 1-d hybrid algorithms based on PI basis functions. Referring to Fig. 6.1a–d, it can be seen that the hybrid algorithm based on PEEC basis function is further subdivided into two parts on the branch of leakage current, thus doubling the dimensions of the related mutual impedance matrix.

Mathematical model based on the mutual influence between branch voltage and branch leakage current case

$$[\bar{\bar{\mathbf{Z}}}_b + j\omega\bar{\bar{\mathbf{M}}}_b][\bar{\mathbf{I}}_b] + [\bar{\bar{\mathbf{A}}}]^T[\bar{\mathbf{V}}_p] = 0 \tag{6.64}$$

$$-[\bar{\bar{\mathbf{A}}}][\bar{\mathbf{I}}_b] + [\bar{\bar{\mathbf{K}}}]^T[\bar{\bar{\mathbf{Z}}}_{s_b}]^{-1}[\bar{\bar{\mathbf{K}}}][\bar{\mathbf{V}}_p] = [\bar{\mathbf{F}}] \tag{6.65}$$

where the connotation of these matrices or vectors ($[\bar{\bar{\mathbf{Z}}}_b]$, $[\bar{\bar{\mathbf{M}}}_b]$, $[\bar{\bar{\mathbf{A}}}]$, $[\bar{\mathbf{V}}_p]$, $[\bar{\mathbf{F}}]$ and $[\bar{\mathbf{I}}_b]$) is the same as the corresponding matrices or vectors in the original formula (6.48) (6.49). Except for matrices $[\bar{\bar{\mathbf{Z}}}_{s_b}]$ and $[\bar{\bar{\mathbf{K}}}]$, where $[\bar{\bar{\mathbf{Z}}}_{s_b}]$ is a $2N_b \times 2N_b$ matrix, and $[\bar{\bar{\mathbf{K}}}]$ is a $2N_b \times N_p$ matrix.

- Further Simplify Method a
  Cancel $[\bar{\mathbf{I}}_b]$ items, combined Eq. (6.64) with Eq. (6.65), equation about hybrid method based on PEEC basic function about voltage unknown can obtained as below

$$\left[[\bar{\bar{\mathbf{K}}}]^T[\bar{\bar{\mathbf{Z}}}_{s_b}]^{-1}[\bar{\bar{\mathbf{K}}}] + [\bar{\bar{\mathbf{A}}}]\left[[\bar{\bar{\mathbf{Z}}}_b] + j\omega[\bar{\bar{\mathbf{M}}}_b]\right]^{-1}[\bar{\bar{\mathbf{A}}}]^T\right][\bar{\mathbf{V}}_p] = [\bar{\mathbf{F}}] \tag{6.66}$$

  For the convenience of explaining various numerical algorithms in the future, this formula is named PEEC 1 U case.
- Further Simplify Method b
  Cancel $[\bar{\mathbf{V}}_p]$ items, combined Eq. (6.64) with Eq. (6.65), equation about hybrid method based on PEEC basic function about current unknown can obtained as below

$$\left[[\bar{\bar{\mathbf{Z}}}_b + j\omega\bar{\bar{\mathbf{M}}}_b] + [\bar{\bar{\mathbf{A}}}]^T\left[[\bar{\bar{\mathbf{K}}}]^T[\bar{\bar{\mathbf{Z}}}_{s_b}]^{-1}[\bar{\bar{\mathbf{K}}}]\right]^{-1}[\bar{\bar{\mathbf{A}}}]\right][\bar{\mathbf{I}}_b] = -[\bar{\bar{\mathbf{A}}}]^T\left[[\bar{\bar{\mathbf{K}}}]^T[\bar{\bar{\mathbf{Z}}}_{s_b}]^{-1}[\bar{\bar{\mathbf{K}}}]\right]^{-1}[\bar{\mathbf{F}}] \tag{6.67}$$

  For the convenience of explaining various numerical algorithms in the future, this formula is named PEEC 1 I case.
  It must be pointed out that due to the limitations of computer calculation accuracy, the double inversion of the mutual impedance matrix $[\bar{\bar{\mathbf{Z}}}_{s_b}]$ will result in unpredictable calculation errors. Therefore, Eq. (6.51) cannot be used for calculation.

The vector of nodal voltage $[\bar{\mathbf{V}}_p]$ may be calculated through solving the Eq. (6.66). The average voltage $[\bar{\mathbf{U}}_a]$ and leakage currents $[\bar{\mathbf{I}}_s]$, the branch voltage $[\bar{\mathbf{U}}_b]$, branch current $[\bar{\mathbf{I}}_b]$ can also be calculated [20–25].

Mathematical model based on the interaction between node voltage and leakage current of the branch connected to the node

$$[\bar{\bar{\mathbf{Z}}}_b + j\omega\bar{\bar{\mathbf{M}}}_b][\bar{\mathbf{I}}_b] + [\bar{\bar{\mathbf{A}}}]^T[\bar{\mathbf{V}}_p] = 0 \tag{6.68}$$

$$-[\bar{\bar{\mathbf{A}}}][\bar{\mathbf{I}}_b] + [\bar{\bar{\mathbf{Z}}}_{s_p}]^{-1}[\bar{\mathbf{V}}_p] = [\bar{\mathbf{F}}] \tag{6.69}$$

where all matrices and vectors here have exactly the same form and content as the corresponding matrices and vectors of the original Eqs. (6.54) (6.55). However, when calculating the matrix $[\bar{\bar{\mathbf{Z}}}_{s_p}]$, due to the need to further subdivide the leakage current branch in the hybrid algorithm based on PEEC basis functions, compared with the hybrid algorithm based on PI basis functions, the length of the leakage current branch is half smaller.

- Further Simplify Method a
  Cancel $[\bar{\mathbf{I}}_b]$ items, combined Eq. (6.68) with Eq. (6.69), equation about hybrid method based on PEEC basic function about voltage unknown can obtained as below

$$[\bar{\bar{\mathbf{Z}}}_{s_p}]^{-1}[\bar{\mathbf{V}}_p] + [\bar{\bar{\mathbf{A}}}]\left[[\bar{\bar{\mathbf{Z}}}_b] + j\omega[\bar{\bar{\mathbf{M}}}_b]\right]^{-1}[\bar{\bar{\mathbf{A}}}]^T[\bar{\mathbf{V}}_p] = [\bar{\mathbf{F}}] \tag{6.70}$$

  For the convenience of explaining various numerical algorithms in the future, this formula is named PEEC 2 U case.
- Further Simplify Method b
  Cancel $[\bar{\mathbf{V}}_p]$ items, combined Eq. (6.68) with Eq. (6.69), equation about hybrid method based on PEEC basic function about current unknown can obtained as below

$$\left[[\bar{\bar{\mathbf{Z}}}_b + j\omega\bar{\bar{\mathbf{M}}}_b] + [\bar{\bar{\mathbf{A}}}]^T[\bar{\bar{\mathbf{Z}}}_{s_p}][\bar{\bar{\mathbf{A}}}]\right][\bar{\mathbf{I}}_b] = -[\bar{\bar{\mathbf{A}}}]^T[\bar{\bar{\mathbf{Z}}}_{s_p}][\bar{\mathbf{F}}] \tag{6.71}$$

  For the convenience of explaining various numerical algorithms in the future, this formula is named PEEC 2 I case.

The vector of nodal voltage $[\bar{\mathbf{V}}_p]$ may be calculated through solving the Eq. (6.70). The average voltage $[\bar{\mathbf{U}}_a]$ and leakage currents $[\bar{\mathbf{I}}_s]$, the branch voltage $[\bar{\mathbf{U}}_b]$, branch current $[\bar{\mathbf{I}}_b]$ can also be calculated [20–25].

Once branch currents and leakage currents have known, the voltage, VMP (A), EFI (E) and MFI (B) at any point can be calculated [20–25].

Computation of $[\bar{\bar{\mathbf{Z}}}_b]$, $[\bar{\bar{\mathbf{M}}}_b]$, $[\bar{\bar{\mathbf{Z}}}_{s_b}]$ and $[\bar{\bar{\mathbf{Z}}}_{s_p}]$ matrices

The calculation method of matrix elements for $[\bar{\bar{\mathbf{Z}}}_b]$ and $[\bar{\bar{\mathbf{M}}}_b]$ can be found in the previous section on hybrid algorithms based on PI basis functions.

The calculation method of matrix elements for $[\bar{\bar{\mathbf{Z}}}_{s_b}]$ and $[\bar{\bar{\mathbf{Z}}}_{s_p}]$ can also be seen in the previous section of the hybrid algorithm based on PI basis functions. However, due to the need to further subdivide the leakage current branch in the hybrid algorithm based on PEEC basis functions, the length of the conductor section in the calculation formula is only half.

6.3.5 Mathematical Model of the Grounding System for the Spatial 1-d Vector Hybrid Method Based on T Basic Function

As shown in Fig. 6.1a, e, Eq. (6.39) in the hybrid method based on PI basis functions can be rewritten as follows,

$$[\bar{\bar{\mathbf{Z}}}_b + j\omega\bar{\bar{\mathbf{M}}}_b][\bar{\mathbf{I}}_b] + [\bar{\bar{\mathbf{A}}}]^T[\bar{\mathbf{V}}_n] = 0 \tag{6.72}$$

here, due to the need to further subdivide the branches of the axial current into two based on the T-basis function hybrid algorithm, the dimensions of the matrices and vectors here need to be doubled. both $[\bar{\bar{\mathbf{Z}}}_b]$ and $[\bar{\bar{\mathbf{M}}}_b]$ are $2N_b \times 2N_b$ matrices. $[\bar{\bar{\mathbf{A}}}]^T$ is $2N_b \times (N_b + N_p)$ rectangular matrix. $[\bar{\mathbf{I}}_b]$ is $2N_b \times 1$ vector, and $[\bar{\mathbf{V}}_n]$ is $(N_b + N_p) \times 1$ vector of all voltage of nodes including end and middle points.

The Eq. (6.72) can be further written as below

$$[\bar{\mathbf{I}}_b] = -[\bar{\bar{\mathbf{Z}}}_b + j\omega\bar{\bar{\mathbf{M}}}_b]^{-1}[\bar{\bar{\mathbf{A}}}]^T[\bar{\mathbf{V}}_n] \tag{6.73}$$

$$[\bar{\mathbf{I}}_b] = [\bar{\bar{\mathbf{Y}}}_b][\bar{\bar{\mathbf{A}}}]^T[\bar{\mathbf{V}}_n] \tag{6.74}$$

where $[\bar{\bar{\mathbf{Y}}}_b] = -[\bar{\bar{\mathbf{Z}}}_b + j\omega\bar{\bar{\mathbf{M}}}_b]^{-1}$ is a $2N_b \times 2N_b$ admittance matrix who includes mutual induction and internal impedance of conductors.

Then $[\bar{\bar{\mathbf{A}}}][\bar{\mathbf{I}}_b] = 0$ was considered, and $[\bar{\mathbf{V}}_n] = \begin{bmatrix} \bar{\mathbf{V}}_p \\ \bar{\mathbf{V}}_b \end{bmatrix}$ and $[\bar{\bar{\mathbf{A}}}] = \begin{bmatrix} \bar{\bar{\mathbf{A}}}_{pb} & \bar{\bar{\mathbf{A}}}_{pb} \\ \bar{\bar{\mathbf{A}}}_{bb} & \bar{\bar{\mathbf{A}}}_{bb} \end{bmatrix}$ and $[\bar{\bar{\mathbf{Y}}}_b] = \begin{bmatrix} \bar{\bar{\mathbf{Y}}}_{bb} & \bar{\bar{\mathbf{Y}}}_{bb} \\ \bar{\bar{\mathbf{Y}}}_{bb} & \bar{\bar{\mathbf{Y}}}_{bb} \end{bmatrix}$ and $[\bar{\bar{\mathbf{A}}}]^T = \begin{bmatrix} \bar{\bar{\mathbf{A}}}_{bp}^T & \bar{\bar{\mathbf{A}}}_{bb}^T \\ \bar{\bar{\mathbf{A}}}_{bp}^T & \bar{\bar{\mathbf{A}}}_{bb}^T \end{bmatrix}$

So it can be obtained as below

$$\begin{bmatrix} \bar{\mathbf{F}} \\ -\bar{\mathbf{I}}_{s_b} \end{bmatrix} = \begin{bmatrix} \bar{\bar{\mathbf{A}}}_{pb}\bar{\bar{\mathbf{Y}}}_{bb}\bar{\bar{\mathbf{A}}}_{bp}^T & \bar{\bar{\mathbf{A}}}_{pb}\bar{\bar{\mathbf{Y}}}_{bb}\bar{\bar{\mathbf{A}}}_{bb}^T \\ \bar{\bar{\mathbf{A}}}_{bb}\bar{\bar{\mathbf{Y}}}_{bb}\bar{\bar{\mathbf{A}}}_{bp}^T & \bar{\bar{\mathbf{A}}}_{bb}\bar{\bar{\mathbf{Y}}}_{bb}\bar{\bar{\mathbf{A}}}_{bb}^T \end{bmatrix} \begin{bmatrix} \bar{\mathbf{V}}_p \\ \bar{\mathbf{V}}_b \end{bmatrix} \tag{6.75}$$

where $[\bar{\mathbf{F}}]$ is a $N_p \times 1$ vector of external currents sources; $[\bar{\bar{\mathbf{Y}}}_{bb}]$ is the $N_b \times N_b$ branch admittance sub-matrix of the matrix $[\bar{\bar{\mathbf{Y}}}_b]$; $[\bar{\mathbf{I}}_{s_b}]$ is a $N_b \times 1$ vector of leakage currents along discrete conductor segments; $\bar{\mathbf{V}}_p$ is a $N_p \times 1$ vector of voltage of the end points; $\bar{\mathbf{V}}_b$ is a $N_b \times 1$ vector of voltage of the middle points; the sub-matrices

$[\bar{\bar{\mathbf{A}}}_{pb}]$ and $[\bar{\bar{\mathbf{A}}}_{bb}]$ are, respectively, $N_p \times N_b$ and $N_b \times N_b$ matrices; the sub-matrices $[\bar{\bar{\mathbf{A}}}_{bp}^T]$ and $[\bar{\bar{\mathbf{A}}}_{bb}^T]$ are, respectively, $N_b \times N_p$ and $N_b \times N_b$ matrices.

As shown in Fig. 6.1a, e, Eq. (6.41) in the hybrid method based on PI basis functions can be rewritten as follows,

$$[\bar{\bar{\mathbf{Z}}}_{s_b}][\bar{\mathbf{I}}_{s_b}] = [\bar{\mathbf{V}}_b] \tag{6.76}$$

where in Eq. (6.76), $[\bar{\bar{\mathbf{Z}}}_{s_b}]$ was $N_b \times N_b$ mutual impedance matrix; $[\bar{\mathbf{V}}_b]$ is $N_b \times 1$ vector of average voltage of discrete conductor segments.

Combining Equations (6.75) and (6.76), we have

$$\begin{bmatrix} \bar{\mathbf{F}} \\ 0 \end{bmatrix} = \begin{bmatrix} \bar{\bar{\mathbf{A}}}_{pb}\bar{\bar{\mathbf{Y}}}_{bb}\bar{\bar{\mathbf{A}}}_{bp}^T & \bar{\bar{\mathbf{A}}}_{pb}\bar{\bar{\mathbf{Y}}}_{bb}\bar{\bar{\mathbf{A}}}_{bb}^T \\ \bar{\bar{\mathbf{A}}}_{bb}\bar{\bar{\mathbf{Y}}}_{bb}\bar{\bar{\mathbf{A}}}_{bp}^T & \bar{\bar{\mathbf{A}}}_{bb}\bar{\bar{\mathbf{Y}}}_{bb}\bar{\bar{\mathbf{A}}}_{bb}^T + \bar{\bar{\mathbf{Z}}}_{s_b}^{-1} \end{bmatrix} \begin{bmatrix} \bar{\mathbf{V}}_p \\ \bar{\mathbf{V}}_b \end{bmatrix} \tag{6.77}$$

At last we have

$$\left[\left[\bar{\bar{\mathbf{A}}}_{pb}\bar{\bar{\mathbf{Y}}}_{bb}\bar{\bar{\mathbf{A}}}_{bp}^T\right] - \left[\bar{\bar{\mathbf{A}}}_{pb}\bar{\bar{\mathbf{Y}}}_{bb}\bar{\bar{\mathbf{A}}}_{bb}^T\left[\bar{\bar{\mathbf{A}}}_{bb}\bar{\bar{\mathbf{Y}}}_{bb}\bar{\bar{\mathbf{A}}}_{bb}^T + \bar{\bar{\mathbf{Z}}}_{s_b}^{-1}\right]^{-1}\bar{\bar{\mathbf{A}}}_{bb}\bar{\bar{\mathbf{Y}}}_{bb}\bar{\bar{\mathbf{A}}}_{bp}^T\right]\right][\bar{\mathbf{V}}_p] = [\bar{\mathbf{F}}] \tag{6.78}$$

For clarity in the following description, this formula can be abbreviated as T case.

The vector of nodal voltage $[\bar{\mathbf{V}}_p]$ may be calculated through solving the Eq. (6.77). The voltage of middle point $[\bar{\mathbf{V}}_b]$ can be calculated by

$$[\bar{\mathbf{V}}_b] = -\left[\bar{\bar{\mathbf{A}}}_{bb}\bar{\bar{\mathbf{Y}}}_{bb}\bar{\bar{\mathbf{A}}}_{bb}^T + \bar{\bar{\mathbf{Z}}}_{s_b}^{-1}\right]^{-1}\left[\bar{\bar{\mathbf{A}}}_{bb}\bar{\bar{\mathbf{Y}}}_{bb}\bar{\bar{\mathbf{A}}}_{bp}^T\right][\bar{\mathbf{V}}_p] \tag{6.79}$$

The leakage current $[\bar{\mathbf{I}}_{s_b}]$, and that the branch voltage $[\bar{\mathbf{U}}_b]$, branch current $[\bar{\mathbf{I}}_b]$ can also be calculated [23–25].

Once branch currents and leakage currents are known, the voltage φ, VMP (A), EFI (E), and MFI (B) at any point can be calculated, which can be referred in papers [23–25].

The study of the grounding systems performance for spatial 1-d hybrid method based on T basic function has been reduced to the computation of $[\bar{\bar{\mathbf{Z}}}_b]$, $[\bar{\bar{\mathbf{M}}}_b]$ and $[\bar{\bar{\mathbf{Z}}}_{s_b}]$ matrices.

Computation of $[\bar{\bar{\mathbf{Z}}}_b]$, $[\bar{\bar{\mathbf{M}}}_b]$ and $[\bar{\bar{\mathbf{Z}}}_{s_b}]$ matrices

The calculation method of matrix elements for $[\bar{\bar{\mathbf{Z}}}_{s_b}]$ can be found in the previous section on hybrid algorithms based on PI basis functions.

The calculation method of matrix elements for $[\bar{\bar{\mathbf{Z}}}_b]$ and $[\bar{\bar{\mathbf{M}}}_b]$ can also be seen in the previous section of the hybrid algorithm based on PI basis functions. However, due to the need to further subdivide the axial current branch in the hybrid algorithm based on PEEC basis functions, the length of the conductor section in the calculation formula is only half.

6.3.6 Mathematical Model of the Grounding System for the Spatial 1-d Vector Hybrid Method Based on TPEEC Basic Function

The Eq. (6.39) in the hybrid method based on PI basis functions can be rewritten as follows,

$$[\bar{\bar{\mathbf{Z}}}_b + j\omega\bar{\bar{\mathbf{M}}}_b][\bar{\mathbf{I}}_b] + [\bar{\bar{\mathbf{A}}}]^T[\bar{\mathbf{V}}_p] = 0 \tag{6.80}$$

where, both $[\bar{\bar{\mathbf{Z}}}_b]$ and $[\bar{\bar{\mathbf{M}}}_b]$ are $N_b \times N_b$ matrices. $[\bar{\bar{\mathbf{A}}}]^T$ is $N_b \times N_p$ rectangular matrix. $[\bar{\mathbf{I}}_b]$ is $N_b \times 1$ vector, and $[\bar{\mathbf{V}}_p]$ is $N_p \times 1$ vector of all voltage of nodes.

As shown in Fig. 6.1a, f, since the hybrid algorithm introduced in this section is based on TPEEC basis functions, it is required that all axial current branches be further subdivided into two parts. Therefore, the axial current is also divided into two parts, so Eq. (6.79) can be written as

$$\frac{1}{2}[\bar{\bar{\mathbf{Z}}}_b + j\omega\bar{\bar{\mathbf{M}}}_b][\bar{\mathbf{I}}_b^1] + \frac{1}{2}[\bar{\bar{\mathbf{Z}}}_b + j\omega\bar{\bar{\mathbf{M}}}_b][\bar{\mathbf{I}}_b^2] + [\bar{\bar{\mathbf{A}}}]^T[\bar{\mathbf{V}}_p] = 0 \tag{6.81}$$

where $[\bar{\mathbf{I}}_b^1]$ and $[\bar{\mathbf{I}}_b^2]$ are $N_b \times 1$ vector.

The Eq. (6.42) in the hybrid method based on PI basis functions can be rewritten as follows,

$$[\bar{\bar{\mathbf{Z}}}_{s_b}][\bar{\mathbf{I}}_{s_b}] = [\bar{\bar{\mathbf{K}}}][\bar{\mathbf{V}}_p] \tag{6.82}$$

where in Eq. (6.82), $[\bar{\bar{\mathbf{Z}}}_{s_b}]$ was $N_b \times N_b$ mutual impedance matrix; $[\bar{\mathbf{I}}_{s_b}]$ was $N_b \times 1$ vector of leakage currents along discrete conductor segments; $[\bar{\bar{\mathbf{K}}}]$ was rectangular matrix of order $N_b \times N_p$.

As shown in Fig. 6.1a, b, as the hybrid algorithm introduced in this section is based on the TPEEC basis function, it is required that all axial current branches be further subdivided into two parts. Therefore, $[\bar{\mathbf{I}}_s] = [\bar{\mathbf{I}}_b^1] - [\bar{\mathbf{I}}_b^2]$, in this way, it can be obtained as below

$$[\bar{\bar{\mathbf{K}}}][\bar{\mathbf{V}}_p] = [\bar{\bar{\mathbf{Z}}}_{s_b}][\bar{\mathbf{I}}_b^1] - [\bar{\bar{\mathbf{Z}}}_{s_b}][\bar{\mathbf{I}}_b^2] \tag{6.83}$$

As shown in Fig. 6.1a, f, Eq. (6.84) can be obtained according to KCL.

$$[\bar{\bar{\mathbf{C}}}][\bar{\mathbf{I}}_b^1] + [\bar{\bar{\mathbf{D}}}][\bar{\mathbf{I}}_b^2] = [\bar{\mathbf{F}}] \tag{6.84}$$

where in Eq. (6.84), $[\bar{\bar{\mathbf{C}}}]$ and $[\bar{\bar{\mathbf{D}}}]$ were rectangular incidence matrices of order $N_p \times N_b$, which are used to relate to branches and nodes, whose element $C_{\theta\vartheta}$ and $D_{\theta\vartheta}$ were defined as follows: $C_{\theta\vartheta} = 1$ if ϑ-th branch is directed from θ-th node, $D_{\theta\vartheta} = -1$ if

ϑ-th branch is directed toward θ-th node, otherwise, $C_{\theta\vartheta} = D_{\theta\vartheta} = 0$. $[\bar{\mathbf{F}}]$ is a $N_p \times 1$ vector of external current sources;

Combined Eq. (6.81), Eq. (6.83) and Eq. (6.84), equation about hybrid method basing on TPEEC basic function can obtained as below

$$\left[[\bar{\bar{\mathbf{D}}} - \bar{\bar{\mathbf{C}}}]\frac{1}{2}[\bar{\bar{\mathbf{Z}}}]_{s_b}^{-1}[\bar{\bar{\mathbf{K}}}] - [\bar{\bar{\mathbf{D}}} + \bar{\bar{\mathbf{C}}}][\bar{\bar{\mathbf{Z}}}_b + j\omega\bar{\bar{\mathbf{M}}}_b]^{-1}[\bar{\bar{\mathbf{A}}}]^T\right][\bar{\mathbf{V}}_p] = [\bar{\mathbf{F}}_p] \quad (6.85)$$

For clarity in the following description, this formula can be abbreviated as TPEEC case.

The vector of nodal voltage $[\bar{\mathbf{V}}_p]$ may be calculated through solving the Eq. (6.84). The average voltage $[\bar{\mathbf{U}}_a]$ and leakage currents $[\bar{\mathbf{I}}_s]$, the branch voltage $[\bar{\mathbf{U}}_b]$, branch current $[\bar{\mathbf{I}}_b]$ can also be calculated [20–25].

Once branch currents and leakage currents have known, the voltage, VMP (A), EFI (E) and MFI (B) at any point can be calculated [20–25].

The study of the grounding systems performance for spatial 1-d hybrid method based on TPEEC basic function has been reduced to the computation of $[\bar{\bar{\mathbf{Z}}}_b]$, $[\bar{\bar{\mathbf{M}}}_b]$ and $[\bar{\bar{\mathbf{Z}}}_{s_b}]$ matrices.

Computation of $[\bar{\bar{\mathbf{Z}}}_b]$, $[\bar{\bar{\mathbf{M}}}_b]$ and $[\bar{\bar{\mathbf{Z}}}_{s_b}]$ matrices

The calculation method of matrix elements for $[\bar{\bar{\mathbf{Z}}}_b]$, $[\bar{\bar{\mathbf{M}}}_b]$ and $[\bar{\bar{\mathbf{Z}}}_{s_b}]$ can be found in the previous section on hybrid algorithms based on PI basis functions.

6.4 The Closed Form of the Green's Function

The important task of simulating a grounding system is calculating the elements (a) $Z_{\theta,\vartheta}$ and $M_{\theta,\vartheta}$ of matrix $[\bar{\bar{\mathbf{Z}}}]$ in (6.7), (b) $Z_{\theta,\vartheta}^{\varsigma\upsilon}$ and $M_{\theta,\vartheta}^{\varsigma\upsilon}$ of matrices $[\bar{\bar{\mathbf{Z}}}_b^{00}]$, $[\bar{\bar{\mathbf{Z}}}_b^{01}]$, $[\bar{\bar{\mathbf{Z}}}_b^{10}]$ and $[\bar{\bar{\mathbf{Z}}}_b^{11}]$ in (6.24), and (c) $M_{\theta,\vartheta}$ and $Z_{s_b}^{\theta,\vartheta}$ and $Z_{s_p}^{\varsigma\upsilon}$ of matrices $[\bar{\bar{\mathbf{M}}}_b]$, $[\bar{\bar{\mathbf{Z}}}_{s_b}]$ and $[\bar{\bar{\mathbf{Z}}}_{s_p}]$ in (6.49) and (6.55). When the earth model is taken to be a horizontal multilayered conductivity media, the influences from the interface must be considered, which will lead to an infinite integral for the Bessel function associated with the Green's function of a scalar monopole or a vector dipole. However, $Z_{\theta\vartheta}$(6.16), $M_{\theta\vartheta}$ (6.13), $Z_{\theta,\vartheta}^{\varsigma\upsilon}$ (6.33), $M_{\theta,\vartheta}^{\varsigma\upsilon}$ (6.30),$M_{\theta,\vartheta}$ (6.57), $Z_{s_b}^{\theta,\vartheta}$ (6.59) and $Z_{s_p}^{\varsigma\upsilon}$ (6.61), which includes the Green's function, can also be quickly calculated by using the QSCIM. The QSCIM and the corresponding closed form Green's function for point current sources in multi-layer horizontally layered soil have been thoroughly studied and discussed in Chap. 3 of this book, and will not be elaborated in this chapter.

6.5 General Analytical Formulas of Mutual Impedance and Induction

Once the closed form of Green's function of a scalar monopole or vector dipole in the horizontal multilayered earth model has been obtained, the mutual induction coefficient (6.14) (6.31) (6.58) or mutual impedance coefficient (6.17) (6.34) (6.60) (6.62) can be quickly analytically calculated, which have been thoroughly studied and discussed in Chap. 2 of this book, and will not be elaborated in this chapter.

6.6 Numerical Results and Discussions for the Various Algorithms

This chapter will conduct numerical analysis on several typical examples to demonstrate the advantages and disadvantages of each numerical algorithm. It should be noted that the letters "PE" in the table are an abbreviation of the word "PEEC", while the letters "TPE" are an abbreviation of the word "TPEEC".

6.6.1 Verification of the Various Algorithms

In order to verify the correctness of the various numerical algorithms introduced in this chapter, numerical examples and test results from public literature were compared and analyzed.

1. case a

   The literature [38] provides three comparative examples, the first two being a comparison with other publicly available examples, and the last one being a numerical example from the literature for the validation of various electromagnetic field numerical algorithms in this chapter.

   First example presented in [38] is a 40m $\times$ 40 m square steel grid with 5 $\times$ 5 uniformly distributed conductors in two directions, whose radius is 5 mm, buried under the ground with a depth of 1m and a soil resistivity of 50 Ωm. The current is injected to the grid from the center of the grid. The hybrid algorithm based on T-basis function is used in reference [38], which can simulate and calculate the grounding problem of substation grounding grid in multi-layer horizontal layered soil. The method presented in [43] is based on the transmission line theory and an unequal-potential model, which is only capable of dealing with uniform earth structure. The grounding resistance in the reference literature [38, 43] and the grounding resistance obtained through various numerical simulation calculation methods can be seen in Table 6.1. The text "(eq)" in the table represents the equipotential electromagnetic model. From this table, it can be seen that the grounding resistance obtained by various numerical calculation methods is basically consistent with the grounding resistance in the literature. The BEM using the equipotential model has the shortest calculation time, and among all the various numerical calculation methods using the inequipotential electromagnetic model, the PI 1 U case takes the shortest time.

Second example also coming from [38]. A commercial software for calculating grounding grids is introduced in [44], which employs an unequal-potential model. An example described in it is a 60 m × 40 m rectangular copper grid with 7 × 5 uniformly distributed conductors in two directions, whose radius is 10 mm, buried under the ground with a depth of 0.5 m and a soil resistivity of 100 Ωm. The injecting current is 1000 A. The calculation results is given in Table 6.2, in which the text "(eq)" in the table represents the equipotential electromagnetic model. As can be seen from Table 6.2, the calculation results of various numerical algorithms in this chapter are basically consistent with those of algorithms in public literature [38, 44].

Last example presented in [38] is a square grid of side 100 m with 7 × 7 uniformly distributed steel conductors of different radius, which is calculated by GGRIDS. Its burying depth is 0.5 m, and the soil is a two-layer structure of top-layer thickness 30 m and top-layer soil resistivity 100 Ωm and bottom-layer resistivity 50Ωm. A current of 1000 A is injected to the grid from one of its corners. The calculation results is given in Table 6.3, where φ_{max} is the maximum potential of the grid, R_u and R_e are the grounding resistance obtained by the unequal-potential and equal-potential models respectively. The text "(eq)" in the table represents the equipotential electromagnetic model. The grounding resistance calculated by various algorithms in this chapter and the calculation results in public literature can be seen in Table 6.3 . It can be seen that the grounding resistance calculated by various algorithms in this chapter is basically consistent with the calculation results in public literature [38]. The spatial 1-d vector Galerkin MM is slightly larger, but it is still acceptable.

2. case b

   The second example comes from measurement data from public literature [45]. Four multi-layer soil models are presented in this study, as shown in Table (6.4). The grounding grids modelled in the study are the following: (1) S4–20 m × 20 m square four-mesh grid. (2) S16–20 m × 20 m square sixteen-mesh grid. The radius of all grounding grid conductors is set to 9.8 mm. The calculated grounding resistance with various algorithms and data from [45] can be seen in Tables (6.5)–(6.8). By referring to these tables, it can be seen that the calculation results of various algorithms are basically consistent with the data in the literature [45] (Tables 6.4, 6.5, 6.6, 6.7 and 6.8).

3. case c

   The last example comes from measurement data from public literature [46]. There are two types of grounding grids used for simulation testing, which are 100 m by 100 m and 50 m by 50 m, and both grounding grids are squares composed of 16 grids. These two types of grounding grids are called grid a and grid b. The burial depth is 0.5 m and the radius of the grid conduc-

Table 6.1 Comparison with publication paper results

	Reference [38]	Reference [43]	MM(e)	BEM(e)	MM	BEM	PI 1 U case	PI 2 U case	PI 2 I case	PE 1 U case	PE 2 U case	PE 2 I case	T case	TPE case
R	0.613	0.610	0.604	0.606	0.652	0.659	0.617	0.617	0.616	0.614	0.617	0.616	0.604	0.617
$Time$	\	\	10.27	10.23	27.30	37.18	17.28	29.50	28.72	31.00	29.00	28.95	22.23	17.59

Table 6.2 Comparison with publication paper results

	Reference [38]	Reference [44]	MM(e)	BEM(e)	MM	BEM	PI 1 U case	PI 2 U case	PI 2 I case	PE 1 U case	PE 2 U case	PE 2 I case	T case	TPE case
GPR	963.3 V	960.9 V	961.0 V	963.7 V	950.5 V	995.9 V	975.5 V	950.5 V	962.7 V	961.8 V	950.5 V	965.1 V	973.3 V	975.7 V
Time	\	\	8.02	8.08	21.57	29.14	13.23	26.46	27.34	28.41	26.22	26.81	20.67	13.57

Table 6.3 Comparison with publication paper results

R(mm)	$R_u(\Omega)$ [38]	$R_e(\Omega)$ [38]	MM(e)	BEM(e)	MM	BEM	PI 1 U case	PI 2 U case	PI 2 I case	PE 1 U case	PE 2 U case	PE 2 I case	T case	TPE case
2	0.531	0.406	0.406	0.407	0.593	0.507	0.539	0.538	0.538	0.538	0.538	0.538	0.539	0.539
5	0.415	0.386	0.394	0.396	0.491	0.410	0.416	0.416	0.416	0.416	0.416	0.416	0.416	0.416
10	0.391	0.386	0.384	0.386	0.465	0.389	0.391	0.391	0.391	0.391	0.391	0.391	0.392	0.391
20	0.379	0.377	0.377	0.379	0.448	0.378	0.378	0.378	0.378	0.378	0.378	0.379	0.379	0.378
30	0.373	0.372	0.372	0.373	0.439	0.379	0.372	0.371	0.373	0.372	0.371	0.372	0.370	0.372
50	0.364	0.364	0.365	0.367	0.428	0.365	0.357	0.359	0.366	0.363	0.359	0.366	0.372	0.357
80	0.360	0.360	0.359	0.361	0.417	0.359	0.353	0.353	0.359	0.351	0.353	0.359	0.382	0.353

Table 6.4 Four multi-layer soil models

Model	Layer	Resistivity (Ωm)	Thickness (m)
(a)	1	2000	3.0
	2	100	3.0
	3	1000	∞
(b)	1	50	3.0
	2	1000	3.0
	3	100	∞

tors is 0.5 cm. The current source tested has two frequencies, namely 100 Hz and 80 Hz. There are two types of uniform soil models used for testing the earth, with earth resistivity of 100 Ωm and 10 Ωm, called uniform earth model 1 and 2, respectively. There are also two horizontal layered soil models with resistivity parameters as (1) $\sigma_1 = 25^{-1}$ Ωm, $\sigma_2 = 120^{-1}$ Ωm, $h = 5$ m; (2) $\sigma_1 = 120^{-1}$ Ωm, $\sigma_2 = 25^{-1}$ Ωm, $h = 5$ m; also known as two-layer earth model 1 and 2, respectively. There are also three horizontal layered soil models with resistivity parameters as (1) $\sigma_1 = 120^{-1}$ Ωm, $\sigma_2 = 25^{-1}$ Ωm, $\sigma_3 = 250^{-1}$ Ωm, $h_1 = 5$m, $h_1 = 9$m; (2) $\sigma_1 = 25^{-1}$ Ωm, $\sigma_2 = 120^{-1}$ Ωm, $\sigma_3 = 10^{-1}$ Ωm, $h_1 = 5$ m, $h_2 = 9$ m; (3) $\sigma_1 = 25^{-1}$ Ω m, $\sigma_2 = 120^{-1}$ Ωm, $\sigma_3 = 10^{-1}$ Ωm, $h_1 = 5$ m, $h_2 = 100$ m; also known as three-layer earth model 1 , 2 and 3, respectively. In all calculation examples, the relative dielectric constant of each layer of the earth is taken as 10. The numerical simulation calculation results can be seen in Tables 6.9, 6.10, 6.11, 6.12 and 6.13. From the comparison of data in these tables, it can be seen that the calculation results of various numerical methods are basically consistent with the data in the literature, indicating the correctness of various algorithms.

6.6.2 *Results and Discussions for the Various Algorithms*

A general grounding system can be seen in Fig. 6.2, which consists of 6 × 6 square grids and the burial depth of the grounding grid is 0.5 m. The size of the grounding grid in the figure is 100m × 100 m. The earth is modelled as horizontal four layers conductive media, whose resistivities and permittivities are $\rho_1 = 100$ Ωm and $\rho_2 = 200$ Ωm, $\rho_3 = 300$ Ωm, $\rho_4 = 400$ Ωm, $\varepsilon_1 = 10\varepsilon_0$, $\varepsilon_2 = 10\varepsilon_0$, $\varepsilon_3 = 10\varepsilon_0$ and $\varepsilon_4 = 10\varepsilon_0$, respectively. The first, second and third layer thickness are $h_1 = 5$ m, $h_2 = 6$ m and $h_3 = 10$ m. All the conductor radius of grids are 5 mm. The external excited a.c. current 1000 A with power frequency 0 Hz, 500 Hz and 50000 Hz is injected from the corner of the grounding system. In this calculation example, there is also a relatively small grounding grid with a size of 10m × 10 m, which consists of 5 × 5

Table 6.5 Soil model (a) and grid type : S4 case

Depth (m)	R_e (Ω) [45]	MM(e)	BEM(e)	MM	BEM	PI 1 U case	PI 2 U case	PI 2 I case	PE 1 U case	PE 2 U case	PE 2 I case	T case	TPE case
1.5	24.46	24.45	24.45	32.61	24.41	24.72	24.55	24.44	24.63	24.55	24.45	24.92	24.72
4.5	7.99	8.02	8.02	8.70	8.02	8.02	8.02	8.02	8.01	8.02	8.02	8.05	8.02
7.5	14.63	14.71	14.71	18.68	14.69	14.67	14.72	14.70	14.67	14.72	14.70	14.83	14.67
15	16.16	16.24	16.24	20.64	16.23	16.22	16.25	16.24	16.31	16.25	16.24	16.44	16.22
50	16.32	16.40	16.40	20.83	16.39	16.43	16.38	16.40	16.44	16.38	16.40	16.40	16.43

Table 6.6 Soil model (a) and grid type : S16 case

Depth (m)	[45]	MM(e)	BEM(e)	MM	BEM	PI 1 U case	PI 2 U case	PI 2 I case	PE 1 U case	PE 2 U case	PE 2 I case	T case	TPE case
1.5	19.40	19.43	19.45	25.19	19.44	19.24	18.93	19.44	19.20	18.93	19.44	20.25	19.24
4.5	7.86	8.02	8.02	8.70	8.02	8.02	8.02	8.02	8.01	8.02	8.02	8.05	8.02
7.5	12.23	12.31	12.32	15.12	12.32	12.44	12.13	12.32	11.94	12.13	12.32	12.66	12.44
15	13.86	13.95	13.97	16.92	13.95	13.86	13.59	13.95	13.56	13.59	13.95	13.87	13.86
50	14.05	14.13	14.16	17.10	14.13	14.00	13.81	14.14	13.75	13.81	14.14	13.97	14.00

Table 6.7 Soil model (b) and grid type : S4 case

Depth (m)	[45]	MM(e)	BEM(e)	MM	BEM	PI 1 U case	PI 2 U case	PI 2 I case	PE 1 U case	PE 2 U case	PE 2 I case	T case	TPE case
1.5	2.57	2.59	2.59	2.93	2.59	2.59	2.59	2.59	2.58	2.59	2.59	2.60	2.59
4.5	8.67	8.75	8.75	12.37	8.74	8.76	8.79	8.76	8.76	8.79	8.76	8.82	8.76
7.5	2.13	2.14	2.14	2.63	2.14	2.14	2.14	2.14	2.14	2.14	2.14	2.14	2.14
15	1.85	1.86	1.86	2.30	1.86	1.86	1.86	1.86	1.86	1.86	1.86	1.86	1.86
50	1.66	1.67	1.67	2.11	1.67	1.67	1.67	1.67	1.67	1.67	1.67	1.67	1.67

Table 6.8 Soil model (b) and grid type : S16 case

Depth (m)	[45]	MM(e)	BEM(e)	MM	BEM	PI 1 U case	PI 2 U case	PI 2 I case	PE 1 U case	PE 2 U case	PE 2 I case	T case	TPE case
1.5	2.51	2.52	2.53	2.71	2.52	2.51	2.50	2.52	2.51	2.50	2.52	2.54	2.51
4.5	6.15	6.23	6.24	8.94	6.25	6.29	6.17	6.24	6.23	6.17	6.24	6.32	6.29
7.5	1.91	1.92	1.93	2.24	1.92	1.92	1.92	1.92	1.92	1.92	1.92	1.92	1.92
15	1.63	1.64	1.64	1.94	1.64	1.63	1.64	1.64	1.64	1.64	1.64	1.65	1.63
50	1.44	1.44	1.44	1.74	1.44	1.44	1.44	1.44	1.44	1.44	1.44	1.44	1.44

Table 6.9 Uniform soil model, frequency = 80 Hz and grid type S16 with $100m \times 100$ m case

$\rho(\Omega m)$	[46]	MM(e)	BEM(e)	MM	BEM	PI 1 U case	PI 2 U case	PI 2 I case	PE 1 U case	PE 2 U case	PE 2 I case	T case	TPE case
100	0.523	$0.531\angle\frac{-3}{10^4}$	$0.532\angle\frac{-3}{10^4}$	$0.571\angle 2.2$	$0.565\angle 2.2$	$0.531\angle 3.8$	$0.531\angle 3.8$	$0.537\angle 2.6$	$0.531\angle 3.8$	$0.531\angle 3.8$	$0.537\angle 2.6$	$0.525\angle 4.6$	$0.531\angle 3.8$
10	0.052	$0.053\angle\frac{-3}{10^5}$	$0.053\angle\frac{-3}{10^5}$	$0.065\angle 1.9$	$0.065\angle 2.0$	$0.064\angle 2.2$	$0.064\angle 2.2$	$0.064\angle 2.1$	$0.064\angle 2.2$	$0.064\angle 2.2$	$0.064\angle 2.1$	$0.064\angle 2.2$	$0.064\angle 2.2$

Table 6.10 Uniform soil model, frequency = 80Hz and grid type S16 with $50m \times 50$ m case

$\rho(\Omega m)$	[46]	MM(e)	BEM(e)	MM	BEM	PI 1 U case	PI 2 U case	PI 2 I case	PE 1 U case	PE 2 U case	PE 2 I case	T case	TPE case
100	0.998	$1.013\angle\frac{-3}{10^4}$	$1.014\angle\frac{-3}{10^4}$	$1.066\angle 0.2$	$1.055\angle 0.6$	$0.997\angle 5.4$	$0.965\angle 5.4$	$1.016\angle 0.6$	$0.966\angle 5.5$	$0.965\angle 5.4$	$1.016\angle 0.6$	$0.901\angle 8.7$	$0.967\angle 5.3$
10	0.100	$0.101\angle\frac{-3}{10^5}$	$0.101\angle\frac{-3}{10^5}$	$0.109\angle 5.4$	$0.109\angle 5.3$	$0.104\angle 6.6$	$0.104\angle 6.6$	$0.104\angle 6.1$	$0.104\angle 6.6$	$0.104\angle 6.6$	$0.104\angle 6.1$	$0.103\angle 7.1$	$0.104\angle 6.6$

Table 6.11 Uniform soil model 1, frequency = 500Hz

Grid	[46]	MM(e)	BEM(e)	MM	BEM	PI 1 U case	PI 2 U case	PI 2 I case	PE 1 U case	PE 2 U case	PE 2 I case	T case	TPE case
a	0.523	$0.531\angle\frac{-2}{10^3}$	$0.532\angle\frac{-2}{10^3}$	$0.573\angle 12.8$	$0.591\angle 12.6$	$0.563\angle 15.5$	$0.563\angle 15.5$	$0.563\angle 15.3$	$0.563\angle 15.5$	$0.563\angle 15.5$	$0.563\angle 15.3$	$0.563\angle 15.8$	$0.563\angle 15.5$
b	1.025	$1.013\angle\frac{-2}{10^3}$	$1.015\angle\frac{-2}{10^3}$	$1.056\angle 3.6$	$1.052\angle 3.4$	$1.018\angle 5.0$	$1.019\angle 5.0$	$1.019\angle 3.9$	$1.018\angle 5.0$	$1.019\angle 5.0$	$1.019\angle 3.9$	$1.017\angle 4.6$	$1.018\angle 5.0$

Table 6.12 Two-layer soil model, frequency = 80Hz and grid type S16 with $100m \times 100$ m case

Model	[46]	MM(e)	BEM(e)	MM	BEM	PI 1 U case	PI 2 U case	PI 2 I case	PE 1 U case	PE 2 U case	PE 2 I case	T case	TPE case
1	0.390	$0.403\angle\frac{-3}{10^4}$	$0.404\angle\frac{-3}{10^4}$	$0.411\angle 3.1$	$0.410\angle 3.1$	$0.406\angle 4.2$	$0.406\angle 4.2$	$0.409\angle 3.3$	$0.406\angle 4.2$	$0.406\angle 4.2$	$0.409\angle 3.3$	$0.402\angle 4.9$	$0.405\angle 4.2$
2	0.276	$0.296\angle\frac{-2}{10^4}$	$0.296\angle\frac{-2}{10^4}$	$0.339\angle 3.9$	$0.330\angle 3.8$	$0.300\angle 5.3$	$0.300\angle 5.3$	$0.301\angle 4.6$	$0.300\angle 5.3$	$0.300\angle 5.3$	$0.301\angle 4.6$	$0.298\angle 5.8$	$0.300\angle 5.3$

Table 6.13 Three-layer soil model, frequency = 80Hz and grid type S16 with $100m \times 100$ m case

Model	[46]	MM(e)	BEM(e)	MM	BEM	PI 1 U case	PI 2 U case	PI 2 I case	PE 1 U case	PE 2 U case	PE 2 I case	T case	TPE case
1	0.658	$0.675\angle\frac{-4}{10^4}$	$0.675\angle\frac{-4}{10^4}$	$0.718\angle 1.7$	$0.713\angle 1.9$	$0.671\angle 3.6$	$0.671\angle 3.6$	$0.680\angle 2.0$	$0.670\angle 3.6$	$0.671\angle 3.6$	$0.680\angle 2.0$	$0.659\angle 4.7$	$0.671\angle 3.6$
2	0.130	$0.132\angle\frac{-1}{10^4}$	$0.133\angle\frac{-1}{10^4}$	$0.148\angle 8.5$	$0.145\angle 8.7$	$0.139\angle 10.1$	$0.140\angle 10.1$	$0.140\angle 9.9$	$0.139\angle 10.1$	$0.140\angle 10.1$	$0.140\angle 9.9$	$0.139\angle 10.3$	$0.139\angle 10.1$
3	0.300	$0.300\angle\frac{-2}{10^4}$	$0.301\angle\frac{-2}{10^4}$	$0.311\angle 4.1$	$0.309\angle 4.1$	$0.304\angle 5.1$	$0.304\angle 5.1$	$0.306\angle 4.4$	$0.304\angle 5.1$	$0.304\angle 5.1$	$0.306\angle 4.4$	$0.302\angle 5.6$	$0.304\angle 5.1$

Fig. 6.2 Grounding system

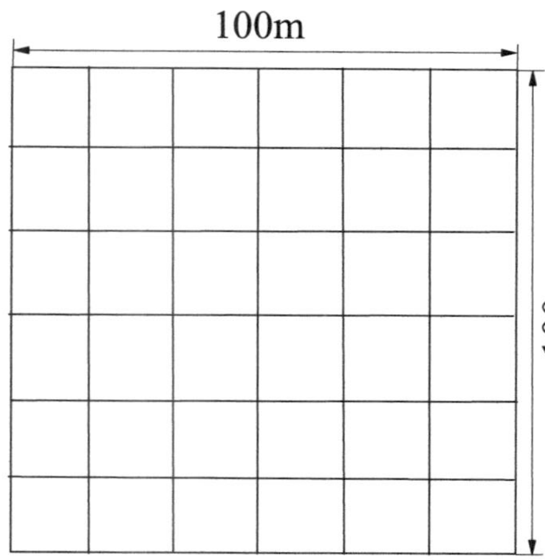

square grids . The remaining parameters of the grounding grid are the same as those shown in Fig. (6.2).

The grounding impedance results and discussions

The grounding impedance values calculated through numerical simulation can be seen in Tables 6.14 and 6.15. Note that the text in the table contains (e) or (eq) subscripts indicating scalar numerical methods. From Tables 6.14 and 6.15, it can be seen that the scalar numerical methods (scalar MM and scalar BEM) ignore the mutual inductance effect between currents on the conductor section of the grounding grid. Therefore, when the frequency is not zero, the imaginary part of the calculated grounding impedance value is relatively small. On the other hand, the absolute values of grounding impedance calculated by various numerical methods are relatively consistent.

For the comparison between scalar numerical algorithms and vector numerical algorithms, this chapter mainly focuses on the Galerkin MM and BEM. As for vector hybrid algorithms based on various basis functions, as this algorithm is a variant of the vector Galerkin MM, no further comparison will be made.

Analysis of 3-d distribution plots of voltage, leakage and branch currents in frequency domains

This chapter presents the 3-d distribution map of leakage current and voltage in the grounding grid based on scalar and vector Galerkin MM and BEM simulation calculations. These 3-d maps can be seen in Figs. 6.3 and 6.4. From Figs. 6.3 and 6.4, it can be seen that the scalar algorithm ignores the influence of current mutual inductance between the conductors of the grounding grid, resulting in a similar voltage distribution on the grounding grid. This is also the source of the equipotential model. The vector algorithm calculates that the voltage distribution on the grounding grid decreases as the distance from the incident point increases, and the voltage distribution on the entire grounding grid is completely different, which is also the

Table 6.14 Calculated grounding impedance (Ω) of grounding grid with $100m \times 100$ m case

Freq	MM(e)	BEM(e)	MM	BEM	PI 1 U case	PI 2 U case	PI 2 I case	PE 1 U case	PE 2 U case	PE 2 I case	T case	TPE case
0Hz	$1.23\angle 0$	$1.23\angle 0$	$1.26\angle 0$	$1.25\angle 0$	$1.23\angle 0$	$1.20\angle 0$	$1.23\angle 0$	$1.28\angle 0$	$1.20\angle 0$	$1.23\angle 0$	$1.23\angle 0$	$1.23\angle 0$
500Hz	$1.23\angle \frac{-1}{10^2}$	$1.23\angle \frac{-1}{10^2}$	$1.29\angle 5.3$	$1.28\angle 5.3$	$1.24\angle 8.2$	$1.24\angle 8.2$	$1.24\angle 6.0$	$1.24\angle 8.2$	$1.24\angle 8.2$	$1.24\angle 6.0$	$1.24\angle 8.1$	$1.24\angle 8.2$
50kHz	$1.23\angle \frac{-5}{10}$	$1.23\angle \frac{-5}{10}$	$6.42\angle 38.3$	$6.42\angle 38.3$	$6.61\angle 39.8$	$6.61\angle 39.7$	$6.61\angle 39.7$	$6.60\angle 39.7$	$6.61\angle 39.7$	$6.61\angle 39.7$	$6.60\angle 39.8$	$6.61\angle 39.8$

Table 6.15 Calculated grounding impedance (Ω) of grounding grid with $10m \times 10$ m case

Freq	MM(e)	BEM(e)	MM	BEM	PI 1 U case	PI 2 U case	PI 2 I case	PE 1 U case	PE 2 U case	PE 2 I case	T case	TPE case
0Hz	$6.26\angle 0$	$6.27\angle 0$	$6.92\angle 0$	$6.45\angle 0$	$6.06\angle 0$	$6.46\angle 0$	$6.26\angle 0$	$6.26\angle 0$	$6.46\angle 0$	$6.26\angle 0$	$6.68\angle 0$	$6.06\angle 0$
500Hz	$6.24\angle \frac{-3}{10^3}$	$6.26\angle \frac{-3}{10^3}$	$6.34\angle 0.1$	$6.25\angle 0.1$	$6.22\angle 3.3$	$6.22\angle 3.4$	$6.25\angle 0.1$	$6.21\angle 3.7$	$6.22\angle 3.4$	$6.25\angle 0.1$	$6.25\angle 1.2$	$6.21\angle 3.9$
50kHz	$6.24\angle \frac{-3}{10}$	$6.25\angle \frac{-3}{10}$	$6.58\angle 9.0$	$6.57\angle 8.9$	$6.41\angle 10.5$	$6.42\angle 10.5$	$6.41\angle 10.0$	$6.41\angle 10.6$	$6.42\angle 10.5$	$6.41\angle 10.0$	$6.42\angle 11.2$	$6.41\angle 10.5$

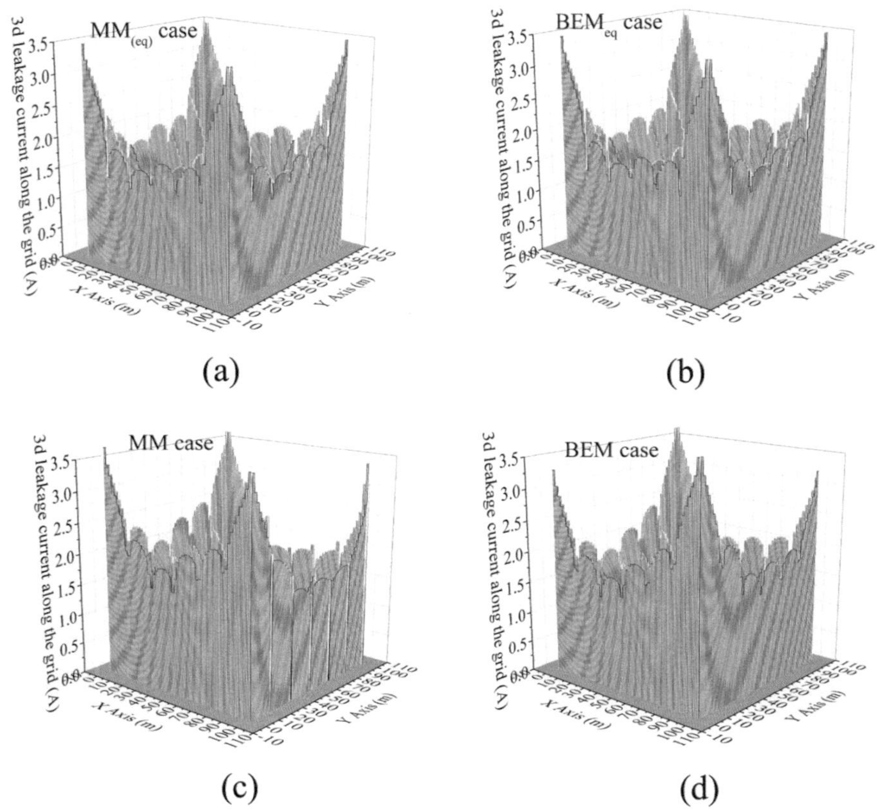

Fig. 6.3 3-d distribution of leakage current along the grid (eq and un case)

source of the unequal potential model. After comparing the voltage distribution, let's compare the leakage current distribution. It can be seen that the distribution of leakage current in the grounding grid obtained by the scalar algorithm is basically based on the complete symmetry of the center point of the grounding grid. The leakage current distribution on the grounding grid obtained by the vector algorithm also undergoes an increase in the degree of deformation of the symmetry relative to the center point as the distance from the incident point increases.

After comparing the differences and similarities in the settlement results between scalar and vector algorithms, this chapter further compares the calculation results between various vector algorithms. Due to the consistency of algorithms between the vector Galerkin MM and the BEM, the calculation results are basically consistent. Therefore, this chapter only considers the comparison of the results obtained by the vector Galerkin MM and various vector hybrid algorithms. Figures 6.5, 6.6, and 6.7 respectively show the 3-d distribution of voltage, branch current, and leakage current on the grounding grid calculated by various vector algorithms.

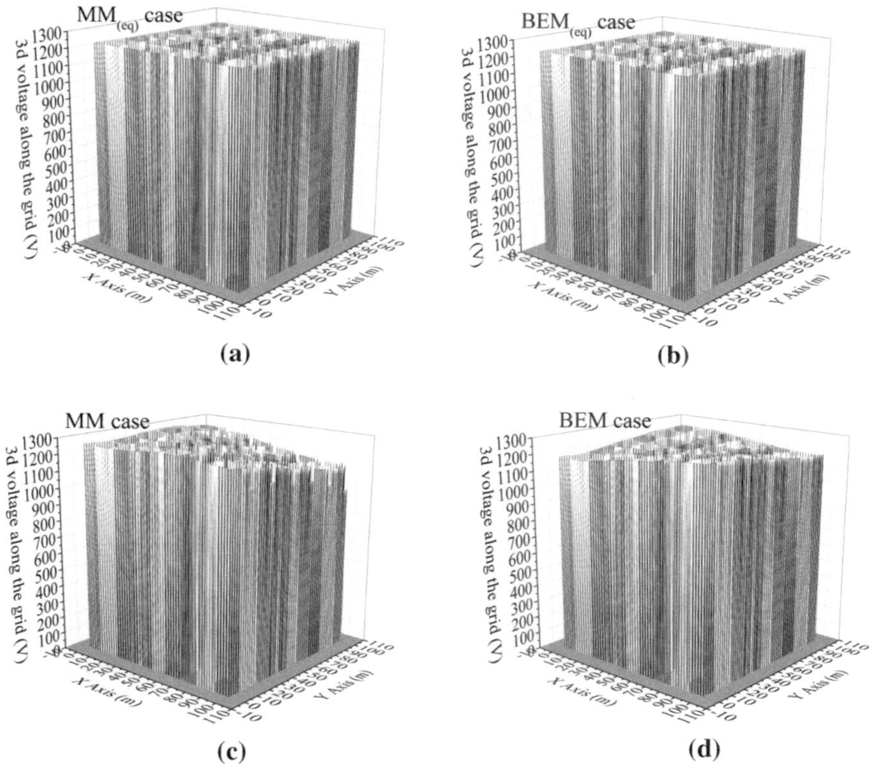

Fig. 6.4 3-d distribution of voltage along the grid (eq and un case)

From Fig. 6.5, it can be seen that although the voltage distribution on the grounding grid is generally consistent among various vector algorithms, there are also differences. It is obvious that the voltage distribution on the grounding grid calculated by the vector Galerkin MM is significantly different from that obtained by various vector hybrid algorithms. The former has extremely obvious unequal distribution, while the latter is relatively less obvious. This is because various vector hybrid algorithms ignore the mutual inductance effect between branch current and leakage current, resulting in a relative reduction in the calculated voltage distribution of the grounding grid. From Fig. 6.6, it can be seen that the distribution of branch currents on the grounding grid calculated by various vector algorithms is basically consistent, except for numerical differences. It can also be seen from Fig. 6.7 that the distribution of leakage current on the grounding grid calculated by various vector algorithms is also basically consistent, except for numerical differences.

The comparison of the above numerical results is based on the situation of a large-sized grounding grid. In order to further illustrate the situation of equipotential and unequal models, the calculation results of small-scale grounding grids are further compared as follows.

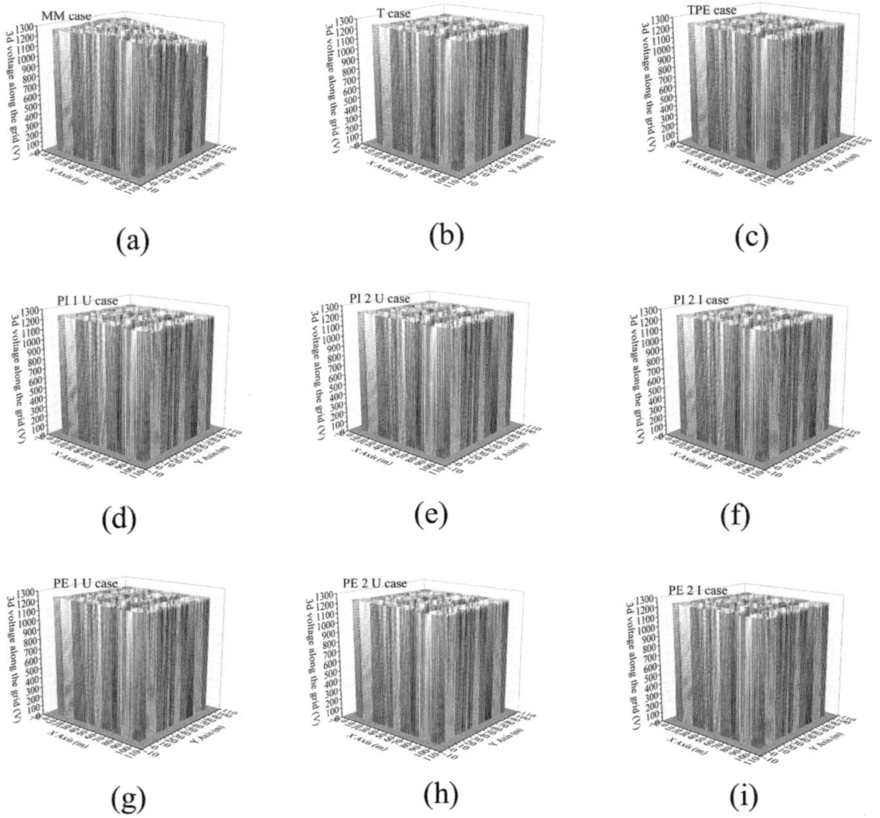

Fig. 6.5 3-d distribution of voltage along the grid

Figures 6.8 and 6.9 respectively provide 3-d distribution maps of voltage and leakage current on small-scale grounding grid, based on scalar and vector (Galerkin MM and BEM) algorithms. Clearly, the results of the two algorithms differ significantly, regardless of size. This also indicates from this perspective that the scalar algorithm based equipotential model has significantly larger computational errors compared to the vector algorithm based equipotential model. Therefore, it is not recommended to use the equipotential model scalar algorithm for the grounding problem of the grounding grid under steady-state conditions.

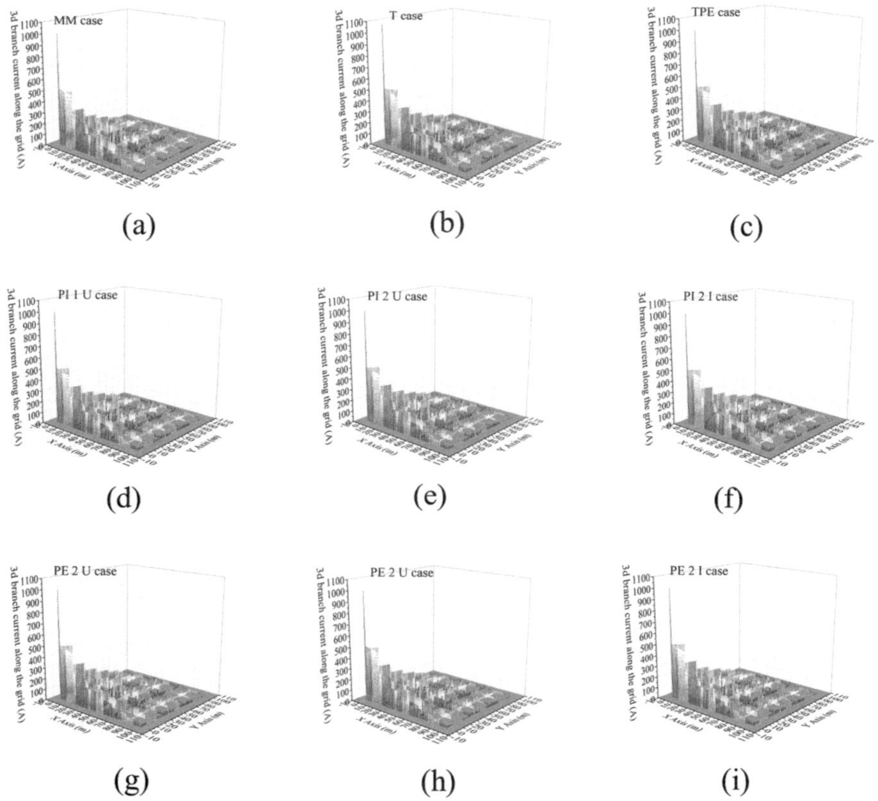

Fig. 6.6 3-d distribution of branch current along the grid

6.7 Summary

This chapter discusses in detail the equipotential grounding grid model based on scalar algorithm and the equipotential grounding grid model based on vector algorithm. By analyzing and discussing numerical simulation examples, it is recommended to use an unequal position grounding grid model based on vector algorithm for steady-state grounding problems.

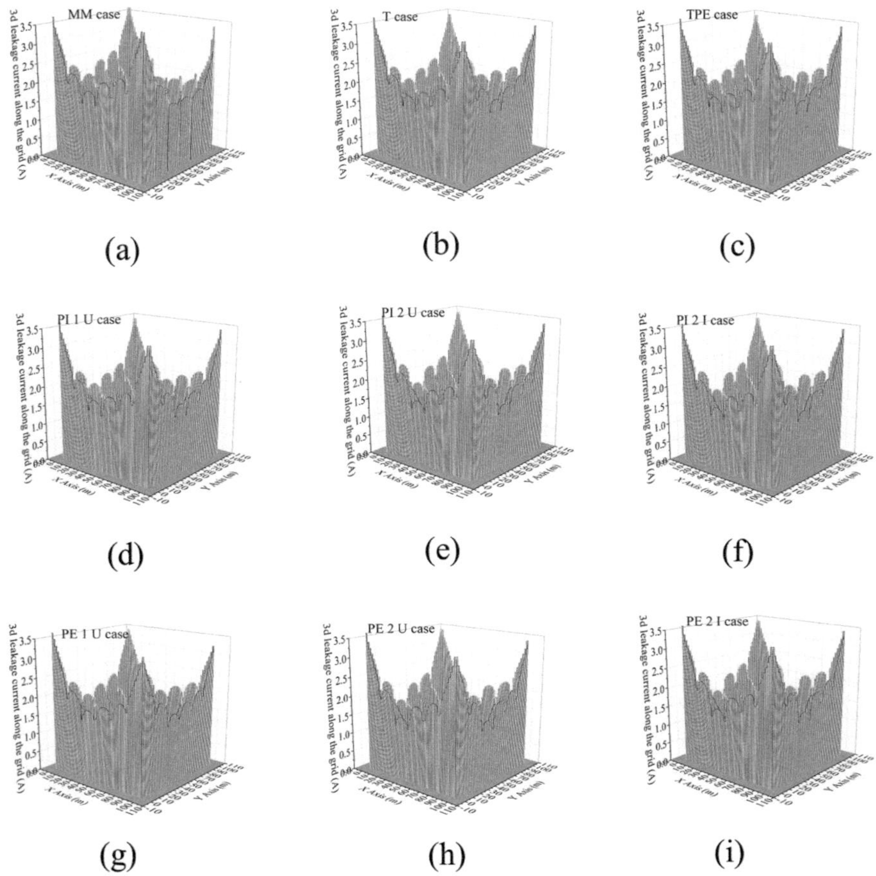

Fig. 6.7 3-d distribution of leakage current along the grid

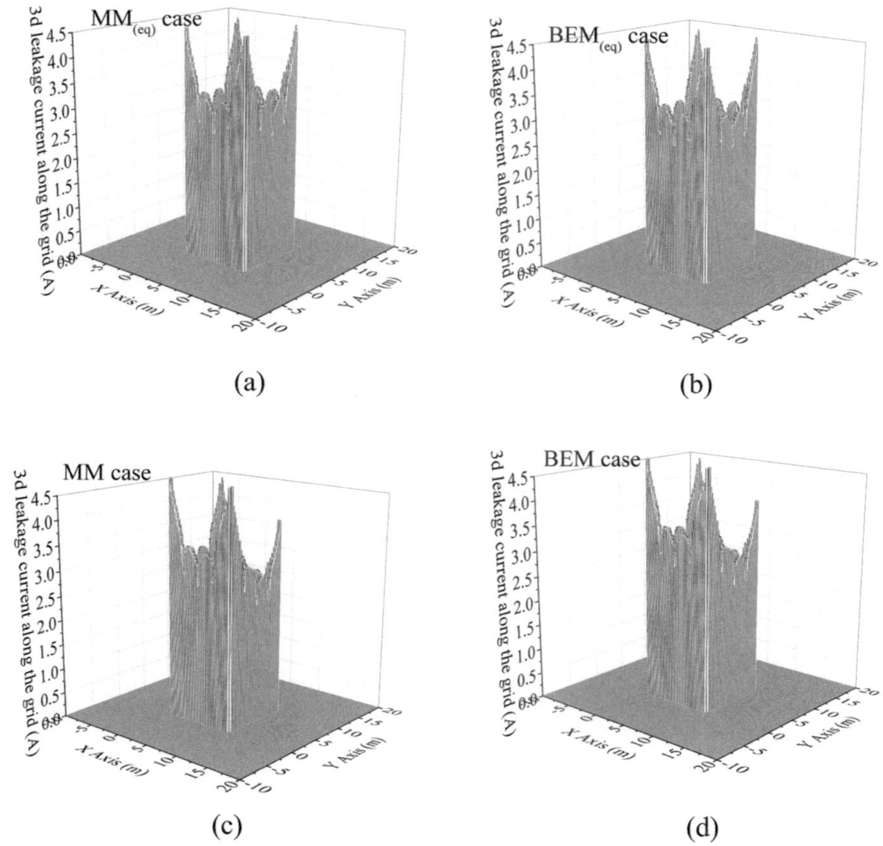

Fig. 6.8 3-d distribution of leakage current along the grid (eq and un case)

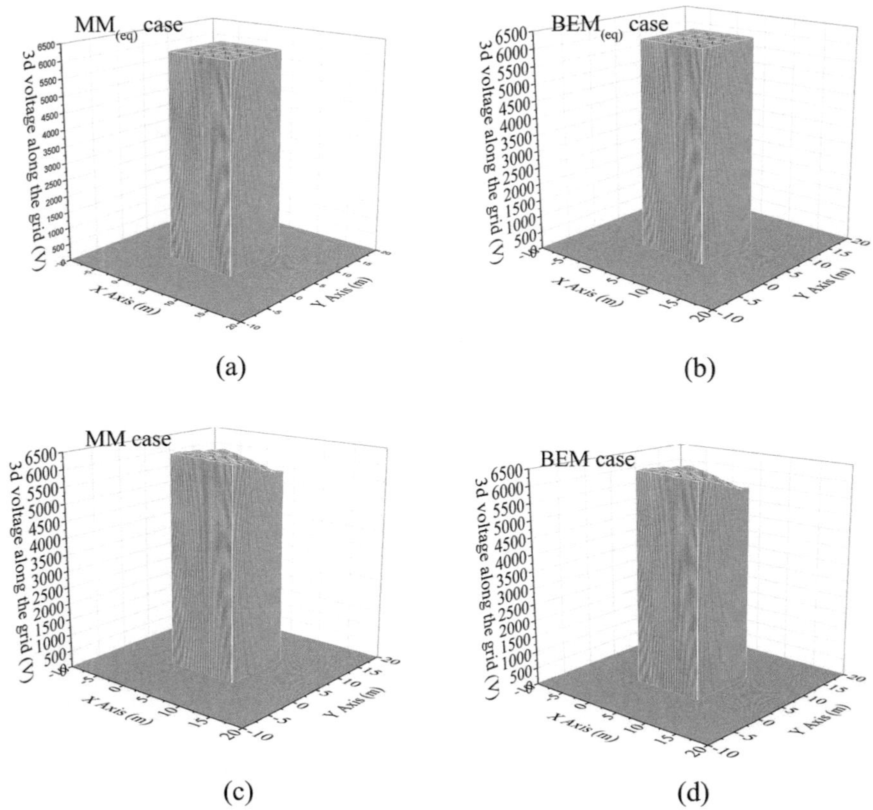

Fig. 6.9 3-d distribution of voltage along the grid (eq and un case) 3-d distribution of voltage along the grid (eq and un case)

References

1. L. Grcev, F.E. Menter, Transient electromagnetic fields near large earthing systems. IEEE Trans. Magn. **32**(3), 1525–1528 (1996)
2. L. Grcev, Computer analysis of transient voltages in large grounding systems. IEEE Trans. Power Delivery **11**(2), 815–823 (1996)
3. L. Grcev, M. Heimbach, Frequency dependent and transient characteristics of substation grounding systems. IEEE Trans. Power Delivery **12**(1), 172–178 (1997)
4. V. Arnautovski-Toseva, L.D. Grcev, Electromagnetic analysis of horizontal wire in two-layered soil. J. Comput. Appl. Mathem. **168**, 21–29 (2004)
5. A.-T. Vesna, G. Leonid, Image and exact models of a vertical wire penetrating a two-layered earth. IEEE Trans. Electromagn. Compat. **53**(4), 968–976 (2011)
6. D. Cavka, B. Harrat, D. Poljak, B. Nekhoul, K. Kerroum, K. El Khamlichi, Drissi, Wire antenna versus modified transmission line approach to the transient analysis of grounding grid. Eng. Anal. Boundary Elements **35**, 1101–1108 (2011)
7. D. Cavka, D. Poljak, On the evaluation of input impedance and transient impedance for grounding electrodes using antenna theory. COMPEL Int. J. Comput. Mathem. Electr. Electron. Eng. **32**(6), 2045–2062 (2013)
8. C. Makhlouf, N. Bachir, P. Dragan, K. Kamal, E.K.D. Khalil, Simplified dipole concept for the assessment of transient electromagnetic field in the vicinity of grounding grid. Int. J. Numer. Model. Electron. Netw. Dev. Fields **28**(4), 404–418 (2015)
9. B. Markovski, L. Grcev, V. Arnautovski-Toseva, Fast and accurate transient analysis of large grounding systems in multilayer soil. IEEE Trans. Power Delivery **36**(2), 589–606 (2021)
10. L. Zhong-Xin, Yu. Yin, W. Ke-Chao, Lightning response of a grounding system buried in multilayered earth based on the galerkin's boundary element and quasi-static complex image methods. IEEE Trans. Electromagn. Compat. **63**(4), 1118–1127 (2021)
11. L. Grcev, F. Dawalibi, An electromagnetic model for transients in grounding systems. IEEE Trans. Power Delivery **5**(4), 1773–1781 (1990)
12. R. Andolfato, L. Bernardi, Fellin, Aerial and grounding system analysis by the shifting complex images method. IEEE Trans. Power Delivery **15**(3), 1001–1009 (2000)
13. H. Karami, K. Sheshyekani, F. Rachidi, Mixed-potential integral equation for full-wave modeling of grounding systems buried in a lossy multilayer stratified ground. IEEE Trans. Electromagn. Compat. **59**(5), 1505–1513 (2017)
14. H. Karami, K. Sheshyekani, F. Rachidi, Harmonic impedance of grounding electrodes buried in a horizontally stratified multilayer ground: A full-wave approach. IEEE Trans. Electromagn. Compat. **60**(4), 899–906 (2018)
15. A.R.J. de Arao, S. Kurokawa, C.M. Seixas, B. Kordi, Optimization of tower-footing grounding impedance for guyed-v transmission towers. Electr. Power Syst. Res. **177**, 105947 (2019)
16. N. Moein, M. Rouzbeh, F. Simon, P. Dawalibi Farid, R. Farhad, Impact of frequency-dependent soil models on grounding system performance for direct and indirect lightning strikes. IEEE Trans. Electromagn. Compat. **63**(1), 134–144 (2021)
17. F. Dawalibi, Electromagnetic fields generated by overhead and buried short conductors. part 2 ground networks. IEEE Power Eng. Rev. PER-6(10), 33–34 (1986)
18. A.D. Papalexopoulos, A.P. Meliopoulos, Frequency dependent characteristics of grounding systems. IEEE Trans. Power Delivery **2**(4), 1073–1081 (1987)
19. F.P. Dawalbi, R.D. Southey, Analysis of electrical interference from power lines to gas pipelines part i: Computation methods. IEEE Power Eng. Rev. **9**(7), 70–70 (1989)
20. A.F. Otero, J. Cidras, J.L. del Alamo, Frequency-dependent grounding system calculation by means of a conventional nodal analysis technique. IEEE Trans. Power Delivery **14**(3), 873–878 (1999)
21. L. Huang, D.G. Kasten, Modeling of ground grid and metallic structure currents in high voltage a.c. substations for the computation of electromagnetic fields. Electr. Power Syst. Res. **59**(1):31–37 (2001)

22. S. Visacro, A. Soares, Hem: a model for simulation of lightning-related engineering problems. IEEE Trans. Power Delivery **20**(2), 1206–1208 (2005)
23. Z.-X. Li, W.-J. Chen, J.-B. Fan, L. Jia-Yu, A novel mathematical modeling of grounding system buried in multilayer earth. IEEE Trans. Power Delivery **21**(3), 1267–1272 (2006)
24. Z.-X. Li, W.-J. Chen, Numerical simulation grounding system buried within horizontal multilayer earth in frequency domain. Commun. Numer. Methods Eng. **23**, 11–27 (2006)
25. Z.X. Li, B.J. Fan, Numerical calculation of grounding system in low-frequency domain based on the boundary element method. Int. J. Numer. Methods Eng. **73**, 685–705 (2008)
26. S. Vujevic, P. Sarajcev, Potential distribution for a harmonic current point source in horizontally stratified multilayer medium. COMPEL Int. J. Comput. Mathem. Electr. Electron. Eng. **27**(3), 624–637 (2008)
27. P. Sarajcev, S. Vujević, D. Lovrić, Time-harmonic current distribution on conductor grid in horizontally stratified multilayer medium. Progress Electromagn. Res. B **31**, 67–87 (2011)
28. S. Vujevic, P. Sarajcev, D. Lovric, Time-harmonic analysis of grounding system in horizontally stratified multilayer medium. Electr. Power Syst. Res. **83**(1), 28–34 (2012)
29. Z.X. Li, G.F. Li, J.B. Fan, Y. Yin, A novel mathematical model for the lightning response of a grounding system buried in multilayered earth based on the quasi-static complex image method. Europ. Ttrans. Electr. Power **22**, 831–845 (2012)
30. Z.X. Li, Y. Yin, C.-X. Zhang, L.-C. Zhang, A mathematical model for the transient lightning response from grounding systems. Progress Electromagn. Res. B **57**, 47–61 (2014)
31. Z. X. Li, Y. Yin, C.-X. Zhang, L.-C. Zhang, Numerical simulation of currents distribution along grounding grid buried in horizontal multi-layered earth in frequency domain based dynamic state complex image method. Electr. Power Components Syst. **43**(14), 1573–1582 (2015)
32. A. Rodolfo, C. Salvatore, Transient behavior of wind towers grounding systems under lightning strikes. Int. J. Energy Environ. Eng. **7**, 235–247 (2016)
33. A. Rodolfo, A. Marco, C. Antonio, H. Pedro, A. Rafael, Closed-form approximation for grounding grids transient analysis. J. Control Autom. Electr. Syst. **32**, 796–806 (2021)
34. P. Kaushal, C. Kandasamy, Transient behaviour of grounding systems in multilayer soil under lightning strikes. Electr. Eng. **104**, 1205–1218 (2022)
35. P. Yutthagowith, A. Ametani, N. Nagaoka, Y. Baba, Application of the partial element equivalent circuit method to analysis of transient potential rises in grounding systems. IEEE Trans. Electromagn. Compat. **53**(3), 726–736 (2011)
36. H. Chen, Y. Du, Lightning grounding grid model considering both the frequency-dependent behaviorand ionization phenomenon. IEEE Trans. Electromagn. Compat. **61**(1), 157–165 (2019)
37. Z.-X. Li, K.-L. Gao, S.-W. Rao, Lightning response of a grounding system buried in multiple layers in earth using the peec method based on the quasi-static complex image method. Electr. Power Components Syst. **48**(3), 291–303 (2020)
38. J.S. Yuan, H.N. Yang, L.P. Zhang, X. Cui, X.S. Ma, Simulation of substation grounding grids with unequal-potential. IEEE Trans. Magn. **36**(4), 1468–1472 (2000)
39. Z.X. Li, K.L. Gao, Y. Yin, D. Ge, Transient lightning responses of grounding systems buried in horizontal multilayered earth with a hybrid method. J. Electrost. **72**(5), 381–386 (2014)
40. Z.X. Li, Y. Yin, C.X. Zhang, L.C. Zhang, Numerical simulation of a grounding grid buried in horizontal multilayered earth based on "t" typical basic elements in the frequency domain. IET Sci. Measur. Technol. **9**(6), 645–653 (2015)
41. Y. Dan, Z. Zhang, H. Zhao, Y. Li, H. Ye, J. Deng, A novel segmented sampling numerical calculation method for grounding parameters in horizontally multilayered soil. Int. J. Electr. Power Energy Syst. **126**, 106586 (2021)
42. J. Salari, C. Portela, Grounding systems modeling including soil ionization. IEEE Trans. Power Delivery **23**(4), 1939–1945 (2008)
43. E.K.N. Yung, An innovative analysis of earthing grids for power substations. IEE Proc. **132**(5), 251–256 (1985)
44. J. Ma, F.P. Dawalibi, Modern computational methods for the design and analysis of power system grounding, in *POWERCON '98. 1998 International Conference on Power System Technology. Proceedings (Cat. No.98EX151)*, vol. 1, pp. 122–126 (1998)

45. F.P. Dawalibi, J. Ma, R.D. Southey, Behaviour of grounding systems in multilayer soil: a parameteric analysis. IEEE Trans. Power Delivery **9**(1), 334–342 (1994)
46. J.X. Ma, F.P. Dawalibi, Influence of inductive coupling between leads on ground impedance measurements using the fall-of-potential method. IEEE Trans. Power Delivery **16**(4), 739–743 (2001)

Chapter 7
Numerical Methods for Grounding Problem at Transient State I

Various frequency-domain numerical methods for grounding problems in the case of lightning transient shock response based on equipotential electromagnetic field modeling.

7.1 Preface

The frequency domain 1-d spatial scalar Galerkin MM and BEM, which are used for the steady-state grounding problem of grounding grid, can be further used for the transient grounding problem of grounding grid through Fourier transform. The scientific and effective nature of this method will be discussed in detail in this chapter.

7.2 Introduction

Although the scalar spatial 1-d Galerkin BEM and the MM combined with the Fourier transform method can be used to numerically simulate the lightning transient response of substation grounding grids in multi-layered horizontally stratified soils, the numerical algorithms ignore the mutual inductance effects between discrete conductor segments of substation grounding grids, which will lead to significant computational errors. As a result, no relevant algorithmic literature has appeared in the public domain. The reason why this chapter is organized in this book is due to the need for the evolution of the numerical algorithms developed in this book.

Z.-X. Li, *Numerical Simulation Algorithm of Electromagnetic Field for Grounding Problems in Power System Substation Grounding Grids*,
https://doi.org/10.1007/978-981-96-0904-8_7

7.3 The Frequency Domain Spatial 1-d Scalar Galerkin BEM for Transient Grounding Problems

In this chapter, we discuss in detail how the 1-d spatial scalar Galerkin BEM simulates the transient grounding of grounding grid buried in horizontal N_e-layer earth model.

7.3.1 *The Frequency Domain Spatial 1-d Scalar Galerkin BEM*

Referring to the formula (5.13) related to the 1-d spatial scalar Galerkin BEM in Chap. 5, this formula can be rewritten as follows.

$$[\bar{\bar{\mathbf{Z}}}_p][\bar{\mathbf{I}}_p] = [\bar{\mathbf{V}}_p] \tag{7.1}$$

where $[\bar{\bar{\mathbf{Z}}}_p]$ is the $N_p \times N_p$ matrix about mutual impedance coefficient between a pair of segments in grounding system, $[\bar{\mathbf{I}}_p]$ is a $N_p \times 1$ vector for leakage currents emanating from grounding system's discrete conductor's node, $[\bar{\mathbf{V}}_p]$ is a $N_p \times 1$ vector for voltage of grounding system's discrete conductor's node. It must be pointed out that the coefficients matrix in linear system Eq. (7.1) is symmetric and positive definite [1].

How to effectively solve this formula will be described in detail in the relevant section of the previous Chap. 5, so this chapter will not go into detail. However, it must be pointed out that once this formula is solved, the leakage current on all conductors of the entire grounding grid can be obtained at a certain frequency. Moreover, the ground impedance value and the voltage value of the incident point can be further solved at this frequency.

7.3.2 *Spatial 1-d Scalar BEM for Frequency Domain Equipotential Electromagnetic Field Mathematical Model Based on Fourier Transform for Transient Grounding Problems*

Traditionally, the transient grounding problem of substation grounding grid can be simulated and calculated by using Fourier transform to convert the transient grounding problem in time domain into the steady-state grounding problem in frequency domain. The main steps are:

- First, the continuous lightning wave time domain function is reasonably evolved into a discrete lightning wave time domain function.

- Then the discrete lightning wave time domain function is converted into a discrete lightning wave frequency domain function through Fourier transform method
- The steady-state response of lightning current at various discrete frequencies was achieved using the 1-d spatial scalar Galerkin BEM to obtain the electromagnetic field response of the grounding grid at each frequency point.
- Finally, the frequency domain response of the electromagnetic field on the grounding grid at each frequency point was transformed into the time domain response of the grounding grid field at each time point through inverse Fourier transform.

The above content is the solution idea for the response of the entire time-domain lightning wave current on the grounding grid of the substation. Specifically, the 1-d spatial scalar Galerkin BEM used in this chapter mainly solves the transient voltage response of lightning inflow into the grounding grid.

7.4 The Frequency Domain Spatial 1-d Scalar Galerkin MM for Transient Grounding Problems

This chapter will discuss in detail how the 1-d spatial scalar Galerkin MM can simulate the transient grounding problem of grounding grid.

7.4.1 *The Frequency Domain Spatial 1-d Scalar Galerkin MM*

This chapter will review and summarize the content related to the fourth chapter and the scalar MM, as detailed below.

Referring to the formula (5.13) related to the 1-d spatial scalar Galerkin MM in Chap. 5, this formula can be rewritten as follows.

$$[\bar{\bar{\mathbf{Z}}}_b][\bar{\mathbf{I}}_b] = [\bar{\mathbf{V}}_b] \tag{7.2}$$

where $[\bar{\bar{\mathbf{Z}}}_b]$ is the $N_b \times N_b$ matrix about mutual impedance coefficient between a pair of segments in grounding system, $[\bar{\mathbf{I}}_b]$ is a $N_b \times 1$ vector for leakage currents emanating from grounding system's discrete conductor's node, $[\bar{\mathbf{V}}_b]$ is a $N_b \times 1$ vector for voltage of grounding system's discrete conductor's node. It must be pointed out that the coefficients matrix in linear system Eq. (7.2) is symmetric and positive definite [1].

How to effectively solve this formula will be described in detail in the relevant section of the previous Chap. 5, so this chapter will not go into detail. However, it must be pointed out that once this formula is solved, the leakage current on all conductors of the entire grounding grid can be obtained at a certain frequency. Moreover, the ground impedance value and the voltage value of the incident point can be further solved at this frequency.

7.4.2 Spatial 1-d Scalar MM for Frequency Domain Equipotential Electromagnetic Field Mathematical Model Based on Fourier Transform for Transient Grounding Problems

The approach to solving the 1-d spatial scalar Galerkin BEM in the previous chapter is exactly the same. Traditionally, the transient grounding problem of a substation grounding grid is also simulated and calculated through Fourier transform, transforming the transient grounding grid problem in the time domain into a steady-state grounding grid problem in the frequency domain. The main steps are also roughly the same and will not be repeated. It should be pointed out that the steady-state frequency domain response of the grounding grid is solved using the 1-d spatial scalar Galerkin MM.

Similarly, specifically, the 1-d spatial scalar Galerkin MM used in this chapter mainly solves the transient voltage response of lightning flowing into the grounding grid.

7.5 Simulation Result and Analysis

According to the approach introduced in this chapter, a program has been implemented. The program can simulate a lightning response of grounding system buried in horizontal up to four-layer earth model, in which the fast convergent 1-d spatial scalar Galerkin MM and BEM with zero order basic functions has been used.

7.5.1 Verification of Accuracy and Effectiveness for Spatial 1-d Scalar Galerkin MM and BEM

To verify the accuracy and effectiveness of the method proposed in this chapter, some cases solved by other authors are studied.

The first example is from [2]. The grounding grid configuration can be seen in Fig. 7.1a. The grounding grid was made of round copper conductors with a cross-sectional area of 50 mm^2 and was buried at 0.5 m. The injected lightning current parameters were set to $T_1 = 65\,\mu s$, $T_2 = 185\,\mu s$, and $I_m = 6.86$ A, and the feed point was set at the corner of the grid. For the earth model, the conductivities and permittivities of the two-layer earth model were set to $\frac{\sigma_1=50^{-1} S/m}{\sigma_2=20^{-1} S/m}$, $\frac{\varepsilon_1=10\varepsilon_0}{\varepsilon_2=10\varepsilon_0}$, and $h = 0.6$m. Figure 7.2a

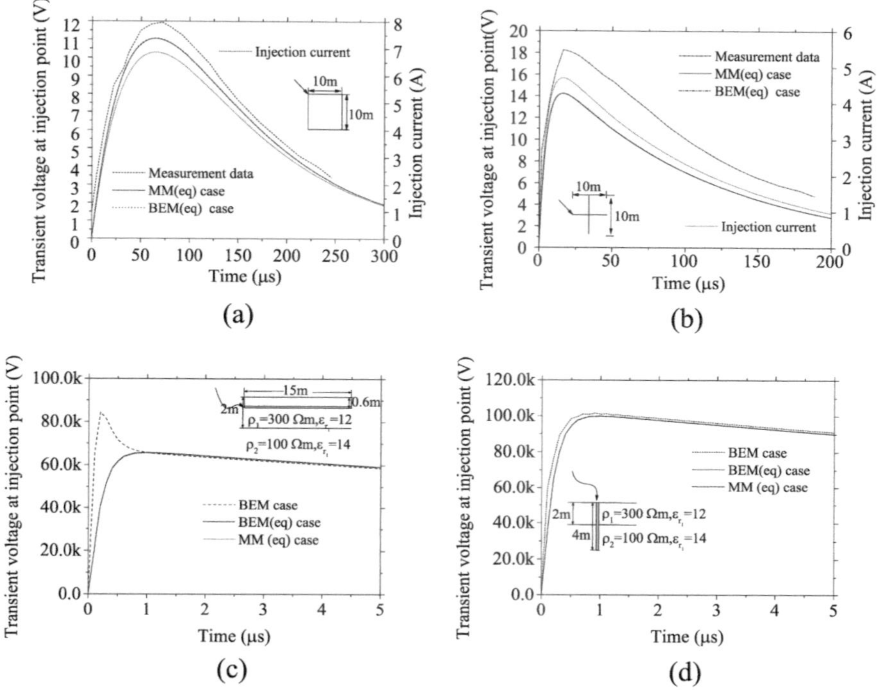

Fig. 7.1 Verified case a

shows the transient voltage curve simulated based on frequency-domain 1-d spatial scalar Galerkin MM and BEM and the measured data [2]. The calculated voltage curves were in good agreement with the measured curve in [2]. It should be noted that in this example, the lightning wave is relatively gentle and the size of the grounding grid is relatively small. Therefore, the equipotential model algorithm developed in this chapter is relatively effective and accurate.

The second verification case came from [3]. The grounding grid configuration can be seen in Fig. 7.1b. The grounding grid was also made of round copper conductors with cross-sectional area 50 mm^2 and was buried at 0.5 m. The injected lightning current parameters were set to $T_1 = 16\,\mu s$, $T_2 = 99\,\mu s$, and $I_m = 4.7A$. For the earth model, the conductivities and permittivities of the two-layer earth model were set to $\frac{\sigma_1=50^{-1}S/m}{\sigma_2=20^{-1}S/m}$, $\frac{\varepsilon_1=10\varepsilon_0}{\varepsilon_2=10\varepsilon_0}$, and $h = 0.6$m. Figure 7.2b shows the transient voltage curve simulated based on 1-d spatial scalar Galerkin MM and BEM, and the data [3]. The calculated voltage curve is very close to the development trend of the measured curve, but there are certain errors. The calculated voltage curve is very close to the development trend of the measured curve, but there are certain errors. This is because the lightning wave waveform in this example is slightly steeper compared to the previous example, and the size of the grounding grid is also relatively small.

The last test case comes from [4, 5]. Figure 7.1c, d shows, respectively, horizontal and vertical grounding electrodes with radius of 12 mm, and the length of horizontal and vertical grounding electrodes is 15m and 4m, respectively. The horizontal grounding electrode was buried at a depth of 0.6m, and both of the electrodes were within in two-layer horizontal earth where the resistivity and relative permittivity ratios for the upper and the lower soil layers were $\frac{\rho_1=300\Omega m}{\rho_2=100\Omega m}$, $\frac{\varepsilon_{r_1}=12}{\varepsilon_{r_2}=14}$, and the upper layer thickness was $H = 2m$. The injected lightning current was set with parameters of: $T_1 = 0.36\,\mu s$, $T_2 = 26.8\,\mu s$, and $I_0 = 3.0$ kA. Figure 7.1c, d show the transient voltage curves of horizontal and vertical grounding rods under lightning current impact, respectively. These curves are derived from literature [4] and the equipotential model algorithm developed in this chapter. From Fig. 7.1c, it can be seen that for horizontal grounding rods, the curve obtained by the equipotential model algorithm developed in this chapter is significantly different from the curve given in Ref. [4], with significant errors. As shown in Fig. 7.1d, for vertical grounding rods, the curve obtained by the equipotential model algorithm developed in this chapter is basically consistent with the curve given in Ref. [4], with relatively small errors. This is because the lightning wave waveform in this calculation example is relatively steep, and the correctness of the equipotential model algorithm developed in this chapter is questionable for horizontal grounding rods, while for vertical grounding rods, the correctness of the equipotential model algorithm developed in this chapter is still acceptable.

From the above three examples, it can be seen that due to the neglect of the mutual inductance effect between the axial branch currents between the grounding grid conductors in the substation, the spatial 1-d scalar Galerkin MM and BEM in the time domain may cause significant deviation or even error in the numerical simulation results of lightning currents with relatively high steepness. This is because lightning currents with relatively high steepness contain more high-frequency components, and in high-frequency situations, the influence of mutual inductance is relatively more significant. Therefore, ignoring the influence of mutual inductance can lead to calculation errors. So it can be seen that the frequency domain 1d spatial scalar Galerkin MM and BEM developed in this chapter have certain limitations and a relatively small scope of application for lightning transient impulse grounding problems in substation grounding grids.

7.5.2 *Numerical Results Analysis*

A general grounding system can be seen in Fig. 7.2a, which consists of 4 × 4 square grids and the burial depth of the grounding grid is 0.5 meters. The earth is modelled as horizontal four layers conductive media, whose resistivities and permittivities are $\rho_1 = 100\,\Omega$m and $\rho_2 = 220\,\Omega$m, $\rho_3 = 310\,\Omega$m, $\rho_4 = 430\,\Omega$m, $\varepsilon_1 = 10\varepsilon_0$, $\varepsilon_2 = 12\varepsilon_0$, $\varepsilon_3 = 13\varepsilon_0$ and $\varepsilon_4 = 14\varepsilon_0$, respectively. The first, second and third layer thickness are $h_1 = 6$ m, $h_2 = 12$ m and $h_3 = 8$ m. All the conductor radius of grids are

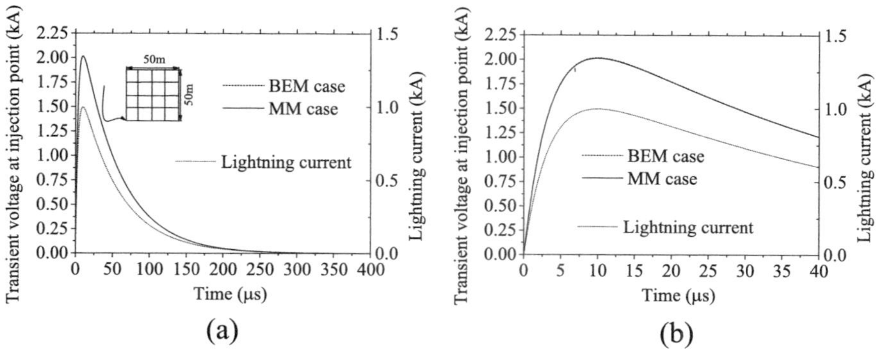

Fig. 7.2 Transient voltage at injection point

5 mm. The external excited lightning current with $1.29 \times (e^{-0.01910t} - e^{-0.292288t})kA$ is injected from the corner of the grounding system. According to the formula of lightning current, T_1 is $10\,\mu s$ and T_2 is $50\,\mu s$.

7.5.3 Analysis and Discussion of Transient Surge Voltages

The transient impulse voltage curve at the incident point obtained through numerical simulation using two scalar algorithms can be seen in Fig. 7.2. It can be known that the two curves overlap with each other, and the calculation accuracy is basically the same, and their correctness is mutually verified.

7.5.4 The Impulse Impedance Results and Discussions

The impulse impedance values of the incident point calculated by the same two scalar algorithms can be seen in Table (7.1). It can also be seen that the impulse grounding impedance of the grounding grid calculated by the two methods are completely the same, indicating the correctness of the two algorithms from another perspective (Table 7.1).

Table 7.1 Calculate impulse impedance

Methods	Impulse impedance
BEM	(2.023, −1.004)
MM	(2.018, −1.002)

Analysis of 3-d distribution plots of voltage and leakage currents in frequency domains

The following two large figures are composed of nine small figures, which respectively provide the 3-d distribution of voltage and leakage current on the grounding grid of the substation at critical frequencies.

Observing Fig. 7.3, it can be seen that the 3-d voltage distribution diagram of the key frequency on the grounding grid of the substation shows the maximum voltage value at zero frequency. As the frequency increases, the voltage gradually weakens until it disappears. At the same time, it can be observed that the amplitude of all frequency voltages is the same, which is because the frequency domain scalar algorithm developed in this chapter is based on the equipotential electromagnetic model.

Observing Fig. 7.4, it can be seen that the 3-d leakage current distribution map of the key frequency on the grounding grid of the substation shows the maximum leakage current value at zero frequency. As the frequency increases, the leakage

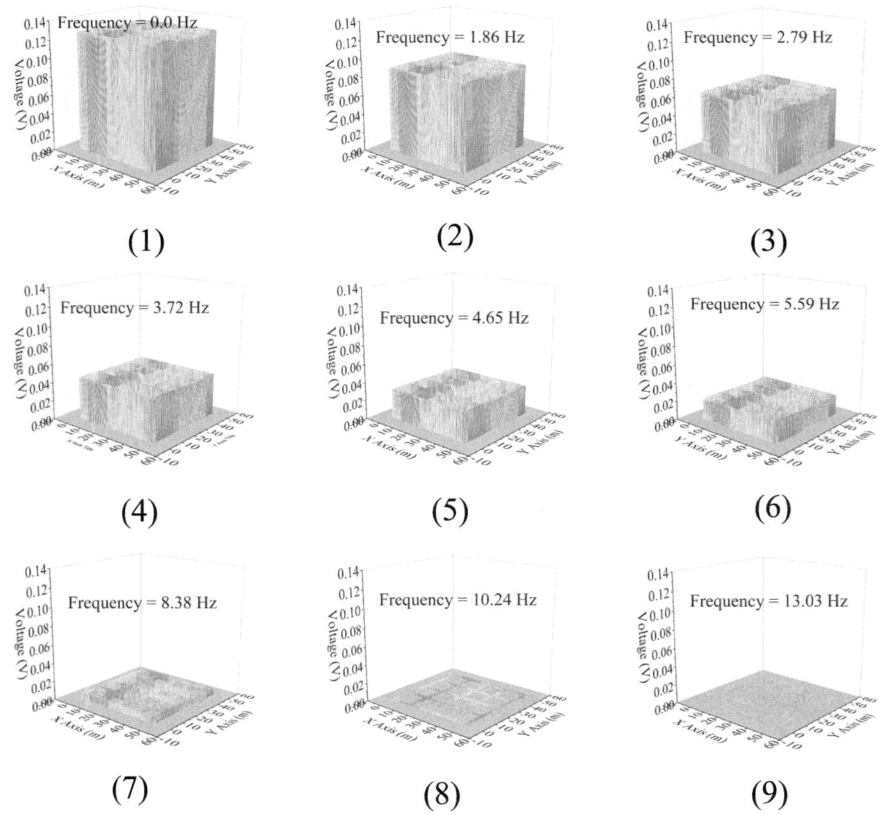

Fig. 7.3 3-d plot for voltage along grounding grid

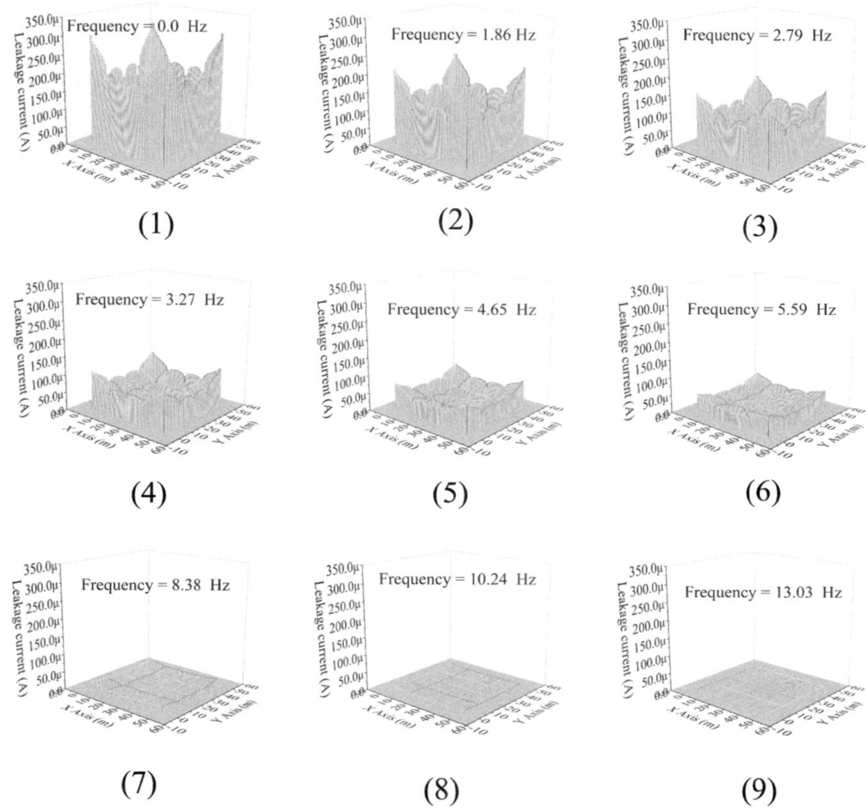

Fig. 7.4 3-d plot for leakage current along grounding grid

current gradually weakens until it disappears. At the same time, it can be observed that the amplitude of leakage current at all frequencies is the same, and the leakage current has the highest amplitude at the four corners of the grounding grid in the substation. From the corners to the middle point of the edge conductors on the four sides of the grounding grid, the leakage current gradually decreases to the relatively lowest. The leakage current on the outermost conductor of the leakage current on the entire grounding grid is relatively large, and then weakens towards the middle point of the grounding grid until it reaches the lowest. This is because the frequency domain scalar algorithm developed in this chapter is based on the equipotential electromagnetic model.

7.6 Summary

The frequency domain spatial 1-d scalar Galerkin MM and BEM developed in this chapter have certain limitations and a relatively small scope of application for lightning transient impulse grounding problems in substation grounding grids.

References

1. C. Johnson, *Numerical Solution of Partial Differential Equations by the Finite Element Method* (Cambridge University Press, Cambridge, U.K., 1987)
2. Z. Stojkovic et al., Experimental investigation of grounding grid impulse characteristic. Europ. Trans. Electr. Power **8**(6), 417–425 (1998)
3. Z. Stojkovic et al., Sensitivity analysis of experimentally determined grounding grid impulse characteristic. IEEE Trans. Power Delivery **13**(3), 1136–1142 (1998)
4. Z.-X. Li, Y. Yu, K.-C. Wang, Lightning response of a grounding system buried in multilayered earth based on the galerkin's boundary element and quasi-static complex image methods. IEEE Trans. Electromagn. Compat. **63**(4), 1118–1127 (2021)
5. M. Jamali, M. Niasati, M. Jazaeri, A two-layer soil model for the calculation of electrical parameters of grounding systems under lightning strikes. Electr. Power Components Syst. **47**(1–2), 181–191 (2019)

Chapter 8
Numerical Methods for Grounding Problem at Transient State II

Various frequency-domain numerical methods for grounding problems in the case of lightning transient shock response based on unequal potential electromagnetic field modeling.

8.1 Preface

Like the scalar numerical method in the previous chapter, the frequency domain 1-d spatial vector Galerkin MM and BEM, as well as the vector hybrid algorithm based on various basis functions, which are used for the steady-state grounding problem of the grounding grid, can also be further used for the transient grounding problem of the grounding grid through Fourier transform. This chapter will discuss in detail the scientificity and effectiveness of these methods.

8.2 Introduction

This chapter mainly studies and discusses the frequency domain numerical methods that combine Fourier transform to simulate the lightning transient impact grounding problem of substation grounding grid, including spatial 1-d vector Galerkin BEM, vector Galerkin MM, and vector Galerkin hybrid algorithm based on various basis functions.

In fact, many literature has published articles [1–11] on the BEM. Among them, articles [1–8] mainly simulate the lightning transient impulse grounding problem of the substation grounding grid in a uniform earth model, while recently published articles [10, 11] simulate the lightning transient impulse grounding problem of the substation grounding grid in a multi-layer horizontal layered soil model that is closer to engineering practice.

Z.-X. Li, *Numerical Simulation Algorithm of Electromagnetic Field for Grounding Problems in Power System Substation Grounding Grids*,
https://doi.org/10.1007/978-981-96-0904-8_8

Regarding the Galerkin MM, the following articles [12–18] have numerically simulated the lightning transient impulse grounding problem of the substation grounding grid. Among them, references [12–16] numerically simulated the lightning transient impulse grounding problem of the substation grounding grid in a uniform earth, while reference [17, 18] numerically simulated the lightning transient impulse grounding problem of the substation grounding grid in a multi-layer horizontal layered earth model that is closer to engineering practice.

There is also a lot of literature [19–22] on numerical algorithms based on the FEM for numerical simulation of lightning transient impulse response of substation grounding grids in a uniform earth, while literature [23] is about moment of line in numerical methods for electromagnetic fields. Meanwhile, there are many literature [24–39] studies discussing the numerical algorithms based on the transmission line method for numerically modeling the lightning transient impulse response of substation grounding grids in a uniform earth. As well as the literature [40] study discusses the problem of numerical algorithms based on the EMPT method for the numerical simulation of the lightning transient shock response of substation grounding grids in a uniform earth. And some literature [41–44] based on the circuit approach for numerical simulation of the lightning transient shock response problem of substation grounding grids in a uniform earth.

More literature [45–60] is on vector Galerkin hybrid algorithms based on various basis functions for simulating lightning transient impulse grounding problems in substation grounding grids. These literature include Galerkin hybrid algorithms based on PI basis functions [45–55], PEEC basis functions [56–58], T basis functions [59], and TPEEC basis functions [60]. Similarly, the hybrid numerical method introduced in references [45–49, 56, 57, 60] is based on the lightning transient impulse grounding problem of the substation grounding grid in a uniform earth, while the hybrid numerical method presented in references [50–55, 58, 59] is based on the lightning transient impulse grounding problem of the substation grounding grid in multi-layer horizontal layered soil that is closer to engineering practice. It should be noted here that the articles [11, 49–51, 58, 59] published by the authors of this book are about numerical algorithms of frequency-domain electromagnetic fields for numerical simulation of the grounding problem of lightning transient shock response of substation grounding grids in multilayer horizontally layered soils, whereas the literature [49–51, 58, 59] is a hybrid algorithm based on various basis functions in the numerical algorithms of frequency-domain electromagnetic fields, and literature [11] is a hybrid algorithm in the numerical algorithms of frequency-domain Galerkin BEM algorithm in EMF numerical algorithms.

8.3 The Frequency Domain Spatial 1-d Vector Galerkin BEM for Transient Grounding Problems

This chapter will discuss in detail how to scientifically simulate lightning transient grounding of substation grounding grid buried in horizontal N_e-layer earth model by 1-d spatial vector Galerkin BEM.

8.3.1 The Frequency Domain Spatial 1-d Vector Galerkin BEM

This chapter will review and summarize the relevant content of the 1-d spatial vector Galerkin BEM in Chap. 6, as follows.

Referring to the formula (6.7) related to the 1-d spatial vector Galerkin BEM in Chap. 6, this formula can be rewritten as follows.

$$[\overline{\overline{\mathbf{Z}}}_p][\bar{\mathbf{I}}_p] = 0 \tag{8.1}$$

where $[\overline{\overline{\mathbf{Z}}}_p]$ is $N_p \times N_p$ matrix, who includes the mutual impedance and induction interaction betweens a parts of conductor segments currents. $[\bar{\mathbf{I}}_p]$ is $N_p \times 1$ vector of unknown currents.

How to effectively solve this formula will be described in detail in the relevant chapters of the first five chapters, so this chapter will not elaborate further.

However, it must be pointed out that once this formula is solved, the leakage current, branch current, and node voltage on all conductor sections of the entire discrete grounding grid can be obtained at a certain frequency. In addition, the grounding impedance and voltage values at the incident point of the grounding grid can be further calculated at this frequency.

8.3.2 Spatial 1-d Vector BEM for Frequency Domain Unequal Potential Electromagnetic Field Mathematical Model Based on Fourier Transform for Transient Grounding Problems

Traditionally, the lightning transient impulse grounding problem of substation grounding grid can be simulated and calculated through Fourier transform, transforming the time-domain lightning transient impulse grounding grid problem into the frequency-domain steady-state grounding grid problem.

The main steps include:

Firstly, the continuous lightning wave time domain function is reasonably evolved into a discrete lightning wave time domain function.

Then, the discrete lightning wave time domain function is transformed into a discrete lightning wave frequency domain function through Fourier transform.

Subsequently, the spatial 1-d vector Galerkin BEM was used to obtain the steady-state response of lightning current at different discrete frequencies, in order to obtain the electromagnetic field response of the grounding grid at each frequency point.

Finally, the frequency domain response of the electromagnetic field at each frequency point on the grounding grid is converted into the time domain response of the grounding grid field at each time point through the inverse Fourier transform.

The above content is a solution to the entire time domain lightning wave current response of the substation grounding grid. Specifically, the spatial 1-d vector Galerkin BEM used in this chapter mainly solves the transient voltage response of lightning flowing into the grounding grid.

8.4 The Frequency Domain Spatial 1-d Vector Galerkin MM for Transient Grounding Problems

This chapter will discuss in detail how to scientifically simulate lightning transient grounding of substation grounding grid by 1-d spatial vector Galerkin MM.

8.4.1 The Frequency Domain Spatial 1-d Vector Galerkin MM

This chapter will review and summarize the relevant content of the spatial 1-d vector Galerkin MM in Chap. 6. The core content is as follows.

Referring to the formula (6.24) related to the spatial 1-d vector Galerkin MM in Chap. 6, the formula can be rewritten as follows.

$$\begin{bmatrix} \overline{\overline{\mathbf{Z}}}_b^{00} & \overline{\overline{\mathbf{Z}}}_b^{01} \\ \overline{\overline{\mathbf{Z}}}_b^{10} & \overline{\overline{\mathbf{Z}}}_b^{11} \end{bmatrix} \begin{bmatrix} \bar{\mathbf{I}}_b^{0} \\ \bar{\mathbf{I}}_b^{1} \end{bmatrix} = 0 \tag{8.2}$$

where the $[\bar{\mathbf{I}}^{0}]_b$ is $N_b \times 1$ vector of unknown branch currents and the $[\bar{\mathbf{I}}_b^{1}]$ is $N_b \times 1$ vector of unknown leakage currents; $[\overline{\overline{\mathbf{Z}}}_b^{00}]$ is $N_b \times N_b$ matrix, who presents the mutual induction interaction betweens a parts of conductor segments branch currents, $[\overline{\overline{\mathbf{Z}}}_b^{01}]$ is $N_b \times N_b$ matrix, who presents the mutual impedance and induction interaction betweens a parts of conductor segments branch and leakage currents, $[\overline{\overline{\mathbf{Z}}}_b^{10}]$ is $N_b \times N_b$ matrix, who presents the mutual impedance and induction interaction betweens a parts of conductor segments branch and leakage currents, $[\overline{\overline{\mathbf{Z}}}_b^{11}]$ is $N_b \times N_b$ matrix, who presents the mutual impedance and induction interaction betweens a parts of conductor segments leakage currents.

How to accurately and effectively solve this formula has been described in detail in the relevant chapters of Chap. 6, so it will not be elaborated in this chapter.

It should be pointed out that by solving this formula, the leakage current, branch current, and node voltage of all conductor segments of the entire discrete grounding grid can be obtained at a certain frequency. At the same time, it is possible to calculate the grounding impedance and voltage value of the lightning inflow into the grounding grid of the substation at this frequency.

8.4.2 Spatial 1-d Vector MM for Frequency Domain Unequal Potential Electromagnetic Field Mathematical Model Based on Fourier Transform for Transient Grounding Problems

It is completely the same as using the 1-d spatial vector Galerkin BEM in the above chapter to solve the transient impulse grounding problem of lightning waves on the substation grounding grid. Traditionally, the lightning transient impulse grounding problem of substation grounding grid is simulated and calculated through Fourier transform, transforming the time-domain lightning transient impulse grounding grid problem into the frequency-domain steady-state grounding grid grounding problem. The main steps are also roughly the same and will not be repeated. It should be noted that the steady-state frequency domain response of the grounding grid is solved using the spatial 1-d vector Galerkin MM.

Therefore, by combining forward and inverse Fourier transform methods, the spatial 1-d Galerkin MM developed in this chapter can effectively simulate the lightning transient impulse grounding problem of substation grounding grid.

8.5 The Frequency Domain Spatial 1-d Vector Hybrid Methods Based on Different Basic Function (PI, PEEC, T and TPEEC) for Transient Grounding Problems

This chapter will discuss in detail how to scientifically simulate the lightning transient grounding problem of substation grounding grid based on 1-d spatial vector hybrid algorithm based on various basis functions.

8.5.1 *The Frequency Domain Spatial 1-d Vector Hybrid Methods Based on Different Basic Function (PI, PEEC, T and TPEEC)*

The application of hybrid algorithms based on various basis functions in the steady-state grounding problem of substation grounding grids is the foundation for solving the lightning transient impulse grounding problem of substation grounding grids. Below, the relevant content is reviewed and summarized as follows.

The frequency domain spatial 1-d vector hybrid methods based on PI basic function

This chapter will review and summarize the relevant content of the spatial 1-d vector hybrid algorithm based on PI basis functions in Chap. 6, as shown below.

For the spatial 1-d vector hybrid algorithm based on PI basis function, there are two solutions to the impedance matrix: (1) considering the impedance matrix of the mutual influence between leakage currents on two conductor segments, and (2) considering the impedance matrix of the mutual influence between leakage currents on two conductor segments connected by two nodes. It can be abbreviated as the branch matrix approach and the node matrix approach, respectively. The relevant concatenation formulas can be given separately as follows.

- The branch matrix approach for the spatial 1-d vector hybrid algorithm based on PI basis function

$$[\bar{\bar{\mathbf{Z}}}_b + j\omega\bar{\bar{\mathbf{M}}}_b][\bar{\mathbf{I}}_b] + [\bar{\bar{\mathbf{A}}}]^T[\bar{\mathbf{V}}_p] = 0 \tag{8.3}$$

$$-[\bar{\bar{\mathbf{A}}}][\bar{\mathbf{I}}_b] + [\bar{\bar{\mathbf{K}}}]^T[\bar{\bar{\mathbf{Z}}}_{s_b}]^{-1}[\bar{\bar{\mathbf{K}}}][\bar{\mathbf{V}}_p] = [\bar{\mathbf{F}}] \tag{8.4}$$

- The node matrix approach for the spatial 1-d vector hybrid algorithm based on PI basis function

$$[\bar{\bar{\mathbf{Z}}}_b + j\omega\bar{\bar{\mathbf{M}}}_b][\bar{\mathbf{I}}_b] + [\bar{\bar{\mathbf{A}}}]^T[\bar{\mathbf{V}}_p] = 0 \tag{8.5}$$

$$-[\bar{\bar{\mathbf{A}}}][\bar{\mathbf{I}}_b] + [\bar{\bar{\mathbf{Z}}}_{s_p}]^{-1}[\bar{\mathbf{V}}_p] = [\bar{\mathbf{F}}] \tag{8.6}$$

In order to reduce the unknown quantity and thus reduce the computational complexity, the two consecutive formulas mentioned above can be further simplified as shown below.

- The branch matrix approach

$$\left[[\bar{\bar{\mathbf{K}}}]^T[\bar{\bar{\mathbf{Z}}}_{s_b}]^{-1}[\bar{\bar{\mathbf{K}}}]+[\bar{\bar{\mathbf{A}}}]\left[[\bar{\bar{\mathbf{Z}}}_b]+j\omega[\bar{\bar{\mathbf{M}}}_b]\right]^{-1}[\bar{\bar{\mathbf{A}}}]^T\right][\bar{\mathbf{V}}_p]=[\bar{\mathbf{F}}] \tag{8.7}$$

  For the convenience of explaining various numerical algorithms in the future, this formula is named PI 1 U case.
- The node matrix approach

$$[\bar{\bar{\mathbf{Z}}}_{s_p}]^{-1}[\bar{\mathbf{V}}_p]+[\bar{\bar{\mathbf{A}}}]\left[[\bar{\bar{\mathbf{Z}}}_b]+j\omega[\bar{\bar{\mathbf{M}}}_b]\right]^{-1}[\bar{\bar{\mathbf{A}}}]^T[\bar{\mathbf{V}}_p]=[\bar{\mathbf{F}}] \tag{8.8}$$

  For the convenience of explaining various numerical algorithms in the future, this formula is named PI 2 U case.

$$\left[[\bar{\bar{\mathbf{Z}}}_b+j\omega\bar{\bar{\mathbf{M}}}_b]+[\bar{\bar{\mathbf{A}}}]^T[\bar{\bar{\mathbf{Z}}}_{s_p}][\bar{\bar{\mathbf{A}}}]\right][\bar{\mathbf{I}}_b]=-[\bar{\bar{\mathbf{A}}}]^T[\bar{\bar{\mathbf{Z}}}_{s_p}][\bar{\mathbf{F}}] \tag{8.9}$$

  For the convenience of explaining various numerical algorithms in the future, this formula is named PI 2 I case.

It must be pointed out that formulas (8.7) and (8.8) both take node voltage as an unknown quantity, while formula (8.9) takes branch current as an unknown quantity. How to accurately and effectively solve these three formulas is discussed and introduced in Chap. 6, and will not be elaborated here.

The frequency domain spatial 1-d vector hybrid methods based on PEEC basic function

This chapter will review and summarize the relevant content of the spatial 1-d vector hybrid algorithm based on PEEC basis functions in Chap. 6, as shown below.

For the spatial 1-d vector hybrid algorithm based on PEEC basis function, there are also two solutions to the impedance matrix: (1) considering the impedance matrix of the mutual influence between leakage currents on two conductor segments, and (2) considering the impedance matrix of the mutual influence between leakage currents on two conductor segments connected by two nodes. It can also be abbreviated as the branch matrix approach and the node matrix approach, respectively. The relevant concatenation formulas can be given separately as follows.

- The branch matrix approach for the spatial 1-d vector hybrid algorithm based on PEEC basis function

$$[\bar{\bar{\mathbf{Z}}}_b+j\omega\bar{\bar{\mathbf{M}}}_b][\bar{\mathbf{I}}_b]+[\bar{\bar{\mathbf{A}}}]^T[\bar{\mathbf{V}}_p]=0 \tag{8.10}$$

$$-[\bar{\bar{\mathbf{A}}}][\bar{\mathbf{I}}_b]+[\bar{\bar{\mathbf{K}}}]^T[\bar{\bar{\mathbf{Z}}}_{s_b}]^{-1}[\bar{\bar{\mathbf{K}}}][\bar{\mathbf{V}}_p]=[\bar{\mathbf{F}}] \tag{8.11}$$

where the connotation of these matrices or vectors ($[\bar{\bar{\mathbf{Z}}}_b]$, $[\bar{\bar{\mathbf{M}}}_b]$, $[\bar{\bar{\mathbf{A}}}]$, $[\bar{\mathbf{V}}_p]$, $[\bar{\mathbf{F}}]$ and $[\bar{\mathbf{I}}_b]$) is the same as the corresponding matrices or vectors in the original formula (8.3) (8.4). Except for matrices $[\bar{\bar{\mathbf{Z}}}_{s_b}]$ and $[\bar{\bar{\mathbf{K}}}]$, where $[\bar{\bar{\mathbf{Z}}}_{s_b}]$ is a $2N_b \times 2N_b$ matrix, and $[\bar{\bar{\mathbf{K}}}]$ is a $2N_b \times N_p$ matrix.

- The node matrix approach for the spatial 1-d vector hybrid algorithm based on PEEC basis function

$$[\bar{\bar{\mathbf{Z}}}_b + j\omega\bar{\bar{\mathbf{M}}}_b][\bar{\mathbf{I}}_b] + [\bar{\bar{\mathbf{A}}}]^T[\bar{\mathbf{V}}_p] = 0 \tag{8.12}$$

$$-[\bar{\bar{\mathbf{A}}}][\bar{\mathbf{I}}_b] + [\bar{\bar{\mathbf{Z}}}_{s_p}]^{-1}[\bar{\mathbf{V}}_p] = [\bar{\mathbf{F}}] \tag{8.13}$$

where all matrices and vectors here have exactly the same form and content as the corresponding matrices and vectors of the original Eq. (8.5) (8.4). However, when calculating the matrix $[\bar{\bar{\mathbf{Z}}}_{s_p}]$, due to the need to further subdivide the leakage current branch in the hybrid algorithm based on PEEC basis functions, compared with the hybrid algorithm based on PI basis functions, the length of the leakage current branch is half smaller.

In order to reduce the unknown quantity and thus reduce the computational complexity, the two consecutive formulas mentioned above can be further simplified as shown below.

- The branch matrix approach

$$\left[[\bar{\bar{\mathbf{K}}}]^T[\bar{\bar{\mathbf{Z}}}_{s_b}]^{-1}[\bar{\bar{\mathbf{K}}}] + [\bar{\bar{\mathbf{A}}}]\left[[\bar{\bar{\mathbf{Z}}}_b] + j\omega[\bar{\bar{\mathbf{M}}}_b]\right]^{-1}[\bar{\bar{\mathbf{A}}}]^T\right][\bar{\mathbf{V}}_p] = [\bar{\mathbf{F}}] \tag{8.14}$$

For the convenience of explaining various numerical algorithms in the future, this formula is named PEEC 1 U case.

- The node matrix approach

$$[\bar{\bar{\mathbf{Z}}}_{s_p}]^{-1}[\bar{\mathbf{V}}_p] + [\bar{\bar{\mathbf{A}}}]\left[[\bar{\bar{\mathbf{Z}}}_b] + j\omega[\bar{\bar{\mathbf{M}}}_b]\right]^{-1}[\bar{\bar{\mathbf{A}}}]^T[\bar{\mathbf{V}}_p] = [\bar{\mathbf{F}}] \tag{8.15}$$

For the convenience of explaining various numerical algorithms in the future, this formula is named PEEC 2 U case.

$$\left[[\bar{\bar{\mathbf{Z}}}_b + j\omega\bar{\bar{\mathbf{M}}}_b] + [\bar{\bar{\mathbf{A}}}]^T[\bar{\bar{\mathbf{Z}}}_{s_p}][\bar{\bar{\mathbf{A}}}]\right][\bar{\mathbf{I}}_b] = -[\bar{\bar{\mathbf{A}}}]^T[\bar{\bar{\mathbf{Z}}}_{s_p}][\bar{\mathbf{F}}] \tag{8.16}$$

For the convenience of explaining various numerical algorithms in the future, this formula is named PEEC 2 I case.

It must be pointed out that formulas (8.14) and (8.15) both take node voltage as an unknown quantity, while formula (8.16) takes branch current as an unknown quantity. How to accurately and effectively solve these three formulas is discussed and introduced in Chap. 6, and will not be elaborated here.

The frequency domain spatial 1-d vector hybrid methods based on T basic function

This chapter will review and summarize the relevant content of the spatial 1-d vector Galerkin hybrid algorithm based on T-basis function in Chap. 6, as shown below.

There is only one solution for the T-based spatial 1-d vector Galerkin hybrid algorithm, as shown below.

$$\left[\left[\bar{\bar{\mathbf{A}}}_{pb}\bar{\bar{\mathbf{Y}}}_{bb}\bar{\bar{\mathbf{A}}}_{bp}^{T}\right]-\left[\bar{\bar{\mathbf{A}}}_{pb}\bar{\bar{\mathbf{Y}}}_{bb}\bar{\bar{\mathbf{A}}}_{bb}^{T}\left[\bar{\bar{\mathbf{A}}}_{bb}\bar{\bar{\mathbf{Y}}}_{bb}\bar{\bar{\mathbf{A}}}_{bb}^{T}+\bar{\bar{\mathbf{Z}}}_{s_b}^{-1}\right]^{-1}\bar{\bar{\mathbf{A}}}_{bb}\bar{\bar{\mathbf{Y}}}_{bb}\bar{\bar{\mathbf{A}}}_{bp}^{T}\right]\right][\bar{\mathbf{V}}_{p}]=[\bar{\mathbf{F}}] \tag{8.17}$$

For clarity in the following description, this formula can be abbreviated as T case.

How to accurately and effectively solve this formula has been discussed and introduced in Chap. 6, and will not be repeated here.

The frequency domain spatial 1-d vector Galerkin hybrid methods based on TPEEC basic function

This chapter will review and summarize the relevant content of the spatial 1-d vector Galerkin hybrid algorithm based on TPEEC-basis function in Chap. 6, as shown below.

There is also only one solution for the TPEEC-based spatial 1-d vector Galerkin hybrid algorithm as below.

$$\left[[\bar{\bar{\mathbf{D}}}-\bar{\bar{\mathbf{C}}}]\frac{1}{2}[\bar{\bar{\mathbf{Z}}}]_{s_b}^{-1}[\bar{\bar{\mathbf{K}}}]-[\bar{\bar{\mathbf{D}}}+\bar{\bar{\mathbf{C}}}][\bar{\bar{\mathbf{Z}}}_{b}+j\omega\bar{\bar{\mathbf{M}}}_{b}]^{-1}[\bar{\bar{\mathbf{A}}}]^{T}\right][\bar{\mathbf{V}}_{p}]=[\bar{\mathbf{F}}_{p}] \tag{8.18}$$

For clarity in the following description, this formula can be abbreviated as TPEEC case.

How to accurately and effectively solve this formula has also been discussed and introduced in Chap. 6, and will not be repeated here.

8.5.2 Spatial 1-d Vector Galerkin Hybrid Methods Based on Different Basic Function (PI, PEEC, T and TPEEC) for Frequency Domain Unequal Potential Electromagnetic Field Mathematical Model Based on Fourier Transform for Transient Grounding Problems

It is also completely the same as using the 1-d spatial vector Galerkin BEM in the above chapter to solve the transient impulse grounding problem of lightning waves on the substation grounding grid.

Traditionally, the lightning transient impulse grounding problem of substation grounding grid is simulated and calculated through Fourier transform, transforming the time-domain lightning transient impulse grounding grid problem into the frequency-domain steady-state grounding grid grounding problem. The main steps are also roughly the same and will not be repeated. It should be noted that the steady-state frequency domain response of the grounding grid is solved using the spatial 1-d vector Galerkin hybrid methods based on different basic function (PI, PEEC, T and TPEEC).

Therefore, by combining forward and backward Fourier transform methods, the spatial 1-d vector Galerkin hybrid methods based on different basic function (PI, PEEC, T and TPEEC) developed in this chapter can effectively simulate the lightning transient impulse grounding problem of substation grounding grid.

8.6 Numerical Results and Discussions for the Various Algorithms

Based on the numerical algorithms of various unequal electromagnetic models introduced above, a transient impulse response program for the substation grounding grid has been developed, which can simulate the lightning transient impulse grounding problem of the substation grounding grid in a four layer horizontal layered earth.

8.6.1 Verification of the Various Algorithms

To verify the accuracy and effectiveness of the method proposed in this chapter, some cases solved by other authors are studied.

The first example is from [61]. The grounding grid configuration can be seen in Fig. 8.1a. The grounding grid was made of round copper conductors with a cross-sectional area of 50 mm^2 and was buried at 0.5 m. The injected lightning current parameters were set to $T_1 = 65\,\mu\text{s}$, $T_2 = 185\,\mu\text{s}$, and $I_m = 6.86$ A, which was generated by a

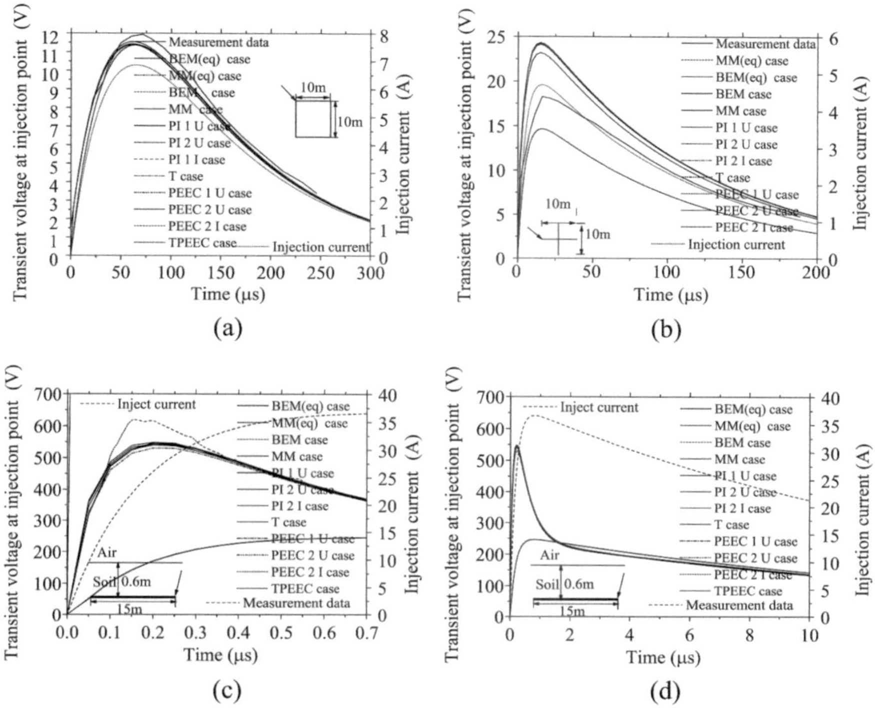

Fig. 8.1 Validation of the impact response of four simple geometries of grounding grids or rods

double exponential function, and the feed point was set at the corner of the grid. For the earth model, the conductivities and permittivities of the two-layer earth model were set to $\frac{\sigma_1=50^{-1}S/m}{\sigma_2=20^{-1}S/m}$, $\frac{\varepsilon_1=10\varepsilon_0}{\varepsilon_2=10\varepsilon_0}$, and $h = 0.6$ m. Figure 8.2a shows the transient voltage curve simulated based on frequency-domain 1-d spatial scalar and vector Galerkin MM and BEM, and vector Galerkin hybrid methods dependant on different basic functions (including PI, PEEC, T an TPEEC and etc), and the measured data [61]. The calculated voltage curves were in good agreement with the measured curve in [61]. It should be noted that in this example, the lightning wave is relatively gentle and the size of the grounding grid is relatively small. Therefore, the equipotential model algorithm developed in this chapter is relatively effective and accurate.

The second verification case came from [62]. The grounding grid configuration can be seen in Fig. 8.1b. The grounding grid was also made of round copper conductors with cross-sectional area 50 mm^2 and was buried at 0.5 m. The injected lightning current parameters were set to $T_1 = 16\,\mu$s, $T_2 = 99\,\mu$s, and $I_m = 4.7$ A, which was generated by a double exponential function. For the earth model, the conductivities and permittivities of the two-layer earth model were set to $\frac{\sigma_1=50^{-1}S/m}{\sigma_2=20^{-1}S/m}$, $\frac{\varepsilon_1=10\varepsilon_0}{\varepsilon_2=10\varepsilon_0}$, and $h = 0.6$ m. Figure 8.2b shows the transient voltage curve simulated based on 1-d spatial scalar and vector Galerkin MM and BEM, and vector Galerkin hybrid methods

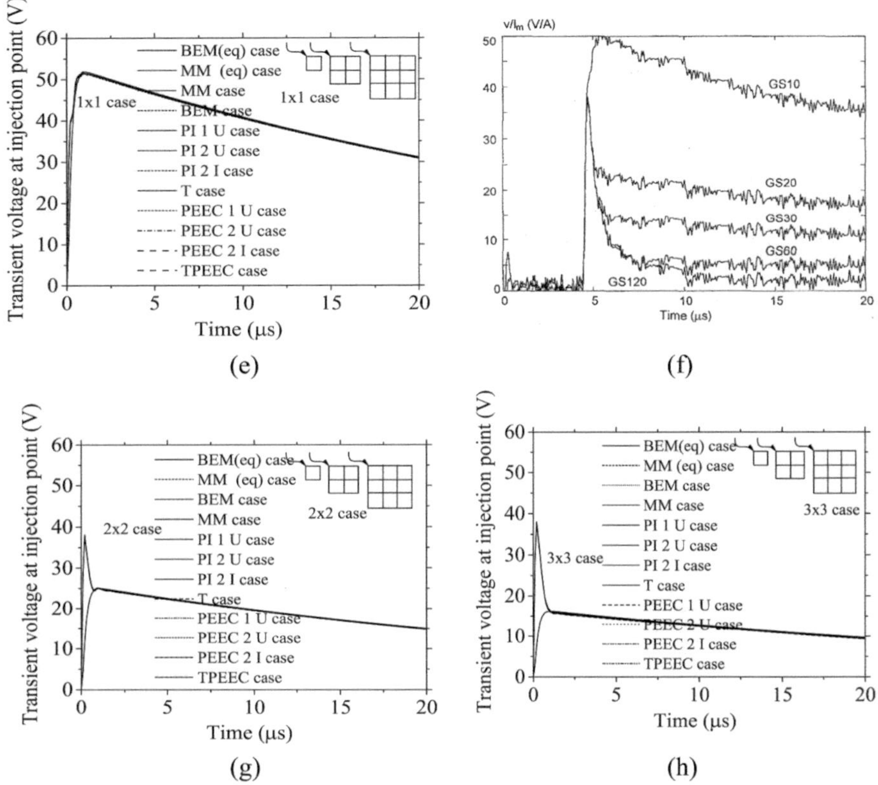

Fig. 8.2 Validation of different lightning wave impact responses for three grounding grids

based on different basic functions (including PI, PEEC, T and TPEEC and etc.), and the data [62]. The calculated voltage curve is very close to the development trend of the measured curve, but there are certain errors. The calculated voltage curve is very close to the development trend of the measured curve, but there are certain errors. This is because the lightning wave waveform in this example is slightly steeper compared to the previous example, and the size of the grounding grid is also relatively small.

Because the steepness of the lightning wave considered in the above two examples were not high, the third example considered the shock of a lightning wave with high steepness. In the second verification case from [1, 11], the length of horizontal conductor was 15 m, and the conductor was buried at 0.6 m. The soil parameters relating to conductivity and relativity permittivity were 70 Ωm and 15, respectively. The conductor radius was 12 mm. The injected lightning current parameters were set to $T_1 = 0.8\,\mu\text{s}$, $T_2 = 12.5\,\mu\text{s}$, and $I_m = 36.5$ A, which was generated by a double exponential function, and the feed point was set at the end of the conductor. Figure 8.1c, d show the transient voltage curves simulated based on 1-d spatial scalar and vector Galerkin MM and BEM, and vector Galerkin hybrid methods dependant on different basic functions (including PI, PEEC, T , TPEEC and etc.), and the measured data [1]. From Fig. 8.1c, it can be seen that the curves come from unequal potential models are basically consistent with the measurement curves, while the equipotential model algorithm developed in this chapter yields curves that are completely different. Figure 8.1d further provides a longer comparison curve to further illustrate this issue. This example demonstrates that for lightning waves with relatively high steepness, despite the relatively small impact size of the grounding electrode, the calculation results of the equipotential model algorithm developed in this chapter are completely incorrect.

The fourth test case comes from [2, 11]. Three different grounding grid sizes were considered: 10 × 10 m, 20 × 20 m, and 30 × 30 m, with 1 × 1, 2 × 2, and 3 × 3 grids, respectively, as shown in Fig. 8.2e–h. All of the grids were buried at a depth of 0.5 m in homogeneous earth with resistivity = 1000 Ωm and relative permittivity = 9. The injected lightning current was set to $T_1 = 0.36\,\mu s$, $T_2 = 26.8\,\mu s$, and $I_0 = 1$ A, which was generated by a double exponential function, with the feed point, as shown in Fig. 8.2e–h. Figure 8.2e–h show that the calculated transient voltage curves obtained by using the proposed method spatial 1-d scalar and vector Galerkin BEM and MM, and vector Galerkin hybrid methods dependant on different basic functions (PI, PEEC, T, TPEEC and etc) [11]. The corresponding measured curves given in Fig. 8.2f by [2]. From 8.2e–h, it can be seen that for small-scale grounding grids, the transient voltage curve given in reference [11] is basically consistent with the corresponding curve obtained by the equipotential model algorithm developed in this chapter, and it is also basically consistent with the measurement curve. However, for the other two grounding grids with relatively large sizes, the transient voltage curve given in reference [11] is completely different from the corresponding curve obtained by the equipotential model algorithm developed in this chapter, and there is a significant difference. This is because the lightning wave steepness in this example is relatively large, which poses a great challenge to the correctness of the algorithm in this chapter.

The fifth test case comes from [11, 63]. Figure 8.3g, h shows, respectively, horizontal and vertical grounding electrodes with radius of 12 mm, and the length of horizontal and vertical grounding electrodes is 15 m and 4 m, respectively. The horizontal grounding electrode was buried at a depth of 0.6 m, and both of the electrodes were within in two-layer horizontal earth where the resistivity and relative permittivity ratios for the upper and the lower soil layers were $\frac{\rho_1=300\Omega m}{\rho_2=100\Omega m}$, $\frac{\varepsilon_{r_1}=12}{\varepsilon_{r_2}=14}$, and the upper layer thickness was $H = 2$ m. The injected lightning current was set with parameters of: (a) $T_1 = 0.36\,\mu s$, $T_2 = 26.8\,\mu s$, and $I_0 = 3.0$ kA, which was generated by a double exponential function. Figure 8.3g, h show the transient voltage curves of horizontal and vertical grounding rods under lightning current impact, respectively. These curves are derived from literature [11] and the equipotential model algorithm developed in this chapter. From Fig. 8.3g, it can be seen that for horizontal grounding rods, the curve obtained by the equipotential model algorithm developed in this

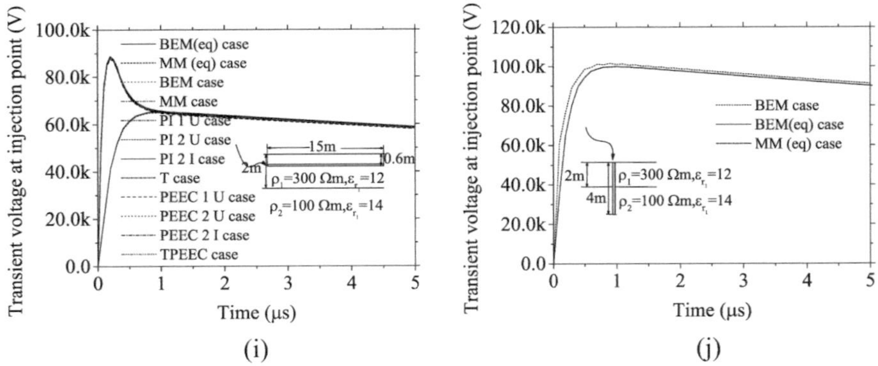

Fig. 8.3 Validation of the impact response of grounding rods in multilayer horizontally stratified soils

chapter is significantly different from the curve given in reference [11], with significant errors. As shown in Fig. 8.3h, for vertical grounding rods, the curve obtained by the equipotential model algorithm developed in this chapter is basically consistent with the curve given in reference [11], with relatively small errors. This is because the lightning wave waveform in this calculation example is relatively steep, and the correctness of the equipotential model algorithm developed in this chapter is questionable for horizontal grounding rods, while for vertical grounding rods, the correctness of the equipotential model algorithm developed in this chapter is still acceptable.

From the above five examples, it can be seen that the 1d spatial scalar Galerkin MM and BEM developed in this chapter have certain limitations and a relatively small scope of application for lightning transient impulse grounding problems in substation grounding grids.

Observing the two subfigures of Fig. 8.5a, b, it can be seen that the curves simulated by the numerical algorithms developed in this book in the frequency and time domains are basically the same as those obtained by the algorithms for electromagnetic field principles (EM) in the original literature [64].

To further validate the various EMF numerical methods in this section, additional experimental validation examples are provided below:

A sixth validation example is given below, where literature [62] describes lightning impulse tests performed against a square grounding grid in two horizontally stratified soil layers. The excitation source for all shock tests was a lightning current waveform formed by a double exponential function. The literature [62] provides transient voltage measurement curves for a total of 19 tests. However, most of the transient voltage curves calculated by the numerical method developed in this book simulated under the same test data do not match each other, which may be due to the lack of accurate measurements of the parameters of the inverse layered soil model. Three of the tests are selected here, and the test-measured transient voltage curves are in general agreement with the numerically simulated transient voltage curves.

Table 8.1 Three choices of experimental biexponential function lightning current parameters

	T_1/T_2 (appointed)	T_1/T_2 (computed)	$\alpha(\mu s^{-1})$	$\beta(\mu s^{-1})$	η	$I_o(A)$
Case 1	7.4/76.0 µs	7.0/76.25 µs	0.0116947	0.2572883	0.823896	10.4
Case 2	10.5/82.0 µs	9.93/82.37 µs	0.0118826	0.165206	0.756818	9.7
Case 3	10.0/81.0 µs	9.44/81.35 µs	0.011854	0.175987	0.767529	9.7

A test grounding grid is given in literature [62] with dimensions of 10 m × 10 m. The test grounding grid consists of four small grids of equal size i.e. $5m \times 5$ m with a cross-sectional area of 50 mm^2 of the grounding conductor. The soil is modeled as a horizontally stratified 2-layer soil with an upper soil resistivity of 50 Ωm and a thickness of 0.6 m and a lower soil resistivity of 20 Ωm. The relative dielectric constant of both layers is 10. The Three selected parameters for the experimental biexponential function of the lightning current are listed in Table 8.1.

The three subplots (a), (b) and (c) of Fig. 8.4 are given to show the transient voltage profiles of the response of the injection point under a three times double exponential function lightning current shock numerically simulated by the six numerical

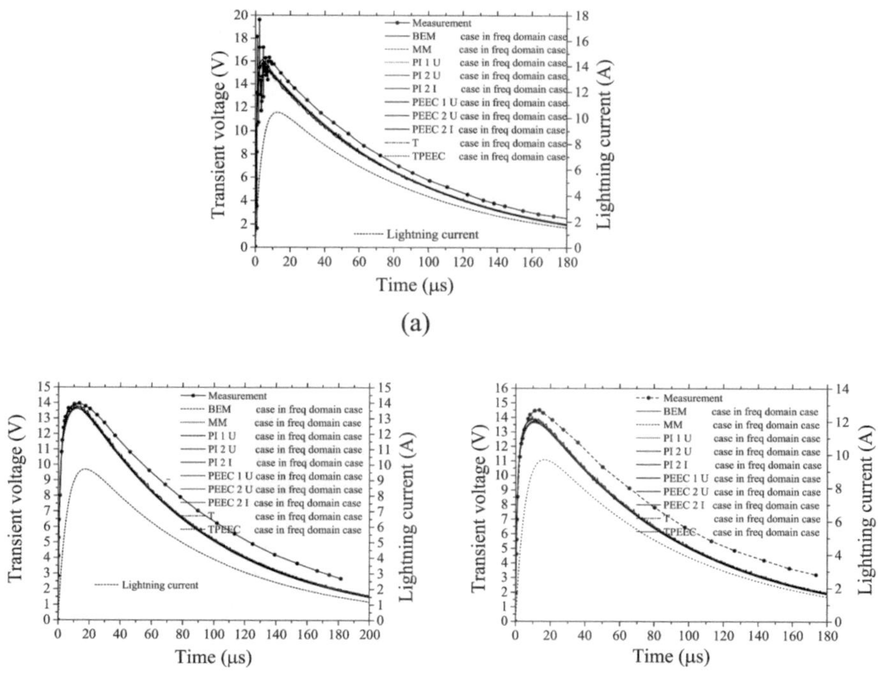

Fig. 8.4 Numerical validation of comparative test results

methods developed in this chapter and the transient voltage profiles obtained from experimental measurements, respectively.

Observing the three subfigures (a), (b) and (c) of Fig. 8.4, it can be known that the transient voltage curves simulated by the six numerical methods developed in this chapter are basically the same as the transient voltage curves of the experimental tests.

The seventh validation example is from the literature [64] and considers a horizontal copper grounding electrode of 30 m length and 0.01 m radius. The electrode is buried in soil at a depth of 1 m with a soil resistivity of 200 Ωm and a relative dielectric constant of 10. Two double-exponential lightning currents are defined as the fast-front current and the slow-front current. The waveform of the fast-front current is $i(t) = 10137.47(e^{-0.02356t} - e^{-12.58497t})$ A. This current waveform has an amplitude of 10 kA, a maximum steepness of $127.4kA/\mu s$, a front end time of 0.5 μs, and a tail end time of 30 μs. The slow-front current is defined as $i(t) = 25869.92(e^{-0.01900989t} - e^{-0.2922879t})$ A, with an amplitude of 20 kA, a maximum steepness of $7kA/\mu s$, a leading-edge time of 10 μs, and a trailing-edge time of 50 μs. Both the fast-edge and the slow-edge currents are applied at one end of the electrode.

The transient voltage curves of two fast and slow lightning current impulse responses of the grounding bar are numerically simulated by various frequency domain numerical methods developed in this book. The transient voltage curves obtained by numerical simulation and those simulated by other numerical methods in the original literature [64] are shown in the two subplots of Fig. 8.5, respectively.

To illustrate the comparisons made, the results of the EM calculations obtained with CDEGS (HIFREQ module) used in the original literature [64] are shown as EM in the figure. It should be noted that CDEGS calculates the transient voltage response numerically with minimal assumptions by means of the full Maxwell's equations and the MOM EM method. Figure 8.5b shows a comparison of the voltage response from the Improved Circuit Model (IM) and the electromagnetic (EM) calculations.

Observing the two subfigures of the comparison diagram, it is known that the transient voltage profiles numerically simulated by the frequency domain numerical algorithms developed in this chapter, the EM and IM in the original literature are basically the same.

All of the above examples use a biexponential function to generate lightning currents, and since the Heidler function for simulating lightning currents is currently popular, the following examples use the Heidler function to generate lightning currents instead.

The eighth calculation example is from reference [65], where the following cases were considered in the analysis according to the arrangement of the values of the following parameters: (a) electrode length: 3 m for the case of " short electrodes" and 30 m for the case of "long electrodes"; (b) earth resistivity: 30 Ωm for the case of "conductive" and 300 Ωm for the case of "resistive"; and (c) lightning current waveforms: typical of the " first" and " second" strokes of the waveforms. The electrode is a horizontal copper conductor of 1.4 cm diameter buried in homogeneous

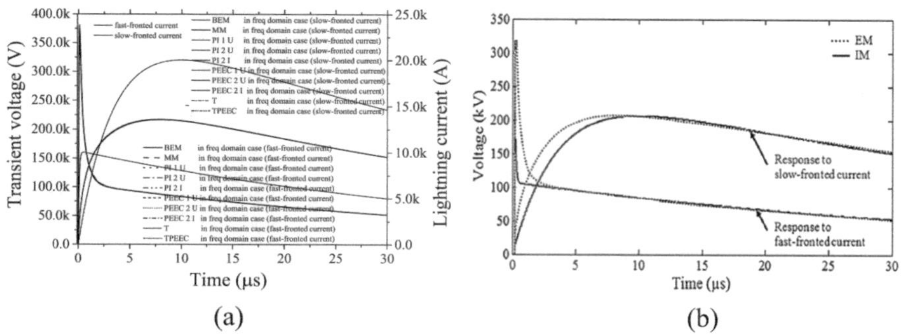

Fig. 8.5 Examining the validation of two types of fast- and slow-varying lightning shock responses generated by a double-exponential function

soil at a depth of 0.8 m. The earth has a relative dielectric constant of 10 and a permeability comparable to air.

Based on the observations of Berger et al. [66], Fig. 8.4a, c illustrates two lightning current waveforms corresponding to a typical first and subsequent return stroke.

The current pulses had the following characteristics: (a) the first return current pulse had a peak value of 30 kA and a zero-to-peak time of about 8 μs; and (b) the subsequent return currents had a peak value of 12 kA and a zero-to-peak time of about 0.8 μs. These waveforms aptly reproduced the concave rising portion of the typical lightning current pulses observed.

Figures 8.6 and 8.9 show the total potential at the injection point relative to the distant ground and its resistive and reactive components in response to a typical injection first stroke and subsequent return stroke current pulse. Figures 8.6 and 8.7 are examples of very conductive soil with a resistivity of 30 Ωm. Figure 8.6 is for a short electrode with length 3 m; and Fig. 8.7 is for a longer electrode with length 30 m. Figures 8.6 and 8.7a, b show responses of the typical first return stroke. Figure 8.6 and 8.7c, d show responses of the typical subsequent return stroke. Figures 8.8 and 8.9 are organized in a similar way, but Figs. 8.8 and 8.9 refer to more resistive soil with a resistivity of 300 m.

Observing Figs. 8.6 and 8.9, it can be seen that the transient voltage curves simulated by the numerical method established in this paper are consistent with the transient voltage curves simulated in the original literature. It can be seen that all the corresponding curves overlap basically exactly. This comparison method proves the accuracy and scientific validity of the various frequency domain EMF numerical methods developed in this chapter.

It should also be noted that the excitation lightning currents in the validation example above were generated using a double exponential function and a Heidler function, whereas the validation algorithm below was generated using a composite Heidler function.

The last validation example is taken from the literature [67], in which a 30-meter-long horizontal grounding rod is buried in a homogeneous soil with a depth of 0.5 m

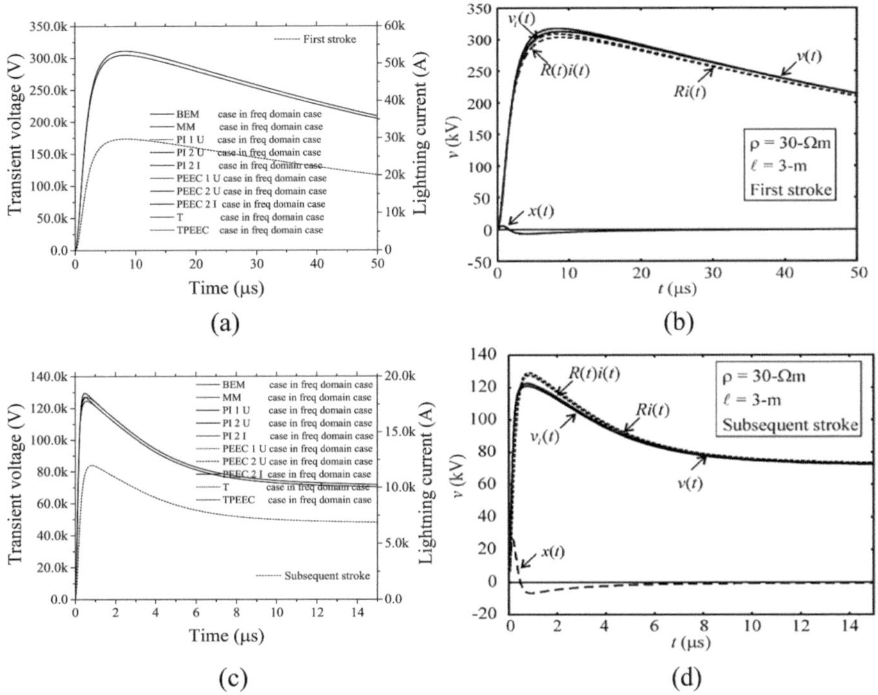

Fig. 8.6 3*m* long horizontal conductor grounding rod buried in 30 Ωm earth resistivity soil

and a resistivity of 300 ohm m. The earth dielectric constant is 10 and the radius of the ground rod is 7 mm.

Representative lightning current waveforms for the first and subsequent strokes used for the excitation of this grounding rod can be seen in Fig. 8.10. Median peak currents and front times: (a) $I_p = 31.1$ kA, $T_{d30} = 3.8\,\mu$s, $T_{d10} = 5.6\,\mu$s and (b) $I_p = 11.8$ kA, $T_{d30} = 0.67\,\mu$s, $T_{d10} = 0.75\,\mu$s. Adapted from [68, 69].

The ground potential rise at the injection point of a horizontal grounding rod under the action of the first and subsequent lightning current impacts can be found in Figs. 8.11 and 8.12, respectively.

Referring to subfigures (a) and (b) of Fig. 8.11, it is important to point out here that subfigure (a) is the voltage profile of the numerical simulation of the algorithm developed in this book, and subfigure (b) is the voltage profile given in the original reference [67]. Observing and comparing the two subplots, it can be seen that: the two sets of transient voltage curves are basically overlapped and consistent, which further illustrates the correctness and scientificity of the various frequency-domain algorithms developed in this book.

The same can be further observed in the two subfigures (a) and (b) of Fig. 8.12, which differ from the two subfigures above in that the former is the response under the action of the first lightning current stroke, while the latter is the response under

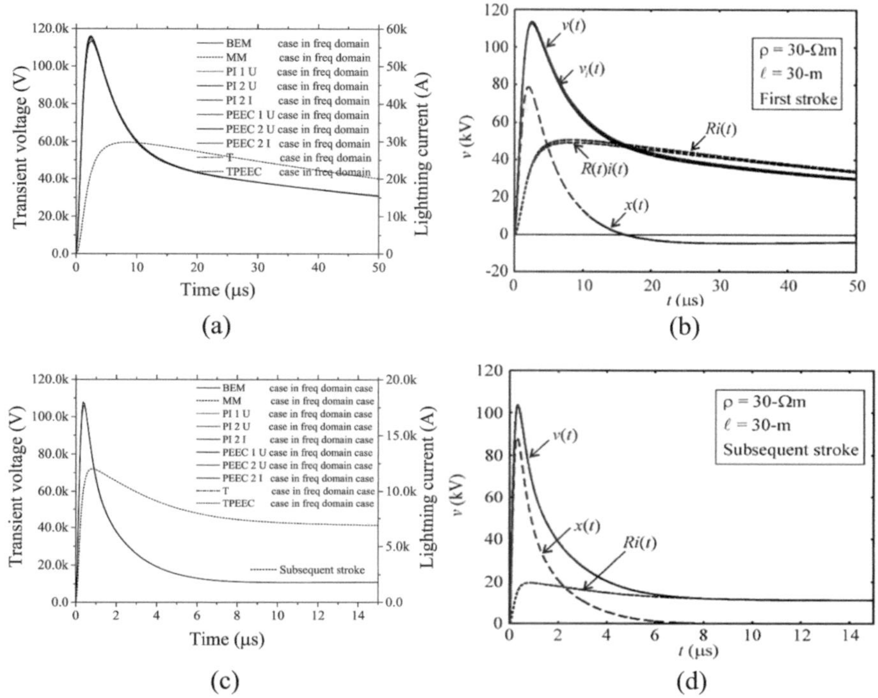

Fig. 8.7 30*m* long horizontal conductor grounding rod buried in 30 Ωm earth resistivity soil

the action of the subsequent lightning current stroke. Comparing the transient voltage curves in the latter two subfigures, it can be seen that the two sets of curves are also basically coincident, which likewise further illustrates the correctness and scientific validity of the various frequency-domain algorithms developed in this chapter.

8.6.2 Results and Discussions for the Various Algorithms

This chapter provides a detailed analysis and discussion of numerical examples, which will be divided into two numerical examples based on the lightning currents generated by two different mathematical formulas.

Using a double exponential function to generate excitation source lightning current

A general grounding system can be seen in Fig. 8.13, which consists of 4×4 square grids and the burial depth of the grounding grid is 0.5 m. The earth is modelled as horizontal four layers conductive media, whose resistivities and permittivities are $\rho_1 = 100\,\Omega\text{m}$ and $\rho_2 = 220\,\Omega\text{m}$, $\rho_3 = 310\,\Omega\text{m}$, $\rho_4 = 430\,\Omega\text{m}$, $\varepsilon_1 = 10\varepsilon_0$,

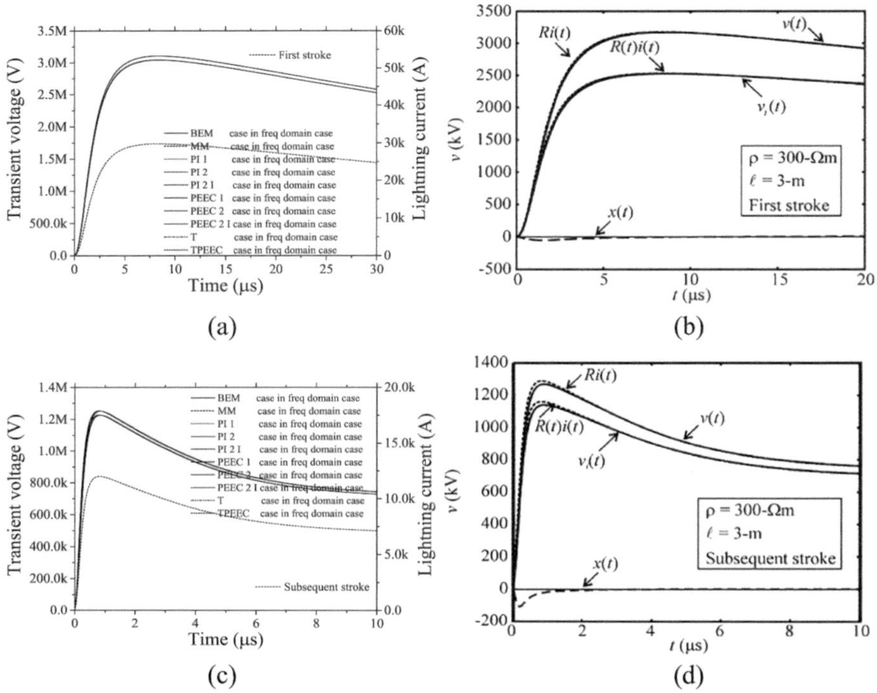

Fig. 8.8 3*m* long horizontal conductor grounding rod buried in 300 Ωm earth resistivity soil

$\varepsilon_2 = 12\varepsilon_0$, $\varepsilon_3 = 13\varepsilon_0$ and $\varepsilon_4 = 14\varepsilon_0$, respectively. The first, second and third layer thickness are $h_1 = 6m$, $h_2 = 12m$ and $h_3 = 8m$. All the conductor radius of grids are 5 mm. The external excited lightning current with (8.19) is injected from the corner of the grounding system. According to the formula of lightning current, T_1 is 10 μs and T_2 is 50 μs.

$$i(t) = 1.29 \times (e^{-0.01910t} - e^{-0.292288t})A \tag{8.19}$$

1. Analysis and discussion of transient surge voltages
   Figure 8.13a shows the transient voltage curve at the injection point of lightning current in the grounding grid simulated by frequency domain equipotential and unequal point electromagnetic models, while Fig. 8.13b shows a locally enlarged view of the peak period in the transient voltage curve. Among them, the numerical models of equipotential electromagnetic fields include spatial 1-d scalar Galerkin MM and BEM, while the numerical models of non equipotential electromagnetic fields include spatial 1-d vector Galerkin MM and BEM, and spatial 1-d vector Galerkin hybrid algorithm based on PI, PEEC, T and TPEEC basis functions.
   From the distribution of transient voltage curves in the Fig. 8.13a, b, it can be seen that the voltage curves simulated and calculated based on the numerical

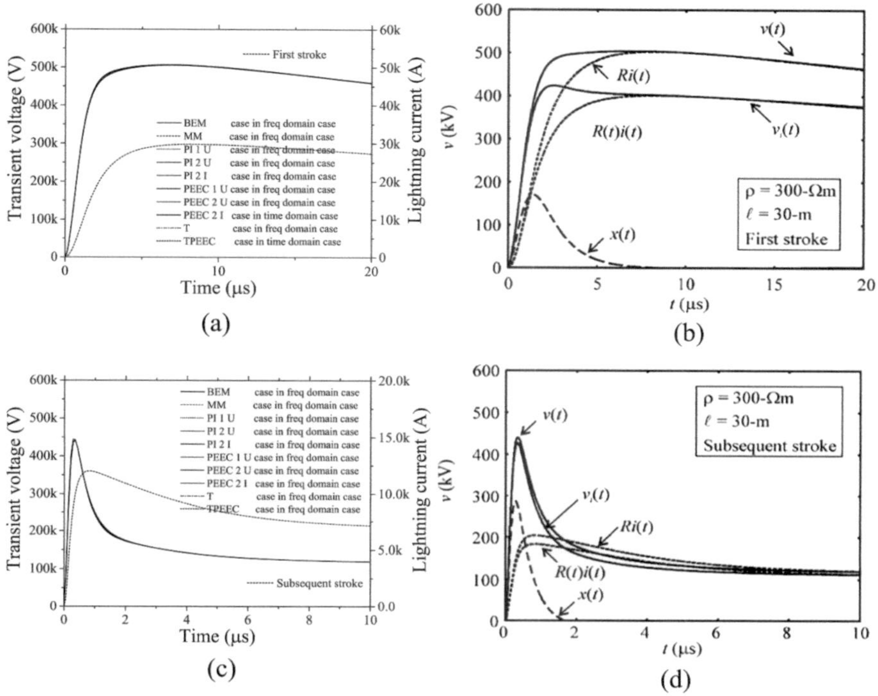

Fig. 8.9 30*m* long horizontal conductor grounding rod buried in 300 Ωm earth resistivity soil

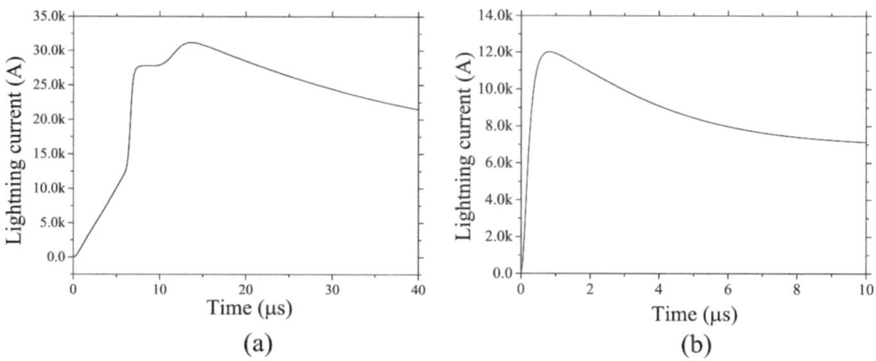

Fig. 8.10 Representative current waveforms of **a** first and **b** subsequent strokes impressed on the electrodes

models of equipotential and non equipotential electromagnetic fields overlap with each other using different algorithms. This includes the overlapping of the two voltage curves calculated using the spatial 1-d scalar Galerkin MM and BEM, the overlapping of the six voltage curves calculated by the spatial 1-d vector Galerkin MM, BEM and hybrid algorithm based on different basis functions.

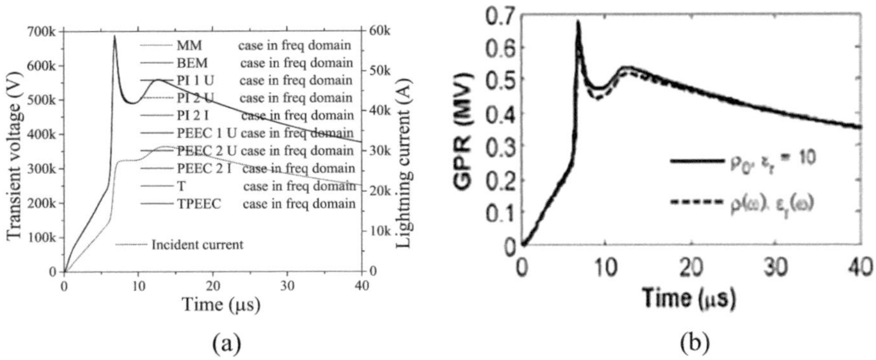

Fig. 8.11 Representative lightning current waveforms for the first strokes

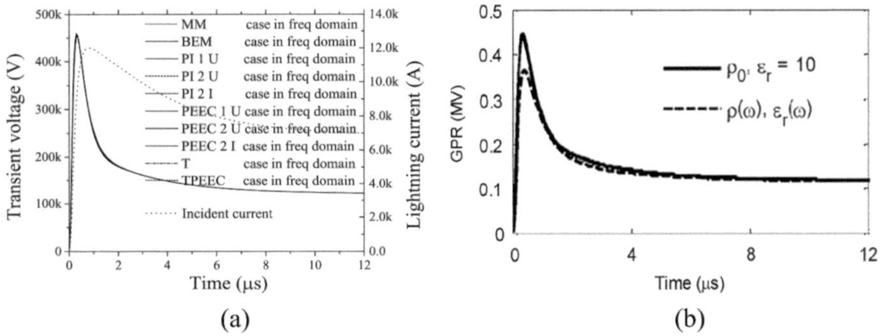

Fig. 8.12 Representative lightning current waveforms for the subsequent strokes

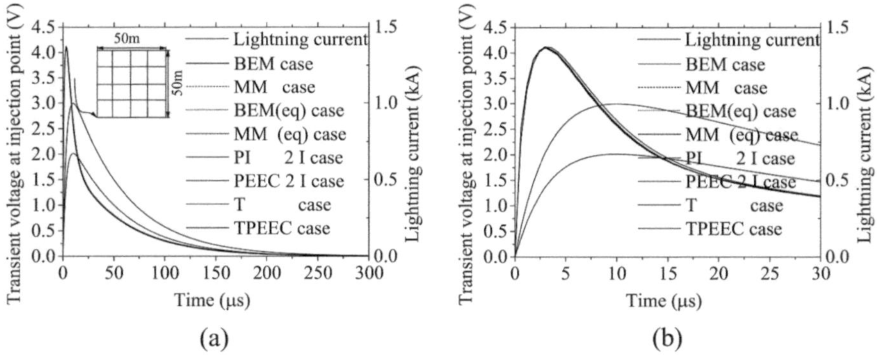

Fig. 8.13 Transient voltage at injection point (biexponential function case)

At the same time, it can also be seen that there is a significant difference between the two curves calculated by the two algorithms based on the equipotential and non equipotential electromagnetic field models. This indicates that the algorithm based on the equipotential electromagnetic field model has a large error in simulating and calculating lightning transient impulse problems. As a calculation tool for predicting lightning transient impulse problems, it is not scientific and accurate.

2. Analysis of 3-d distribution plots of voltage, leakage and branch currents in frequency domains
   The following three large images, consisting of nine small images, respectively, display the 3-d distribution of frequency domain branch current, leakage current, and voltage on the grounding grid at critical frequencies.
   Observing Fig. 8.14, it can be seen that the distribution of branch currents on the entire grounding grid is basically consistent at various frequencies, with only the amplitude decreasing from large to small until it disappears. For a single frequency, the lightning current injected into the grounding grid has the highest current value in the electrical branch, while the current value in the diagonal branch is the smallest. The branch current in the entire grounding grid gradually weakens and decreases as the conductor flows from the injection point to the

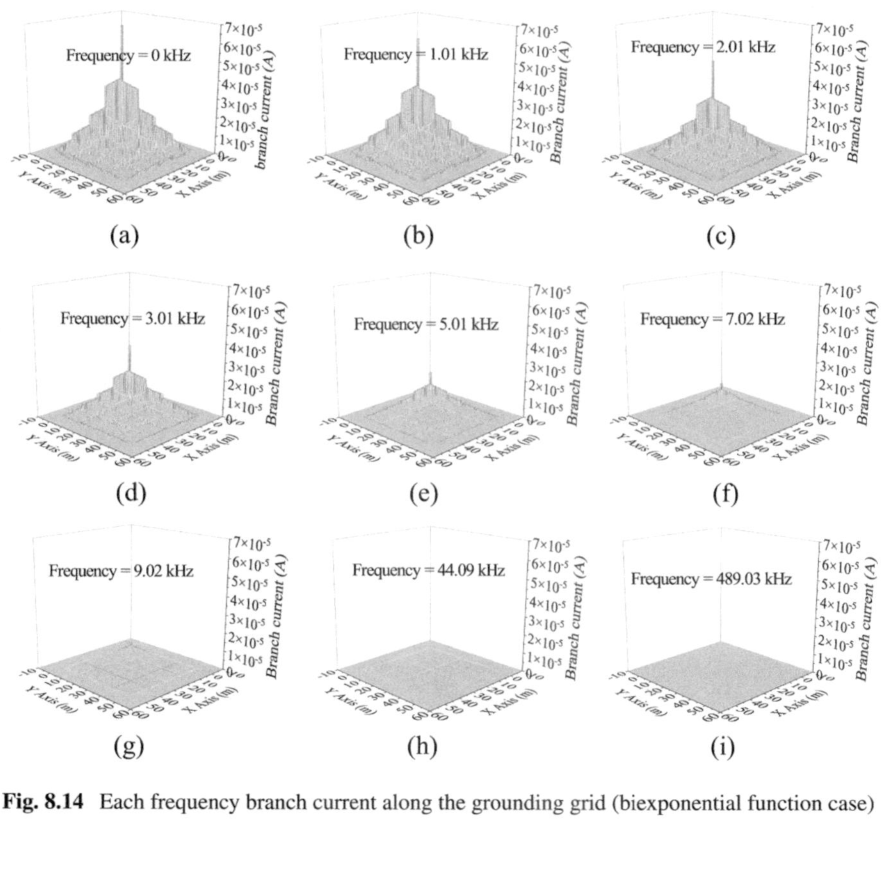

Fig. 8.14 Each frequency branch current along the grounding grid (biexponential function case)

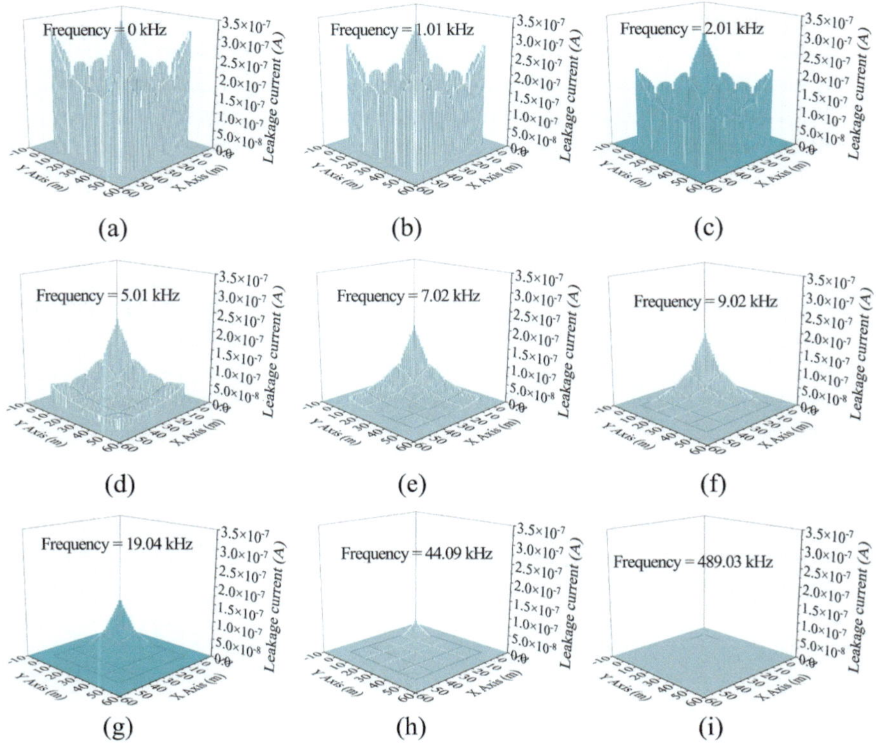

Fig. 8.15 Each frequency leakage current along the grounding grid (biexponential function case)

diagonal point, until the minimum current at the diagonal point disappears. It is obvious that when the frequency is zero, the overall branch current value is maximum, gradually weakening and decreasing until it disappears with increasing frequency. Firstly, the branch current value starts to decrease and disappear from the diagonal point, and then gradually extends towards the injection point. Observing Fig. 8.15, it can be seen that the distribution of leakage current on the entire grounding grid is basically consistent at various frequencies, with only the amplitude decreasing from large to small until it disappears. For a single frequency, the maximum leakage current value is injected into the grounding grid by lightning current, while the leakage current value at the diagonal point is relatively small compared to the other three grounding grid corner points. The leakage current value of the entire grounding grid is relatively large at the four corner points of the grounding grid, and then gradually decreases towards the midpoint of the conductor on the four side conductors; The leakage current gradually weakens and decreases as the grounding grid conductor moves towards the midpoint of the grounding grid, until it reaches its minimum. It is obvious that when the frequency is zero, the overall leakage current value is maximum,

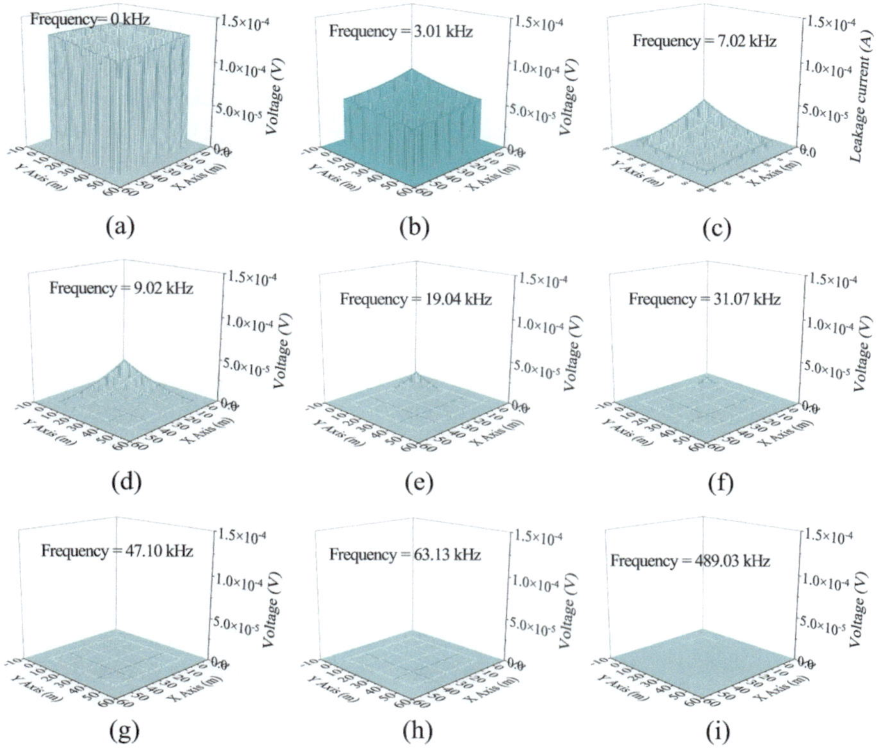

Fig. 8.16 Each frequency voltage along the grounding grid (biexponential function case)

and gradually decreases and disappears as the frequency increases. Firstly, the branch current value starts to decrease and disappear from the diagonal point, and then gradually extends towards the injection point.

Observing Fig. 8.16, it can be seen that the voltage distribution on the entire grounding grid is completely different from the previous current distribution, with completely different frequency distributions, except for the amplitude from large to small, until it disappears. For a single frequency, the voltage value at the lightning current injection point of the grounding grid is the highest, while the voltage value at the diagonal point is relatively small compared to the other three grounding grid corner points. The voltage on the entire grounding grid gradually weakens and decreases as the grounding grid conductor moves towards the diagonal point of the grounding grid, until it reaches its minimum. It is obvious that when the frequency is zero, the overall voltage value reaches its maximum, gradually weakening and decreasing until it disappears as the frequency increases. Firstly, the branch voltage starts to decrease and disappear from the diagonal point, and then gradually extends towards the injection point.

Using the Heidler function to generate excitation source lightning current

The substation grounding grid of this example is buried in five layers of horizontally stratified soil, the grounding grid is buried at a depth of 0.5 m. The substation grounding grid is composed of a square-shaped grounding grid, the entire grid size of 60 m by 60 m, consisting of 12 m by 12 m of small squares, a total of 5 by 5 a total of 25 small grids, which can be seen in Fig. 8.17a.

The parameters of the five layers of horizontally stratified soil, i.e., resistivity and relative permittivity, are as follows: the first layer of soil has a resistivity of 300 Ωm, the first layer of soil has a resistivity of 196.57 Ωm, the third layer of soil has a resistivity of 100 Ωm, the fourth layer has a resistivity of 400 Ωm and the last layer of The resistivity of the soil is 600 Ωm; the relative dielectric constant of the first to third and last layers of soil is 10, and only the fourth layer of soil has a relative dielectric constant of 50.

The excitation source lightning current is generated by the Heidler function with the following equation:

$$i(t) = \frac{I_0}{\eta}\frac{(\frac{t}{\tau_1})^n}{1+(\frac{t}{\tau_1})^n}e^{-\frac{t}{\tau_2}} \tag{8.20}$$

where $I_0 = 14.862\ A$, $\eta = 0.7833$, $\tau_1 = 2.4411\ \mu\text{s}$, $\tau_2 = 18.761\ \mu\text{s}$ and $n = 5$. The choice of these parameters implies a wavehead time of $T_1 = 3.26\ \mu\text{s}$ and a half-wave time of $T_2 = 21.34\ \mu\text{s}$ for the lightning current.

1. Analysis and discussion of transient surge voltages
   Observing the two subfigures (a) and (b) in Fig. 8.17, it can be known that the whole process of the transient voltage curve at the injection point under the lightning impact of the horizontal grounding grid buried in first layer in five

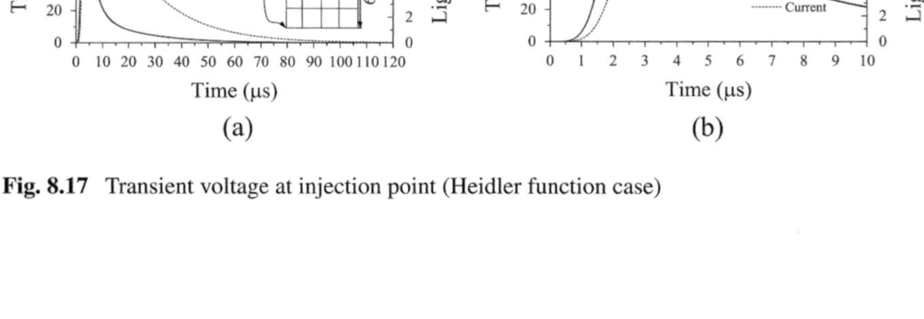

Fig. 8.17 Transient voltage at injection point (Heidler function case)

horizontally stratified layers of soil simulated by various numerical algorithms of frequency-domain unequal potential electromagnetic fields appears in subfigure (a), and subfigure (b) is the local zoom-in view near the peak of the numerically simulated transient voltage curve.

Meanwhile, after observing the two subfigures (a) and (b) in Fig. 8.17, it can be seen that the transient voltage profiles simulated and computed by various different unequal potential EMF numerical methods such as vector Galerkin BEM, vector Galerkin MM, as well as vector hybrid algorithms based on various basis functions (PI and PEEC) basically converge and composite with each other. These mutually overlapping transient voltage curves illustrate that the various numerical algorithms for electromagnetic fields with unequal potentials developed in this chapter mutually validate the correctness and accuracy of their own algorithms.

2. Analysis of 3-d distribution plots of voltage, leakage and branch currents in frequency domains

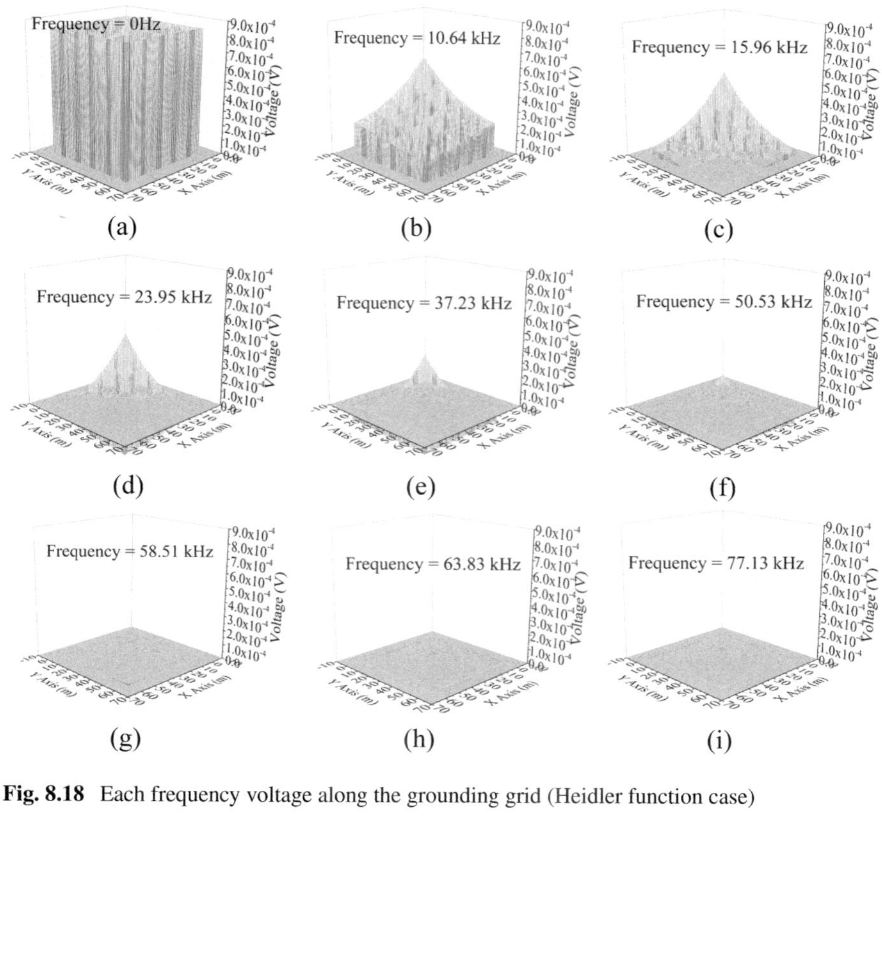

Fig. 8.18 Each frequency voltage along the grounding grid (Heidler function case)

Each of the following large graphs consists of four large graphs made up of nine smaller graphs showing the 3-d distributions of frequency domain voltages, branch currents, and leakage currents in the grounding grid at key frequencies. Both of the last two large plots, which require expenditures, show the leakage currents in question.

Observation of Fig. 8.18 can be seen, the voltage distribution of the entire grounding grid is closely related to the frequency, the frequency of zero, the entire grounding grid distribution of voltage amplitude is the largest, with the increase in frequency, the distribution of voltage on the grounding grid as a whole with the increase in frequency and gradually weakened. Obviously, regardless of the frequency, the voltage distribution on the entire grounding grid to meet the law that the voltage amplitude on the injection point is the largest, and and the injection point on the grounding grid and the injection point on the diagonal of the voltage amplitude is the lowest.

Observation of Fig. 8.21 shows that the 3-d distribution of branch currents on the whole grounding grid is basically similar at different frequencies, with the current amplitude at the injection point of the grounding grid being the largest, and the current amplitude at the diagonal point of the injection point of the

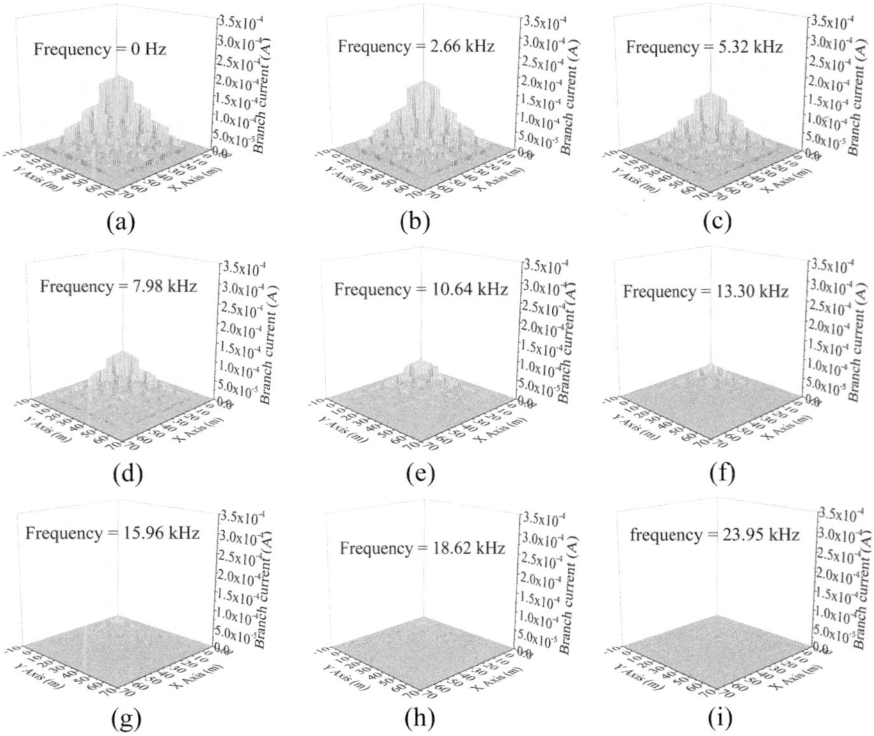

Fig. 8.19 Each frequency branch current along the grounding grid (Heidler function case)

grounding grid being the smallest. The 3-d distribution of branch currents on the whole grounding grid is gradually weakened with the increase of frequency until it disappears.

Note that this large graph consists of nine small graphs showing the 3-d distribution of branch currents on the grounding grid at various frequencies, generated in the frequency domain by the lightning current excitation produced by the Heidler function, while the 3-d distribution of branch currents on the grounding grid at various frequencies previously generated in the frequency domain by the double-exponential function generating the lightning current excitation is basically similar, and will not be repeated here (Fig. 8.19).

Similarly, by observing the 3-d distribution of leakage current on the grounding grid as a function of frequency on the nine small graphs in Fig. 8.20, it can be concluded that the leakage current on the grounding grid is slightly different from the 3-d distribution of leakage current in the frequency domain generated by the lightning current excitation generated by the double exponential function. Although they are basically consistent with the 3-d distribution of voltage and branch current on the grounding grid, that is, the leakage current amplitude at

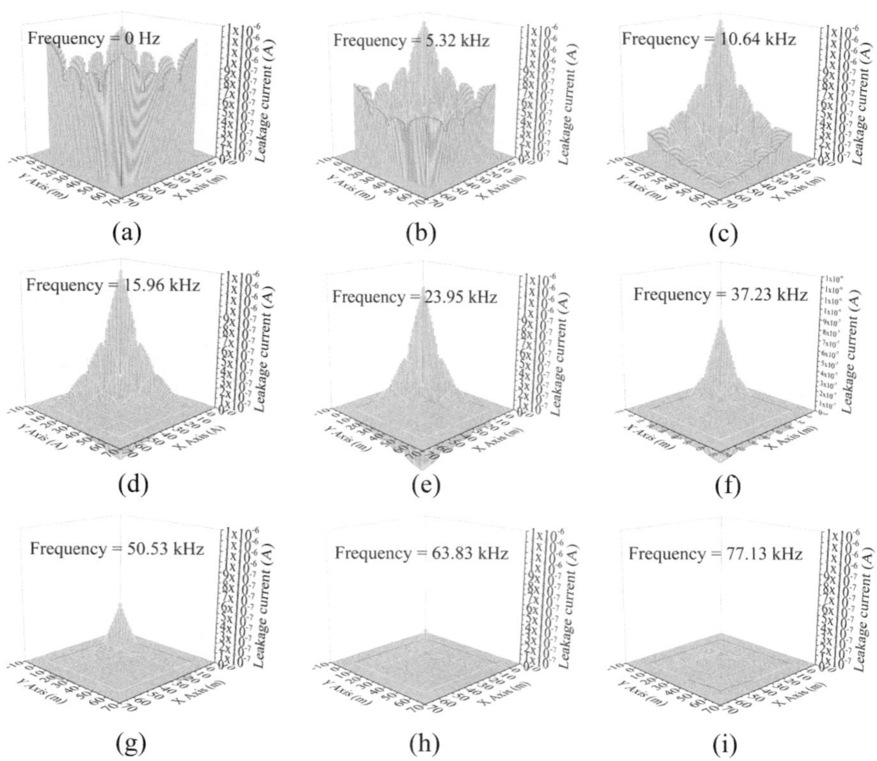

Fig. 8.20 Each frequency leakage current along the grounding grid (Heidler function case)

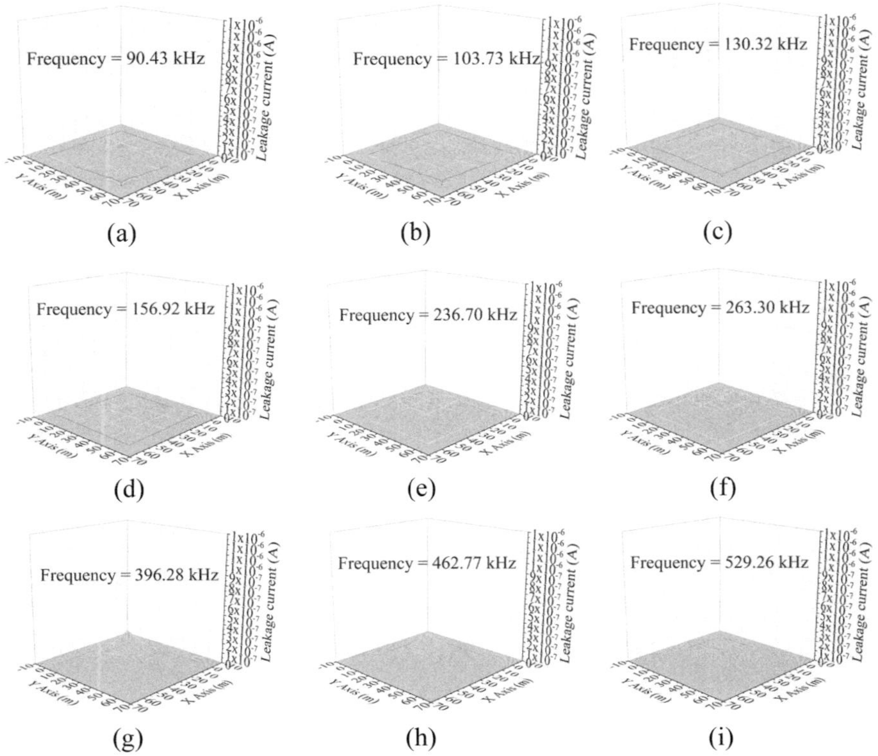

Fig. 8.21 Each frequency leakage current along the grounding grid (Heidler function case)

the injection point on the grounding grid is the largest, and the leakage current amplitude at the diagonal point of the injection point on the grounding grid is the smallest; But as the frequency continues to increase, the three-dimensional distribution of leakage current on the grounding grid here weakens and disappears extremely slowly, to the point where a large graph consisting of nine small graphs is added to show the trend of leakage current on the grounding grid as the frequency increases.

8.7 Summary

Combining forward and inverse Fourier transform, frequency domain 1-d spatial vector Galerkin BEM, vector MM, and vector hybrid algorithm based on various basis functions can effectively simulate the lightning transient impulse grounding response of substation grounding grids.

References

1. L. Grcev, Computer analysis of transient voltages in large grounding systems. IEEE Trans. Power Delivery **11**(2), 815–823 (1996)
2. M. Heimbach, L. Grcev, Grounding system analysis in transient programs applying electromagnetic field approach. IEEE Trans. Power Delivery **12**(1), 186–193 (1997)
3. D. Poljak, V. Doric, Wire antenna mopdel for transient analyais of simple grounding systems, part i: the vertical grounding electrode. Progress Electromagn. Res. **64**, 149–166 (2006)
4. D. Poljak, V. Doric, Wire antenna mopdel for transient analyais of simple grounding systems, part i: the horizontal grounding electrode. Progress Electromagn. Res. **64**, 167–189 (2006)
5. T. Topa, A. Karwowski, A. Noga, Using gpu with cuda to accelerate mom-based electromagnetic simulation of wire-grid models. IEEE Antennas Wirel. Propagation Lett. **10**, 342–345 (2011)
6. D. Cavka, B. Harrat, D. Poljak, B. Nekhoul, K. Kerroum, K.E.K. Drissi, Wire antenna versus modified transmission line approach to the transient analysis of grounding grid. Eng. Anal. Boundary Elements **35**, 1101–1108 (2011)
7. K. Sheshyekani, S.H.H. Sadeghi, R. Moini, F. Rachidi, Frequency-domain analysis of ground electrodes buried in an ionized soil when subjected to surge currents: A mom-aom approach. Electr. Power Syst. Res. **81**, 290–296 (2011)
8. D. Cavka, D. Poljak, On the evaluation of input impedance and transient impedance for grounding electrodes using antenna theory. COMPEL Int. J. Comput. Mathem. Electr. Electron. Eng. **32**(6), 2045–2062 (2013)
9. M. Chouki, B. Nekhoul, D. Poljak, K. Kerroum, K.E.K. Drissi, Simplified dipole concept for the assessment of transient electromagnetic field in the vicinity of grounding grid. Int. J. Numer. Model. Electron. Netw. Dev. Fields **28**, 404–418 (2015)
10. B. Markovski, L. Grcev, V. Arnautovski-Toseva, Fast and accurate transient analysis of large grounding systems in multilayer soil. IEEE Trans. Power Delivery **36**(2), 598–606 (2021)
11. Z.-X. Li, Y. Yu, K.-C. Wang, Lightning response of a grounding system buried in multilayered earth based on the galerkin's boundary element and quasi-static complex image methods. IEEE Trans. Electromagn. Compat. **63**(4), 1118–1127 (2021)
12. L. Grcev, F. Dawalibi, An electromagnegtic model for transient in grounding systems. IEEE Trans. Power Delivery **5**(4), 1773–1781 (1990)
13. G. Ala, M.L.D. Silvestre, A simulation model for electromagnetic transients in lightning protection systems. IEEE Trans. Electromagn. Compat. **44**(4), 539–554 (2002)
14. G. Ala, M.L.D. Silvestre, E. Francomano, A. Tortorici, Wavelet-based efficient simulation of electromagnetic transients in a lightning protection system. IEEE Trans. Magn. **39**(3), 1257–1260 (2003)
15. G. Ala, M.L.D. Silvestre, E. Francomano, A. Tortorici, An advanced numerical model in solving thin-wire integral equations by using semi-orthogonal compactly supported spline wavelets. IEEE Trans. Electromagn. Compat. **45**(2), 218–228 (2003)
16. T. Rouibaha, A. Bayadia, K. Kerroum, Accelerating the frequency-domain response calculation of complexgrounding system using wavelet based mbpe technique. Electr. Power Syst. Res. **121**, 287–294 (2015)
17. M. Nazari, R. Moini, S. Fortin, F.P. Dawalibi, F. Rachidi, Impact of frequency-dependent soil models on grounding system performance for direct and indirect lightning strikes. IEEE Trans. Electromagn. Compat. **63**(1), 134–144 (2021)
18. M. Shafieipour, M. Nazari, F.P. Dawalibi, S. Fortin, A. Tatematsu, J.D. Silva, P. Gmez, Full-wave 3-d transient analysis with method of moments and numerical laplace transform including resistive non-linear elements. IEEE Trans. Power Delivery **38**(1), 299–308 (2023)
19. L. Qi, X. Cui, Z. Zhao, H. Li, Grounding performance analysis of the substation grounding grids by finite element method in frequency domain. IEEE Trans. Magn. **43**(4), 1181–1184 (2007)

20. M. Akbari, K. Sheshyekani, M.R. Alemi, The effect of frequency dependence of soil electrical parameters on the lightning performance of grounding systems. IEEE Trans. Electromagn. Compat. **55**(4), 739–746 (2013)
21. B. Salarieh, J.D. Silva, B. Kordi, High frequency response of grounding electrodes: effect of soil dielectric constant. IET Gen. Transmission Distrib. **14**(15), 2915–2921 (2020)
22. N. Yu, Z. Cai, R. Li, Q. Zhang, L. Gao, Z. Sun, Calculation of earth surface potential and neutral current caused by hvdc considering three-dimensional complex soil structure. IEEE Trans. Electromagn. Compat. **63**(5), 1480–1490 (2021)
23. H. Zamani, K. Sheshyekani, Method of line for modeling of grounding electrodes buried in stratified multilayer soil. IEEE Trans. Electromagn. Compat. **64**(6), 2131–2140 (2022)
24. Y. Liu, M. Zitnik, R. Thottappillil, An improved transmission-line model of grounding system. IEEE Trans. Electromagn. Compat. **43**(3), 348–355 (2001)
25. Y. Liu, N. Theethayi, R. Thottappillil, An engineering model for transient analysis of grounding system under lightning strikes: Nonuniform transmission-line approach. IEEE Trans. Power Delivery **20**(2), 722–730 (2005)
26. N. Harid, H. Griffiths, A. Haddad, A new frequency-dependent surge impedance calculation method for high-voltage towers. IEEE Trans. Power Delivery **21**(3), 1430–1437 (2006)
27. M. Al-Shaher, M. Saied, s-domain analysis of grounding electrodes impulse response. Electr. Power Components Syst. **34**, 985–999 (2006)
28. N. Theethayi, R. Thottappillil, M. Paolone, C.A. Nucci, F. Rachidi, External impedance and admittance of buried horizontal wires for transient studies using transmission line analysis. IEEE Trans. Dielectrics Electr. Insulation **14**(3), 751–761 (2007)
29. J.-H. Choi, B.-H. Lee, An analysis on the frequency-dependent grounding impedance based on the ground current dissipation of counterpoises in the two-layered soils. J. Electrost. **70**, 184–191 (2012)
30. A. Jardines, J.L. Guardado, J. Torres, J.J. Chez, M. Herndez, A multiconductor transmission line model for grounding grids. Electr. Power Energy Syst. **60**, 24–33 (2014)
31. R. Shariatinasab, J. Gholinezhad, The effect of grounding system modeling on lightning-related studies of transmission lines. J. Appl. Res. Technol. **15**, 545–554 (2017)
32. S. Yassin, A.Z. El Dein, Harmonic impedance of horizontal grounding electrode under soil ionization effect. Wseas Trans. Power Syst. **13**, 200–208 (2018)
33. S.S. Sajjadi, S.R. Ostadzadeh, Lightning response of multi-port grids buried in dispersive soils: An approximation versus full-wave methods and experiment. Adva. Electromagn. **8**(1), 43–50 (2019)
34. C.M. de Seixas, S. Kurokawa, Using circuit elements to represent the distributed parameters of a grounding system under lightning strokes. Electr. Power Syst. Res. **172**, 213–220 (2019)
35. R. Batista, J.O.S. Paulino, A practical approach to estimate grounding impedance of a vertical rod in a two-layer soil. Electr. Power Syst. Res. **177**, 105973 (2019)
36. R. Batista, C.E.F. Caetano, J.O.S. Paulino, W.C. Boaventur, I.J.S. Lopes, E.N. Cardoso, A study of grounding arrangements composed by vertical electrodes for two layered stratified soil models. Electr. Power Syst. Res. **180**, 106129 (2020)
37. S.S. Sajadi, S.R. Ostadzadeh, S.H.H. Sadeghi, Parametric dependence of lightning impulse behavior of grounding electrodes buried in a dispersive and ionized lossy soil under first- and subsequent-stroke currents. COMPEL Int. J. Computation Mathem. Electr. Electron. Eng. **39**(4), 757–773 (2020)
38. R. Batista, J.O.S. Paulino, Computing grounding resistance and impulse impedance of horizontal electrodes parallel or perpendicular to the interface of a vertically stratified soil using transmission line theory. Electr. Power Syst. Res. **194**, 107060 (2021)
39. D. Tashiro, K. Sameshima, T. Hisakado, A.K.M.M. Islam, O. Wada, Single-conductor transmission-line model for bent wire structures. IEEE Trans. Electromagn. Compat. **65**(5), 1432–1445 (2023)
40. F.E. Menter, L. Grcev, Empt-based model for grounding system analysis. IEEE Trans. Power Delivery **9**(4), 1838–1849 (1994)

41. A.C. Liew, M. Darveniza, Dynamic model of impulse characteristics of concentrated earths. Proc. IEE **121**(2), 124–135 (1974)
42. A. Geri, E. Garbagnati, G. Sartorio, Non-linear behaviour of ground electrodes under lightning surge currents: Computer modelling and comparison with experimental results. IEEE Trans. Magn. **28**(2), 1442–1445 (1992)
43. A. Geri, Behaviour of grounding systems excited by high impulse currents: the model and its validation. IEEE Trans. Power Delivery **14**(3), 1008–1017 (1999)
44. J. Wang, A.C. Liew, M. Darveniza, Extension of dynamic model of impulse behavior of concentrated grounds at high currents. IEEE Trans. Power Delivery **20**(3), 2160–2165 (2005)
45. A.F. Otero, J. Cidras, J.L.D. Alamo, Frequency-depantant grounding system calculation by means of a conventional nodal analysis technique. IEEE Trans. Power Delivery **14**(3), 873–878 (1999)
46. J. Cidras, A.F. Otero, C. Garrido, Nodal frequency analysis of grounding systems considering the soil ionization effect. IEEE Trans. Power Delivery **15**(1), 103–107 (2000)
47. L. Huang, D.G. Kasten, Modeling of ground grid and metallic structure currents in high voltage a.c. substations for the computation of electromagnetic fields. Electr. Power Syst. Res. **59**, 31–37 (2001)
48. S. Visacro, Jr. A. Soares. Hem: a model for simulation of lightning-related engineering problems. IEEE Trans. Power Delivery **20**(2), 1206–1208 (2005)
49. Z.-X. Li, Y. Yin, C.-X. Zhang, L.-C. Zhang, A mathematical model for the transient lightning response from grounding systems. Progress Electromagn. Res. B **57**, 47–61 (2014)
50. Z.-X. Li, G.-F. Li, J.-B. Fan, Y. Yin, A novel mathematical model for the lightning response of a grounding system buried in multilayered earth based on the quasi-static complex image method. Euro. Trans. Electr. Power **22**, 831–845 (2012)
51. Z.-X. Li, Y. Yin, C.-X. Zhang, L.-C. Zhang, A mathematical model for the transient lightning response from grounding systems. Progress Electromagn. Res. B **57**, 47–61 (2014)
52. A. Rodolfo, C. Salvatore, Transient behavior of wind towers grounding systems under lightning strikes. Int. J. Energy Environ. Eng. **7**, 235–247 (2016)
53. R.A.R. de Moura, M.A. de Oliveira Schroeder, A.C.S. de Lima, P.H.N. Vieira, R. Alipio, Closed-form approximation for grounding grids transient analysis. J. Control Autom. Electr. Syst. **32**, 796–806 (2021)
54. K.P. Sengar, K. Chandrasekaran, Transient behaviour of grounding systems in multilayer soil under lightning strikes. Electr. Eng. **104**, 1205–1218 (2022)
55. K.P. Sengar, K. Chandrasekaran, Analysis of grounding system transient with consideration of soil ionization and mutual coupling effect. J. Electr. Syst. Inf. Technol. **10**, 53 (2023)
56. P. Yutthagowith, A. Ametani, N. Nagaoka, Y. Baba, Application of the partial element equivalent circuit method to analysis of transient potential rises in grounding systems. IEEE Trans. Electromagn. Compat. **53**(3), 726–736 (2011)
57. H. Chen, Y. Du, Lightning grounding grid model considering both the frequency-dependent behavior and ionization phenomenon. IEEE Trans. Electromagn. Compat. **61**(1), 157–165 (2019)
58. Z.-X. Li, K.-L. Gao, S.-W. Rao, Lightning response of a grounding system buried in multiple layers in earth using the peec method based on the quasi-static complex image method. Electr. Power Components Syst. **48**(3), 291–303 (2020)
59. Z.-X. Li, K.-L. Gao, Y. Yin, D. Ge, Transient lightning responses of grounding systems buried in horizontal multilayered earth with a hybrid method. J. Electrost. **73**(5), 381–386 (2014)
60. J.C. Salari, C. Portela, Grounding systems modeling including soil ionization. IEEE Trans. Power Delivery **23**(4), 1939–1945 (2008)
61. Z. Stojkovic et al., Experimental investigation of grounding grid impulse characteristic. Europ. Trans. Electr. Power **8**(6), 417–425 (1998)
62. Z. Stojkovic et al., Sensitivity analysis of experimentally determined grounding grid impulse characteristic. IEEE Trans. Power Delivery **13**(3), 1136–1142 (1998)
63. M. Jamali, M. Niasati, M. Jazaeri, A two-layer soil model for the calculation of electrical parameters of grounding systems under lightning strikes. Electr. Power Components Syst. **47**(1), 181–191 (2019)

64. M. Mokhtari, Z. Abdul-Malek, Z. Salam, An improved circuit-based model of a grounding electrode by considering the current rate of rise and soil ionization factors. IEEE Trans. Power Delivery **30**(1), 211–219 (2015)
65. L. Grcev, Time- and frequency-dependent lightning surge characteristics of grounding electrodes. IEEE Trans. Power Delivery **24**(4), 2186–2196 (2009)
66. K. Berger, R.B. Anderson, H. Kroninger, Parameters of lightning flashes. Electra **41**, 23–37 (1975)
67. R. Alipio, S. Visacro, Impulse efficiency of grounding electrodes: Effect of frequency-dependent soil parameters. IEEE Trans. Electromagn. Compat. **29**(2), 716–723 (2014)
68. S. Visacro, A representative curve for lightning current waveshape of first negative stroke. Geophys. Res. Lett. **31**, L07112 (2004)
69. A. De Conti, S. Visacro, Analytical representation of single- and double-peaked lightning current waveforms. IEEE Trans. Electromagn. Compat. **49**(2), 448–451 (2007)

Chapter 9
Numerical Methods for Grounding Problem at Transient State III

Various time-domain numerical methods for grounding problems in the case of lightning transient shock response based on equipotential electromagnetic field modeling.

9.1 Preface

The frequency domain 1-d spatial scalar Galerkin MM and BEM used for steady-state grounding problems of grounding grids can be transformed into time-domain 1-d spatial scalar Galerkin MM and BEM through Fourier transform, and can be directly used to analyze and simulate lightning transient grounding problems of substation grounding grids. In this chapter, the frequency-domain scalar numerical method is transformed into a time-domain scalar numerical method, which can directly simulate the lightning transient impact problem of substation grounding grid in the time domain without resorting to the Fourier transform method. This chapter will discuss in detail the scientificity and effectiveness of this method.

9.2 Introduction

This chapter mainly discusses the time domain spatial 1-d scalar Galerkin BEM and the MM, the method from the previous chapter of the frequency domain spatial 1-d scalar Galerkin BEM and the MM evolved from the development of the algorithms of the literature published, the reason why this chapter is so arranged, is to consider the continuation of the development of the whole book's research content of the science and rationality.

Z.-X. Li, *Numerical Simulation Algorithm of Electromagnetic Field for Grounding Problems in Power System Substation Grounding Grids*,
https://doi.org/10.1007/978-981-96-0904-8_9

9.3 The Time Domain Spatial 1-d Scalar Galerkin MM for Transient Grounding Problems

In this chapter, we discuss in detail how the time domain spatial 1-d scalar Galerkin MM simulates the transient grounding of grounding grid buried in horizontal N_e-layer earth model.

9.3.1 The Time Domain Spatial 1-d Scalar Galerkin's MM Obtained With Fourier Transform

The frequency domain spatial 1-d scalar Galerkin's MM

Referring to the frequency domain formula (5.4) related to the 1-d spatial scalar Galerkin's MM in Chap. 5, this formula can be rewritten as follows.

$$[\bar{\bar{\mathbf{Z}}}_b][\bar{\mathbf{I}}_b] = [\bar{\mathbf{V}}_b] \tag{9.1}$$

where $[\bar{\bar{\mathbf{Z}}}_b]$ is the $N_b \times N_b$ matrix about mutual impedance coefficient between a pair of segments in grounding system, $[\bar{\mathbf{I}}_b]$ is a $N_b \times 1$ vector for leakage currents emanating from grounding system's conductor segment, $[\bar{\mathbf{V}}_b]$ is a $N_b \times 1$ vector for voltage of grounding system's conductor segment. It must be pointed out that the coefficients matrix in linear system Eq. (9.1) is symmetric and positive definite [1].

This frequency domain equation (9.1) is assembled from the following equations.

$$\sum_{i=1}^{N_b} \frac{1}{l_j} \int_{l_j} \frac{1}{l_i} \int_{l_i} G_{\xi\zeta}^{\varphi^f}(\bar{r}_j, \bar{r}_i, \omega) dr_i dr_j I_{l_i}(\bar{r}_i) = V_{l_j}(\bar{r}_j) \tag{9.2}$$

where $\xi = 0, 1, \ldots, N_e$; $\zeta = 0, 1, \ldots, N_e$. $j = 1, \ldots, N_b$. $G_{\xi\zeta}^{\varphi^f}(\bar{r}_j, \bar{r}_i, \omega)$ is the frequency domain Green's function of point current source in multi-layer horizontal layered soil where the field point is located at $\bar{r}_j$ and the source point is located at $\bar{r}_i$, $I_{l_i}(\bar{r}_i)$ is unknown amount of current located at $\bar{r}_i$, $V_{l_j}(\bar{r}_j)$ is unknown amount of voltage located at $\bar{r}_j$.

According to reference [2, 3] and the Chap. 3 related content, the frequency domain Green's function of point current source in multi-layer horizontal layered soil can be further written as the following formula.

$$G_{\xi\zeta}^{\varphi^f}(\bar{r}_j, \bar{r}_i, \omega) = G_{\xi\zeta}^{\varphi_s^f}(\bar{r}_j, \bar{r}_i) + G_{\xi\zeta}^{\varphi_q^f}(\bar{r}_j, \bar{r}_i, \omega) \tag{9.3}$$

The time domain spatial 1-d scalar Galerkin's MM obtained with Fourier transform

Substituting formula (9.3) into the formula (9.2) and considering the Fourier transform, the following equation can be obtained.

$$\frac{1}{2\pi}\int_{-\infty}^{\infty}\sum_{i=1}^{N_b}\frac{1}{l_j}\int_{l_j}\frac{1}{l_i}\int_{l_i}\left(G_{\xi\zeta}^{\varphi_s^f}(\bar{r}_j,\bar{r}_i)+G_{\xi\zeta}^{\varphi_q^f}(\bar{r}_j,\bar{r}_i,\omega)\right)dr_i dr_j I_{l_i}(\bar{r}_i)e^{j\omega t}d\omega=\frac{1}{2\pi}\int_{-\infty}^{\infty}V_{l_j}(\bar{r}_j)e^{j\omega t}d\omega \tag{9.4}$$

In this way, the time-domain formula can be obtained as follows

$$\sum_{i=1}^{N_b}\frac{1}{l_j}\int_{l_j}\frac{1}{l_i}\int_{l_i}\left(G_{\xi\zeta}^{\varphi_s^t}(\bar{r}_j,\bar{r}_i)I_{l_i}(\bar{r}_i,t)+G_{\xi\zeta}^{\varphi_q^t}(\bar{r}_j,\bar{r}_i,t)I_{l_i}(\bar{r}_i,t-t_\varphi)\right)dr_i dr_j=V_{l_j}(\bar{r}_j,t) \tag{9.5}$$

This formula can be further written in matrix form as follows.

$$[\bar{\bar{\mathbf{Z}}}_b^s][\bar{\mathbf{I}}_b(t)]+[\bar{\bar{\mathbf{Z}}}_b^q][\bar{\mathbf{I}}_b(t-t_\varphi)]=[\bar{\mathbf{V}}_b(t)] \tag{9.6}$$

where $[\bar{\bar{\mathbf{Z}}}_b^s]$ and $[\bar{\bar{\mathbf{Z}}}_b^q]$ are the $N_b\times N_b$ equation matrices without delay effects and with delay effects, respectively; $[\bar{\mathbf{I}}_b(t)]$ and $[\bar{\mathbf{V}}_b(t)]$ are $N_b\times 1$ vectors representing the current and voltage of the conductor section of the grounding grid, respectively.

9.3.2 *The Time Domain Spatial 1-d Scalar Galerkin's MM Obtained with Laplace Transform*

The s domain spatial 1-d scalar Galerkin's MM based on laplace transform

Referring to the frequency domain formula (5.3) related to the 1-d spatial scalar Galerkin's MM in Chap. 5, this formula can be rewritten as follows.

$$[\bar{\bar{\mathbf{Z}}}_b][\bar{\mathbf{I}}_b]=[\bar{\mathbf{V}}_b] \tag{9.7}$$

where $[\bar{\bar{\mathbf{Z}}}_b]$ is the $N_b\times N_b$ matrix about mutual impedance coefficient between a pair of segments in grounding system, $[\bar{\mathbf{I}}_b]$ is a $N_b\times 1$ vector for leakage currents emanating from grounding system's conductor segment, $[\bar{\mathbf{V}}_b]$ is a $N_b\times 1$ vector for voltage of grounding system's conductor segment. It must be pointed out that the coefficients matrix in linear system Eq. (9.7) is symmetric and positive definite [1].

This s domain equation (9.7) is assembled from the following equations.

$$\sum_{i=1}^{N_b} \frac{1}{l_j} \int_{l_j} \frac{1}{l_i} \int_{l_i} G_{\xi\zeta}^{\varphi^s}(\bar{r}_j, \bar{r}_i, s) dr_i dr_j I_{l_i}(\bar{r}_i) = V_{l_j}(\bar{r}_j) \tag{9.8}$$

where $\xi = 0, 1, \ldots, N_e$; $\zeta = 0, 1, \ldots, N_e$. $j = 1, \ldots, N_b$. $G_{\xi\zeta}^{\varphi^s}(\bar{r}_j, \bar{r}_i, s)$ is the s domain Green's function of point current source in multi-layer horizontal layered soil where the field point is located at $\bar{r}_j$ and the source point is located at $\bar{r}_i$, $I_{l_i}(\bar{r}_i)$ is unknown amount of current located at $\bar{r}_i$, $V_{l_j}(\bar{r}_j)$ is unknown amount of voltage located at $\bar{r}_j$.

According to Chap. 3, the s domain Green's function of point current source in multi-layer horizontal layered soil can be further written as the following formula.

$$G_{\xi\zeta}^{\varphi^s}(\bar{r}_j, \bar{r}_i, s) = G_{\xi\zeta}^{\varphi_s^s}(\bar{r}_j, \bar{r}_i) + G_{\xi\zeta}^{\varphi_q^s}(\bar{r}_j, \bar{r}_i, s) \tag{9.9}$$

The time domain spatial 1-d scalar Galerkin's MM obtained with Laplace transform

Substituting formula (9.9) into the formula (9.8) and considering the laplace transform, the following equation can be obtained.

$$\frac{1}{2\pi j} \int_{\sigma - j\infty}^{\sigma + j\infty} \sum_{i=1}^{N_b} \frac{1}{l_j} \int_{l_j} \frac{1}{l_i} \int_{l_i} \left(G_{\xi\zeta}^{\varphi_s^s}(\bar{r}_j, \bar{r}_i) + G_{\xi\zeta}^{\varphi_q^s}(\bar{r}_j, \bar{r}_i, s) \right) dr_i dr_j I_{l_i}(\bar{r}_i) e^{st} ds$$
$$= \frac{1}{2\pi j} \int_{\sigma - j\infty}^{\sigma + j\infty} V_{l_j}(\bar{r}_j) e^{st} ds \tag{9.10}$$

In this way, the time-domain formula can be obtained as follows

$$\sum_{i=1}^{N_b} \frac{1}{l_j} \int_{l_j} \frac{1}{l_i} \int_{l_i} \left(G_{\xi\zeta}^{\varphi_s^t}(\bar{r}_j, \bar{r}_i) I_{l_i}(\bar{r}_i, t) + G_{\xi\zeta}^{\varphi_q^t}(\bar{r}_j, \bar{r}_i) I_{l_i}(\bar{r}_i, t - t_q) \right) dr_i dr_j = V_{l_j}(\bar{r}_j, t) \tag{9.11}$$

This formula can be further written in matrix form as follows.

$$[\bar{\bar{\mathbf{Z}}}_b^s][\bar{\mathbf{I}}_b(t)] + [\bar{\bar{\mathbf{Z}}}_b^q][\bar{\mathbf{I}}_b(t - t_q)] = [\bar{\mathbf{V}}_b(t)] \tag{9.12}$$

where $[\bar{\bar{\mathbf{Z}}}_b^s]$ and $[\bar{\bar{\mathbf{Z}}}_b^q]$ are the $N_b \times N_b$ equation matrices without delay effects and with delay effects, respectively; $[\bar{\mathbf{I}}_b(t)]$ and $[\bar{\mathbf{V}}_b(t)]$ are $N_b \times 1$ vectors representing the current and voltage of the conductor section of the grounding grid, respectively and $t_q = j\beta_q$.

9.3.3 Spatial 1-d Scalar MM for Time Domain Equipotential Electromagnetic Field Mathematical Model for Transient Grounding Problems

Traditionally, the transient grounding problem of a substation grounding grid can be simulated and calculated using Fourier transform, transforming the transient grounding grid problem in the time domain into a steady-state grounding grid problem in the frequency domain. Then, the steady-state grounding problem is solved using the 1-d spatial scalar MM in the frequency domain, and the frequency domain response of the steady-state grounding problem is transformed into the time domain response of the transient grounding problem using inverse Fourier & Laplace transform.

The time-domain 1D spatial scalar MM developed in this chapter first converts the frequency-domain 1-d spatial scalar MM into the time-domain 1-d spatial scalar MM through Fourier transformation. It can directly solve the lightning transient impulse grounding problem of the substation grounding grid in the time domain without the need for inverse Fourier & Laplace transform.

By discretizing continuous time, the discrete time domain matrix formula can be obtained as follows

$$[\bar{\bar{\mathbf{Z}}}_b^s][\bar{\mathbf{I}}_b(t_k)] + [\bar{\bar{\mathbf{Z}}}_b^q][\bar{\mathbf{I}}_b(t_k - t_n)] = [\bar{\mathbf{V}}_b(t_k)] \tag{9.13}$$

where $t_k = k\Delta, k = 1, 2, \ldots, N_t$. Δt is the discrete time interval. N_t is the maximum discrete time value. $t_n = \frac{t_\varphi}{\Delta t}$.

Equation (9.13) is a marching-on-in-time formulation that can be quickly solved with an iterative procedure. By solving formula (9.13), the response of lightning waves on the grounding grid of the substation can be obtained at any time.

9.4 The Time Domain Spatial 1-d Scalar Galerkin BEM for Transient Grounding Problems

This chapter will discuss in detail how the time domain 1-d spatial scalar Galerkin BEM can simulate the transient grounding problem of grounding grid.

9.4.1 The Time Domain Spatial 1-d Scalar Galerkin's BEM Obtained With Fourier Transform

The frequency domain spatial 1-d scalar Galerkin BEM

This chapter will review and summarize the content related to the fourth chapter and the scalar BEM, as detailed below.

Referring to the formula (5.13) related to the 1-d spatial scalar Galerkin's BEM in Chap. 5, this formula can be rewritten as follows.

$$[\bar{\bar{\mathbf{Z}}}_p][\bar{\mathbf{I}}_p] = [\bar{\mathbf{V}}_p] \tag{9.14}$$

where $[\bar{\bar{\mathbf{Z}}}_p]$ is the $N_p \times N_p$ matrix about mutual impedance coefficient between a pair of segments in grounding system, $[\bar{\mathbf{I}}_p]$ is a $N_p \times 1$ vector for leakage currents emanating from grounding system's discrete conductor's node, $[\bar{\mathbf{V}}_p]$ is a $N_p \times 1$ vector for voltage of grounding system's discrete conductor's node. It must be pointed out that the coefficients matrix in linear system Eq. (9.14) is symmetric and positive definite [1].

his frequency domain equation (9.14) is assembled from the following equations.

$$\sum_{i=1}^{N_b} \int_{l_j} \int_{l_i} G_{\varphi}^{f}(\bar{r}_j, \bar{r}_i, \omega) \begin{bmatrix} \frac{1}{l_i}\frac{1}{l_j} & -\frac{1}{l_i}\frac{1}{l_j} \\ -\frac{1}{l_i}\frac{1}{l_j} & \frac{1}{l_i}\frac{1}{l_j} \end{bmatrix} dr_i dr_j \begin{bmatrix} I_{l_j}^{k_1}(\bar{r}_i) \\ I_{l_j}^{k_2}(\bar{r}_i) \end{bmatrix} = \begin{bmatrix} 1 \\ -1 \end{bmatrix} \varphi_{l_j}(\bar{r}_j) \tag{9.15}$$

where $j = 1, \ldots, N_b$. k_1 and k_2 are the numbers of the two nodes in the i-th conductor segment, respectively, $k_1, k_2 = 1, \ldots, N_p$.

The time domain spatial 1-d scalar Galerkin BEM obtained with Fourier transform

Substituting formula (9.3) into the formula (9.15) and considering the Fourier transform, the following equation can be obtained.

$$\begin{aligned} &\frac{1}{2\pi}\int_{-\infty}^{\infty} \sum_{i=1}^{N_b} \int_{l_j} \int_{l_i} \left(G_{\xi\zeta}^{\varphi_s^f}(\bar{r}_j, \bar{r}_i) + G_{\xi\zeta}^{\varphi_q^f}(\bar{r}_j, \bar{r}_i, \omega) \right) \begin{bmatrix} \frac{1}{l_i}\frac{1}{l_j} & -\frac{1}{l_i}\frac{1}{l_j} \\ -\frac{1}{l_i}\frac{1}{l_j} & \frac{1}{l_i}\frac{1}{l_j} \end{bmatrix} dr_i dr_j \begin{bmatrix} I_{l_j}^{k_1}(\bar{r}_i) \\ I_{l_j}^{k_2}(\bar{r}_i) \end{bmatrix} e^{j\omega t} d\omega \\ &= \frac{1}{2\pi}\int_{-\infty}^{\infty} \begin{bmatrix} 1 \\ -1 \end{bmatrix} \varphi_{l_j}(\bar{r}_j) e^{j\omega t} d\omega \end{aligned} \tag{9.16}$$

This formula can be further written in matrix form as follows.

$$\sum_{i=1}^{N_b} \int_{l_j} \int_{l_i} \left[G_{\xi\zeta}^{\varphi_s^t}(\bar{r}_j, \bar{r}_i) \begin{bmatrix} \frac{1}{l_i}\frac{1}{l_j} & -\frac{1}{l_i}\frac{1}{l_j} \\ -\frac{1}{l_i}\frac{1}{l_j} & \frac{1}{l_i}\frac{1}{l_j} \end{bmatrix} dr_i dr_j \begin{bmatrix} I_{l_j}^{k_1}(\bar{r}_i, t) \\ I_{l_j}^{k_2}(\bar{r}_i, t) \end{bmatrix} + G_{\xi\zeta}^{\varphi_q^t}(\bar{r}_j, \bar{r}_i, t) \begin{bmatrix} \frac{1}{l_i}\frac{1}{l_j} & -\frac{1}{l_i}\frac{1}{l_j} \\ -\frac{1}{l_i}\frac{1}{l_j} & \frac{1}{l_i}\frac{1}{l_j} \end{bmatrix} dr_i dr_j \begin{bmatrix} I_{l_j}^{k_1}(\bar{r}_i, t - t_\varphi) \\ I_{l_j}^{k_2}(\bar{r}_i, t - t_\varphi) \end{bmatrix} \right] = \begin{bmatrix} 1 \\ -1 \end{bmatrix} \varphi_{l_j}(\bar{r}_j, t) \tag{9.17}$$

This formula can be further written in matrix form as follows.

$$[\bar{\bar{\mathbf{Z}}}_p^s][\bar{\mathbf{I}}_p(t)] + [\bar{\bar{\mathbf{Z}}}_p^q][\bar{\mathbf{I}}_p(t - t_\varphi)] = [\bar{\mathbf{V}}_p(t)] \tag{9.18}$$

where $[\bar{\bar{\mathbf{Z}}}_p^s]$ and $[\bar{\bar{\mathbf{Z}}}_p^q]$ are the $N_p \times N_p$ equation matrices without delay effects and with delay effects, respectively; $[\bar{\mathbf{I}}_p(t)]$ and $[\bar{\mathbf{V}}_p(t)]$ are $N_p \times 1$ vectors representing the current and voltage of the conductor's node of the grounding grid, respectively.

9.4.2 The Time Domain Spatial 1-d Scalar Galerkin's BEM Obtained With Laplace Transform

The s domain spatial 1-d scalar Galerkin's BEM

This chapter will review and summarize the content related to the fourth chapter and the scalar BEM, as detailed below.

Referring to the formula (5.3) related to the 1-d spatial scalar Galerkin's BEM in Chap. 5, this formula can be rewritten as follows.

$$[\bar{\bar{\mathbf{Z}}}_p][\bar{\mathbf{I}}_p] = [\bar{\mathbf{V}}_p] \tag{9.19}$$

where $[\bar{\bar{\mathbf{Z}}}_p]$ is the $N_p \times N_p$ matrix about mutual impedance coefficient between a pair of segments in grounding system, $[\bar{\mathbf{I}}_p]$ is a $N_p \times 1$ vector for leakage currents emanating from grounding system's discrete conductor's node, $[\bar{\mathbf{V}}_p]$ is a $N_p \times 1$ vector for voltage of grounding system's discrete conductor's node. It must be pointed out that the coefficients matrix in linear system Eq. (9.19) is symmetric and positive definite [1].

The s domain equation (9.19) is assembled from the following equations.

$$\sum_{i=1}^{N_b} \int_{l_j} \int_{l_i} G_{\xi\zeta}^{\varphi^s}(\bar{r}_j, \bar{r}_i, s) \begin{bmatrix} \frac{1}{l_i}\frac{1}{l_j} & -\frac{1}{l_i}\frac{1}{l_j} \\ -\frac{1}{l_i}\frac{1}{l_j} & \frac{1}{l_i}\frac{1}{l_j} \end{bmatrix} dr_i dr_j \begin{bmatrix} I_{l_j}^{k_1}(\bar{r}_i) \\ I_{l_j}^{k_2}(\bar{r}_i) \end{bmatrix} = \begin{bmatrix} 1 \\ -1 \end{bmatrix} \varphi_{l_j}(\bar{r}_j) \tag{9.20}$$

where $j = 1, \ldots, N_b$. k_1 and k_2 are the numbers of the two nodes in the i-th conductor segment, respectively, $k_1, k_2 = 1, \ldots, N_p$.

The time domain spatial 1-d scalar Galerkin's BEM obtained with Laplace transform

Substituting formula (9.9) into the formula (9.20) and considering the laplace transform, the following equation can be obtained.

$$\frac{1}{2\pi j} \int_{\sigma - j\infty}^{\sigma + j\infty} \sum_{i=1}^{N_b} \int_{l_j} \int_{l_i} \left(G_{\xi\zeta}^{\varphi_s^s}(\bar{r}_j, \bar{r}_i) + G_{\xi\zeta}^{\varphi_q^s}(\bar{r}_j, \bar{r}_i, s) \right) \begin{bmatrix} \frac{1}{l_i}\frac{1}{l_j} & -\frac{1}{l_i}\frac{1}{l_j} \\ -\frac{1}{l_i}\frac{1}{l_j} & \frac{1}{l_i}\frac{1}{l_j} \end{bmatrix} dr_i dr_j \begin{bmatrix} I_{l_j}^{k_1}(\bar{r}_i) \\ I_{l_j}^{k_2}(\bar{r}_i) \end{bmatrix} e^{st} ds$$

$$= \frac{1}{2\pi j} \int_{\sigma - j\infty}^{\sigma + j\infty} \begin{bmatrix} 1 \\ -1 \end{bmatrix} \varphi_{l_j}(\bar{r}_j) e^{st} ds \tag{9.21}$$

This formula can be further written in matrix form as follows.

$$\sum_{i=1}^{N_b}\int_{l_j}\int_{l_i}[G_{\xi\zeta}^{\varphi_s^t}(\bar{r}_j,\bar{r}_i)\begin{bmatrix}\frac{1}{l_i}\frac{1}{l_j} & -\frac{1}{l_i}\frac{1}{l_j}\\ -\frac{1}{l_i}\frac{1}{l_j} & \frac{1}{l_i}\frac{1}{l_j}\end{bmatrix}dr_i dr_j\begin{bmatrix}I_{l_j}^{k_1}(\bar{r}_i,t)\\ I_{l_j}^{k_2}(\bar{r}_i,t)\end{bmatrix}$$

$$+G_{\xi\zeta}^{\varphi_q^t}(\bar{r}_j,\bar{r}_i)\begin{bmatrix}\frac{1}{l_i}\frac{1}{l_j} & -\frac{1}{l_i}\frac{1}{l_j}\\ -\frac{1}{l_i}\frac{1}{l_j} & \frac{1}{l_i}\frac{1}{l_j}\end{bmatrix}dr_i dr_j\begin{bmatrix}I_{l_j}^{k_1}(\bar{r}_i,t-j\beta_e)\\ I_{l_j}^{k_2}(\bar{r}_i,t-j\beta_e)\end{bmatrix}]=\begin{bmatrix}1\\ -1\end{bmatrix}\varphi_{l_j}(\bar{r}_j,t) \tag{9.22}$$

This formula can be further written in matrix form as follows.

$$[\bar{\bar{\mathbf{Z}}}_p^s][\bar{\mathbf{I}}_p(t)]+[\bar{\bar{\mathbf{Z}}}_p^q][\bar{\mathbf{I}}_p(t-t_q)]=[\bar{\mathbf{V}}_p(t)] \tag{9.23}$$

where $[\bar{\bar{\mathbf{Z}}}_p^s]$ and $[\bar{\bar{\mathbf{Z}}}_p^q]$ are the $N_p \times N_p$ equation matrices without delay effects and with delay effects, respectively; $[\bar{\mathbf{I}}_p(t)]$ and $[\bar{\mathbf{V}}_p(t)]$ are $N_p \times 1$ vectors representing the current and voltage of the conductor's node of the grounding grid, respectively.

9.4.3 Spatial 1-d Scalar Galerkin BEM for Time Domain Equipotential Electromagnetic Field Mathematical Model for Transient Grounding Problems

The time-domain 1-d spatial scalar Galerkin BEM developed in this chapter can directly solve the lightning transient impulse grounding problem of the substation grounding grid in the time domain without the need for inverse Fourier & Laplace transform.

By discretizing continuous time, the discrete time domain matrix formula can be obtained as follows

$$[\bar{\bar{\mathbf{Z}}}_p^s][\bar{\mathbf{I}}_p(t_k)]+[\bar{\bar{\mathbf{Z}}}_p^q][\bar{\mathbf{I}}_p(t_k-t_n)]=[\bar{\mathbf{V}}_p(t_k)] \tag{9.24}$$

where $t_k = k\Delta, k = 1, 2, \ldots, N_t$. Δt is the discrete time interval. N_t is the maximum discrete time value.

Equation (9.24) is a marching-on-in-time formulation that can be quickly solved with an iterative procedure. By solving formula (9.24), the response of lightning waves on the grounding grid of the substation can be obtained at any time.

9.5 Simulation Result and Analysis

According to the approach introduced in this chapter, a program has been implemented. The program can simulate a lightning response of grounding system buried in horizontal up to four-layer earth model, in which the fast convergent 1-d spatial scalar Galerkin MM and BEM with zero order basic functions has been used.

9.5.1 *Verification of Accuracy and Effectiveness for Spatial 1-d Scalar Galerkin MM and BEM*

To verify the accuracy and effectiveness of the method proposed in this chapter, some cases solved by other authors are studied. In the following arithmetic diagrams, the abbreviations MM(eq) and BEM(eq) denote the scalar Galerkin MM and the scalar Galerkin BEM for equipotential mathematical models, respectively. The abbreviations MM and BEM denote the vector Galerkin MM and the vector Galerkin BEM of the unequal potential mathematical model, respectively.

The first example is from [4]. The grounding grid configuration can be seen in Fig. 9.1a. The grounding grid was made of round copper conductors with a cross-sectional

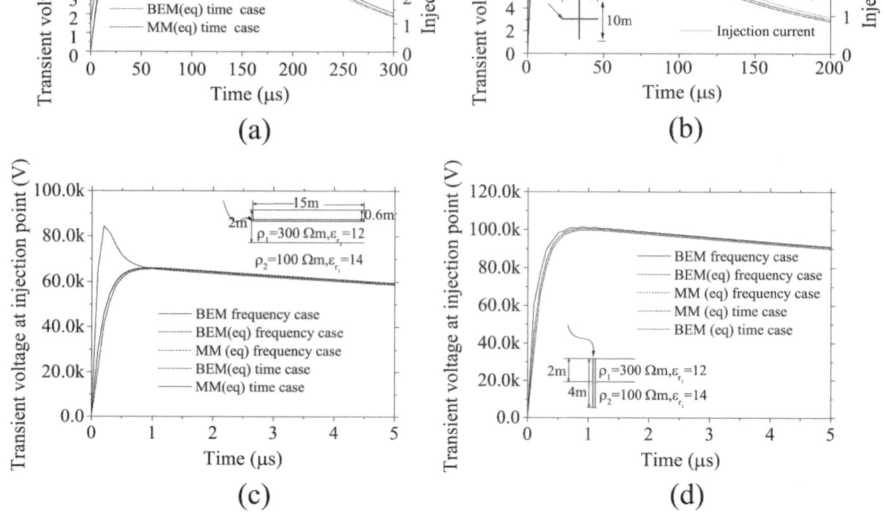

Fig. 9.1 Verified case

area of 50 mm^2 and was buried at 0.5 m. The injected lightning current parameters were set to $T_1 = 65\,\mu\text{s}$, $T_2 = 185\,\mu\text{s}$, and $I_m = 6.86A$, and the feed point was set at the corner of the grid. For the earth model, the conductivities and permittivities of the two-layer earth model were set to $\frac{\sigma_1=50^{-1}S/m}{\sigma_2=20^{-1}S/m}$, $\frac{\varepsilon_1=10\varepsilon_0}{\varepsilon_2=10\varepsilon_0}$, and $h = 0.6$ m. Figure 9.1a shows the transient voltage curve simulated based on frequency-domain 1-d spatial scalar Galerkin's MM and BEM and the measured data [4]. The calculated voltage curves were in good agreement with the measured curve in [4]. It should be noted that in this example, the lightning wave is relatively gentle and the size of the grounding grid is relatively small. Therefore, the equipotential model algorithm developed in this chapter is relatively effective and accurate.

The second verification case came from [5]. The grounding grid configuration can be seen in Fig. 9.1b. The grounding grid was also made of round copper conductors with cross-sectional area 50 mm^2 and was buried at 0.5 m. The injected lightning current parameters were set to $T_1 = 16\,\mu\text{s}$, $T_2 = 99\,\mu\text{s}$, and $I_m = 4.7A$. For the earth model, the conductivities and permittivities of the two-layer earth model were set to $\frac{\sigma_1=50^{-1}S/m}{\sigma_2=20^{-1}S/m}$, $\frac{\varepsilon_1=10\varepsilon_0}{\varepsilon_2=10\varepsilon_0}$, and $h = 0.6$ m. Figure 9.1b shows the transient voltage curve simulated based on 1-d spatial scalar Galerkin's MM and BEM, and the data [5]. The calculated voltage curve is very close to the development trend of the measured curve, but there are certain errors. The calculated voltage curve is very close to the development trend of the measured curve, but there are certain errors. This is because the lightning wave waveform in this example is slightly steeper compared to the previous example, and the size of the grounding grid is also relatively small.

The last test case comes from [3, 6]. Figure 9.1c, d shows, respectively, horizontal and vertical grounding electrodes with radius of 12 mm, and the length of horizontal and vertical grounding electrodes is 15 m and 4 m, respectively. The horizontal grounding electrode was buried at a depth of 0.6 m, and both of the electrodes were within in two-layer horizontal earth where the resistivity and relative permittivity ratios for the upper and the lower soil layers were $\frac{\rho_1=300\,\Omega m}{\rho_2=100\,\Omega m}$, $\frac{\varepsilon_{r_1}=12}{\varepsilon_{r_2}=14}$, and the upper layer thickness was $H = 2$ m. The injected lightning current was set with parameters of: (a) $T_1 = 0.36\,\mu\text{s}$, $T_2 = 26.8\,\mu\text{s}$, and $I_0 = 3.0\,\text{kA}$. Figure 9.1c, d show the transient voltage curves of horizontal and vertical grounding rods under lightning current impact, respectively. These curves are derived from literature [3] and the equipotential model algorithm developed in this chapter. From Fig. 9.1c, it can be seen that for horizontal grounding rods, the curve obtained by the equipotential model algorithm developed in this chapter is significantly different from the curve given in reference [3], with significant errors. As shown in Fig. 9.1d, for vertical grounding rods, the curve obtained by the equipotential model algorithm developed in this chapter is basically consistent with the curve given in reference [3], with relatively small errors. This is because the lightning wave waveform in this calculation example is relatively steep, and the correctness of the equipotential model algorithm developed in this chapter is questionable for horizontal grounding rods, while for vertical grounding rods, the correctness of the equipotential model algorithm developed in this chapter is still acceptable.

From the above five examples, it can be seen that due to the neglect of the mutual inductance effect between the axial branch currents between the conductors of the substation grounding grid, the spatial 1-d scalar Galerkin MM and BEM in the time domain may result in significant deviations or even errors in the numerical simulation results for lightning currents with relatively high steepness. This is because lightning currents with relatively high steepness contain more high-frequency components, and in high-frequency situations, the influence of mutual inductance is relatively increased. Therefore, ignoring the influence of mutual inductance will lead to calculation errors. Therefore, it can be known that the time domain 1d spatial scalar Galerkin MM and BEM developed in this chapter have certain limitations and a relatively small scope of application for lightning transient impulse grounding problems in substation grounding grids.

9.5.2 *Numerical Results Analysis*

A general grounding system can be seen in Fig. 9.2, which consists of 4×4 square grids and the burial depth of the grounding grid is 0.5 meters. The dimensions of this grounding grid can be found in the figure. The earth is modelled as horizontal four layers conductive media, whose resistivities and permittivities are $\rho_1 = 100\,\Omega$m and $\rho_2 = 220\,\Omega$m, $\rho_3 = 310\,\Omega$m, $\rho_4 = 430\,\Omega$m, $\varepsilon_1 = 10\varepsilon_0$, $\varepsilon_2 = 12\varepsilon_0$, $\varepsilon_3 = 13\varepsilon_0$ and $\varepsilon_4 = 14\varepsilon_0$, respectively. The first, second and third layer thickness are $h_1 = 6$ m, $h_2 = 12$ m and $h_3 = 8$ m. All the conductor radius of grids are 5 mm. The external excited lightning current with $1.29 \times (e^{-0.01910t} - e^{-0.292288t})kA$ is injected from the corner of the grounding system. According to the formula of lightning current, T_1 is 10 μs and T_2 is 50 μs.

Impulse impedance case

Referring to Table 9.1, it can be seen that the impulse impedance calculated by the two scalar algorithms (MM and BEM) in the frequency and time domains are basically the same, except that the two scalar algorithms in the frequency domain are slightly higher than the two scalar scalar algorithms in the time domain. This can be seen from the comparison of the impulse impedance values calculated by the four algorithms to see the correctness of the two scalar algorithms in both the frequency and time domains.

Table 9.1 Calculate impulse impedance

	Impulse impedance	
Methods	Time domain case	Frequency domain case
BEM	(1.980, −0.184)	(2.023, −1.004)
MM	(1.976, −0.184)	(2.018, −1.002)

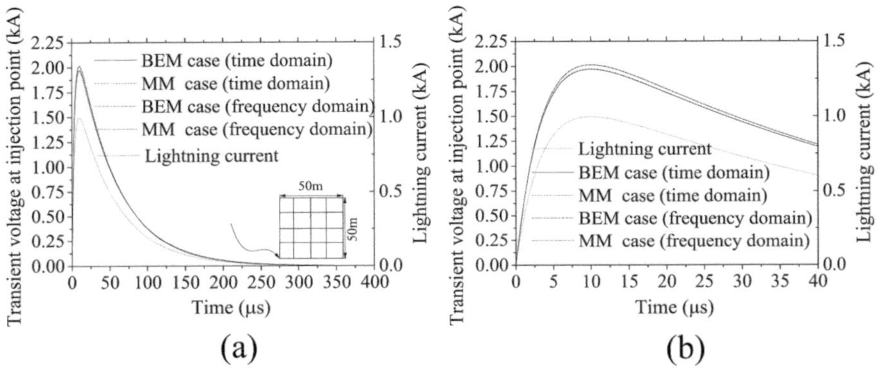

Fig. 9.2 Transient voltage at injection point

Transient voltage case

The transient voltage curve at the injection point of the lightning current is shown in Fig. 9.2a, b is a partial enlargement of figure (a). Observing Fig. 9.2, we can know that the transient voltage curves at the injection point of the lightning current calculated by the scalar Galerkin MM and the BEM in the frequency and time domains are basically the same. The curves computed by the scalar MM and the scalar BEM in the time and frequency domains are in perfect coincidence, respectively, while the curves computed by the scalar MM and the scalar BEM in the frequency domain are slightly higher than those computed by the scalar MM and the scalar BEM in the time domain. This situation is basically consistent with the impulse impedance values analyzed above.

3-d plot for transient voltage and leakage current case

In the following, the 3-d distribution of transient voltage and transient leakage current on the grounding grid of the substation will be numerically simulated from the two methods of scalar Galerkin BEM in the frequency and time domains, and the two algorithms will simulate the 3-d distribution of transient voltage and transient leakage current on the grounding grid at the same time, which can be compared with each other, and verified by each other.

Firstly, Figs. 9.3 and 9.4 show the 3-d distribution of the transient voltage on the grounding grid, which is obtained by numerical simulation of the time-domain and frequency-domain algorithms, and Figs. 9.5 and 9.6 show the 3-d distribution of the transient leakage current on the grounding grid, which is obtained by numerical simulation of the time-domain and frequency-domain algorithms, respectively.

Observe Fig. 9.2 can know the whole lightning current impact process occurs within 400 μs, therefore, first observe Fig. 9.3, can see: from 0 μs, gradually induced 3-d distribution of transient voltage, from Fig. 9.3 (2) can be seen in the 1.02 μs induced 3-d distribution of transient voltage, in with the increase of time, the induced 3-d distribution of the transient voltage gradually reaches the maximum value, here

Fig. 9.3 (4) shows the maximum transient voltage of the 3-d distribution in 10.23 μs. After that, with the change of time, the 3-d distributed transient voltage gradually weakens until it disappears. Here Fig. 9.3 (9) shows the transient voltage of the 3-d distribution that basically disappears at 305.91 μs. Observing the nine subplots (1)-(9) of Fig. 9.3, the 3-d distributed transient voltages induced on this grounding grid during the entire lightning current impact can be observed. It can also be seen that the 3-d distributed transient voltages are basically square, regardless of the moment, which is due to the fact that the numerical algorithm for the electromagnetic field developed in this chapter is the scalar Galerkin BEM of the equipotential electromagnetic model. Therefore, the 3-d transient voltage induced on the grounding grid at any moment throughout the lightning strike is plane distributed.

A detailed analysis and discussion of the evolution of the 3-d distribution of the transient voltage induced on the substation grounding grid during lightning strikes for the numerical simulation of the time-domain algorithm is presented above. In the following, the frequency domain algorithm is further analyzed to numerically simulate the evolution of the induced 3-d distributed transient voltage on the grounding grid during lightning strikes. In fact, the two algorithms numerically simulate the 3-d distribution of transient voltages induced on the grounding grid in the lightning strike process is basically the same. Here we should observe the 3-d distribution of transient voltages induced on the grounding grid of the two algorithms at critical moments, for example, observe the 3-d distribution of the induced transient voltages just appeared at the moment of 102 microseconds, and it can be known from the comparison of the two 3-d diagrams in Figs. 9.3(2) and 9.4(2) that the two are basically the same. Secondly, observing and comparing the 3-d distribution of transient voltage in Figs. 9.3(4) and 9.4(4) at the moment of maximum value, it can also be known that: the two are basically the same. In the same way, respectively, the observation of Figs. 9.1 and 9.2 corresponds to the same moment of the 3-d transient voltage 3-d distribution diagrams, the same can be obtained from the two algorithms of numerical simulation of transient voltage on the grounding grid in the process of lightning impact evolution process and the value of the full compliance with the same, from this point of view to respond to the correctness of the two electromagnetic field numerical algorithms.

The above numerical algorithms for electromagnetic fields in the frequency and time domains are analyzed and evaluated for the evolution of transient voltages induced on substation grounding grids during lightning strikes. In the following, the evolution of the transient leakage current induced on the substation grounding grid in the lightning impact process is analyzed and judged by the frequency domain and time domain electromagnetic field numerical algorithms.

First, the evolution of the transient leakage currents on the grounding grid during the lightning impact is observed, and the 3-d distribution of the transient leakage currents presented here at various moments of the lightning impact is simulated by the numerical algorithm of the electromagnetic field in the time domain. First, the evolution of the transient leakage currents on the grounding grid during the lightning impact is observed, and the 3-d distribution of the transient leakage currents presented here at various moments of the lightning impact is simulated by the numerical algorithm of the time-domain electromagnetic field. From observing the subfigures

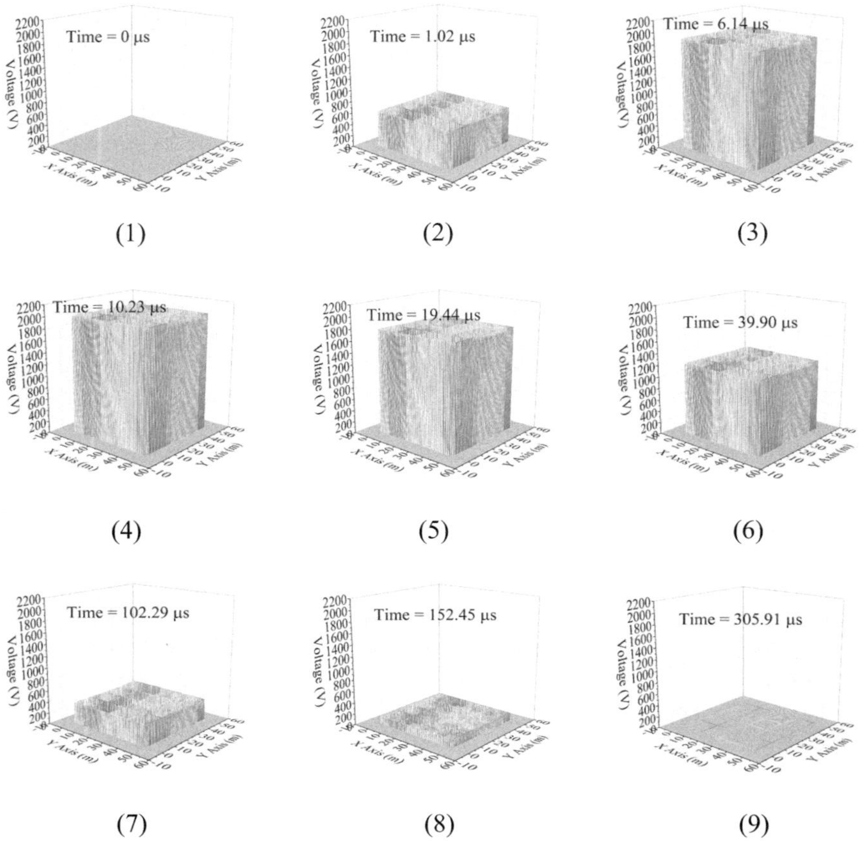

Fig. 9.3 3-d plot for voltage along grounding grid (time domain case)

(1) to (9) of Fig. (9.5), it can be known that: under the lightning impact, consistent with the transient voltage case above, the 3-d distribution of transient leakage current appears at 1.02 μs, and after that, with the change of time, it reaches the maximum value of the distribution of 3-d transient leakage current at 10.23 μs. After that, with the change of time, it gradually weakened until disappeared. The evolution of 3-d transient leakage current on the grounding grid during the whole lightning impact process is basically the same as the evolution of 3-d transient voltage on the grounding grid above. But the difference is: 3-d transient leakage distribution is no longer flat, but in the grounding grid around the leakage current value is the largest and gradually to the center of the grounding grid weakened until the leakage current in the middle of the grounding grid is the smallest. This is due to the use of time-domain scalar Galerkin BEM, numerical simulation of the grounding grid induced transient voltage is distributed in a planar manner, but due to the grounding grid at any location of the grounding grid to the ground resistance is different and caused by the

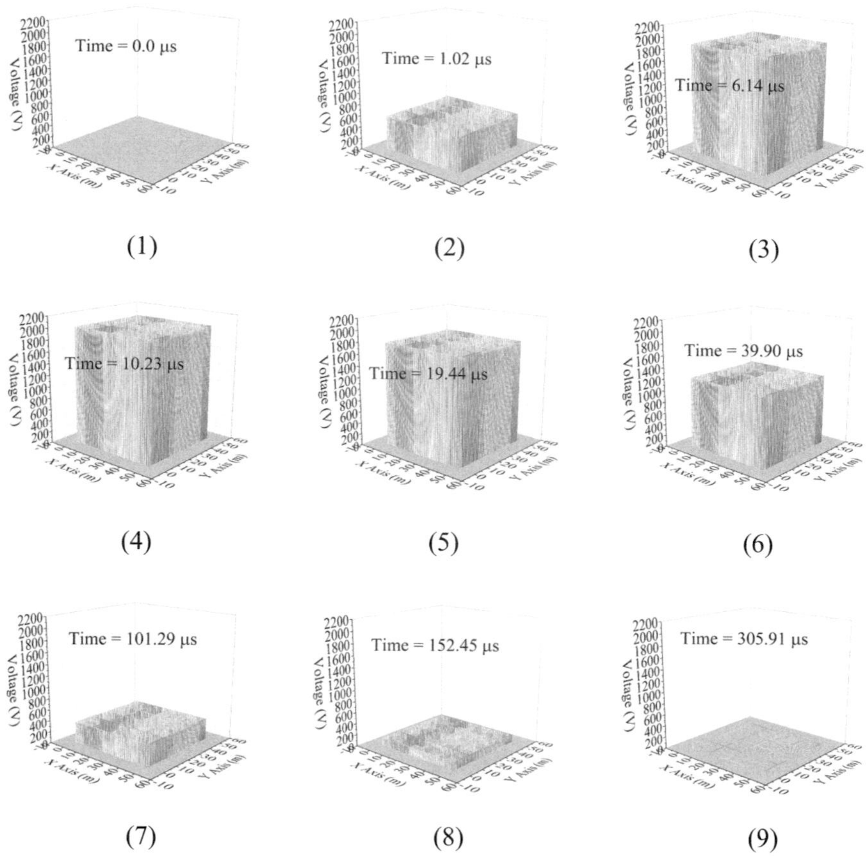

Fig. 9.4 3-d plot for voltage along grounding grid (frequency domain case)

distribution of transient leakage current is no longer distributed in a planar manner of the 3-d distribution.

The evolution of the 3-d distribution of the transient leakage currents appearing on the grounding grid during lightning strikes numerically simulated by the time-domain electromagnetic field method has been analyzed and discussed previously. In the following, the evolution of the 3-d distribution of the transient leakage current on the grounding grid during lightning impact simulated numerically by the frequency-domain electromagnetic field method is further observed and discussed.

And the previous for the time domain and frequency domain electromagnetic field numerical algorithms for simulation and calculation of the two groups of grounding grid induced transient voltage of the 3-d distribution of the evolution process of comparison and analysis of the same idea, the following for the time and frequency domain electromagnetic field numerical algorithms for simulation and calculation of

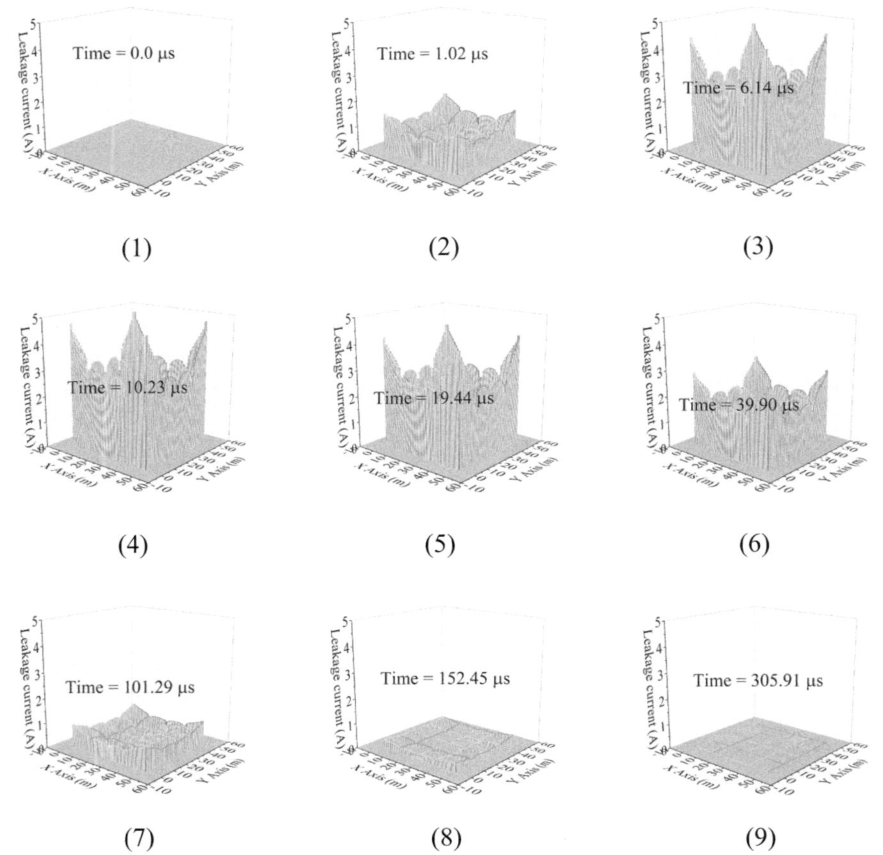

Fig. 9.5 3-d plot for leakage current along grounding grid (time domain case)

the two groups of time and frequency domain electromagnetic field numerical algorithms for simulation and calculation of the two groups of grounded grid induced leakage current of the 3-d distribution of evolution of the process of comparison and analysis of the 3-d distribution, the same idea. The same idea, observation and comparison of Figs. 9.5 and 9.6 in each moment corresponding to the sub-diagram can be known: the evolution of the two groups of induced leakage current on the grounding grid of the 3-d distribution of the map in the amount of value and distribution trend is basically the same. As in the case of transient voltages, the correctness and accuracy of the two different numerical algorithms can also be illustrated from the point of view of this correctly obtained numerical result.

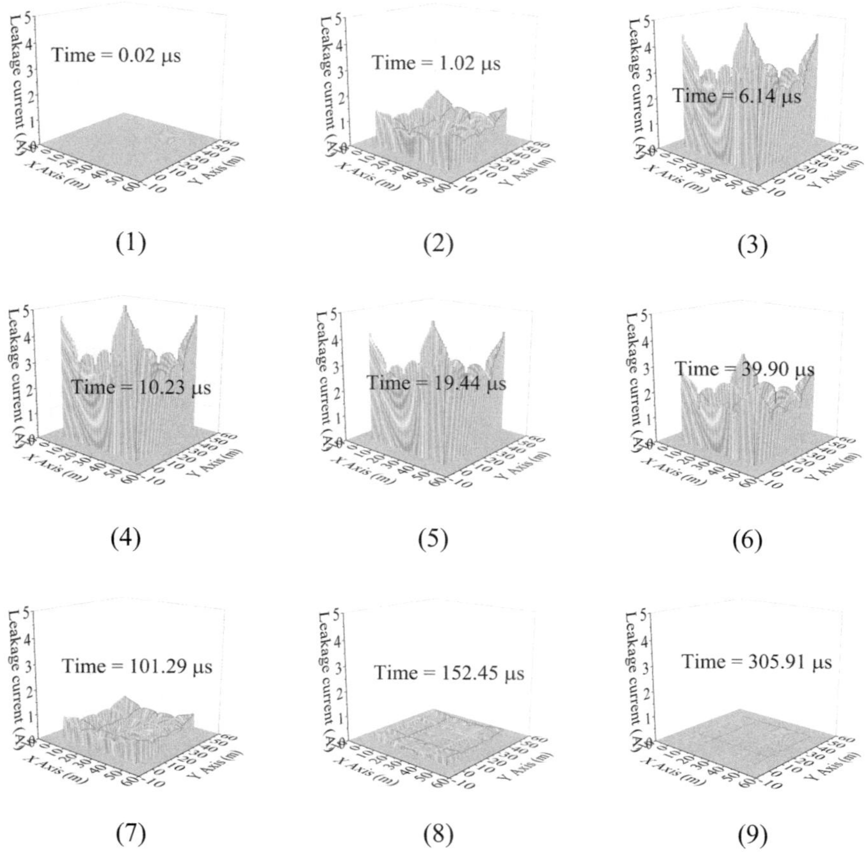

Fig. 9.6 3-d plot for leakage current along grounding grid (frequency domain case)

9.6 Summary

Due to the neglect of mutual inductance between the conductors of the substation grounding grid by the time-domain scalar Galerkin MM and BEM, the numerical simulation results are basically incorrect for lightning waves with high steepness. So the spatial 1d scalar Galerkin MM and BEM developed in this chapter have certain limitations and a relatively small scope of application for lightning transient impulse grounding problems in substation grounding grids.

References

1. C. Johnson, *Numerical Solution of Partial Differential Equations by the Finite Element Method* (Cambridge University Press, Cambridge, U.K., 1987)
2. Z.-X. Li, S.-W. Rao, Transient lightning response of grounding grids in a horizontal multi-layered soil model that considers soil ionization effects with time-domain quasi-static complex images. Int. J. Numer. Model. Electron. Network Dev. Fields **32**(1), e2463 (2019)
3. Z.-X. Li, Y. Yu, K.-C. Wang, Lightning response of a grounding system buried in multilayered earth based on the galerkin's boundary element and quasi-static complex image methods. IEEE Trans. Electromagn. Compat. **63**(4), 1118–1127 (2021)
4. Z. Stojkovic et al., Experimental investigation of grounding grid impulse characteristic. Europ. Transa. Electr. Power **8**(6), 417–425 (1998)
5. Z. Stojkovic et al., Sensitivity analysis of experimentally determined grounding grid impulse characteristic. IEEE Trans. Power Delivery **13**(3), 1136–1142 (1998)
6. M. Jamali, M. Niasati, M. Jazaeri, A two-layer soil model for the calculation of electrical parameters of grounding systems under lightning strikes. Electr. Power Components Syst. **47**(1–2), 181–191 (2019)

Chapter 10
Numerical Methods for Grounding Problem at Transient State IV

Various time-domain numerical methods for grounding problems in the case of lightning transient shock response based on unequal potential electromagnetic field modeling.

10.1 Preface

Like the scalar numerical method in the previous chapter, the frequency domain 1-d spatial vector Galerkin MM and BEM, as well as the vector hybrid algorithm based on various basis functions, which are used for the steady-state grounding problem of the grounding grid, can also be further used for the transient grounding problem of the grounding grid through Fourier transform. This chapter will discuss in detail the scientificity and effectiveness of these methods.

10.2 Introduction

This chapter mainly discusses the time-domain spatial 1-d vector Galerkin BEM, time-domain vector Galerkin MM, and time-domain vector Galerkin hybrid algorithm based on PI and PEEC basis functions.

For numerical simulation of substation grounding grid lightning transient impact grounding problem of electromagnetic field numerical methods can be divided into two categories, respectively, discrete solution field algorithms and discrete solution source algorithms, the former can be directly solved for the required field points of the electromagnetic field amount of information, but the need for the calculation of

Z.-X. Li, *Numerical Simulation Algorithm of Electromagnetic Field for Grounding Problems in Power System Substation Grounding Grids*,
https://doi.org/10.1007/978-981-96-0904-8_10

the field of the relatively large and caused by the problem of the amount of huge calculations; The latter requires two steps to solve the electromagnetic field information of the required field point, i.e., to solve the electromagnetic field information of the distributed source point first, and then to solve the electromagnetic field information of the required field point through the known source point information, which is relatively troublesome but reduces the computational amount relative to the former due to the fact that it only discretely solves the electromagnetic field information of a limited number of source points.

At present, the development of FDTD methods for numerical algorithms for electromagnetic fields, which belong to the first category, is currently very lush. There is also a lot of literature [1–19] on the use of FDTD methods to numerically simulate the lightning transient shock response of substation grounding grids in uniform as well as multilayered layered earth.

Currently belonging to the second category of electromagnetic field numerical algorithms of various types of algorithms are also developed very lush, and the variety is relatively large compared to the first category of electromagnetic field numerical algorithms. The literature [23–26] has discussed the use of time-domain BEM to simulate the lightning transient impulse grounding problem of substation grounding grid in a uniform earth. In contrast, literature [27] describes the time-domain Galerkin BEM for numerical simulation of the lightning transient shock response of substation grounding grids in multilayer horizontally stratified soils. In the literature [20–22], the time-domain FEM has also been used to numerically simulate the lightning transient shock response of substation grounding grids in a uniform earth. At the same time, some literature [28–36] introduces time-domain hybrid algorithms for numerical simulation of lightning transient impulse grounding problems in substation grounding grids. The time-domain hybrid algorithms introduced in these literature [28–34] are based on PI basis functions, while some [35, 36] are based on PEEC basis functions. There is also a lot of literature [37–54] on the numerical simulation of the lightning transient impulse response of substation grounding grids in a uniform earth using the time-domain transmission line method. Finally, it should be further pointed out that there are a lot of literatures that first establish the system of equations of frequency-domain finite-element method [55], boundary-element method [56, 57], method of moments [58, 59], and hybrid algorithm [60, 61], and then transform the system of equations of frequency-domain into the system of equations of time-domain through the method of vector line fitting, so as to obtain the corresponding numerical method of time-domain for numerical simulation of lightning transient response of the substation grounding grid in a homogeneous earth. It should be noted that the articles published by the authors of this book [27, 29–31, 36] are on the numerical simulation of the grounding problem of lightning transient impacts on the substation grounding grid in a multilayer horizontally layered soil using the time-domain Galerkin BEM as well as hybrid algorithms based on various basis functions.

10.3 The Time Domain Spatial 1-d Vector Galerkin BEM for Transient Grounding Problems

This chapter will discuss in detail how to scientifically simulate lightning transient grounding of substation grounding grid buried in horizontal N_e-layer earth model by time domain 1-d spatial vector Galerkin BEM.

10.3.1 The Time Domain Spatial 1-d Vector Galerkin BEM Obtained with Fourier Transform

In this chapter, we will discuss in detail how the frequency domain 1-d spatial vector Galerkin BEM can be converted to a time domain 1-d spatial vector Galerkin BEM.

The frequency domain spatial 1-d vector Galerkin BEM

This chapter will review and summarize the relevant content of the 1-d spatial vector Galerkin BEM in Chap. 6, as follows.

Referring to the formula (6.7) related to the 1-d spatial vector Galerkin BEM in Chap. 6, this formula can be rewritten as follows.

$$[\bar{\bar{\mathbf{Z}}}_p][\bar{\mathbf{I}}_p] = 0 \tag{10.1}$$

where $[\bar{\bar{\mathbf{Z}}}_p]$ is $N_p \times N_p$ matrix, who includes the mutual impedance and induction interaction betweens a parts of conductor segments currents. $[\bar{\mathbf{I}}_p]$ is $N_p \times 1$ vector of unknown currents.

Matrix equation (10.1) is assembled from the following equations.

$$Z_{l_\theta} \int_{l_\theta} [\bar{q}(\bar{r}_\theta)][\bar{p}(\bar{r}_\theta)] dr_\theta [\bar{I}_\theta^\chi]$$

$$+ \sum_{\vartheta=1}^{N_b} \int_{l_\theta} \int_{l_\vartheta} \hat{r}_\theta \cdot j\omega \bar{\bar{G}}_{\xi\zeta}^{Af}(\bar{r}_\theta, \bar{r}_\vartheta, \omega) \cdot \hat{r}_\vartheta^T [\bar{q}(\bar{r}_\theta)][\bar{p}(\bar{r}_\vartheta)] dr_\vartheta dr_\theta [\bar{I}_\vartheta^\chi]$$

$$+ \sum_{\vartheta=1}^{N_b} \int_{l_\theta} \int_{l_\vartheta} G_{\xi\zeta}^{\varphi f}(\bar{r}_\theta, \bar{r}_\vartheta, \omega) \frac{\partial[\bar{q}(\bar{r}_\theta)]}{\partial r_\theta} \frac{\partial[\bar{p}(\bar{r}_\vartheta)]}{\partial r_\vartheta} dr_\vartheta dr_\theta [\bar{I}_\vartheta^\chi] = 0 \tag{10.2}$$

where $\xi = 0, 1, \ldots, N_e$; $\zeta = 0, 1, \ldots, N_e$. $\theta = 1, 2, \ldots, N_b$; $\vartheta = 1, 2, \ldots, N_b$; $\chi = 1, 2, \ldots, N_p$; $\bar{r}_\theta$ and $\bar{r}_\vartheta$ is location vector along θ-th and ϑ-th conductors segment, respectively; r_θ and r_ϑ, respectively, parameter variable along the θ-th and

ϑ-th thin conductor axis. Z_{l_θ} is the internal impedance of the θ-th thin conductor segment. $\hat{r}_\theta$ and $\hat{r}_\vartheta$ are, respectively, the 1×3 dimensional unit director vector along the axis of the θ-th and ϑ-th cylindrical conductor segments. The symbol T represents matrix transposition. j is a imaginary unit, ω is angular frequency. $G_{\xi\zeta}^{\varphi^f}(\bar{r}_\theta, \bar{r}_\vartheta, \omega)$ and $\bar{\bar{G}}_{\xi\zeta}^{A^f}(\bar{r}_\theta, \bar{r}_\vartheta, \omega)$ are, respectively, frequency domain scalar monopole Green's function and vector dipole dyadic Green's function within horizontal multilayered earth model, which had been fully discussed in front Chap. 3. $[\bar{p}(\bar{r}_\vartheta)]$ and $[\bar{q}(\bar{r}_\theta)]$ are 1×2 dimensional 1 order basis functions and 2×1 dimensional 1 order test functions, respectively. $[\bar{p}(\bar{r}_\vartheta)] = \begin{bmatrix} \frac{l_\vartheta - r_\vartheta}{l_\vartheta} & \frac{r_\vartheta}{l_\vartheta} \end{bmatrix}$ and $[\bar{q}(\bar{r}_\theta)] = \begin{bmatrix} \frac{l_\theta - r_\theta}{l_\theta} \\ \frac{r_\theta}{l_\theta} \end{bmatrix}$.

Reference [29] shows that the Green's function of point current sources in multilayered horizontal layered soil can be further classified as:

$$\bar{\bar{G}}_{\xi\zeta}^{A^f}(\bar{r}_\theta, \bar{r}_\vartheta; \omega) = \bar{\bar{G}}_{\xi\zeta}^{A_s^f}(\bar{r}_\theta, \bar{r}_\vartheta) + \bar{\bar{G}}_{\xi\zeta}^{A_q^f}(\bar{r}_\theta, \bar{r}_\vartheta; \omega) \tag{10.3}$$

$$G_{\xi\zeta}^{\varphi^f}(\bar{r}_\theta, \bar{r}_\vartheta; \omega) = G_{\xi\zeta}^{\varphi_s^f}(\bar{r}_\theta, \bar{r}_\vartheta) + G_{\xi\zeta}^{\varphi_q^f}(\bar{r}_\theta, \bar{r}_\vartheta; \omega) \tag{10.4}$$

Reference [30] provides the expansion formula for internal impedance as follows

$$Z_{l_\theta} = \frac{l_\theta}{\pi r_c^2 \sigma_c} + \sum_{c=1}^{M_c} \alpha_c e^{\beta_c \omega} = Z_{l_0} + Z_l(\omega) \tag{10.5}$$

where M_c was total number of complex images, α_c and β_c were coefficients of complex images, r_c, l_θ and σ_c are the conductor radius, length and conductivity of the grounding grid, respectively.

The time domain spatial 1-d vector Galerkin BEM obtained with Fourier transform

Substituting formulas (10.3) , (10.4) and (10.5) into formula (10.2) and considering the Fourier transform, there is the following formula

$$\frac{1}{2\pi} \int_{-\infty}^{\infty} [\big(Z_{l_0} + Z_l(\omega)\big) \int_{l_\theta} [\bar{q}(\bar{r}_\theta)][\bar{p}(\bar{r}_\theta)] dr_\theta [\bar{I}_\theta^\chi]$$

$$+ \sum_{\vartheta=1}^{N_b} \int_{l_\theta} \int_{l_\vartheta} \hat{r}_\theta \cdot j\omega \left(\bar{\bar{G}}_{\xi\zeta}^{A_s^f}(\bar{r}_\theta, \bar{r}_\vartheta) + \bar{\bar{G}}_{\xi\zeta}^{A_q^f}(\bar{r}_\theta, \bar{r}_\vartheta; \omega) \right) \cdot \hat{r}_\vartheta^T [\bar{q}(\bar{r}_\theta)][\bar{p}(\bar{r}_\vartheta)] dr_\vartheta dr_\theta [\bar{I}_\vartheta^\chi]$$

$$+\sum_{\vartheta=1}^{N_b}\int_{l_\theta}\int_{l_\vartheta}\left(G_{\xi\zeta}^{\varphi_s^f}(\bar{r}_\theta,\bar{r}_\vartheta)+G_{\xi\zeta}^{\varphi_q^f}(\bar{r}_\theta,\bar{r}_\vartheta;\omega)\right)\frac{\partial[\bar{q}(\bar{r}_\theta)]}{\partial r_\theta}\frac{\partial\bar{p}(\bar{r}_\vartheta)}{\partial r_\vartheta}dr_\vartheta dr_\theta[\bar{I}_\vartheta^X]]e^{j\omega t}d\omega=0 \tag{10.6}$$

where $\theta = 1, 2, \ldots, N_b$.

By using Fourier transform, the following formulas in the time domain can be obtained.

$$\left(Z_{l_0}\int_{l_\theta}[\bar{q}(\bar{r}_\theta)][\bar{p}(\bar{r}_\theta)]dr_\theta[\bar{I}_\theta^X(t)]+Z_l\int_{l_\theta}[\bar{q}(\bar{r}_\theta)][\bar{p}(\bar{r}_\theta)]dr_\theta[\bar{I}_\theta^X(t-t_c)]\right)$$

$$+\sum_{\vartheta=1}^{N_b}(\int_{l_\theta}\int_{l_\vartheta}\hat{r}_\theta\cdot\bar{\bar{G}}_{\xi\zeta}^{A_s^t}(\bar{r}_\theta,\bar{r}_\vartheta)\cdot\hat{r}_\vartheta^T[\bar{q}(\bar{r}_\theta)][\bar{p}(\bar{r}_\vartheta)]dr_\vartheta dr_\theta\frac{d[\bar{I}_\vartheta^X(t)]}{dt}$$

$$+\int_{l_\theta}\int_{l_\vartheta}\hat{r}_\theta\cdot\bar{\bar{G}}_{\xi\zeta}^{A_q^t}(\bar{r}_\theta,\bar{r}_\vartheta,t)\cdot\hat{r}_\vartheta^T[\bar{q}(\bar{r}_\theta)][\bar{p}(\bar{r}_\vartheta)]dr_\vartheta dr_\theta\frac{d[\bar{I}_\vartheta^X(t-t_A)]}{dt})$$

$$+\sum_{\vartheta=1}^{N_b}(\int_{l_\theta}\int_{l_\vartheta}G_{\xi\zeta}^{\varphi_s^t}(\bar{r}_\theta,\bar{r}_\vartheta)\frac{\partial[\bar{q}(\bar{r}_\theta)]}{\partial r_\theta}\frac{d\bar{p}(\bar{r}_\vartheta)}{dr_\vartheta}dr_\vartheta dr_\theta[\bar{I}_\vartheta^X(t)]$$

$$+\int_{l_\theta}\int_{l_\vartheta}G_{\xi\zeta}^{\varphi_q^t}(\bar{r}_\theta,\bar{r}_\vartheta,t)\frac{\partial[\bar{q}(\bar{r}_\theta)]}{\partial r_\theta}\frac{\partial\bar{p}(\bar{r}_\vartheta)}{\partial r_\vartheta}dr_\vartheta dr_\theta[\bar{I}_\vartheta^X(t-t_\varphi)]) = 0 \tag{10.7}$$

where $\theta = 1, 2, \ldots, N_b$. $G_{\xi\zeta}^{\varphi_q^t}(\bar{r}_\theta,\bar{r}_\vartheta,t)$ and $\bar{\bar{G}}_{\xi\zeta}^{A_q^t}(\bar{r}_\theta,\bar{r}_\vartheta,t)$ are, respectively, time domain scalar monopole Green's function and vector dipole dyadic Green's function within horizontal multilayered earth model, which had been fully discussed in front Chap. 3.

This formula (10.7) can also be further written as matrix formation as shown below.

$$[\bar{\bar{\mathbf{Z}}}_b^s][I_p(t)]+[\bar{\bar{\mathbf{M}}}_b^s]\frac{d[I_b(t)]}{dt}+[\bar{\bar{\mathbf{Z}}}_s^s][I_p(t)]=-[\bar{\bar{\mathbf{Z}}}_b^q][I_p(t-t_c)]-[\bar{\bar{\mathbf{M}}}_b^q]\frac{d[I_p(t-t_A)]}{dt}-[\bar{\bar{\mathbf{Z}}}_s^q][I_p(t-t_\varphi)] \tag{10.8}$$

where $[\bar{\bar{\mathbf{Z}}}_b^s]$ and $[\bar{\bar{\mathbf{Z}}}_b^q]$ are the self impedance matrices of grounding grid conductors without and with delay effects in the $N_p \times N_p$ dimension, respectively. $[\bar{\bar{\mathbf{M}}}_b^s]$ and $[\bar{\bar{\mathbf{M}}}_b^q]$ are the mutual induction matrices of grounding grid conductors without and with delay effects in the $N_p \times N_p$ dimension, respectively. $[\bar{\bar{\mathbf{Z}}}_s^s]$ and $[\bar{\bar{\mathbf{Z}}}_s^q]$ are the mutual impedance matrices of grounding grid conductors without and with delay effects in the $N_p \times N_p$ dimension, respectively.

10.3.2 The Time Domain Spatial 1-d Vector Galerkin BEM Obtained with Laplace Transform

In this chapter, we will discuss in detail how the s domain 1-d spatial vector Galerkin BEM can be converted to a time domain 1-d spatial vector Galerkin BEM .

The s domain spatial 1-d vector Galerkin BEM

This chapter will review and summarize the relevant content of the 1-d spatial vector Galerkin BEM in Chap. 6, as follows.

Referring to the formula (6.7) related to the 1-d spatial vector Galerkin BEM in Chap. 6, this formula can be rewritten as follows.

$$[\bar{\bar{\mathbf{Z}}}_p][\bar{\mathbf{I}}_p] = 0 \tag{10.9}$$

where $[\bar{\bar{\mathbf{Z}}}_p]$ is $N_p \times N_p$ matrix, who includes the mutual impedance and induction interaction betweens a parts of conductor segments currents. $[\bar{\mathbf{I}}_p]$ is $N_p \times 1$ vector of unknown currents.

Matrix equation (10.9) is assembled from the following equations.

$$\sum_{\theta=1}^{N_b} Z_{l_\theta} \int_{l_\theta} [\bar{q}(\bar{r}_\theta)][\bar{p}(\bar{r}_\theta)]dr_\theta[\bar{I}_\theta^\chi]$$

$$+\sum_{\vartheta=1}^{N_b} \int_{l_\theta} \int_{l_\vartheta} \hat{r}_\theta \cdot j\omega \bar{\bar{G}}_{\xi\zeta}^{A^s}(\bar{r}_\theta, \bar{r}_\vartheta, s) \cdot \hat{r}_\vartheta^T [\bar{q}(\bar{r}_\theta)][\bar{p}(\bar{r}_\vartheta)]dr_\vartheta dr_\theta[\bar{I}_\vartheta^\chi]$$

$$+\sum_{\vartheta=1}^{N_b} \int_{l_\theta} \int_{l_\vartheta} G_{\xi\zeta}^{\varphi^s}(\bar{r}_\theta, \bar{r}_\vartheta, s) \frac{\partial[\bar{q}(\bar{r}_\theta)]}{\partial r_\theta} \frac{\partial[\bar{p}(\bar{r}_\vartheta)]}{\partial r_\vartheta} dr_\vartheta dr_\theta[\bar{I}_\vartheta^\chi] = 0 \tag{10.10}$$

where $\xi = 0, 1, \ldots, N_e$, $\zeta = 0, 1, \ldots, N_e$. $\theta = 1, 2, \ldots, N_b$; $\chi = 1, 2, \ldots, N_p$; $\bar{r}_\theta$ and $\bar{r}_\vartheta$ is location vector along θ-th and ϑ-th conductors segment, respectively; r_θ and r_ϑ, respectively, parameter variable along the θ-th and ϑ-th thin conductor axis. Z_{l_θ} is the internal impedance of the θ-th thin conductor segment. $\hat{r}_\theta$ and $\hat{r}_\vartheta$ are, respectively, the 1×3 dimensional unit director vector along the axis of the θ-th and ϑ-th cylindrical conductor segments. The symbol T represents matrix transposition. j is a imaginary unit, ω is angular frequency. $G_{\xi\zeta}^{\varphi^s}(\bar{r}_\theta, \bar{r}_\vartheta, s)$ and $\bar{\bar{G}}_{\xi\zeta}^{A^s}(\bar{r}_\theta, \bar{r}_\vartheta, s)$ are, respectively, s domain scalar monopole Green's function and vector dipole dyadic Green's function within horizontal multilayered earth model, which had been fully discussed in front Chap. 3. $[\bar{p}(\bar{r}_\vartheta)]$ and $[\bar{q}(\bar{r}_\theta)]$ are 1×2 dimensional 1 order basis functions and 2×1 dimensional 1 order test functions, respectively. $[\bar{p}(\bar{r}_\vartheta)] = \left[\begin{array}{cc} \frac{l_\vartheta - r_\vartheta}{l_\vartheta} & \frac{r_\vartheta}{l_\vartheta} \end{array}\right]$ and $[\bar{q}(\bar{r}_\theta)] = \left[\begin{array}{c} \frac{l_\theta - r_\theta}{l_\theta} \\ \frac{r_\theta}{l_\theta} \end{array}\right]$.

The Chap. 3 shows that the s domain Green's function of point current sources in multilayered horizontal layered soil can be further classified as:

$$\bar{\bar{G}}_{\xi\zeta}^{A^s}(\bar{r}_\theta, \bar{r}_\vartheta, s) = \bar{\bar{G}}_{\xi\zeta}^{A_s^s}(\bar{r}_\theta, \bar{r}_\vartheta) + \bar{\bar{G}}_{\xi\zeta}^{A_q^s}(\bar{r}_\theta, \bar{r}_\vartheta, s) \tag{10.11}$$

$$G_{\xi\zeta}^{\varphi^s}(\bar{r}_\theta, \bar{r}_\vartheta, s) = G_{\xi\zeta}^{\varphi_s^s}(\bar{r}_\theta, \bar{r}_\vartheta) + G_{\xi\zeta}^{\varphi_q^s}(\bar{r}_\theta, \bar{r}_\vartheta, s) \tag{10.12}$$

Reference [30] provides the expansion formula for internal impedance as follows

$$Z_{l_\theta} = \frac{l_\theta}{\pi r_c^2 \sigma_c} + \sum_{c=1}^{M_c} \alpha_c e^{\beta_c \omega} = \frac{l_\theta}{\pi r_c^2 \sigma_c} + \sum_{c=1}^{M_c} \alpha_c e^{-j\beta_c j\omega} = \frac{l_\theta}{\pi r_c^2 \sigma_c} + \sum_{c=1}^{M_c} \alpha_c e^{-j\beta_c s} = Z_{l_0} + Z_l(s) \tag{10.13}$$

where $s = j\omega$, M_c was total number of complex images, α_c and β_c were coefficients of complex images, r_c, l_θ and σ_c are the conductor radius, length and resistivity of the grounding grid, respectively.

The time domain spatial 1-d vector Galerkin BEM obtained with Laplace transform

Substituting formulas (10.11) , (10.12) and (10.13) into formula (10.10) and considering the laplace transform, there is the following formula

$$\frac{1}{2\pi j} \int_{\sigma - j\infty}^{\sigma + j\infty} [\sum_{\theta=1}^{N_b} \left(Z_{l_0} + Z_l(s)\right) \int_{l_\theta} [\bar{q}(\bar{r}_\theta)][\bar{p}(\bar{r}_\theta)] dr_\theta [\bar{I}_\theta^\chi]$$

$$+ \sum_{\vartheta=1}^{N_b} \int_{l_\theta} \int_{l_\vartheta} \hat{r}_\theta \cdot s \left(\bar{\bar{G}}_{\xi\zeta}^{A_s^s}(\bar{r}_\theta, \bar{r}_\vartheta) + \bar{\bar{G}}_{\xi\zeta}^{A_q^s}(\bar{r}_\theta, \bar{r}_\vartheta, s) \right) \cdot \hat{r}_\vartheta^T [\bar{q}(\bar{r}_\theta)][\bar{p}(\bar{r}_\vartheta)] dr_\vartheta dr_\theta [\bar{I}_\vartheta^\chi]$$

$$+\sum_{\vartheta=1}^{N_b}\int_{l_\theta}\int_{l_\vartheta}\left(G_{\xi\zeta}^{\varphi_s^s}(\bar{r}_\theta,\bar{r}_\vartheta)+G_{\xi\zeta}^{\varphi_q^s}(\bar{r}_\theta,\bar{r}_\vartheta,s)\right)\frac{\partial[\bar{q}(\bar{r}_\theta)]}{\partial r_\theta}\frac{\partial\bar{p}(\bar{r}_\vartheta)}{\partial r_\vartheta}dr_\vartheta dr_\theta[\bar{I}_\vartheta^X]]e^{st}ds=0 \tag{10.14}$$

where $\theta = 1, 2, \ldots, N_b$.

By using laplace transform, the following formulas in the time domain can be obtained.

$$[\sum_{\theta=1}^{N_b}\left(Z_{l_0}\int_{l_\theta}[\bar{q}(\bar{r}_\theta)][\bar{p}(\bar{r}_\theta)]dr_\theta[\bar{I}_\theta^X(t)]+Z_l\int_{l_\theta}[\bar{q}(\bar{r}_\theta)][\bar{p}(\bar{r}_\theta)]dr_\theta[\bar{I}_\theta^X(t-t_c)]\right)$$

$$+\sum_{\vartheta=1}^{N_b}(\int_{l_\theta}\int_{l_\vartheta}\hat{r}_\theta\cdot\bar{\bar{G}}_{\xi\zeta}^{A_s^t}(\bar{r}_\theta,\bar{r}_\vartheta)\cdot\hat{r}_\vartheta^T[\bar{q}(\bar{r}_\theta)][\bar{p}(\bar{r}_\vartheta)]dr_\vartheta dr_\theta\frac{d[\bar{I}_\vartheta^X(t)]}{dt}$$

$$+\int_{l_\theta}\int_{l_\vartheta}\hat{r}_\theta\cdot\bar{\bar{G}}_{\xi\zeta}^{A_q^t}(\bar{r}_\theta,\bar{r}_\vartheta)\cdot\hat{r}_\vartheta^T[\bar{q}(\bar{r}_\theta)][\bar{p}(\bar{r}_\vartheta)]dr_\vartheta dr_\theta\frac{d[\bar{I}_\vartheta^X(t-t_A)]}{dt})$$

$$+\sum_{\vartheta=1}^{N_b}(\int_{l_\theta}\int_{l_\vartheta}G_{\xi\zeta}^{\varphi_s^t}(\bar{r}_\theta,\bar{r}_\vartheta)\frac{\partial[\bar{q}(\bar{r}_\theta)]}{\partial r_\theta}\frac{d\bar{p}(\bar{r}_\vartheta)}{dr_\vartheta}dr_\vartheta dr_\theta[\bar{I}_\vartheta^X(t)]$$

$$+\int_{l_\theta}\int_{l_\vartheta}G_{\xi\zeta}^{\varphi_q^t}(\bar{r}_\theta,\bar{r}_\vartheta)\frac{\partial[\bar{q}(\bar{r}_\theta)]}{\partial r_\theta}\frac{\partial\bar{p}(\bar{r}_\vartheta)}{\partial r_\vartheta}dr_\vartheta dr_\theta[\bar{I}_\vartheta^X(t-t_\varphi)]])=0 \tag{10.15}$$

where $\theta = 1, 2, \ldots, N_b$. $G_{\xi\zeta}^{\varphi_q^t}(\bar{r}_\theta,\bar{r}_\vartheta,t_\varphi)$ and $\bar{\bar{G}}_{\xi\zeta}^{A_q^t}(\bar{r}_\theta,\bar{r}_\vartheta,t_A)$ are, respectively, time domain scalar monopole Green's function and vector dipole dyadic Green's function within horizontal multilayered earth model with delay effects, which had been fully discussed in front Chap. 3.

This formula (10.15) can also be further written as matrix formation as shown below.

$$[\bar{\bar{\mathbf{Z}}}_b^s][I_p(t)]+[\bar{\bar{\mathbf{M}}}_b^s]\frac{d[I_b(t)]}{dt}+[\bar{\bar{\mathbf{Z}}}_s^s][I_p(t)]=-[\bar{\bar{\mathbf{Z}}}_b^q][I_p(t-t_c)]-[\bar{\bar{\mathbf{M}}}_b^q]\frac{d[I_p(t-t_A)]}{dt}-[\bar{\bar{\mathbf{Z}}}_s^q][I_p(t-t_\varphi)] \tag{10.16}$$

where $[\bar{\bar{\mathbf{Z}}}_b^s]$ and $[\bar{\bar{\mathbf{Z}}}_b^q]$ are the self impedance matrices of grounding grid conductors without and with delay effects in the $N_p \times N_p$ dimension, respectively. $[\bar{\bar{\mathbf{M}}}_b^s]$ and $[\bar{\bar{\mathbf{M}}}_b^q]$ are the mutual induction matrices of grounding grid conductors without and with delay effects in the $N_p \times N_p$ dimension, respectively. $[\bar{\bar{\mathbf{Z}}}_s^s]$ and $[\bar{\bar{\mathbf{Z}}}_s^q]$ are the mutual impedance matrices of grounding grid conductors without and with delay effects in the $N_p \times N_p$ dimension, respectively.

10.3.3 The Time Domain Spatial 1-d Vector Galerkin BEM and its Solution

Equations (10.8) and (10.16) can be unified to be written as below

$$[\bar{\bar{\mathbf{Z}}}_b^s][I_p(t)] + [\bar{\bar{\mathbf{M}}}_b^s]\frac{d[I_b(t)]}{dt} + [\bar{\bar{\mathbf{Z}}}_s^s][I_p(t)] = -[\bar{\bar{\mathbf{Z}}}_b^q][I_p(t-t_c)] - [\bar{\bar{\mathbf{M}}}_b^q]\frac{d[I_p(t-t_A)]}{dt} - [\bar{\bar{\mathbf{Z}}}_s^q][I_p(t-t_\varphi)] \tag{10.17}$$

The excitation is incorporated into the formulation through the boundary condition:

$$I_{p_1}(t) = I_g(t) \tag{10.18}$$

where $I_g(t)$ denotes the lightning current generator and $I_{p_1}(t)$ current in the injection point.

At a junction consisting of two or more segments the continuity properties of the electric field must be satisfied, which is ensured by applying the Kirchhoff current law

$$\sum_{m=1}^{n} I_{p_m}(t) = 0 \tag{10.19}$$

where n is the total number of conductor segments connecting the node.

And continuity equation

$$[\frac{\partial I_{p_1}(t)}{\partial r_1}]_{at\ junction} = [\frac{\partial I_{p_2}(t)}{\partial r_2}]_{at\ junction} = \cdots = [\frac{\partial I_{p_n}(t)}{\partial r_n}]_{at\ junction} \tag{10.20}$$

The condition (10.20) ensures the discontinuities in charge per unit length to be ruled out in passing from one conductor to another across the junction.

At the conductor free ends, the total current vanishes, i.e. is forced to be zero.

Once Eq. (10.8) is solved, the unknown currents were known, the branch and leakage currents can be obtained easily. The voltage (zero potential reference point at infinity), the electrical field intensity (E) and the magnetic field intensity (B) at any point can be calculated by [29–31, 36, 62].

The study of the performance of the grounding system for time domain spatial 1-d Galerkin BEM has been reduced to the computation of $[\bar{\bar{\mathbf{Z}}}_b^s]$, $[\bar{\bar{\mathbf{Z}}}_b^q]$, $[\bar{\bar{\mathbf{M}}}_b^s]$, $[\bar{\bar{\mathbf{M}}}_b^q]$, $[\bar{\bar{\mathbf{Z}}}_s^s]$ and $[\bar{\bar{\mathbf{Z}}}_s^q]$ matrices.

Computation of $[\overline{\overline{\mathbf{Z}}}_b^s]$, $[\overline{\overline{\mathbf{Z}}}_b^q]$, $[\overline{\overline{\mathbf{M}}}_b^s]$, $[\overline{\overline{\mathbf{M}}}_b^q]$, $[\overline{\overline{\mathbf{Z}}}_s^s]$ and $[\overline{\overline{\mathbf{Z}}}_s^q]$ matrices

From [29–31, 36, 62], it can be seen that each segment owns internal impedance. Mutual inductances or impedances between segments' currents are also included in the model; To discuss the calculation technique of computation of element for $[\overline{\overline{\mathbf{Z}}}_b^s]$, $[\overline{\overline{\mathbf{Z}}}_b^q]$, $[\overline{\overline{\mathbf{M}}}_b^s]$, $[\overline{\overline{\mathbf{M}}}_b^q]$, $[\overline{\overline{\mathbf{Z}}}_s^s]$ and $[\overline{\overline{\mathbf{Z}}}_s^q]$ matrices, the Eq. (10.7) has been used as reference.

1. The first term of the Eq. (10.7) for element of $[\overline{\overline{\mathbf{Z}}}_b^s]$ matrix:

$$Z_\theta^s = Z_{l_0} \int_{l_\theta} [\bar{q}(\bar{r}_\theta)][\bar{p}(\bar{r}_\theta)] dr_\theta = Z_{l_0} \int_{l_\theta} \begin{bmatrix} \frac{r_\theta^2 - 2l_\theta r_\theta + l_\theta^2}{l_\theta^2} & \frac{-r_\theta^2 + l_\theta r_\theta}{l_\theta} \\ \frac{-r_\theta^2 + l_\theta r_\theta}{l_\theta^2} & \frac{r_\theta^2}{l_\theta^2} \end{bmatrix} dr_\theta \quad (10.21)$$

   where Z_{l_0} consists of internal resistance, the formula for internal resistance can be found in [30, 31, 36, 62].

2. The core of second term of the Eq. (10.7) for element of $[\overline{\overline{\mathbf{Z}}}_b^q]$ matrix:

$$Z_\theta^q = Z_l \int_{l_\theta} [\bar{q}(\bar{r}_\theta)][\bar{p}(\bar{r}_\theta)] dr_\theta = \sum_{c=1}^{M_c} \alpha_c \int_{l_\theta} \begin{bmatrix} \frac{r_\theta^2 - 2l_\theta r_\theta + l_\theta^2}{l_\theta^2} & \frac{-r_\theta^2 + l_\theta r_\theta}{l_\theta} \\ \frac{-r_\theta^2 + l_\theta r_\theta}{l_\theta^2} & \frac{r_\theta^2}{l_\theta^2} \end{bmatrix} dr_\theta \quad (10.22)$$

3. The core of third term of the Eq. (10.7) for element of $[\overline{\overline{\mathbf{M}}}_b^s]$ matrix: The core is formula of mutual induction interaction betweens a pars of currents along conductor segments without delay effects.

$$M_{\theta,\vartheta}^s = \int_{l_\theta} \int_{l_\vartheta} \hat{r}_\theta \cdot \bar{\bar{G}}_{\xi\zeta}^{A_s^t}(\bar{r}_\theta, \bar{r}_\vartheta, t) \cdot \hat{r}_\vartheta^T [\bar{p}(\bar{r}_\theta)][\bar{q}(\bar{r}_\vartheta)] dr_\vartheta dr_\theta \quad (10.23)$$

$$= \int_{l_\theta} \int_{l_\vartheta} \hat{r}_\theta \cdot \bar{\bar{G}}_{\xi\zeta}^{A_s^t}(\bar{r}_\theta, \bar{r}_\vartheta) \cdot \hat{r}_\vartheta^T \begin{bmatrix} \frac{r_\theta r_\vartheta - l_\vartheta r_\theta - l_\theta r_\vartheta + l_\theta l_\vartheta}{l_\theta l_\vartheta} & \frac{-r_\theta r_\vartheta + l_\theta r_\vartheta}{l_\theta r_\vartheta} \\ \frac{-r_\theta r_\vartheta + l_\vartheta r_\theta}{l_\theta l_\vartheta} & \frac{r_\theta r_\vartheta}{l_\theta l_\vartheta} \end{bmatrix} dr_\vartheta dr_\theta \quad (10.24)$$

   For an half infinite homogeneous conductivity medium, one has

$$\bar{\bar{G}}_{11}^{A_s^t}(\bar{r}_\vartheta, \bar{r}_\theta) = \frac{\mu}{4\pi} \begin{bmatrix} \frac{1}{R_{\theta\vartheta}} & 0 & -\hat{k}_{01}^{TM} \frac{\partial(\ln(-(z_\theta + z_\vartheta) + R'_{\theta\vartheta})}{\partial x} \\ 0 & \frac{1}{R_{\theta\vartheta}} & -\hat{k}_{01}^{TM} \frac{\partial(\ln(-(z_\theta + z_\vartheta) + R'_{\theta\vartheta})}{\partial y} \\ 0 & 0 & \frac{1}{R_{\theta\vartheta}} - \frac{\hat{k}_{01}^{TM}}{R'_{\theta\vartheta}} \end{bmatrix}, \quad (10.25)$$

where $R_{\theta\vartheta} = |\bar{r}_\theta - \bar{r}_\vartheta|$, $R'_{\theta\vartheta} = |\bar{r}_\theta + \bar{r}_\vartheta|$, $\hat{k}_{01}^{TM} = \frac{\sigma_1 - \sigma_0}{\sigma_1 + \sigma_0}$.

$$M^s_{\theta,\vartheta} = \frac{\mu}{4\pi} \int_{l_\theta} \int_{l_\vartheta} \hat{r}_\theta \cdot \begin{bmatrix} \frac{1}{R_{\theta\vartheta}} & 0 & -\hat{k}_{01}^{TM} \frac{\partial(\ln(-(z_\theta+z_\vartheta)+R'_{\theta\vartheta})}{\partial x} \\ 0 & \frac{1}{R_{\theta\vartheta}} & -\hat{k}_{01}^{TM} \frac{\partial(\ln(-(z_\theta+z_\vartheta)+R'_{\theta\vartheta})}{\partial y} \\ 0 & 0 & \frac{1}{R_{\theta\vartheta}} - \frac{\hat{k}_{01}^{TM}}{R'_{\theta\vartheta}} \end{bmatrix}$$

$$\cdot \hat{r}_\vartheta^T \begin{bmatrix} \frac{r_\theta r_\vartheta - l_\vartheta r_\theta - l_\theta r_\vartheta + l_\theta l_\vartheta}{l_\theta l_\vartheta} & \frac{-r_\theta r_\vartheta + l_\theta r_\vartheta}{l_\theta l_\vartheta} \\ \frac{-r_\theta r_\vartheta + l_\vartheta r_\theta}{l_\theta l_\vartheta} & \frac{r_\theta r_\vartheta}{l_\theta l_\vartheta} \end{bmatrix} dr_\vartheta dr_\theta \tag{10.26}$$

4. The core of fourth term of the Eq. (10.7) for element of $[\overline{\overline{\mathbf{M}}}_b^q]$ matrix: The core is formula of mutual induction interaction betweens a pars of currents along conductor segments with delay effects.

$$M^q_{\theta,\vartheta} = \int_{l_\theta} \int_{l_\vartheta} \hat{r}_\theta \cdot \bar{\bar{G}}_{\xi\zeta}^{A_q^t}(\bar{r}_\theta, \bar{r}_\vartheta) \cdot \hat{r}_\vartheta^T [\bar{p}(\bar{r}_\theta)][\bar{q}(\bar{r}_\vartheta)] dr_\vartheta dr_\theta$$

$$= \int_{l_\theta} \int_{l_\vartheta} \hat{r}_\theta \cdot \bar{\bar{G}}_{\xi\zeta}^{A_q^t}(\bar{r}_\theta, \bar{r}_\vartheta) \cdot \hat{r}_\vartheta^T \begin{bmatrix} \frac{r_\theta r_\vartheta - l_\vartheta r_\theta - l_\theta r_\vartheta + l_\theta l_\vartheta}{l_\theta l_\vartheta} & \frac{-r_\theta r_\vartheta + l_\theta r_\vartheta}{l_\theta r_\vartheta} \\ \frac{-r_\theta r_\vartheta + l_\vartheta r_\theta}{l_\theta l_\vartheta} & \frac{r_\theta r_\vartheta}{l_\theta l_\vartheta} \end{bmatrix} dr_\vartheta dr_\theta \tag{10.27}$$

For an half infinite homogeneous conductivity medium, one has

$$\bar{\bar{G}}_{11}^{A_q^t}(\bar{r}_\vartheta, \bar{r}_\theta) = \frac{\mu}{4\pi} \begin{bmatrix} 0 & 0 & -\sum_{q_A=1}^{M_A} \alpha_{q_A} \frac{\partial(\ln(-(z_\theta+z_\vartheta)+R'_{\theta\vartheta})}{\partial x} \\ 0 & 0 & -\sum_{q_A=1}^{M_A} \alpha_{q_A} \frac{\partial(\ln(-(z_\theta+z_\vartheta)+R'_{\theta\vartheta})}{\partial y} \\ 0 & 0 & -\sum_{q_A=1}^{M_A} \alpha_{q_A} \frac{1}{R'_{\theta\vartheta}} \end{bmatrix}, \tag{10.28}$$

so

$$M^q_{\theta,\vartheta} = \frac{\mu}{4\pi}\int_{l_\theta}\int_{l_\vartheta} \hat{r}_\theta \cdot \begin{bmatrix} 0\;0 - \sum_{q_A=1}^{M_A}\alpha_{q_A}\frac{\partial\left(\ln\left(-(z_\theta+z_\vartheta)+R'_{\theta\vartheta}\right)\right)}{\partial x} \\ 0\;0 - \sum_{q_A=1}^{M_A}\alpha_{q_A}\frac{\partial\left(\ln\left(-(z_\theta+z_\vartheta)+R'_{\theta\vartheta}\right)\right)}{\partial y} \\ 0\;0 \quad -\sum_{q_A=1}^{M_A}\alpha_{q_A}\frac{1}{R'_{\theta\vartheta}} \end{bmatrix}$$

$$\cdot\,\hat{r}^T_\vartheta \begin{bmatrix} \frac{r_\theta r_\vartheta - l_\vartheta r_\theta - l_\theta r_\vartheta + l_\theta l_\vartheta}{l_\theta l_\vartheta} & \frac{-r_\theta r_\vartheta + l_\theta r_\vartheta}{l_\theta l_\vartheta} \\ \frac{-r_\theta r_\vartheta + l_\vartheta r_\theta}{l_\theta l_\vartheta} & \frac{r_\theta r_\vartheta}{l_\theta l_\vartheta} \end{bmatrix} dr_\vartheta dr_\theta \tag{10.29}$$

5. The core of fifth term of the Eq. (10.7): The core is formula of mutual impedance interaction betweens a pars of currents along conductor segments without delay effect.

$$Z^s_{\theta,\vartheta} = \int_{l_\theta}\int_{l_\vartheta} G^{\varphi^t_s}_{\xi\zeta}(\bar{r}_\theta,\bar{r}_\vartheta)\frac{\partial[\bar{q}(\bar{r}_\theta)]}{\partial r_\theta}\frac{\partial[\bar{p}(\bar{r}_\vartheta)]}{\partial r_\vartheta}dr_\vartheta dr_\theta \tag{10.30}$$

$$Z^s_{\theta,\vartheta} = \int_{l_\theta}\int_{l_\vartheta} G^{\varphi^t_s}_{\xi\zeta}(\bar{r}_\theta,\bar{r}_\vartheta)\begin{bmatrix} \frac{1}{l_\theta l_\vartheta} & -\frac{1}{l_\theta l_\vartheta} \\ -\frac{1}{l_\theta l_\vartheta} & \frac{1}{l_\theta l_\vartheta} \end{bmatrix} dr_\vartheta dr_\theta \tag{10.31}$$

For an half infinite homogeneous conductivity medium, one has

$$G^{\varphi^t_s}_{11}(\bar{r}_\vartheta,\bar{r}_\theta) = \frac{1}{4\pi\sigma_1}\frac{1}{R_{\theta\vartheta}}, \tag{10.32}$$

so

$$Z^s_{\theta,\vartheta} = \frac{1}{4\pi\sigma_1 l_\theta l_\vartheta}\int_{l_\theta}\int_{l_\vartheta}\frac{1}{R_{\theta\vartheta}}\begin{bmatrix} 1 & -1 \\ -1 & 1 \end{bmatrix} dr_\theta dr_\vartheta \tag{10.33}$$

6. The core of sixth term of the Eq. (10.7): The core is formula of mutual impedance interaction betweens a pars of currents along conductor segments with delay effect.

$$Z^q_{\theta,\vartheta} = \int_{l_\theta}\int_{l_\vartheta} G^{\varphi^t_q}_{\xi\zeta}(\bar{r}_\theta,\bar{r}_\vartheta)\frac{\partial[\bar{q}(\bar{r}_\theta)]}{\partial r_\theta}\frac{\partial[\bar{p}(\bar{r}_\vartheta)]}{\partial r_\vartheta}dr_\vartheta dr_\theta \tag{10.34}$$

$$Z^q_{\theta,\vartheta} = \int_{l_\theta}\int_{l_\vartheta} G^{\varphi^t_q}_{\xi\zeta}(\bar{r}_\theta, \bar{r}_\vartheta)\begin{bmatrix} \frac{1}{l_\theta l_\vartheta} & -\frac{1}{l_\theta l_\vartheta} \\ -\frac{1}{l_\theta l_\vartheta} & \frac{1}{l_\theta l_\vartheta} \end{bmatrix} dr_\vartheta dr_\theta \tag{10.35}$$

For an half infinite homogeneous conductivity medium, one has

$$G^{\varphi^t_q}_{11}(\bar{r}_\vartheta, \bar{r}_\theta) = \frac{1}{4\pi\sigma_1}\left(\sum_{q_0=1}^{M_0}\alpha_{q_0}\frac{1}{R_{\theta\vartheta}} - \sum_{q_\varphi=1}^{M_\varphi}\alpha_{q_\varphi}\frac{1}{R'_{\theta\vartheta}}\right), \tag{10.36}$$

so

$$Z^q_{\theta,\vartheta} = \frac{1}{4\pi\sigma_1 l_\theta l_\vartheta}\int_{l_\theta}\int_{l_\vartheta}\left[\sum_{q_0=1}^{M_0}\alpha_{q_0}\frac{1}{R_{\theta\vartheta}} - \sum_{q_\varphi=1}^{M_\varphi}\alpha_{q_\varphi}\frac{1}{R'_{\theta\vartheta}}\right]\begin{bmatrix} 1 & -1 \\ -1 & 1 \end{bmatrix} dr_\theta dr_\vartheta \tag{10.37}$$

Note the medium surrounding the point current source here was considered as half infinite homogeneous conductivity medium. However, in any practical case, the earth is represented via a multilayered earth model. The time domain QSCIM can be used to dealt with the multilayered earth model, which has be fully discussed in Chap. 3.

10.3.4 Time Domain Spatial 1-d Vector Galerkin BEM Solution for Transient Grounding Problems

The matrix equation of the time domain spatial 1-d vector Galerkin BEM was given in the previous section as follows.

$$[\bar{\bar{\mathbf{Z}}}^s_b][\bar{\mathbf{I}}_p(t)] + [\bar{\bar{\mathbf{M}}}^s_p]\frac{d[\bar{\mathbf{I}}_p(t)]}{dt} + [\bar{\bar{\mathbf{Z}}}^s_s][\bar{\mathbf{I}}_p(t)] = -[\bar{\bar{\mathbf{Z}}}^q_b][\bar{\mathbf{I}}_p(t-t_c)] - [\bar{\bar{\mathbf{M}}}^q_b]\frac{d[\bar{\mathbf{I}}_p(t-t_A)]}{dt} - [\bar{\bar{\mathbf{Z}}}^q_s][\bar{\mathbf{I}}_p(t-t_\varphi)] \tag{10.38}$$

Here, the time-domain equation is first discretized in time, with a time interval of Δt. Therefore, $t_k = k\Delta t, k = 0, 1, \ldots, N_t$, N_t is is the maximum number of discrete times, so

$$[\bar{\bar{\mathbf{Z}}}_b^s][\bar{\mathbf{I}}_p(t_k)] + [\bar{\bar{\mathbf{M}}}_b^s]\frac{[\bar{\mathbf{I}}_p(t_k)] - [\bar{\mathbf{I}}_p(t_{k-1})]}{\Delta t} + [\bar{\bar{\mathbf{Z}}}_s^s][I_p(t_k)]$$

$$= -[\bar{\bar{\mathbf{Z}}}_b^q][\bar{\mathbf{I}}_p(t_k - t_c^k)] - [\bar{\bar{\mathbf{M}}}_b^q]\frac{[\bar{\mathbf{I}}_p(t_k - t_A^k)] - [\bar{\mathbf{I}}_p(t_{k-1} - t_A^k)]}{\Delta t} - [\bar{\bar{\mathbf{Z}}}_s^q][\bar{\mathbf{I}}_p(t_k - t_\varphi^k)] \tag{10.39}$$

where $t_c^k = \frac{t_c}{\Delta t}$, $t_A^k = \frac{t_A}{\Delta t}$, $t_\varphi^k = \frac{t_\varphi}{\Delta t}$.

The above time domain equation can be further written as the formula shown below.

$$\left[\bar{\bar{\mathbf{Z}}}_b^s + \frac{\bar{\bar{\mathbf{M}}}_b^s}{\Delta t} + \bar{\bar{\mathbf{Z}}}_s^s\right][\bar{\mathbf{I}}_p(t_k)] = [\bar{\bar{\mathbf{M}}}_b^s]\frac{[\bar{\mathbf{I}}_p(t_{k-1})]}{\Delta t} - [\bar{\bar{\mathbf{Z}}}_b^q][\bar{\mathbf{I}}_p(t_k - t_c^k)]$$

$$- [\bar{\bar{\mathbf{M}}}_b^q]\frac{[\bar{\mathbf{I}}_p(t_k - t_A^k)] - [\bar{\mathbf{I}}_p(t_{k-1} - t_A^k)]}{\Delta t} - [\bar{\bar{\mathbf{Z}}}_s^q][\bar{\mathbf{I}}_p(t_k - t_\varphi^k)] \tag{10.40}$$

The excitation is incorporated into the formulation through the boundary condition:

$$I_{p_1}(t_k) = I_g(t_k) \tag{10.41}$$

where $I_g(t_k)$ denotes the lightning current generator and $I_{p_1}(t_k)$ current in the injection point at t_k moment.

At a junction consisting of two or more segments the continuity properties of the electric field must be satisfied, which is ensured by applying the Kirchhoff current law

$$\sum_{m=1}^{n} I_{p_m}(t_k) = 0 \tag{10.42}$$

And continuity equation

$$\left[\frac{\partial I_{p_1}(t_k)}{\partial r_1}\right]_{at\ junction} = \left[\frac{\partial I_{p_2}(t_k)}{\partial r_2}\right]_{at\ junction} = \cdots = \left[\frac{\partial I_{p_n}(t_k)}{\partial r_n}\right]_{at\ junction} \tag{10.43}$$

The condition (10.43) ensures the discontinuities in charge per unit length to be ruled out in passing from one conductor to another across the junction.

At the conductor free ends, the total current vanishes, i.e. is forced to be zero.

Eq. (10.40) represents a marching-on-in-time formulation. Accordingly, it can be quickly solved through iterations.

10.4 The Time Domain Spatial 1-d Vector Galerkin MM for Transient Grounding Problems

This chapter will discuss in detail how to scientifically simulate lightning transient grounding of substation grounding grid by time domain 1-d spatial vector Galerkin MM.

10.4.1 The Time Domain Spatial 1-d Vector Galerkin MM Obtained with Fourier Transform

In this chapter, we will discuss in detail how the frequencuy domain 1-d spatial vector Galerkin MM can be converted to a time domain 1-d spatial vector Galerkin MM .

The frequency domain spatial 1-d vector Galerkin MM

This chapter will review and summarize the relevant content of the frequency domain spatial 1-d vector Galerkin MM in Chap. 6. The core content is as follows.

Referring to the formula (6.24) related to the spatial 1-d vector Galerkin MM in Chap. 6, the formula can be rewritten as follows.

$$\begin{bmatrix} \overline{\overline{\mathbf{Z}}}^{00} & \overline{\overline{\mathbf{Z}}}^{01} \\ \overline{\overline{\mathbf{Z}}}^{10} & \overline{\overline{\mathbf{Z}}}^{11} \end{bmatrix} \begin{bmatrix} \bar{\mathbf{I}}^{0} \\ \bar{\mathbf{I}}^{1} \end{bmatrix} = 0 \tag{10.44}$$

where the $[\bar{\mathbf{I}}^{0}]$ is $N_b \times 1$ vector of unknown branch currents and the $[\bar{\mathbf{I}}^{1}]$ is $N_b \times 1$ vector of unknown leakage currents; $[\overline{\overline{\mathbf{Z}}}^{00}]$ is $N_b \times N_b$ matrix, who presents the mutual induction interaction betweens a parts of conductor segments branch currents, $[\overline{\overline{\mathbf{Z}}}^{01}]$ is $N_b \times N_b$ matrix, who presents the mutual impedance and induction interaction betweens a parts of conductor segments branch and leakage currents, $[\overline{\overline{\mathbf{Z}}}^{10}]$ is $N_b \times N_b$ matrix, who presents the mutual impedance and induction interaction betweens a parts of conductor segments branch and leakage currents, $[\overline{\overline{\mathbf{Z}}}^{11}]$ is $N_b \times N_b$ matrix, who presents the mutual impedance and induction interaction betweens a parts of conductor segments leakage currents.

Matrix equation (10.44) is assembled from the following equations.

$$Z_{l_\theta} \int_{l_\theta} [\bar{q}(\bar{r}_\theta)][\bar{p}(\bar{r}_\theta)] dr_\theta [\bar{I}_\theta^\chi] + \sum_{\vartheta=1}^{N_b} \int_{l_\theta} \int_{l_\vartheta} \hat{r}_\theta \cdot j\omega \bar{\bar{G}}_{\xi\zeta}^{A_q^f}(\bar{r}_\theta, \bar{r}_\vartheta) \cdot \hat{r}_\vartheta^T [\bar{q}(\bar{r}_\theta)][\bar{p}(\bar{r}_\vartheta)] dr_\vartheta dr_\theta [\bar{I}_\vartheta^\chi]$$

$$+\sum_{\vartheta=1}^{N_b}\int_{l_\theta}\int_{l_\vartheta} G_{\xi\zeta}^{\varphi_q^f}(\bar{r}_\theta,\bar{r}_\vartheta)\frac{\partial[\bar{q}(\bar{r}_\theta)]}{\partial r_\theta}\frac{\partial[\bar{p}(\bar{r}_\vartheta)]}{\partial r_\vartheta}dr_\vartheta dr_\theta[\bar{I}_\vartheta^\chi]=0 \tag{10.45}$$

where $\theta = 1, 2, \ldots, N_b$; $\chi = 0, 1$; $\theta = 1, \ldots, N_b$. $\bar{r}_\theta$ and $\bar{r}_\vartheta$ is location vector along θ-th and ϑ-th conductors segment, respectively; r_θ and r_ϑ, respectively, parameter variable along the θ-th and ϑ-th thin conductor axis. Z_{l_θ} is the internal impedance of the θ-th thin conductor segment. $\hat{r}_\theta$ and $\hat{r}_\vartheta$ are, respectively, the 1×3 dimensional unit director vector along the axis of the θ-th and ϑ-th cylindrical conductor segments. j is a imaginary unit, ω is angular frequency. $G_{\xi\zeta}^{\varphi_q^f}(\bar{r}_\theta,\bar{r}_\vartheta)$ and $\bar{\bar{G}}_{\xi\zeta}^{A_q^f}(\bar{r}_\theta,\bar{r}_\vartheta)$ are, respectively, scalar monopole Green's function and vector dipole dyadic Green's function within horizontal multilayered earth model, which had been fully discussed in front Chap. 3. $[\bar{p}(\bar{r}_\vartheta)]$ and $[\bar{q}(\bar{r}_\theta)]$ are 1×2 dimensional 1 order basis functions and 2×1 dimensional 1 order test functions, respectively. $[\bar{p}(\bar{r}_\vartheta)] = \begin{bmatrix} 1 & -\frac{r_\vartheta}{l_\vartheta} \end{bmatrix}$ and $[\bar{q}(\bar{r}_\theta)] = \begin{bmatrix} 1 \\ -\frac{r_\theta}{l_\theta} \end{bmatrix}$.

The time domain spatial 1-d vector Galerkin MM obtained with Fourier transform

Just like the BEM, consider equations (10.3), (10.4), and (10.5) simultaneously, and substituting formulas (10.3) , (10.4) and (10.5) into formula (10.45) and considering the Fourier transform, there is the following formula.

$$\frac{1}{2\pi}\int_{-\infty}^{\infty}[(Z_{l_0}+Z_l(\omega))\int_{l_\theta}[\bar{q}(\bar{r}_\theta)][\bar{p}(\bar{r}_\theta)]dr_\theta[\bar{I}_\theta^\chi]$$

$$+\sum_{\vartheta=1}^{N_b}\int_{l_\theta}\int_{l_\vartheta}\hat{r}_\theta\cdot j\omega\left(\bar{\bar{G}}_{\xi\zeta}^{A_s^f}(\bar{r}_\theta,\bar{r}_\vartheta)+\bar{\bar{G}}_{\xi\zeta}^{A_q^f}(\bar{r}_\theta,\bar{r}_\vartheta,\omega)\right)\cdot\hat{r}_\vartheta^T[\bar{q}(\bar{r}_\theta)][\bar{p}(\bar{r}_\vartheta)]dr_\vartheta dr_\theta[\bar{I}_\vartheta^\chi]$$

$$+\sum_{\vartheta=1}^{N_b}\int_{l_\theta}\int_{l_\vartheta}\left(G_{\xi\zeta}^{\varphi_s^f}(\bar{r}_\theta,\bar{r}_\vartheta)+G_{\xi\zeta}^{\varphi_q^f}(\bar{r}_\theta,\bar{r}_\vartheta,\omega)\right)\frac{\partial[\bar{q}(\bar{r}_\theta)]}{\partial r_\theta}\frac{\partial[\bar{p}(\bar{r}_\vartheta)]}{\partial r_\vartheta}dr_\vartheta dr_\theta[\bar{I}_\vartheta^\chi]e^{j\omega t}d\omega=0 \tag{10.46}$$

where $\theta = 1, 2, \ldots, N_b$, $\chi = 0, 1$.

By using Fourier transform, the following formulas in the time domain can be obtained.

$$\left(Z_{l_0}\int_{l_\theta}[\bar{q}(\bar{r}_\theta)][\bar{p}(\bar{r}_\theta)]dr_\theta[\bar{I}_\theta^\chi(t)]+Z_l\int_{l_\theta}[\bar{q}(\bar{r}_\theta)][\bar{p}(\bar{r}_\theta)]dr_\theta[\bar{I}_\theta^\chi(t-t_c)]\right)$$

$$+\sum_{\vartheta=1}^{N_b}(\int_{l_\theta}\int_{l_\vartheta}\hat{r}_\theta\cdot\bar{\bar{G}}_{\xi\zeta}^{A_s^t}(\bar{r}_\theta,\bar{r}_\vartheta)\cdot\hat{r}_\vartheta^T[\bar{q}(\bar{r}_\theta)][\bar{p}(\bar{r}_\vartheta)]dr_\vartheta dr_\theta\frac{d[\bar{I}_\vartheta^\chi(t)]}{dt}$$

$$+\int_{l_\theta}\int_{l_\vartheta}\hat{r}_\theta\cdot\bar{\bar{G}}_{\xi\zeta}^{A_q^t}(\bar{r}_\theta,\bar{r}_\vartheta)\cdot\hat{r}_\vartheta^T[\bar{q}(\bar{r}_\theta)][\bar{p}(\bar{r}_\vartheta)]dr_\vartheta dr_\theta\frac{d[\bar{I}_\vartheta^\chi(t-t_A)]}{dt})$$

$$+\sum_{\vartheta=1}^{N_b}(\int_{l_\theta}\int_{l_\vartheta}G_{\xi\zeta}^{\varphi_s^t}(\bar{r}_\theta,\bar{r}_\vartheta)\frac{\partial[\bar{q}(\bar{r}_\theta)]}{\partial r_\theta}\frac{d[\bar{p}(\bar{r}_\vartheta)]}{dr_\vartheta}dr_\vartheta dr_\theta[\bar{I}_\vartheta^\chi(t)]$$

$$+\int_{l_\theta}\int_{l_\vartheta}G_{\xi\zeta}^{\varphi_q^t}(\bar{r}_\theta,\bar{r}_\vartheta,t)\frac{\partial[\bar{q}(\bar{r}_\theta)]}{\partial r_\theta}\frac{\partial[\bar{p}(\bar{r}_\vartheta)]}{\partial r_\vartheta}dr_\vartheta dr_\theta[\bar{\mathbf{I}}_\vartheta^\chi(t-t_\varphi)]) = 0 \quad (10.47)$$

where $\theta = 1, 2, \ldots, N_b$; $\chi = 0, 1$; $G_{\xi\zeta}^{\varphi_q^t}(\bar{r}_\theta, \bar{r}_\vartheta)$ and $\bar{\bar{G}}_{\xi\zeta}^{A_q^t}(\bar{r}_\theta, \bar{r}_\vartheta)$ are, respectively, time domain scalar monopole Green's function and vector dipole dyadic Green's function within horizontal multilayered earth model with delay effects, which had been fully discussed in front Chap. 3.

This formula (10.47) can also be further written as matrix formation as shown below.

$$[\bar{\bar{\mathbf{Z}}}_b^s][\mathbf{I}_b^\chi(t)] + [\bar{\bar{\mathbf{Z}}}_b^q][\mathbf{I}_b^\chi(t-t_c)] + [\bar{\bar{\mathbf{M}}}_b^s]\frac{d[\mathbf{I}_b^\chi(t)]}{dt} + [\bar{\bar{\mathbf{M}}}_b^q]\frac{d[\mathbf{I}_b^\chi(t-t_A)]}{dt} + [\bar{\bar{\mathbf{Z}}}_s^s][\mathbf{I}_b^\chi(t)] + [\bar{\bar{\mathbf{Z}}}_s^q][\mathbf{I}_b^\chi(t-t_\varphi)] = 0 \quad (10.48)$$

where $\chi = 0, 1$; $[\bar{\bar{\mathbf{Z}}}_b^s] = \begin{bmatrix}\bar{\bar{\mathbf{Z}}}_b^{s00} & \bar{\bar{\mathbf{Z}}}_b^{s01}\\ \bar{\bar{\mathbf{Z}}}_b^{s10} & \bar{\bar{\mathbf{Z}}}_b^{s11}\end{bmatrix}$ and $[\bar{\bar{\mathbf{Z}}}_b^q] = \begin{bmatrix}\bar{\bar{\mathbf{Z}}}_b^{q00} & \bar{\bar{\mathbf{Z}}}_b^{q01}\\ \bar{\bar{\mathbf{Z}}}_b^{q10} & \bar{\bar{\mathbf{Z}}}_b^{q11}\end{bmatrix}$ are the self impedance matrices of grounding grid conductors without and with delay effects in the $2N_b \times 2N_b$ dimension, respectively. $[\bar{\bar{\mathbf{M}}}_b^s] = \begin{bmatrix}\bar{\bar{\mathbf{M}}}_b^{s00} & \bar{\bar{\mathbf{M}}}_b^{s01}\\ \bar{\bar{\mathbf{M}}}_b^{s10} & \bar{\bar{\mathbf{M}}}_b^{s11}\end{bmatrix}$ and $[\bar{\bar{\mathbf{M}}}_b^q] = \begin{bmatrix}\bar{\bar{\mathbf{M}}}_b^{q00} & \bar{\bar{\mathbf{M}}}_b^{q01}\\ \bar{\bar{\mathbf{M}}}_b^{q10} & \bar{\bar{\mathbf{M}}}_b^{q11}\end{bmatrix}$ are the mutual induction matrices of grounding grid conductors without and with delay effects in the $2N_b \times 2N_b$ dimension, respectively. $[\bar{\bar{\mathbf{Z}}}_s^s] = \begin{bmatrix}0 & 0\\ 0 & \bar{\bar{\mathbf{Z}}}_s^{s11}\end{bmatrix}$ and $[\bar{\bar{\mathbf{Z}}}_s^q] = \begin{bmatrix}0 & 0\\ 0 & \bar{\bar{\mathbf{Z}}}_s^{q11}\end{bmatrix}$ are the mutual impedance matrices of grounding grid conductors without and with delay effects in the $2N_b \times 2N_b$ dimension, respectively. $[\bar{\mathbf{I}}_\vartheta^\chi(t)] = \begin{bmatrix}\bar{\mathbf{I}}_b^0(t)\\ \bar{\mathbf{I}}_b^1(t)\end{bmatrix}$, $[\bar{\mathbf{I}}_b^\chi(t-t_c)] = \begin{bmatrix}\bar{\mathbf{I}}_b^0(t-t_c)\\ \bar{\mathbf{I}}_b^1(t-t_c)\end{bmatrix}$, $[\bar{\mathbf{I}}_b^\chi(t-t_A)] = \begin{bmatrix}\bar{\mathbf{I}}_b^0(t-t_A)\\ \bar{\mathbf{I}}_b^1(t-t_A)\end{bmatrix}$ and $[\bar{\mathbf{I}}_b^\chi(t-t_\varphi)] = \begin{bmatrix}\bar{\mathbf{I}}_b^0(t-t_\varphi)\\ \bar{\mathbf{I}}_b^1(t-t_\varphi)\end{bmatrix}$.

So, this equation can be further written as follows

$$\begin{bmatrix} \bar{\bar{\mathbf{Z}}}_b^{s00} & \bar{\bar{\mathbf{Z}}}_b^{s01} \\ \bar{\bar{\mathbf{Z}}}_b^{s10} & \bar{\bar{\mathbf{Z}}}_b^{s11} + \bar{\bar{\mathbf{Z}}}_s^{s11} \end{bmatrix} \begin{bmatrix} \bar{\mathbf{I}}_b^0(t) \\ \bar{\mathbf{I}}_b^1(t) \end{bmatrix} + \begin{bmatrix} \bar{\bar{\mathbf{M}}}_b^{s00} & \bar{\bar{\mathbf{M}}}_b^{s01} \\ \bar{\bar{\mathbf{M}}}_b^{s10} & \bar{\bar{\mathbf{M}}}_b^{s11} \end{bmatrix} \frac{d}{dt} \begin{bmatrix} \bar{\mathbf{I}}_b^0(t) \\ \bar{\mathbf{I}}_b^1(t) \end{bmatrix} = - \begin{bmatrix} \bar{\bar{\mathbf{Z}}}_b^{q00} & \bar{\bar{\mathbf{Z}}}_b^{q01} \\ \bar{\bar{\mathbf{Z}}}_b^{q10} & \bar{\bar{\mathbf{Z}}}_b^{q11} \end{bmatrix} \begin{bmatrix} \bar{\mathbf{I}}_b^0(t-t_c) \\ \bar{\mathbf{I}}_b^1(t-t_c) \end{bmatrix}$$

$$- \begin{bmatrix} \bar{\bar{\mathbf{M}}}_b^{q00} & \bar{\bar{\mathbf{M}}}_b^{q01} \\ \bar{\bar{\mathbf{M}}}_b^{q10} & \bar{\bar{\mathbf{M}}}_b^{q11} \end{bmatrix} \frac{d}{dt} \begin{bmatrix} \bar{\mathbf{I}}_b^0(t-t_A) \\ \bar{\mathbf{I}}_b^1(t-t_A) \end{bmatrix} - \begin{bmatrix} 0 & 0 \\ 0 & \bar{\bar{\mathbf{Z}}}_s^{q11} \end{bmatrix} \begin{bmatrix} \bar{\mathbf{I}}_b^0(t-t_\varphi) \\ \bar{\mathbf{I}}_b^1(t-t_\varphi) \end{bmatrix} \tag{10.49}$$

10.4.2 *The Time Domain Spatial 1-d Vector Galerkin MM Obtained with Laplace Transform*

In this chapter, we will discuss in detail how the s domain 1-d spatial vector Galerkin MM can be converted to a time domain 1-d spatial vector Galerkin MM .

The s domain spatial 1-d vector Galerkin MM

This chapter will review and summarize the relevant content of the frequency domain spatial 1-d vector Galerkin MM in Chap. 6. The core content is as follows.

Referring to the formula (6.24) related to the spatial 1-d vector Galerkin MM in Chap. 6, the formula can be rewritten as follows.

$$\begin{bmatrix} \overline{\overline{\mathbf{Z}}}^{00} & \overline{\overline{\mathbf{Z}}}^{01} \\ \overline{\overline{\mathbf{Z}}}^{10} & \overline{\overline{\mathbf{Z}}}^{11} \end{bmatrix} \begin{bmatrix} \bar{\mathbf{I}}^0 \\ \bar{\mathbf{I}}^1 \end{bmatrix} = 0 \tag{10.50}$$

where the $[\bar{\mathbf{I}}^0]$ is $N_b \times 1$ vector of unknown branch currents and the $[\bar{\mathbf{I}}^1]$ is $N_b \times 1$ vector of unknown leakage currents; $[\overline{\overline{\mathbf{Z}}}^{00}]$ is $N_b \times N_b$ matrix, who presents the mutual induction interaction betweens a parts of conductor segments branch currents, $[\overline{\overline{\mathbf{Z}}}^{01}]$ is $N_b \times N_b$ matrix, who presents the mutual impedance and induction interaction betweens a parts of conductor segments branch and leakage currents, $[\overline{\overline{\mathbf{Z}}}^{10}]$ is $N_b \times N_b$ matrix, who presents the mutual impedance and induction interaction betweens a parts of conductor segments branch and leakage currents, $[\overline{\overline{\mathbf{Z}}}^{11}]$ is $N_b \times N_b$ matrix, who presents the mutual impedance and induction interaction betweens a parts of conductor segments leakage currents.

Matrix equation (10.50) is assembled from the following equations.

$$\sum_{\theta=1}^{N_b} Z_{l_\theta} \int_{l_\theta} [\bar{q}(\bar{r}_\theta)][\bar{p}(\bar{r}_\theta)] dr_\theta [\bar{I}_\theta^\chi]$$

$$+\sum_{\vartheta=1}^{N_b}\int_{l_\theta}\int_{l_\vartheta}\hat{r}_\theta\cdot s\bar{\bar{G}}_{\xi\zeta}^{A^s}(\bar{r}_\theta,\bar{r}_\vartheta)\cdot\hat{r}_\vartheta^T[\bar{q}(\bar{r}_\theta)][\bar{p}(\bar{r}_\vartheta)]dr_\vartheta dr_\theta[\bar{I}_\vartheta^\chi]$$

$$+\sum_{\vartheta=1}^{N_b}\int_{l_\theta}\int_{l_\vartheta}G_{\xi\zeta}^{\varphi^s}(\bar{r}_\theta,\bar{r}_\vartheta)\frac{\partial[\bar{q}(\bar{r}_\theta)]}{\partial r_\theta}\frac{\partial[\bar{p}(\bar{r}_\vartheta)]}{\partial r_\vartheta}dr_\vartheta dr_\theta[\bar{I}_\vartheta^\chi]=0 \tag{10.51}$$

where $\theta=1,2,\ldots,N_b$; $\chi=0,1$; $\theta=1,\ldots,N_b$. $\bar{r}_\theta$ and $\bar{r}_\vartheta$ is location vector along θ-th and ϑ-th conductors segment, respectively; r_θ and r_ϑ, respectively, parameter variable along the θ-th and ϑ-th thin conductor axis. Z_{l_θ} is the internal impedance of the θ-th thin conductor segment. $\hat{r}_\theta$ and $\hat{r}_\vartheta$ are, respectively, the 1×3 dimensional unit director vector along the axis of the θ-th and ϑ-th cylindrical conductor segments. j is a imaginary unit, ω is angular frequency. $G_{\xi\zeta}^{\varphi^s}(\bar{r}_\theta,\bar{r}_\vartheta)$ and $\bar{\bar{G}}_{\xi\zeta}^{A^s}(\bar{r}_\theta,\bar{r}_\vartheta)$ are, respectively, s domain scalar monopole Green's function and vector dipole dyadic Green's function within horizontal multilayered earth model, which had been fully discussed in front Chap. 3. $[\bar{p}(\bar{r}_\vartheta)]$ and $[\bar{q}(\bar{r}_\theta)]$ are 1×2 dimensional 1 order basis functions and 2×1 dimensional 1 order test functions, respectively. $[\bar{p}(\bar{r}_\vartheta)]=\begin{bmatrix}1 & -\frac{r_\vartheta}{l_\vartheta}\end{bmatrix}$ and $[\bar{q}(\bar{r}_\theta)]=\begin{bmatrix}1\\ -\frac{r_\theta}{l_\theta}\end{bmatrix}$.

The time domain spatial 1-d vector Galerkin MM obtained with Laplace transform

Just like the BEM, consider equations (10.11), (10.12), and (10.13) simultaneously, and substituting formulas (10.11) , (10.12) and (10.13) into formula (10.51) and considering the laplace transform, there is the following formula.

$$\frac{1}{2\pi j}\int_{\sigma-\infty}^{\sigma+\infty}[\sum_{\theta=1}^{N_b}\left(Z_{l_0}+Z_l(s)\right)\int_{l_\theta}[\bar{q}(\bar{r}_\theta)][\bar{p}(\bar{r}_\theta)]dr_\theta[\bar{I}_\theta^\chi]$$

$$+\sum_{\vartheta=1}^{N_b}\int_{l_\theta}\int_{l_\vartheta}\hat{r}_\theta\cdot s\left(\bar{\bar{G}}_{\xi\zeta}^{A_s^s}(\bar{r}_\theta,\bar{r}_\vartheta)+\bar{\bar{G}}_{\xi\zeta}^{A_q^s}(\bar{r}_\theta,\bar{r}_\vartheta,s)\right)\cdot\hat{r}_\vartheta^T[\bar{q}(\bar{r}_\theta)][\bar{p}(\bar{r}_\vartheta)]dr_\vartheta dr_\theta[\bar{I}_\vartheta^\chi]$$

$$+\sum_{\vartheta=1}^{N_b}\int_{l_\theta}\int_{l_\vartheta}\left(G_{\xi\zeta}^{\varphi_s^s}(\bar{r}_\theta,\bar{r}_\vartheta)+G_{\xi\zeta}^{\varphi_q^s}(\bar{r}_\theta,\bar{r}_\vartheta,s)\right)\frac{\partial[\bar{q}(\bar{r}_\theta)]}{\partial r_\theta}\frac{\partial[\bar{p}(\bar{r}_\vartheta)]}{\partial r_\vartheta}dr_\vartheta dr_\theta[\bar{I}_\vartheta^\chi]e^{st}ds=0 \tag{10.52}$$

where $\theta=1,2,\ldots,N_b$, $\chi=0,1$.

By using laplace transform, the following formulas in the time domain can be obtained.

$$\sum_{\theta=1}^{N_b}\left(Z_{l_0}\int_{l_\theta}[\bar{q}(\bar{r}_\theta)][\bar{p}(\bar{r}_\theta)]dr_\theta[\bar{I}_\theta^\chi(t)]+Z_l\int_{l_\theta}[\bar{q}(\bar{r}_\theta)][\bar{p}(\bar{r}_\theta)]dr_\theta[\bar{I}_\theta^\chi(t-t_c)]\right)$$

$$+\sum_{\vartheta=1}^{N_b}\left(\int_{l_\theta}\int_{l_\vartheta}\hat{r}_\theta\cdot\bar{\bar{G}}_{\xi\zeta}^{A_s^t}(\bar{r}_\theta,\bar{r}_\vartheta)\cdot\hat{r}_\vartheta^T[\bar{q}(\bar{r}_\theta)][\bar{p}(\bar{r}_\vartheta)]dr_\vartheta dr_\theta\frac{d[\bar{I}_\vartheta^\chi(t)]}{dt}\right.$$

$$\left.+\int_{l_\theta}\int_{l_\vartheta}\hat{r}_\theta\cdot\bar{\bar{G}}_{\xi\zeta}^{A_q^t}(\bar{r}_\theta,\bar{r}_\vartheta)\cdot\hat{r}_\vartheta^T[\bar{q}(\bar{r}_\theta)][\bar{p}(\bar{r}_\vartheta)]dr_\vartheta dr_\theta\frac{d[\bar{I}_\vartheta^\chi(t-t_A)]}{dt}\right)$$

$$+\sum_{\vartheta=1}^{N_b}\left(\int_{l_\theta}\int_{l_\vartheta}G_{\xi\zeta}^{\varphi_s^t}(\bar{r}_\theta,\bar{r}_\vartheta)\frac{\partial[\bar{q}(\bar{r}_\theta)]}{\partial r_\theta}\frac{d[\bar{p}(\bar{r}_\vartheta)]}{dr_\vartheta}dr_\vartheta dr_\theta[\bar{I}_\vartheta^\chi(t)]\right.$$

$$\left.+\int_{l_\theta}\int_{l_\vartheta}G_{\xi\zeta}^{\varphi_q^t}(\bar{r}_\theta,\bar{r}_\vartheta)\frac{\partial[\bar{q}(\bar{r}_\theta)]}{\partial r_\theta}\frac{\partial[\bar{p}(\bar{r}_\vartheta)]}{\partial r_\vartheta}dr_\vartheta dr_\theta[\bar{\mathbf{I}}_\vartheta^\chi(t-t_\varphi)]\right)=0 \qquad (10.53)$$

where $\theta = 1, 2, \ldots, N_b$; $\chi = 0, 1$; $G_{\xi\zeta}^{\varphi_q^t}(\bar{r}_\theta,\bar{r}_\vartheta)$ and $\bar{\bar{G}}_{\xi\zeta}^{A_q^t}(\bar{r}_\theta,\bar{r}_\vartheta)$ are, respectively, time domain scalar monopole Green's function and vector dipole dyadic Green's function within horizontal multilayered earth model with delay effects, which had been fully discussed in front Chap. 3.

This formula (10.53) can also be further written as matrix formation as shown below.

$$[\bar{\bar{\mathbf{Z}}}_b^s][\mathbf{I}_b^\chi(t)]+[\bar{\bar{\mathbf{Z}}}_b^q][\mathbf{I}_b^\chi(t-t_c)]+[\bar{\bar{\mathbf{M}}}_b^s]\frac{d[\mathbf{I}_b^\chi(t)]}{dt}+[\bar{\bar{\mathbf{M}}}_b^q]\frac{d[\mathbf{I}_b^\chi(t-t_A)]}{dt}$$

$$+[\bar{\bar{\mathbf{Z}}}_s^s][\mathbf{I}_b^\chi(t)]+[\bar{\bar{\mathbf{Z}}}_s^q][\mathbf{I}_b^\chi(t-t_\varphi)]=0 \qquad (10.54)$$

where $\chi = 0, 1$; $[\bar{\bar{\mathbf{Z}}}_b^s]=\begin{bmatrix}\bar{\bar{\mathbf{Z}}}_b^{s_{00}} & \bar{\bar{\mathbf{Z}}}_b^{s_{01}}\\ \bar{\bar{\mathbf{Z}}}_b^{s_{10}} & \bar{\bar{\mathbf{Z}}}_b^{s_{11}}\end{bmatrix}$ and $[\bar{\bar{\mathbf{Z}}}_b^q]=\begin{bmatrix}\bar{\bar{\mathbf{Z}}}_b^{q_{00}} & \bar{\bar{\mathbf{Z}}}_b^{q_{01}}\\ \bar{\bar{\mathbf{Z}}}_b^{q_{10}} & \bar{\bar{\mathbf{Z}}}_b^{q_{11}}\end{bmatrix}$ are the self impedance matrices of grounding grid conductors without and with delay effects in the $2N_b \times 2N_b$ dimension, respectively. $[\bar{\bar{\mathbf{M}}}_b^s]=\begin{bmatrix}\bar{\bar{\mathbf{M}}}_b^{s_{00}} & \bar{\bar{\mathbf{M}}}_b^{s_{01}}\\ \bar{\bar{\mathbf{M}}}_b^{s_{10}} & \bar{\bar{\mathbf{M}}}_b^{s_{11}}\end{bmatrix}$ and $[\bar{\bar{\mathbf{M}}}_b^q]=\begin{bmatrix}\bar{\bar{\mathbf{M}}}_b^{q_{00}} & \bar{\bar{\mathbf{M}}}_b^{q_{01}}\\ \bar{\bar{\mathbf{M}}}_b^{q_{10}} & \bar{\bar{\mathbf{M}}}_b^{q_{11}}\end{bmatrix}$ are the mutual induction matrices of grounding grid conductors without and with delay effects in the $2N_b \times 2N_b$ dimension, respectively. $[\bar{\bar{\mathbf{Z}}}_s^s]=\begin{bmatrix}0 & 0\\ 0 & \bar{\bar{\mathbf{Z}}}_s^{s_{11}}\end{bmatrix}$ and $[\bar{\bar{\mathbf{Z}}}_s^q]=\begin{bmatrix}0 & 0\\ 0 & \bar{\bar{\mathbf{Z}}}_s^{q_{11}}\end{bmatrix}$ are the mutual impedance matrices of grounding grid conductors without and with delay effects in the $2N_b \times 2N_b$ dimension, respectively.

$[\bar{\mathbf{I}}_{\vartheta}^{\chi}(t)] = \begin{bmatrix} \bar{\mathbf{I}}_b^0(t) \\ \bar{\mathbf{I}}_b^1(t) \end{bmatrix}$, $[\bar{\mathbf{I}}_b^{\chi}(t-t_c)] = \begin{bmatrix} \bar{\mathbf{I}}_b^0(t-t_c) \\ \bar{\mathbf{I}}_b^1(t-t_c) \end{bmatrix}$, $[\bar{\mathbf{I}}_b^{\chi}(t-t_A)] = \begin{bmatrix} \bar{\mathbf{I}}_b^0(t-t_A) \\ \bar{\mathbf{I}}_b^1(t-t_A) \end{bmatrix}$ and $[\bar{\mathbf{I}}_b^{\chi}(t-t_{\varphi})] = \begin{bmatrix} \bar{\mathbf{I}}_b^0(t-t_{\varphi}) \\ \bar{\mathbf{I}}_b^1(t-t_{\varphi}) \end{bmatrix}$.

So, this equation can be further written as follows

$$\begin{bmatrix} \bar{\bar{\mathbf{Z}}}_b^{s_{00}} & \bar{\bar{\mathbf{Z}}}_b^{s_{01}} \\ \bar{\bar{\mathbf{Z}}}_b^{s_{10}} & \bar{\bar{\mathbf{Z}}}_b^{s_{11}} + \bar{\bar{\mathbf{Z}}}_s^{s_{11}} \end{bmatrix} \begin{bmatrix} \bar{\mathbf{I}}_b^0(t) \\ \bar{\mathbf{I}}_b^1(t) \end{bmatrix} + \begin{bmatrix} \bar{\bar{\mathbf{M}}}_b^{s_{00}} & \bar{\bar{\mathbf{M}}}_b^{s_{01}} \\ \bar{\bar{\mathbf{M}}}_b^{s_{10}} & \bar{\bar{\mathbf{M}}}_b^{s_{11}} \end{bmatrix} \frac{d}{dt} \begin{bmatrix} \bar{\mathbf{I}}_b^0(t) \\ \bar{\mathbf{I}}_b^1(t) \end{bmatrix}$$
$$= - \begin{bmatrix} \bar{\bar{\mathbf{Z}}}_b^{q_{00}} & \bar{\bar{\mathbf{Z}}}_b^{q_{01}} \\ \bar{\bar{\mathbf{Z}}}_b^{q_{10}} & \bar{\bar{\mathbf{Z}}}_b^{q_{11}} \end{bmatrix} \begin{bmatrix} \bar{\mathbf{I}}_b^0(t-t_c) \\ \bar{\mathbf{I}}_b^1(t-t_c) \end{bmatrix} - \begin{bmatrix} \bar{\bar{\mathbf{M}}}_b^{q_{00}} & \bar{\bar{\mathbf{M}}}_b^{q_{01}} \\ \bar{\bar{\mathbf{M}}}_b^{q_{10}} & \bar{\bar{\mathbf{M}}}_b^{q_{11}} \end{bmatrix} \frac{d}{dt} \begin{bmatrix} \bar{\mathbf{I}}_b^0(t-t_A) \\ \bar{\mathbf{I}}_b^1(t-t_A) \end{bmatrix}$$
$$- \begin{bmatrix} 0 & 0 \\ 0 & \bar{\bar{\mathbf{Z}}}_s^{q_{11}} \end{bmatrix} \begin{bmatrix} \bar{\mathbf{I}}_b^0(t-t_{\varphi}) \\ \bar{\mathbf{I}}_b^1(t-t_{\varphi}) \end{bmatrix} \tag{10.55}$$

10.4.3 The Time Domain Spatial 1-d Vector Galerkin MM and its Solution

The Eqs. 10.49 and 10.55 can be unified to be written as below

$$\begin{bmatrix} \bar{\bar{\mathbf{Z}}}_b^{s_{00}} & \bar{\bar{\mathbf{Z}}}_b^{s_{01}} \\ \bar{\bar{\mathbf{Z}}}_b^{s_{10}} & \bar{\bar{\mathbf{Z}}}_b^{s_{11}} + \bar{\bar{\mathbf{Z}}}_s^{s_{11}} \end{bmatrix} \begin{bmatrix} \bar{\mathbf{I}}_b^0(t) \\ \bar{\mathbf{I}}_b^1(t) \end{bmatrix} + \begin{bmatrix} \bar{\bar{\mathbf{M}}}_b^{s_{00}} & \bar{\bar{\mathbf{M}}}_b^{s_{01}} \\ \bar{\bar{\mathbf{M}}}_b^{s_{10}} & \bar{\bar{\mathbf{M}}}_b^{s_{11}} \end{bmatrix} \frac{d}{dt} \begin{bmatrix} \bar{\mathbf{I}}_b^0(t) \\ \bar{\mathbf{I}}_b^1(t) \end{bmatrix}$$
$$= - \begin{bmatrix} \bar{\bar{\mathbf{Z}}}_b^{q_{00}} & \bar{\bar{\mathbf{Z}}}_b^{q_{01}} \\ \bar{\bar{\mathbf{Z}}}_b^{q_{10}} & \bar{\bar{\mathbf{Z}}}_b^{q_{11}} \end{bmatrix} \begin{bmatrix} \bar{\mathbf{I}}_b^0(t-t_c) \\ \bar{\mathbf{I}}_b^1(t-t_c) \end{bmatrix} - \begin{bmatrix} \bar{\bar{\mathbf{M}}}_b^{q_{00}} & \bar{\bar{\mathbf{M}}}_b^{q_{01}} \\ \bar{\bar{\mathbf{M}}}_b^{q_{10}} & \bar{\bar{\mathbf{M}}}_b^{q_{11}} \end{bmatrix} \frac{d}{dt} \begin{bmatrix} \bar{\mathbf{I}}_b^0(t-t_A) \\ \bar{\mathbf{I}}_b^1(t-t_A) \end{bmatrix}$$
$$- \begin{bmatrix} 0 & 0 \\ 0 & \bar{\bar{\mathbf{Z}}}_s^{q_{11}} \end{bmatrix} \begin{bmatrix} \bar{\mathbf{I}}_b^0(t-t_{\varphi}) \\ \bar{\mathbf{I}}_b^1(t-t_{\varphi}) \end{bmatrix} \tag{10.56}$$

The excitation is incorporated into the formulation through the boundary condition:

$$I_{b_1}^0(t) = I_g(t) \tag{10.57}$$

where $I_g(t)$ denotes the lightning current generator and $I_{b_1}^0(t)$ current in the injection point.

At a junction consisting of two or more segments the continuity properties of the electric field must be satisfied, which is ensured by applying the Kirchhoff current law

$$\sum_{m=1}^{n} I_{b_m}^0(t) = 0 \tag{10.58}$$

where n is the total number of conductor segments connected to the node.

and continuity equation

$$[\frac{\partial I_{b_1}^0(t)}{\partial r_1}]_{at\ junction} = [\frac{\partial I_{b_2}^0(t)}{\partial r_2}]_{at\ junction} = \cdots = [\frac{\partial I_{b_n}^0(t)}{\partial r_n}]_{at\ junction} \tag{10.59}$$

The condition (10.59) ensures the discontinuities in charge per unit length to be ruled out in passing from one conductor to another across the junction.

At the conductor free ends, the total current vanishes, i.e. is forced to be zero.

Once Eq. (10.49) is solved, the unknown currents were known, the branch and leakage currents can be obtained easily. The voltage, the electrical field intensity (E) and the magnetic field intensity (B) at any point can be calculated by [29–31, 36, 62].

The study of the performance of the grounding system for time domain spatial 1-d Galerkin MM has been reduced to the computation of $[\overline{\overline{\mathbf{Z}}}_b^{s00}]$, $[\overline{\overline{\mathbf{Z}}}_b^{s}]$, $[\overline{\overline{\mathbf{Z}}}_b^{q}]$, $[\overline{\overline{\mathbf{M}}}_b^{s}]$, $[\overline{\overline{\mathbf{M}}}_b^{q}]$, $[\overline{\overline{\mathbf{Z}}}_s^{s11}]$ and $[\overline{\overline{\mathbf{Z}}}_s^{q11}]$ matrices.

Computation of $[\overline{\overline{\mathbf{Z}}}_b^{s00}]$, $[\overline{\overline{\mathbf{Z}}}_b^{s}]$, $[\overline{\overline{\mathbf{Z}}}_b^{q}]$, $[\overline{\overline{\mathbf{M}}}_b^{s}]$, $[\overline{\overline{\mathbf{M}}}_b^{q}]$, $[\overline{\overline{\mathbf{Z}}}_s^{s11}]$ and $[\overline{\overline{\mathbf{Z}}}_s^{q11}]$ matrices

From [29–31, 36, 62], it can be seen that each segment owns internal impedance. Mutual inductances or impedances between segments' currents are also included in the model; To discuss the calculation technique of computation of element for $[\overline{\overline{\mathbf{Z}}}_b^{s}]$, $[\overline{\overline{\mathbf{Z}}}_b^{q}]$, $[\overline{\overline{\mathbf{M}}}_b^{s}]$, $[\overline{\overline{\mathbf{M}}}_b^{q}]$, $[\overline{\overline{\mathbf{Z}}}_s^{s11}]$ and $[\overline{\overline{\mathbf{Z}}}_s^{q11}]$ matrices, the Eq. (10.47) has been used as reference.

1. The first term of the Eq. (10.24) for element of $[\overline{\overline{\mathbf{Z}}}_b^{s}]$ matrix:

$$Z_\theta^s = Z_{l_0} \int_{l_\theta} [\bar{q}(\bar{r}_\theta)][\bar{p}(\bar{r}_\theta)] dr_\theta = Z_{l_0} \int_{l_\theta} \begin{bmatrix} 1 & \frac{-r_\theta}{l_\theta} \\ \frac{-r_\theta}{l_\theta} & \frac{r_\theta^2}{l_\theta^2} \end{bmatrix} dr_\theta \tag{10.60}$$

   where Z_{l_0} consists of internal resistance, the formula for internal resistance can be found in [30, 31, 36, 62].

2. The core of second term of the Eq. (10.7) for element of $[\overline{\overline{\mathbf{Z}}}_b^{q}]$ matrix:

$$Z_\theta^q = Z_l \int_{l_\theta} [\bar{q}(\bar{r}_\theta)][\bar{p}(\bar{r}_\theta)]dr_\theta = \sum_{c=1}^{M_c} \alpha_c \int_{l_\theta} \begin{bmatrix} 1 & \frac{-r_\vartheta}{l_\vartheta} \\ \frac{-r_\theta}{l_\theta} & \frac{r_\theta r_\vartheta}{l_\theta l_\vartheta} \end{bmatrix} dr_\theta \tag{10.61}$$

3. The core of third term of the Eq. (10.7) for element of $[\overline{\overline{\mathbf{M}}}_b^s]$ matrix: The core is formula of mutual induction interaction betweens a pars of currents along conductor segments without delay effects.

$$M_{\theta,\vartheta}^s = \int_{l_\theta}\int_{l_\vartheta} \hat{r}_\theta \cdot \bar{\bar{G}}_{\xi\zeta}^{A_s^t}(\bar{r}_\theta, \bar{r}_\vartheta) \cdot \hat{r}_\vartheta^T [\bar{p}(\bar{r}_\theta)][\bar{q}(\bar{r}_\vartheta)]dr_\vartheta dr_\theta \tag{10.62}$$

$$= \int_{l_\theta}\int_{l_\vartheta} \hat{r}_\theta \cdot \bar{\bar{G}}_{\xi\zeta}^{A_s^t}(\bar{r}_\theta, \bar{r}_\vartheta) \cdot \hat{r}_\vartheta^T \begin{bmatrix} 1 & \frac{-r_\vartheta}{l_\vartheta} \\ \frac{-r_\theta}{l_\theta} & \frac{r_\theta r_\vartheta}{l_\theta l_\vartheta} \end{bmatrix} dr_\vartheta dr_\theta \tag{10.63}$$

For an half infinite homogeneous conductivity medium, one has

$$\bar{\bar{G}}_{11}^{A_s^t}(\bar{r}_\vartheta, \bar{r}_\theta) = \frac{\mu}{4\pi} \begin{bmatrix} \frac{1}{R_{\theta\vartheta}} & 0 & -\hat{k}_{01}^{TM} \frac{\partial(\ln(-(z_\theta+z_\vartheta)+R'_{\theta\vartheta})}{\partial x} \\ 0 & \frac{1}{R_{\theta\vartheta}} & -\hat{k}_{01}^{TM} \frac{\partial(\ln(-(z_\theta+z_\vartheta)+R'_{\theta\vartheta})}{\partial y} \\ 0 & 0 & \frac{1}{R_{\theta\vartheta}} - \frac{\hat{k}_{01}^{TM}}{R'_{\theta\vartheta}} \end{bmatrix}, \tag{10.64}$$

where $R_{\theta\vartheta} = |\bar{r}_\theta - \bar{r}_\vartheta|$, $R'_{\theta\vartheta} = |\bar{r}_\theta + \bar{r}_\vartheta|$, $\hat{k}_{01}^{TM} = \frac{\sigma_1 - \sigma_0}{\sigma_1 + \sigma_0}$.
so

$$M_{\theta,\vartheta}^s =$$

$$\frac{\mu}{4\pi}\int_{l_\theta}\int_{l_\vartheta} \hat{r}_\theta \cdot \begin{bmatrix} \frac{1}{R_{\theta\vartheta}} & 0 & -\hat{k}_{01}^{TM} \frac{\partial(\ln(-(z_\theta+z_\vartheta)+R'_{\theta\vartheta})}{\partial x} \\ 0 & \frac{1}{R_{\theta\vartheta}} & -\hat{k}_{01}^{TM} \frac{\partial(\ln(-(z_\theta+z_\vartheta)+R'_{\theta\vartheta})}{\partial y} \\ 0 & 0 & \frac{1}{R_{\theta\vartheta}} - \frac{\hat{k}_{01}^{TM}}{R'_{\theta\vartheta}} \end{bmatrix} \cdot \hat{r}_\vartheta^T \begin{bmatrix} 1 & \frac{-r_\vartheta}{l_\vartheta} \\ \frac{-r_\theta}{l_\theta} & \frac{r_\theta r_\vartheta}{l_\theta l_\vartheta} \end{bmatrix} dr_\vartheta dr_\theta \tag{10.65}$$

4. The core of fourth term of the Eq. (10.7) for element of $[\overline{\overline{\mathbf{M}}}_b^q]$ matrix: The core is formula of mutual induction interaction betweens a pars of currents along conductor segments with delay effects.

$$M_{\theta,\vartheta}^q = \int_{l_\theta}\int_{l_\vartheta} \hat{r}_\theta \cdot \bar{\bar{G}}_{\xi\zeta}^{A_q^t}(\bar{r}_\theta, \bar{r}_\vartheta) \cdot \hat{r}_\vartheta^T [\bar{p}(\bar{r}_\theta)][\bar{q}(\bar{r}_\vartheta)]dr_\vartheta dr_\theta \tag{10.66}$$

$$= \int_{l_\theta}\int_{l_\vartheta} \hat{r}_\theta \cdot \bar{\bar{G}}^{A_q^t}_{\xi\zeta}(\bar{r}_\theta, \bar{r}_\vartheta) \cdot \hat{r}_\vartheta^T \begin{bmatrix} 1 & \frac{-r_\vartheta}{l_\vartheta} \\ \frac{-r_\theta}{l_\theta} & \frac{r_\theta r_\vartheta}{l_\theta l_\vartheta} \end{bmatrix} dr_\vartheta dr_\theta \tag{10.67}$$

For an half infinite homogeneous conductivity medium, one has

$$\bar{\bar{G}}^{A_q^t}_{11}(\bar{r}_\vartheta, \bar{r}_\theta) = \frac{\mu}{4\pi} \begin{bmatrix} 0\ 0 - \sum_{q_A=1}^{M_A} \alpha_{q_A} \frac{\partial\left(\ln\left(-(z_\theta+z_\vartheta)+R'_{\theta\vartheta}\right)\right)}{\partial x} \\ 0\ 0 - \sum_{q_A=1}^{M_A} \alpha_{q_A} \frac{\partial\left(\ln\left(-(z_\theta+z_\vartheta)+R'_{\theta\vartheta}\right)\right)}{\partial y} \\ 0\ 0 \qquad - \sum_{q_A=1}^{M_A} \alpha_{q_A} \frac{1}{R'_{\theta\vartheta}} \end{bmatrix}, \tag{10.68}$$

so

$$M^q_{\theta,\vartheta} =$$

$$\frac{\mu}{4\pi}\int_{l_\theta}\int_{l_\vartheta} \hat{r}_\theta \cdot \begin{bmatrix} 0\ 0 - \sum_{q_A=1}^{M_A} \alpha_{q_A} \frac{\partial\left(\ln\left(-(z_\theta+z_\vartheta)+R'_{\theta\vartheta}\right)\right)}{\partial x} \\ 0\ 0 - \sum_{q_A=1}^{M_A} \alpha_{q_A} \frac{\partial\left(\ln\left(-(z_\theta+z_\vartheta)+R'_{\theta\vartheta}\right)\right)}{\partial y} \\ 0\ 0 \qquad - \sum_{q_A=1}^{M_A} \alpha_{q_A} \frac{1}{R'_{\theta\vartheta}} \end{bmatrix} \cdot \hat{r}_\vartheta^T \begin{bmatrix} 1 & \frac{-r_\vartheta}{l_\vartheta} \\ \frac{-r_\theta}{l_\theta} & \frac{r_\theta r_\vartheta}{l_\theta l_\vartheta} \end{bmatrix} dr_\vartheta dr_\theta \tag{10.69}$$

5. The core of fifth term of the Eq. (10.7): The core is formula of mutual impedance interaction betweens a pars of currents along conductor segments without delay effect.

$$Z^s_{\theta,\vartheta} = \int_{l_\theta}\int_{l_\vartheta} G^{\varphi_s^t}_{\xi\zeta}(\bar{r}_\theta, \bar{r}_\vartheta) \frac{\partial[\bar{q}(\bar{r}_\theta)]}{\partial r_\theta} \frac{\partial[\bar{p}(\bar{r}_\vartheta)]}{\partial r_\vartheta} dr_\vartheta dr_\theta \tag{10.70}$$

$$Z^s_{\theta,\vartheta} = \int_{l_\theta}\int_{l_\vartheta} G^{\varphi_s^t}_{\xi\zeta}(\bar{r}_\theta, \bar{r}_\vartheta) \begin{bmatrix} 0 & 0 \\ 0 & \frac{1}{l_\theta l_\vartheta} \end{bmatrix} dr_\vartheta dr_\theta \tag{10.71}$$

$$Z^{s_{11}}_{\theta,\vartheta} = \frac{1}{l_\theta l_\vartheta}\int_{l_\theta}\int_{l_\vartheta} G^{\varphi_s^t}_{\xi\zeta}(\bar{r}_\theta, \bar{r}_\vartheta) dr_\vartheta dr_\theta \tag{10.72}$$

For an half infinite homogeneous conductivity medium, one has

$$G_{11}^{\varphi_s^t}(\bar{r}_\vartheta, \bar{r}_\theta) = \frac{1}{4\pi\sigma_1}\frac{1}{R_{\theta\vartheta}}, \tag{10.73}$$

so

$$Z_{\theta,\vartheta}^{s_{11}} = \frac{1}{4\pi\bar{\sigma}_1 l_\theta l_\vartheta}\int_{l_\theta}\int_{l_\vartheta}\frac{1}{R_{\theta\vartheta}}dr_\theta dr_\vartheta \tag{10.74}$$

6. The core of sixth term of the Eq. (10.7): The core is formula of mutual impedance interaction betweens a pars of currents along conductor segments with delay effect.

$$Z_{\theta,\vartheta}^{q} = \int_{l_\theta}\int_{l_\vartheta} G_{\xi\zeta}^{\varphi_q^t}(\bar{r}_\theta, \bar{r}_\vartheta, t)\frac{\partial[\bar{q}(\bar{r}_\theta)]}{\partial r_\theta}\frac{\partial[\bar{p}(\bar{r}_\vartheta)]}{\partial r_\vartheta}dr_\vartheta dr_\theta \tag{10.75}$$

$$Z_{\theta,\vartheta}^{q} = \int_{l_\theta}\int_{l_\vartheta} G_{\xi\zeta}^{\varphi_q^t}(\bar{r}_\theta, \bar{r}_\vartheta, t)\begin{bmatrix} 0 & 0 \\ 0 & \frac{1}{l_\theta l_\vartheta}\end{bmatrix}dr_\vartheta dr_\theta \tag{10.76}$$

$$Z_{\theta,\vartheta}^{q_{11}} = \frac{1}{l_\theta l_\vartheta}\int_{l_\theta}\int_{l_\vartheta} G_{\xi\zeta}^{\varphi_q^t}(\bar{r}_\theta, \bar{r}_\vartheta, t)dr_\vartheta dr_\theta \tag{10.77}$$

For an half infinite homogeneous conductivity medium, one has

$$G_{11}^{\varphi_q^t}(\bar{r}_\vartheta, \bar{r}_\theta, t) = \frac{1}{4\pi\sigma_1}\left(\sum_{q_0=1}^{M_0}\alpha_{q_0}\frac{1}{R_{\theta\vartheta}} - \sum_{q_\varphi=1}^{M_\varphi}\alpha_{q_\varphi}\frac{1}{R'_{\theta\vartheta}}\right), \tag{10.78}$$

so

$$Z_{\theta,\vartheta}^{q_{11}} = \frac{1}{4\pi\sigma_1 l_\theta l_\vartheta}\int_{l_\theta}\int_{l_\vartheta}\left[\sum_{q_0}^{M_0}\alpha_{q_0}\frac{1}{R_{\theta\vartheta}} - \sum_{q_\varphi}^{M_\varphi}\alpha_{q_\varphi}\frac{1}{R'_{\theta\vartheta}}\right]dr_\theta dr_\vartheta \tag{10.79}$$

Note the medium surrounding the point current source here was considered as half infinite homogeneous conductivity medium. However, in any practical case, the earth is represented via a multilayered earth model. The QSCIM can be used to dealt with the multilayered earth model, which has be fully discussed in Chap. 3.

10.4.4 Time Domain Spatial 1-d Vector Galerkin MM for Transient Grounding Problems

The matrix equation of the time domain spatial 1-d vector Galerkin MM was given in the previous section as follows.

$$\begin{bmatrix} \bar{\bar{\mathbf{Z}}}_b^{s00} & \bar{\bar{\mathbf{Z}}}_b^{s01} \\ \bar{\bar{\mathbf{Z}}}_b^{s10} & \bar{\bar{\mathbf{Z}}}_b^{s11}+\bar{\bar{\mathbf{Z}}}_s^{s11} \end{bmatrix} \begin{bmatrix} \bar{\mathbf{I}}_b^0(t) \\ \bar{\mathbf{I}}_b^1(t) \end{bmatrix} + \begin{bmatrix} \bar{\bar{\mathbf{M}}}_b^{s00} & \bar{\bar{\mathbf{M}}}_b^{s01} \\ \bar{\bar{\mathbf{M}}}_b^{s10} & \bar{\bar{\mathbf{M}}}_b^{s11} \end{bmatrix} \frac{d}{dt} \begin{bmatrix} \bar{\mathbf{I}}_b^0(t) \\ \bar{\mathbf{I}}_b^1(t) \end{bmatrix}$$

$$= -\begin{bmatrix} \bar{\bar{\mathbf{Z}}}_b^{q00} & \bar{\bar{\mathbf{Z}}}_b^{q01} \\ \bar{\bar{\mathbf{Z}}}_b^{q10} & \bar{\bar{\mathbf{Z}}}_b^{q11} \end{bmatrix} \begin{bmatrix} \bar{\mathbf{I}}_b^0(t-t_c) \\ \bar{\mathbf{I}}_b^1(t-t_c) \end{bmatrix} - \begin{bmatrix} \bar{\bar{\mathbf{M}}}_b^{q00} & \bar{\bar{\mathbf{M}}}_b^{q01} \\ \bar{\bar{\mathbf{M}}}_b^{q10} & \bar{\bar{\mathbf{M}}}_b^{q11} \end{bmatrix} \frac{d}{dt} \begin{bmatrix} \bar{\mathbf{I}}_b^0(t-t_A) \\ \bar{\mathbf{I}}_b^1(t-t_A) \end{bmatrix}$$

$$- \begin{bmatrix} 0 & 0 \\ 0 & \bar{\bar{\mathbf{Z}}}_s^{q11} \end{bmatrix} \begin{bmatrix} \bar{\mathbf{I}}_b^0(t-t_\varphi) \\ \bar{\mathbf{I}}_b^1(t-t_\varphi) \end{bmatrix} \tag{10.80}$$

The time-domain matrix equation is discretized in time with a time interval of Δt, and has $t_k = k\Delta t, k = 0, 1, \ldots, N_t$, N_t is the maximum number of discrete time intervals. In this way, the following time discrete matrix equation can be obtained.

$$\begin{bmatrix} \bar{\bar{\mathbf{Z}}}_b^{s00} & \bar{\bar{\mathbf{Z}}}_b^{s01} \\ \bar{\bar{\mathbf{Z}}}_b^{s10} & \bar{\bar{\mathbf{Z}}}_b^{s11}+\bar{\bar{\mathbf{Z}}}_s^{s11} \end{bmatrix} \begin{bmatrix} \bar{\mathbf{I}}_b^0(t_k) \\ \bar{\mathbf{I}}_b^1(t_k) \end{bmatrix} + \begin{bmatrix} \bar{\bar{\mathbf{M}}}_b^{s00} & \bar{\bar{\mathbf{M}}}_b^{s01} \\ \bar{\bar{\mathbf{M}}}_b^{s10} & \bar{\bar{\mathbf{M}}}_b^{s11} \end{bmatrix} \begin{bmatrix} \frac{\bar{\mathbf{I}}_b^0(t_k)-\bar{\mathbf{I}}_b^0(t_{k-1})}{\Delta t} \\ \frac{\bar{\mathbf{I}}_b^1(t_k)-\bar{\mathbf{I}}_b^1(t_{k-1})}{\Delta t} \end{bmatrix}$$

$$= -\begin{bmatrix} \bar{\bar{\mathbf{Z}}}_b^{q00} & \bar{\bar{\mathbf{Z}}}_b^{q01} \\ \bar{\bar{\mathbf{Z}}}_b^{q10} & \bar{\bar{\mathbf{Z}}}_b^{q11} \end{bmatrix} \begin{bmatrix} \bar{\mathbf{I}}_b^0(t_k-t_c^k) \\ \bar{\mathbf{I}}_b^1(t_k-t_c^k) \end{bmatrix} - \begin{bmatrix} \bar{\bar{\mathbf{M}}}_b^{q00} & \bar{\bar{\mathbf{M}}}_b^{q01} \\ \bar{\bar{\mathbf{M}}}_b^{q10} & \bar{\bar{\mathbf{M}}}_b^{q11} \end{bmatrix} \begin{bmatrix} \frac{\bar{\mathbf{I}}_b^0(t_k-t_A^k)-\bar{\mathbf{I}}_b^0(t_{k-1}-t_A^k)}{\Delta t} \\ \frac{\bar{\mathbf{I}}_b^1(t_k-t_A^k)-\bar{\mathbf{I}}_b^1(t_{k-1}-t_A^k)}{\Delta t} \end{bmatrix}$$

$$- \begin{bmatrix} 0 & 0 \\ 0 & \bar{\bar{\mathbf{Z}}}_s^{q11} \end{bmatrix} \begin{bmatrix} \bar{\mathbf{I}}_b^0(t_k-t_\varphi^k) \\ \bar{\mathbf{I}}_b^1(t_k-t_\varphi^k) \end{bmatrix} \tag{10.81}$$

The above discrete time domain matrix formula can be further written as the following formula.

$$\begin{bmatrix} \bar{\bar{\mathbf{Z}}}_b^{s00}+\frac{\bar{\bar{\mathbf{M}}}_b^{s00}}{\Delta t} & \bar{\bar{\mathbf{Z}}}_b^{s01}+\frac{\bar{\bar{\mathbf{M}}}_b^{s01}}{\Delta t} \\ \bar{\bar{\mathbf{Z}}}_b^{s10}+\frac{\bar{\bar{\mathbf{M}}}_b^{s10}}{\Delta t} & \bar{\bar{\mathbf{Z}}}_b^{s11}+\bar{\bar{\mathbf{Z}}}_s^{s11}+\frac{\bar{\bar{\mathbf{M}}}_b^{s11}}{\Delta t} \end{bmatrix} \begin{bmatrix} \bar{\mathbf{I}}_b^0(t_k) \\ \bar{\mathbf{I}}_b^1(t_k) \end{bmatrix}$$

$$= \frac{1}{\Delta t} \begin{bmatrix} \bar{\bar{\mathbf{M}}}_b^{s00} & \bar{\bar{\mathbf{M}}}_b^{s01} \\ \bar{\bar{\mathbf{M}}}_b^{s10} & \bar{\bar{\mathbf{M}}}_b^{s11} \end{bmatrix} \begin{bmatrix} \bar{\mathbf{I}}_b^0(t_{k-1}) \\ \bar{\mathbf{I}}_b^1(t_{k-1}) \end{bmatrix} - \begin{bmatrix} \bar{\bar{\mathbf{Z}}}_b^{q00} & \bar{\bar{\mathbf{Z}}}_b^{q01} \\ \bar{\bar{\mathbf{Z}}}_b^{q10} & \bar{\bar{\mathbf{Z}}}_b^{q11} \end{bmatrix} \begin{bmatrix} \bar{\mathbf{I}}_b^0(t_k-t_c^k) \\ \bar{\mathbf{I}}_b^1(t_k-t_c^k) \end{bmatrix}$$

$$- \begin{bmatrix} \bar{\bar{\mathbf{M}}}_b^{q00} & \bar{\bar{\mathbf{M}}}_b^{q01} \\ \bar{\bar{\mathbf{M}}}_b^{q10} & \bar{\bar{\mathbf{M}}}_b^{q11} \end{bmatrix} \begin{bmatrix} \frac{\bar{\mathbf{I}}_b^0(t_k-t_A^k)-\bar{\mathbf{I}}_b^0(t_{k-1}-t_A^k)}{\Delta t} \\ \frac{\bar{\mathbf{I}}_b^1(t_k-t_A^k)-\bar{\mathbf{I}}_b^1(t_{k-1}-t_A^k)}{\Delta t} \end{bmatrix} - \begin{bmatrix} 0 & 0 \\ 0 & \bar{\bar{\mathbf{Z}}}_s^{q11} \end{bmatrix} \begin{bmatrix} \bar{\mathbf{I}}_b^0(t_k-t_\varphi^k) \\ \bar{\mathbf{I}}_b^1(t_k-t_\varphi^k) \end{bmatrix} \tag{10.82}$$

The excitation is incorporated into the formulation through the boundary condition:

$$I_{b_1}^0(t_k) = I_g(t_k) \tag{10.83}$$

where $I_g(t_k)$ denotes the lightning current generator and $I_{b_1}^0(t_k)$ current in the injection point at t_k moment.

At a junction consisting of two or more segments the continuity properties of the electric field must be satisfied, which is ensured by applying the Kirchhoff current law

$$\sum_{m=1}^{n} I_{b_m}^0(t_k) = 0 \tag{10.84}$$

where n is the total number of conductor segments connected to the node.

and continuity equation

$$[\frac{\partial I_{b_1}^0(t_k)}{\partial r_1}]_{at\ junction} = [\frac{\partial I_{b_2}^0(t_k)}{\partial r_2}]_{at\ junction} = \cdots = [\frac{\partial I_{b_n}^0(t_k)}{\partial r_n}]_{at\ junction} \tag{10.85}$$

The condition (10.85) ensures the discontinuities in charge per unit length to be ruled out in passing from one conductor to another across the junction.

At the conductor free ends, the total current vanishes, i.e. is forced to be zero.

Equation (10.82) represents a marching-on-in-time formulation. Accordingly, it can be quickly solved through iterations.

10.5 The Time Domain Spatial 1-d Vector Hybrid Methods Based on Different Basic Function (PI and PEEC) for Transient Grounding Problems

This chapter will discuss in detail how to scientifically simulate the lightning transient grounding problem of substation grounding grid based on time domain spatial 1-d vector hybrid algorithm based on various basis functions.

10.5.1 The Time Domain Spatial 1-d Vector Hybrid Methods Based on Different Basic Function (PI and PEEC) Obtained with Fourier Transform

The application of hybrid algorithms based on various basis functions in the steady-state grounding problem of substation grounding grids is the foundation for solving the lightning transient impulse grounding problem of substation grounding grids. Below, the relevant content is reviewed and summarized as follows.

The time domain spatial 1-d vector hybrid methods based on PI basic function

This chapter will review and summarize the relevant content of the frequency domain spatial 1-d vector hybrid algorithm based on PI basis functions in Chap. 6, as shown below.

For the spatial frequency domain 1-d vector hybrid algorithm based on PI basis function, there are two solutions to the impedance matrix: (1) considering the impedance matrix of the mutual influence between leakage currents on two conductor segments, and (2) considering the impedance matrix of the mutual influence between leakage currents on two conductor segments connected by two nodes. It can be abbreviated as the branch matrix approach and the node matrix approach, respectively. The relevant concatenation formulas can be given separately as follows.

- The branch matrix approach for the spatial frequency domain 1-d vector hybrid algorithm based on PI basis function

$$[\bar{\bar{\mathbf{Z}}}_b + j\omega\bar{\bar{\mathbf{M}}}_b][\bar{\mathbf{I}}_b] + [\bar{\bar{\mathbf{A}}}]^T[\bar{\mathbf{V}}_p] = 0 \tag{10.86}$$

$$-[\bar{\bar{\mathbf{A}}}][\bar{\mathbf{I}}_b] + [\bar{\bar{\mathbf{K}}}]^T[\bar{\bar{\mathbf{Z}}}_{s_b}]^{-1}[\bar{\bar{\mathbf{K}}}][\bar{\mathbf{V}}_p] = [\bar{\mathbf{F}}] \tag{10.87}$$

  For the convenience of explaining various numerical algorithms in the future, these two consecutive is named frequency domain PI 1 U case.
- The node matrix approach for the spatial frequency domain 1-d vector hybrid algorithm based on PI basis function

$$[\bar{\bar{\mathbf{Z}}}_b + j\omega\bar{\bar{\mathbf{M}}}_b][\bar{\mathbf{I}}_b] + [\bar{\bar{\mathbf{A}}}]^T[\bar{\mathbf{V}}_p] = 0 \tag{10.88}$$

$$-[\bar{\bar{\mathbf{A}}}][\bar{\mathbf{I}}_b] + [\bar{\bar{\mathbf{Z}}}_{s_p}]^{-1}[\bar{\mathbf{V}}_p] = [\bar{\mathbf{F}}] \tag{10.89}$$

  For the convenience of explaining various numerical algorithms in the future, these two consecutive is named frequency domain PI 2 U case.

In order to reduce the unknown quantity and thus reduce the computational complexity, the two consecutive formulas mentioned above can be further simplified as shown below.

$$\left[[\bar{\bar{\mathbf{Z}}}_b + j\omega\bar{\bar{\mathbf{M}}}_b] + [\bar{\bar{\mathbf{A}}}]^T[\bar{\bar{\mathbf{Z}}}_{s_p}][\bar{\bar{\mathbf{A}}}]\right][\bar{\mathbf{I}}_b] = -[\bar{\bar{\mathbf{A}}}]^T[\bar{\bar{\mathbf{Z}}}_{s_p}][\bar{\mathbf{F}}] \tag{10.90}$$

For the convenience of explaining various numerical algorithms in the future, this formula is named frequency domain PI 2 I case.

It must be pointed out here that although the two consecutive equations of the branch matrix method can further simplify the unknown quantity to obtain two single matrix equations, and although these two single matrix equations can also simulate and calculate the transient and steady-state grounding problems of the substation grounding grid in the frequency domain, it is because the mutual inductance impedance matrix needs to be inverted twice after these two single equations are transformed into a single matrix equation in the time domain by Fourier transform, Therefore, the calculation results are completely incorrect and cannot be simulated correctly, so it will not be mentioned in this chapter. At the same time, although the two consecutive equations of the node matrix method can also be simplified into a single matrix equation of the unknown node voltage, the equation also needs two times of inverse calculation of mutual inductance matrix, which leads to errors and cannot realize correct simulation calculation. It is no longer mentioned in this chapter. Therefore, only three effective equations were mentioned above.

The above three matrix formulas are all assembled from the following three formulas as shown in the following equation

- The branch matrix approach

$$[\sum_{\theta=1}^{N_b} Z_{l_\theta} l_\theta + j\omega \sum_{\theta=1}^{N_b}\sum_{\vartheta=1}^{N_b}\int_{l_\theta}\int_{l_\vartheta} \hat{r}_\theta \cdot \bar{\bar{G}}_{\xi\zeta}^{Af}(\bar{r}_\theta, \bar{r}_\vartheta, \omega)\cdot \hat{r}_\vartheta^T dr_\vartheta dr_\theta][\bar{\mathbf{I}}_b] + [\bar{\bar{\mathbf{A}}}]^T[\bar{\mathbf{V}}_p] = 0 \tag{10.91}$$

$$-[\bar{\bar{\mathbf{A}}}][\bar{\mathbf{I}}_b] + [\bar{\bar{\mathbf{K}}}]^T[\sum_{\theta=1}^{N_b}\sum_{\vartheta=1}^{N_b}\int_{l_\theta}\int_{l_\vartheta} G_{\xi\zeta}^{\varphi f}(\bar{r}_\theta, \bar{r}_\vartheta, \omega)\frac{1}{l_\theta}\frac{1}{l_\vartheta} dr_\vartheta dr_\theta]^{-1}[\bar{\bar{\mathbf{K}}}][\bar{\mathbf{V}}_p] = [\bar{\mathbf{F}}] \tag{10.92}$$

- The node matrix approach

$$[\sum_{\theta=1}^{N_b} Z_{l_\theta} l_\theta + j\omega \sum_{\theta=1}^{N_b}\sum_{\vartheta=1}^{N_b}\int_{l_\theta}\int_{l_\vartheta} \hat{r}_\theta \cdot \bar{\bar{G}}_{\xi\zeta}^{Af}(\bar{r}_\theta, \bar{r}_\vartheta, \omega)\cdot \hat{r}_\vartheta^T dr_\vartheta dr_\theta][\bar{\mathbf{I}}_b] + [\bar{\bar{\mathbf{A}}}]^T[\bar{\mathbf{V}}_p] = 0 \tag{10.93}$$

$$-[\bar{\bar{\mathbf{A}}}][\bar{\mathbf{I}}_b]+[\frac{1}{N_\theta^\varsigma}\sum_{\theta=1}^{N_\theta^\varsigma}\frac{1}{l_\theta^\varsigma}\int_{l_\theta^\varsigma}\frac{1}{N_\vartheta^\upsilon}\sum_{\vartheta=1}^{N_\vartheta^\upsilon}\frac{1}{l_\vartheta^\upsilon}\int_{l_\vartheta^\upsilon}G_{\xi\zeta}^{\varphi f}(\bar{r}_\varsigma,\bar{r}_\upsilon,\omega)dr_\varsigma dr_\upsilon]^{-1}[\bar{\mathbf{V}}_p]=[\bar{\mathbf{F}}] \tag{10.94}$$

where ς, υ are the node numbers of discrete conductor segments in the grounding grid. N_θ^ς and N_ϑ^υ are the total number of conductor segments connecting nodes ς and υ. l_θ^ς and l_ϑ^υ are the lengths of the θ-th and ϑ-th conductor segments connecting nodes ς and υ, respectively.

$$[[\sum_{\theta=1}^{N_b}Z_{l_\theta}l_\theta+j\omega\sum_{\theta=1}^{N_b}\sum_{\vartheta=1}^{N_b}\int_{l_\theta}\int_{l_\vartheta}\hat{r}_\theta\cdot\bar{\bar{G}}_{\xi\zeta}^{A^f}(\bar{r}_\theta,\bar{r}_\vartheta,\omega)\cdot\hat{r}_\vartheta^T dr_\vartheta dr_\theta]$$

$$+[\bar{\bar{\mathbf{A}}}]^T[\frac{1}{N_\theta^\varsigma}\sum_{\theta=1}^{N_\theta^\varsigma}\frac{1}{l_\theta^\varsigma}\int_{l_\theta^\varsigma}\frac{1}{N_\vartheta^\upsilon}\sum_{\vartheta=1}^{N_\vartheta^\upsilon}\frac{1}{l_\vartheta^\upsilon}\int_{l_\vartheta^\upsilon}G_{\xi\zeta}^{\varphi f}(\bar{r}_\varsigma,\bar{r}_\upsilon,\omega)dr_\varsigma dr_\upsilon][\bar{\bar{\mathbf{A}}}]][\bar{\mathbf{I}}_b]$$

$$=-[\bar{\bar{\mathbf{A}}}]^T[\frac{1}{N_\theta^\varsigma}\sum_{\theta=1}^{N_\theta^\varsigma}\frac{1}{l_\theta^\varsigma}\int_{l_\theta^\varsigma}\frac{1}{N_\vartheta^\upsilon}\sum_{\vartheta=1}^{N_\vartheta^\upsilon}\frac{1}{l_\vartheta^\upsilon}\int_{l_\vartheta^\upsilon}G_{\xi\zeta}^{\varphi f}(\bar{r}_\varsigma,\bar{r}_\upsilon,\omega)dr_\varsigma dr_\upsilon][\bar{\mathbf{F}}] \tag{10.95}$$

Refer to formulas (10.3) , (10.4) and (10.5), and substitute these three formulas into formulas (10.91), (10.92), (10.93), (10.94), and (10.95), taking into account the Fourier transform, as follows

$$\frac{1}{2\pi}\int_{-\infty}^{\infty}[[\sum_{\theta=1}^{N_b}(Z_{l_0}+Z_l(\omega))l_\theta$$

$$+j\omega\sum_{\theta=1}^{N_b}\sum_{\vartheta=1}^{N_b}\int_{l_\theta}\int_{l_\vartheta}\hat{r}_\theta\cdot\left(\bar{\bar{G}}_{\xi\zeta}^{A_s^f}(\bar{r}_\theta,\bar{r}_\vartheta)+\bar{\bar{G}}_{\xi\zeta}^{A_q^f}(\bar{r}_\theta,\bar{r}_\vartheta,\omega)\right)\cdot\hat{r}_\vartheta^T dr_\vartheta dr_\theta][\bar{\mathbf{I}}_b]$$

$$+[\bar{\bar{\mathbf{A}}}]^T[\bar{\mathbf{V}}_p]]e^{j\omega t}d\omega=0 \tag{10.96}$$

$$\frac{1}{2\pi}\int_{-\infty}^{\infty}[-[\bar{\bar{\mathbf{A}}}][\bar{\mathbf{I}}_b]$$

$$+[\bar{\bar{\mathbf{K}}}]^T[\sum_{\theta=1}^{N_b}\sum_{\vartheta=1}^{N_b}\int_{l_\theta}\int_{l_\vartheta}\left(G_{\xi\zeta}^{\varphi_s^f}(\bar{r}_\theta,\bar{r}_\vartheta)+G_{\xi\zeta}^{A_q^f}(\bar{r}_\theta,\bar{r}_\vartheta,\omega)\right)\frac{1}{l_\theta}\frac{1}{l_\vartheta}dr_\vartheta dr_\theta]^{-1}[\bar{\bar{\mathbf{K}}}][\bar{\mathbf{V}}_p]]e^{j\omega t}d\omega$$

$$=\frac{1}{2\pi}\int_{-\infty}^{\infty}[\bar{\mathbf{F}}]e^{j\omega t}d\omega \tag{10.97}$$

$$\frac{1}{2\pi}\int_{-\infty}^{\infty}[[\sum_{\theta=1}^{N_b}(Z_{l_0}+Z_l(\omega))l_\theta$$

$$+j\omega\sum_{\theta=1}^{N_b}\sum_{\vartheta=1}^{N_b}\int_{l_\theta}\int_{l_\vartheta}\hat{r}_\theta\cdot\left(\bar{\bar{G}}_{\xi\zeta}^{A_s^f}(\bar{r}_\theta,\bar{r}_\vartheta)+\bar{\bar{G}}_{\xi\zeta}^{A_q^f}(\bar{r}_\theta,\bar{r}_\vartheta,\omega)\right)\cdot\hat{r}_\vartheta^T dr_\vartheta dr_\theta][\bar{\mathbf{I}}_b]$$

$$+[\bar{\bar{\mathbf{A}}}]^T[\bar{\mathbf{V}}_p]]e^{j\omega t}d\omega=0 \tag{10.98}$$

$$\frac{1}{2\pi}\int_{-\infty}^{\infty}[-[\bar{\bar{\mathbf{A}}}][\bar{\mathbf{I}}_b]$$

$$+[\frac{1}{N_\theta^\varsigma}\sum_{\theta=1}^{N_\theta^\varsigma}\frac{1}{l_\theta^\varsigma}\int_{l_\theta^\varsigma}\frac{1}{N_\vartheta^\upsilon}\sum_{\vartheta=1}^{N_\vartheta^\upsilon}\frac{1}{l_\vartheta^\upsilon}\int_{l_\vartheta^\upsilon}\left(G_{\xi\zeta}^{\varphi_s^f}(\bar{r}_\theta,\bar{r}_\vartheta)+G_{\xi\zeta}^{\varphi_q^f}(\bar{r}_\theta,\bar{r}_\vartheta,\omega)\right)dr_\varsigma dr_\upsilon]^{-1}[\bar{\mathbf{V}}_p]]e^{j\omega t}d\omega$$

$$=\frac{1}{2\pi}\int_{-\infty}^{\infty}[\bar{\mathbf{F}}]e^{j\omega t}d\omega \tag{10.99}$$

$$\frac{1}{2\pi}\int_{-\infty}^{\infty}[[[\sum_{\theta=1}^{N_b}(Z_{l_0}+Z_l(\omega))l_\theta$$

$$+j\omega\sum_{\theta=1}^{N_b}\sum_{\vartheta=1}^{N_b}\int_{l_\theta}\int_{l_\vartheta}\hat{r}_\theta\cdot\left(\bar{\bar{G}}_{\xi\zeta}^{A_s^f}(\bar{r}_\theta,\bar{r}_\vartheta)+\bar{\bar{G}}_{\xi\zeta}^{A_q^f}(\bar{r}_\theta,\bar{r}_\vartheta,\omega)\right)\cdot\hat{r}_\vartheta^T dr_\vartheta dr_\theta]$$

$$+[\bar{\bar{\mathbf{A}}}]^T[\frac{1}{N_\theta^\varsigma}\sum_{\theta=1}^{N_\theta^\varsigma}\frac{1}{l_\theta^\varsigma}\int_{l_\theta^\varsigma}\frac{1}{N_\vartheta^\upsilon}\sum_{\vartheta=1}^{N_\vartheta^\upsilon}\frac{1}{l_\vartheta^\upsilon}\int_{l_\vartheta^\upsilon}\left(G_{\xi\zeta}^{\varphi_s^f}(\bar{r}_\theta,\bar{r}_\vartheta)+G_{\xi\zeta}^{\varphi_q^f}(\bar{r}_\theta,\bar{r}_\vartheta,\omega)\right)dr_\varsigma dr_\upsilon][\bar{\bar{\mathbf{A}}}]][\bar{\mathbf{I}}_b]]e^{j\omega t}d\omega$$

$$=-\frac{1}{2\pi}\int_{-\infty}^{\infty}[\bar{\bar{\mathbf{A}}}]^T[\frac{1}{N_\theta^\varsigma}\sum_{\theta=1}^{N_\theta^\varsigma}\frac{1}{l_\theta^\varsigma}\int_{l_\theta^\varsigma}\frac{1}{N_\vartheta^\upsilon}\sum_{\vartheta=1}^{N_\vartheta^\upsilon}\frac{1}{l_\vartheta^\upsilon}\int_{l_\vartheta^\upsilon}\left(G_{\xi\zeta}^{\varphi_s^f}(\bar{r}_\theta,\bar{r}_\vartheta)+G_{\xi\zeta}^{\varphi_q^f}(\bar{r}_\theta,\bar{r}_\vartheta,\omega)\right)dr_\varsigma dr_\upsilon][\bar{\mathbf{F}}]e^{j\omega t}d\omega \tag{10.100}$$

By using the Fourier transform, the following three formulas can be obtained in the time domain.

- The branch matrix approach

$$[\bar{\bar{\mathbf{Z}}}_b^s][\bar{\mathbf{I}}_b(t)]+[\bar{\bar{\mathbf{M}}}_b^s]\frac{d[\bar{\mathbf{I}}_b(t)]}{dt}+[\bar{\bar{\mathbf{A}}}]^T[\bar{\mathbf{V}}_p(t)]=-[\bar{\bar{\mathbf{Z}}}_b^q][\bar{\mathbf{I}}_b(t-t_c)]-[\bar{\bar{\mathbf{M}}}_b^q]\frac{d[\bar{\mathbf{I}}_b(t-t_A)]}{dt} \tag{10.101}$$

$$-[\bar{\bar{\mathbf{A}}}][\bar{\mathbf{I}}_b(t)]+[\bar{\bar{\mathbf{K}}}]^T[\bar{\bar{\mathbf{Z}}}_{s_b}^s]^{-1}[\bar{\bar{\mathbf{K}}}][\bar{\mathbf{V}}_p(t)]=[\bar{\mathbf{F}}(t)]-[\bar{\bar{\mathbf{K}}}]^T[\bar{\bar{\mathbf{Z}}}_{s_b}^s]^{-1}[\bar{\bar{\mathbf{Z}}}_{s_b}^q][\bar{\mathbf{I}}_{s_b}(t-t_\varphi)] \tag{10.102}$$

For the convenience of explaining various numerical algorithms in the future, these two consecutive is named time domain PI 1 U case.

- The node matrix approach

$$[\bar{\bar{\mathbf{Z}}}_b^s][\bar{\mathbf{I}}_b(t)]+[\bar{\bar{\mathbf{M}}}_b^s]\frac{d[\bar{\mathbf{I}}_b(t)]}{dt}+[\bar{\bar{\mathbf{A}}}]^T[\bar{\mathbf{V}}_p]=-[\bar{\bar{\mathbf{Z}}}_b^q][\bar{\mathbf{I}}_b(t-t_c)]-[\bar{\bar{\mathbf{M}}}_b^q]\frac{d[\bar{\mathbf{I}}_b(t-t_A)]}{dt} \tag{10.103}$$

$$[\bar{\bar{\mathbf{A}}}][\bar{\mathbf{I}}_b(t)]+[\bar{\bar{\mathbf{Z}}}_{s_p}^s]^{-1}[\bar{\mathbf{V}}_p(t)]=[\bar{\mathbf{F}}(t)]-[\bar{\bar{\mathbf{Z}}}_{s_p}^s]^{-1}[\bar{\bar{\mathbf{Z}}}_{s_p}^q][\bar{\mathbf{I}}_{s_p}(t-t_\varphi)] \tag{10.104}$$

For the convenience of explaining various numerical algorithms in the future, these two consecutive is named time domain PI 2 U case.

$$[\bar{\bar{\mathbf{Z}}}_b^s][\bar{\mathbf{I}}_b(t)]+[\bar{\bar{\mathbf{M}}}_b^s]\frac{d[\bar{\mathbf{I}}_b(t)]}{dt}+[\bar{\bar{\mathbf{A}}}]^T[\bar{\bar{\mathbf{Z}}}_{s_p}^s][\bar{\bar{\mathbf{A}}}][\bar{\mathbf{I}}_b(t)]=-[\bar{\bar{\mathbf{Z}}}_b^q][\bar{\mathbf{I}}_b(t-t_c)]$$

$$-[\bar{\bar{\mathbf{M}}}_b^q]\frac{d[\bar{\mathbf{I}}_b(t-t_A)]}{dt}-[\bar{\bar{\mathbf{A}}}]^T[\bar{\bar{\mathbf{Z}}}_{s_p}^s][\bar{\mathbf{F}}(t)]+[\bar{\bar{\mathbf{A}}}]^T[\bar{\bar{\mathbf{Z}}}_{s_p}^q][\bar{\mathbf{I}}_{s_p}(t-t_\varphi)] \tag{10.105}$$

For the convenience of explaining various numerical algorithms in the future, this formula is named time domain PI 2 I case.

The time domain spatial 1-d vector hybrid methods based on PEEC basic function

This chapter will review and summarize the relevant content of the frequency domain spatial 1-d vector hybrid algorithm based on PEEC basis functions in Chap. 6, as shown below.

For the frequency domain spatial 1-d vector hybrid algorithm based on PEEC basis function, there are also two solutions to the impedance matrix: (1) considering the impedance matrix of the mutual influence between leakage currents on two conductor segments, and (2) considering the impedance matrix of the mutual influence between leakage currents on two conductor segments connected by two nodes. It can also be abbreviated as the branch matrix approach and the node matrix approach, respectively. The relevant concatenation formulas can be given separately as follows.

- The branch matrix approach for the frequency domain spatial 1-d vector hybrid algorithm based on PEEC basis function

$$[\bar{\bar{\mathbf{Z}}}_b + j\omega\bar{\bar{\mathbf{M}}}_b][\bar{\mathbf{I}}_b] + [\bar{\bar{\mathbf{A}}}]^T[\bar{\mathbf{V}}_p] = 0 \tag{10.106}$$

$$-[\bar{\bar{\mathbf{A}}}][\bar{\mathbf{I}}_b] + [\bar{\bar{\mathbf{K}}}]^T[\bar{\bar{\mathbf{Z}}}_{s_b}]^{-1}[\bar{\bar{\mathbf{K}}}][\bar{\mathbf{V}}_p] = [\bar{\mathbf{F}}] \tag{10.107}$$

where the connotation of these matrices or vectors ($[\bar{\bar{\mathbf{Z}}}_b]$, $[\bar{\bar{\mathbf{M}}}_b]$, $[\bar{\bar{\mathbf{A}}}]$, $[\bar{\mathbf{V}}_p]$, $[\bar{\mathbf{F}}]$ and $[\bar{\mathbf{I}}_b]$) is the same as the corresponding matrices or vectors in the original formula (10.86) (10.87). Except for matrices $[\bar{\bar{\mathbf{Z}}}_{s_b}]$ and $[\bar{\bar{\mathbf{K}}}]$, where $[\bar{\bar{\mathbf{Z}}}_{s_b}]$ is a $2N_b \times 2N_b$ matrix, and $[\bar{\bar{\mathbf{K}}}]$ is a $2N_b \times N_p$ matrix.

For the convenience of explaining various numerical algorithms in the future, there two consecutive formula is named frequency domain PEEC 1 U case.

- The node matrix approach for the frequency domain spatial 1-d vector hybrid algorithm based on PEEC basis function

$$[\bar{\bar{\mathbf{Z}}}_b + j\omega\bar{\bar{\mathbf{M}}}_b][\bar{\mathbf{I}}_b] + [\bar{\bar{\mathbf{A}}}]^T[\bar{\mathbf{V}}_p] = 0 \tag{10.108}$$

$$-[\bar{\bar{\mathbf{A}}}][\bar{\mathbf{I}}_b] + [\bar{\bar{\mathbf{Z}}}_{s_p}]^{-1}[\bar{\mathbf{V}}_p] = [\bar{\mathbf{F}}] \tag{10.109}$$

where all matrices and vectors here have exactly the same form and content as the corresponding matrices and vectors of the original Eqs. (10.88) (10.89). However, when calculating the matrix $[\bar{\bar{\mathbf{Z}}}_{s_p}]$, due to the need to further subdivide the leakage current branch in the hybrid algorithm based on PEEC basis functions, compared with the hybrid algorithm based on PI basis functions, the length of the leakage current branch is half smaller.

For the convenience of explaining various numerical algorithms in the future, there two consecutive formula is named frequency domain PEEC 2 U case.

In order to reduce the unknown quantity and thus reduce the computational complexity, the two consecutive formulas mentioned above can also be further simplified as shown below.

$$\left[[\bar{\bar{\mathbf{Z}}}_b + j\omega\bar{\bar{\mathbf{M}}}_b] + [\bar{\bar{\mathbf{A}}}]^T[\bar{\bar{\mathbf{Z}}}_{s_p}][\bar{\bar{\mathbf{A}}}]\right][\bar{\mathbf{I}}_b] = -[\bar{\bar{\mathbf{A}}}]^T[\bar{\bar{\mathbf{Z}}}_{s_p}][\bar{\mathbf{F}}] \tag{10.110}$$

For the convenience of explaining various numerical algorithms in the future, this formula is named frequency domain PEEC 2 I case.

The entire process of evolving from a spatial 1-d hybrid algorithm based on PEEC basis functions in the frequency domain to a spatial 1-d hybrid algorithm based on PEEC basis functions in the time domain is exactly the same as the evolution process of the spatial 1-d hybrid algorithm based on PI basis functions introduced in the previous section, and will not be repeated in this section. The three relevant formulas for directly providing the spatial 1-d hybrid algorithm based on PEEC basis functions in the time domain are as follows.

- The branch matrix approach

$$[\bar{\bar{\mathbf{Z}}}_b^s][\bar{\mathbf{I}}_b(t)] + [\bar{\bar{\mathbf{M}}}_b^s]\frac{d[\bar{\mathbf{I}}_b(t)]}{dt} + [\bar{\bar{\mathbf{A}}}]^T[\bar{\mathbf{V}}_p] = -[\bar{\bar{\mathbf{Z}}}_b^q][\bar{\mathbf{I}}_b(t-t_c)] - [\bar{\bar{\mathbf{M}}}_b^q]\frac{d[\bar{\mathbf{I}}_b(t-t_A)]}{dt} \tag{10.111}$$

$$-[\bar{\bar{\mathbf{A}}}][\bar{\mathbf{I}}_b(t)] + [\bar{\bar{\mathbf{K}}}]^T[\bar{\bar{\mathbf{Z}}}_{s_b}^s]^{-1}[\bar{\bar{\mathbf{K}}}][\bar{\mathbf{V}}_p(t)] = [\bar{\mathbf{F}}(t)] - [\bar{\bar{\mathbf{K}}}]^T[\bar{\bar{\mathbf{Z}}}_{s_b}^s]^{-1}[\bar{\bar{\mathbf{Z}}}_{s_b}^q][\bar{\mathbf{I}}_{s_b}(t-t_\varphi)] \tag{10.112}$$

For the convenience of explaining various numerical algorithms in the future, these two consecutive is named time domain PEEC 1 U case.

- The node matrix approach

$$[\bar{\bar{\mathbf{Z}}}_b^s][\bar{\mathbf{I}}_b(t)] + [\bar{\bar{\mathbf{M}}}_b^s]\frac{d[\bar{\mathbf{I}}_b(t)]}{dt} + [\bar{\bar{\mathbf{A}}}]^T[\bar{\mathbf{V}}_p] = -[\bar{\bar{\mathbf{Z}}}_b^q][\bar{\mathbf{I}}_b(t-t_c)] - [\bar{\bar{\mathbf{M}}}_b^q]\frac{d[\bar{\mathbf{I}}_b(t-t_A)]}{dt} \tag{10.113}$$

$$-[\bar{\bar{\mathbf{A}}}][\bar{\mathbf{I}}_b(t)]+[\bar{\bar{\mathbf{Z}}}^s_{s_p}]^{-1}[\bar{\mathbf{V}}_p(t)]=[\bar{\mathbf{F}}(t)]-[\bar{\bar{\mathbf{Z}}}^s_{s_p}]^{-1}[\bar{\bar{\mathbf{Z}}}^q_{s_p}][\bar{\mathbf{I}}_{s_p}(t-t_\varphi)] \quad (10.114)$$

For the convenience of explaining various numerical algorithms in the future, these two consecutive is named time domain PEEC 2 U case.

$$[\bar{\bar{\mathbf{Z}}}^s_b][\bar{\mathbf{I}}_b(t)]+[\bar{\bar{\mathbf{M}}}^s_b(t)]\frac{d[\bar{\mathbf{I}}_b(t)]}{dt}+[\bar{\bar{\mathbf{A}}}]^T[\bar{\bar{\mathbf{Z}}}^s_{s_p}][\bar{\bar{\mathbf{A}}}][\bar{\mathbf{I}}_b(t)]=-[\bar{\bar{\mathbf{Z}}}^q_b][\bar{\mathbf{I}}_b(t-t_c)]$$

$$-[\bar{\bar{\mathbf{M}}}^q_b(t)]\frac{d[\bar{\mathbf{I}}_b(t-t_A)]}{dt}-[\bar{\bar{\mathbf{A}}}]^T[\bar{\bar{\mathbf{Z}}}^s_{s_p}][\bar{\mathbf{F}}(t)]+[\bar{\bar{\mathbf{A}}}]^T[\bar{\bar{\mathbf{Z}}}^q_{s_p}][\bar{\mathbf{I}}_{s_p}(t-t_\varphi)] \quad (10.115)$$

For the convenience of explaining various numerical algorithms in the future, this formula is named time domain PEEC 2 I case.

10.5.2 The Time Domain Spatial 1-d Vector Hybrid Methods Based on Different Basic Function (PI and PEEC) Based on Laplace Transform

The application of hybrid algorithms based on various basis functions in the steady-state grounding problem of substation grounding grids is the foundation for solving the lightning transient impulse grounding problem of substation grounding grids. Below, the relevant content is reviewed and summarized as follows.

The time domain spatial 1-d vector hybrid methods based on PI basic function

This chapter will review and summarize the relevant content of the frequency domain spatial 1-d vector hybrid algorithm based on PI basis functions in Chap. 5, as shown below.

For the spatial frequency domain 1-d vector hybrid algorithm based on PI basis function, there are two solutions to the impedance matrix: (1) considering the impedance matrix of the mutual influence between leakage currents on two conductor segments, and (2) considering the impedance matrix of the mutual influence between leakage currents on two conductor segments connected by two nodes. It can be abbreviated as the branch matrix approach and the node matrix approach, respectively. The relevant concatenation formulas can be given separately as follows.

- The branch matrix approach for the spatial frequency domain 1-d vector hybrid algorithm based on PI basis function

$$[\bar{\bar{\mathbf{Z}}}_b+s\bar{\bar{\mathbf{M}}}_b][\bar{\mathbf{I}}_b]+[\bar{\bar{\mathbf{A}}}]^T[\bar{\mathbf{V}}_p]=0 \quad (10.116)$$

$$-[\bar{\bar{\mathbf{A}}}][\bar{\mathbf{I}}_b] + [\bar{\bar{\mathbf{K}}}]^T[\bar{\bar{\mathbf{Z}}}_{s_b}]^{-1}[\bar{\bar{\mathbf{K}}}][\bar{\mathbf{V}}_p] = [\bar{\mathbf{F}}] \tag{10.117}$$

For the convenience of explaining various numerical algorithms in the future, these two consecutive is named frequency domain PI 1 case.

- The node matrix approach for the spatial frequency domain 1-d vector hybrid algorithm based on PI basis function

$$[\bar{\bar{\mathbf{Z}}}_b + s\bar{\bar{\mathbf{M}}}_b][\bar{\mathbf{I}}_b] + [\bar{\bar{\mathbf{A}}}]^T[\bar{\mathbf{V}}_p] = 0 \tag{10.118}$$

$$-[\bar{\bar{\mathbf{A}}}][\bar{\mathbf{I}}_b] + [\bar{\bar{\mathbf{Z}}}_{s_p}]^{-1}[\bar{\mathbf{V}}_p] = [\bar{\mathbf{F}}] \tag{10.119}$$

For the convenience of explaining various numerical algorithms in the future, these two consecutive is named frequency domain PI 2 case.
In order to reduce the unknown quantity and thus reduce the computational complexity, the two consecutive formulas mentioned above can be further simplified as shown below.

$$\left[[\bar{\bar{\mathbf{Z}}}_b + s\bar{\bar{\mathbf{M}}}_b] + [\bar{\bar{\mathbf{A}}}]^T[\bar{\bar{\mathbf{Z}}}_{s_p}][\bar{\bar{\mathbf{A}}}]\right][\bar{\mathbf{I}}_b] = -[\bar{\bar{\mathbf{A}}}]^T[\bar{\bar{\mathbf{Z}}}_{s_p}][\bar{\mathbf{F}}] \tag{10.120}$$

For the convenience of explaining various numerical algorithms in the future, this formula is named frequency domain PI 2 I case.

It must be pointed out here that although the two consecutive equations of the branch matrix method can further simplify the unknown quantity to obtain two single matrix equations, and although these two single matrix equations can also simulate and calculate the transient and steady-state grounding problems of the substation grounding grid in the frequency domain, it is because the mutual inductance impedance matrix needs to be inverted twice after these two single equations are transformed into a single matrix equation in the time domain by laplace transform, Therefore, the calculation results are completely incorrect and cannot be simulated correctly, so it will not be mentioned in this chapter. At the same time, although the two consecutive equations of the node matrix method can also be simplified into a single matrix equation of the unknown node voltage, the equation also needs two times of inverse calculation of mutual inductance matrix, which leads to errors and cannot realize correct simulation calculation. It is no longer mentioned in this chapter. Therefore, only three effective equations were mentioned above.

The above three matrix formulas are all assembled from the following three formulas as shown in the following equation

- The branch matrix approach

$$[\sum_{\theta=1}^{N_b} Z_{l_\theta} l_\theta + s\sum_{\theta=1}^{N_b}\sum_{\vartheta=1}^{N_b}\int_{l_\theta}\int_{l_\vartheta}\hat{r}_\theta\cdot\bar{\bar{G}}_{\xi\zeta}^{A^s}(\bar{r}_\theta,\bar{r}_\vartheta,s)\cdot\hat{r}_\vartheta^T dr_\vartheta dr_\theta][\bar{\mathbf{I}}_b]+[\bar{\bar{\mathbf{A}}}]^T[\bar{\mathbf{V}}_p]=0 \tag{10.121}$$

$$-[\bar{\bar{\mathbf{A}}}][\bar{\mathbf{I}}_b]+[\bar{\bar{\mathbf{K}}}]^T[\sum_{\theta=1}^{N_b}\sum_{\vartheta=1}^{N_b}\int_{l_\theta}\int_{l_\vartheta}G_{\xi\zeta}^{\varphi^s}(\bar{r}_\theta,\bar{r}_\vartheta,s)\frac{1}{l_\theta}\frac{1}{l_\vartheta}dr_\vartheta dr_\theta]^{-1}[\bar{\bar{\mathbf{K}}}][\bar{\mathbf{V}}_p]=[\bar{\mathbf{F}}] \tag{10.122}$$

- The node matrix approach

$$[\sum_{\theta=1}^{N_b} Z_{l_\theta} l_\theta + s\sum_{\theta=1}^{N_b}\sum_{\vartheta=1}^{N_b}\int_{l_\theta}\int_{l_\vartheta}\hat{r}_\theta\cdot\bar{\bar{G}}_{\xi\zeta}^{A^s}(\bar{r}_\theta,\bar{r}_\vartheta,s)\cdot\hat{r}_\vartheta^T dr_\vartheta dr_\theta][\bar{\mathbf{I}}_b]+[\bar{\bar{\mathbf{A}}}]^T[\bar{\mathbf{V}}_p]=0 \tag{10.123}$$

$$-[\bar{\bar{\mathbf{A}}}][\bar{\mathbf{I}}_b]+[\frac{1}{N_\theta^\varsigma}\sum_{\theta=1}^{N_\theta^\varsigma}\frac{1}{l_\theta^\varsigma}\int_{l_\theta^\varsigma}\frac{1}{N_\vartheta^\upsilon}\sum_{\vartheta=1}^{N_\vartheta^\upsilon}\frac{1}{l_\vartheta^\upsilon}\int_{l_\vartheta^\upsilon}G_{\xi\zeta}^{\varphi^s}(\bar{r}_\varsigma,\bar{r}_\upsilon,s)dr_\varsigma dr_\upsilon]^{-1}[\bar{\mathbf{V}}_p]=[\bar{\mathbf{F}}] \tag{10.124}$$

where ς, υ are the node numbers of discrete conductor segments in the grounding grid. N_θ^ς and N_ϑ^υ are the total number of conductor segments connecting nodes ς and υ. l_θ^ς and l_ϑ^υ are the lengths of the θ-th and ϑ-th conductor segments connecting nodes ς and υ, respectively.

$$[[\sum_{\theta=1}^{N_b} Z_{l_\theta} l_\theta + s\sum_{\theta=1}^{N_b}\sum_{\vartheta=1}^{N_b}\int_{l_\theta}\int_{l_\vartheta}\hat{r}_\theta\cdot\bar{\bar{G}}_{\xi\zeta}^{A^s}(\bar{r}_\theta,\bar{r}_\vartheta,s)\cdot\hat{r}_\vartheta^T dr_\vartheta dr_\theta]$$

$$+[\bar{\bar{\mathbf{A}}}]^T[\frac{1}{N_\theta^\varsigma}\sum_{\theta=1}^{N_\theta^\varsigma}\frac{1}{l_\theta^\varsigma}\int_{l_\theta^\varsigma}\frac{1}{N_\vartheta^\upsilon}\sum_{\vartheta=1}^{N_\vartheta^\upsilon}\frac{1}{l_\vartheta^\upsilon}\int_{l_\vartheta^\upsilon}G_{\xi\zeta}^{\varphi^s}(\bar{r}_\varsigma,\bar{r}_\upsilon,s)dr_\varsigma dr_\upsilon][\bar{\bar{\mathbf{A}}}]][\bar{\mathbf{I}}_b]$$

$$=-[\bar{\bar{\mathbf{A}}}]^T[\frac{1}{N_\theta^\varsigma}\sum_{\theta=1}^{N_\theta^\varsigma}\frac{1}{l_\theta^\varsigma}\int_{l_\theta^\varsigma}\frac{1}{N_\vartheta^\upsilon}\sum_{\vartheta=1}^{N_\vartheta^\upsilon}\frac{1}{l_\vartheta^\upsilon}\int_{l_\vartheta^\upsilon}G_{\xi\zeta}^{\varphi^s}(\bar{r}_\varsigma,\bar{r}_\upsilon,s)dr_\varsigma dr_\upsilon][\bar{\mathbf{F}}] \tag{10.125}$$

Refer to formulas (10.11) , (10.12) and (10.13), and substitute these three formulas into formulas (10.120), (10.122), (10.123), (10.124), and (10.125), taking into account the laplace transform, as follows

$$\frac{1}{2\pi j}\int_{\sigma-\infty}^{\sigma+\infty}[[\sum_{\theta=1}^{N_b}(Z_{l_0}+Z_l(s))l_\theta$$

$$+s\sum_{\theta=1}^{N_b}\sum_{\vartheta=1}^{N_b}\int_{l_\theta}\int_{l_\vartheta}\hat{r}_\theta\cdot\left(\bar{\bar{G}}_{\xi\zeta}^{A_s^s}(\bar{r}_\theta,\bar{r}_\vartheta)+\bar{\bar{G}}_{\xi\zeta}^{A_q^s}(\bar{r}_\theta,\bar{r}_\vartheta,s)\right)\cdot\hat{r}_\vartheta^T dr_\vartheta dr_\theta][\bar{\mathbf{I}}_b]+[\bar{\bar{\mathbf{A}}}]^T[\bar{\mathbf{V}}_p]]e^{st}ds=0 \tag{10.126}$$

$$\frac{1}{2\pi j}\int_{\sigma-\infty}^{\sigma+\infty}[-[\bar{\bar{\mathbf{A}}}][\bar{\mathbf{I}}_b]$$

$$+[\bar{\bar{\mathbf{K}}}]^T[\sum_{\theta=1}^{N_b}\sum_{\vartheta=1}^{N_b}\int_{l_\theta}\int_{l_\vartheta}\left(G_{\xi\zeta}^{\varphi_s^s}(\bar{r}_\theta,\bar{r}_\vartheta)+G_{\xi\zeta}^{\varphi_q^s}(\bar{r}_\theta,\bar{r}_\vartheta,s)\right)\frac{1}{l_\theta}\frac{1}{l_\vartheta}dr_\vartheta dr_\theta]^{-1}[\bar{\bar{\mathbf{K}}}][\bar{\mathbf{V}}_p]]e^{st}ds$$

$$=\frac{1}{2\pi j}\int_{\sigma-\infty}^{\sigma+\infty}[\bar{\mathbf{F}}]e^{st}ds \tag{10.127}$$

$$\frac{1}{2\pi j}\int_{\sigma-\infty}^{\sigma+\infty}[[\sum_{\theta=1}^{N_b}(Z_{l_0}+Z_l(s))l_\theta$$

$$+s\sum_{\theta=1}^{N_b}\sum_{\vartheta=1}^{N_b}\int_{l_\theta}\int_{l_\vartheta}\hat{r}_\theta\cdot\left(\bar{\bar{G}}_{\xi\zeta}^{A_s^s}(\bar{r}_\theta,\bar{r}_\vartheta)+\bar{\bar{G}}_{\xi\zeta}^{A_q^s}(\bar{r}_\theta,\bar{r}_\vartheta,s)\right)\cdot\hat{r}_\vartheta^T dr_\vartheta dr_\theta][\bar{\mathbf{I}}_b]+[\bar{\bar{\mathbf{A}}}]^T[\bar{\mathbf{V}}_p]]e^{st}ds=0 \tag{10.128}$$

$$\frac{1}{2\pi j}\int_{\sigma-\infty}^{\sigma+\infty}[-[\bar{\bar{\mathbf{A}}}][\bar{\mathbf{I}}_b]$$

$$+[\frac{1}{N_\theta^\varsigma}\sum_{\theta=1}^{N_\theta^\varsigma}\frac{1}{l_\theta^\varsigma}\int_{l_\theta^\varsigma}\frac{1}{N_\vartheta^\upsilon}\sum_{\vartheta=1}^{N_\vartheta^\upsilon}\frac{1}{l_\vartheta^\upsilon}\int_{l_\vartheta^\upsilon}\left(G_{\xi\zeta}^{\varphi_s^s}(\bar{r}_\theta,\bar{r}_\vartheta)+G_{\xi\zeta}^{\varphi_q^s}(\bar{r}_\theta,\bar{r}_\vartheta,s)\right)dr_\varsigma dr_\upsilon]^{-1}[\bar{\mathbf{V}}_p]]e^{st}ds$$

$$=\frac{1}{2\pi}\int_{\sigma-\infty}^{\sigma+\infty}[\bar{\mathbf{F}}]e^{st}ds \tag{10.129}$$

$$\frac{1}{2\pi j}\int_{\sigma-\infty}^{\sigma+\infty}[[[\sum_{\theta=1}^{N_b}(Z_{l_0}+Z_l(s))l_\theta$$

$$+s\sum_{\theta=1}^{N_b}\sum_{\vartheta=1}^{N_b}\int_{l_\theta}\int_{l_\vartheta}\hat{r}_\theta\cdot\left(\bar{\bar{G}}_{\xi\zeta}^{A_s^s}(\bar{r}_\theta,\bar{r}_\vartheta)+\bar{\bar{G}}_{\xi\zeta}^{A_q^s}(\bar{r}_\theta,\bar{r}_\vartheta,s)\right)\cdot\hat{r}_\vartheta^T dr_\vartheta dr_\theta]$$

$$+[\bar{\bar{\mathbf{A}}}]^T[\frac{1}{N_\theta^\varsigma}\sum_{\theta=1}^{N_\theta^\varsigma}\frac{1}{l_\theta^\varsigma}\int_{l_\theta^\varsigma}\frac{1}{N_\vartheta^\upsilon}\sum_{\vartheta=1}^{N_\vartheta^\upsilon}\frac{1}{l_\vartheta^\upsilon}\int_{l_\vartheta^\upsilon}\left(G_{\xi\zeta}^{\varphi_s^s}(\bar{r}_\theta,\bar{r}_\vartheta)+G_{\xi\zeta}^{\varphi_q^s}(\bar{r}_\theta,\bar{r}_\vartheta,s)\right)dr_\varsigma dr_\upsilon][\bar{\bar{\mathbf{A}}}]][\bar{\mathbf{I}}_b]]e^{st}ds$$

$$=-\frac{1}{2\pi j}\int_{\sigma-\infty}^{\sigma+\infty}[\bar{\bar{\mathbf{A}}}]^T[\frac{1}{N_\theta^\varsigma}\sum_{\theta=1}^{N_\theta^\varsigma}\frac{1}{l_\theta^\varsigma}\int_{l_\theta^\varsigma}\frac{1}{N_\vartheta^\upsilon}\sum_{\vartheta=1}^{N_\vartheta^\upsilon}\frac{1}{l_\vartheta^\upsilon}\int_{l_\vartheta^\upsilon}\left(G_{\xi\zeta}^{\varphi_s^s}(\bar{r}_\theta,\bar{r}_\vartheta)+G_{\xi\zeta}^{\varphi_q^s}(\bar{r}_\theta,\bar{r}_\vartheta,s)\right)dr_\varsigma dr_\upsilon][\bar{\mathbf{F}}]e^{st}ds \tag{10.130}$$

By using the laplace transform, the following three formulas can be obtained in the time domain.

- The branch matrix approach

$$[\bar{\bar{\mathbf{Z}}}_b^s][\bar{\mathbf{I}}_b(t)]+[\bar{\bar{\mathbf{M}}}_b^s]\frac{d[\bar{\mathbf{I}}_b(t)]}{dt}+[\bar{\bar{\mathbf{A}}}]^T[\bar{\mathbf{V}}_p(t)]=-[\bar{\bar{\mathbf{Z}}}_b^q][\bar{\mathbf{I}}_b(t-t_c)]-[\bar{\bar{\mathbf{M}}}_b^q]\frac{d[\bar{\mathbf{I}}_b(t-t_A)]}{dt} \tag{10.131}$$

$$-[\bar{\bar{\mathbf{A}}}][\bar{\mathbf{I}}_b(t)]+[\bar{\bar{\mathbf{K}}}]^T[\bar{\bar{\mathbf{Z}}}_{s_b}^s]^{-1}[\bar{\bar{\mathbf{K}}}][\bar{\mathbf{V}}_p(t)]=[\bar{\mathbf{F}}(t)]-[\bar{\bar{\mathbf{K}}}]^T[\bar{\bar{\mathbf{Z}}}_{s_b}^s]^{-1}[\bar{\bar{\mathbf{Z}}}_{s_b}^q][\bar{\mathbf{I}}_{s_b}(t-t_\varphi)] \tag{10.132}$$

For the convenience of explaining various numerical algorithms in the future, these two consecutive is named time domain PI 1 case.

- The node matrix approach

$$[\bar{\bar{\mathbf{Z}}}_b^s][\bar{\mathbf{I}}_b(t)]+[\bar{\bar{\mathbf{M}}}_b^s]\frac{d[\bar{\mathbf{I}}_b(t)]}{dt}+[\bar{\bar{\mathbf{A}}}]^T[\bar{\mathbf{V}}_p]=-[\bar{\bar{\mathbf{Z}}}_b^q][\bar{\mathbf{I}}_b(t-t_c)]-[\bar{\bar{\mathbf{M}}}_b^q]\frac{d[\bar{\mathbf{I}}_b(t-t_A)]}{dt} \tag{10.133}$$

$$-[\bar{\bar{\mathbf{A}}}][\bar{\mathbf{I}}_b(t)]+[\bar{\bar{\mathbf{Z}}}_{s_p}^s]^{-1}[\bar{\mathbf{V}}_p(t)]=[\bar{\mathbf{F}}(t)]-[\bar{\bar{\mathbf{Z}}}_{s_p}^s]^{-1}[\bar{\bar{\mathbf{Z}}}_{s_p}^q][\bar{\mathbf{I}}_{s_p}(t-t_\varphi)] \tag{10.134}$$

For the convenience of explaining various numerical algorithms in the future, these two consecutive is named time domain PI 2 case.

$$[\bar{\bar{\mathbf{Z}}}_b^s][\bar{\mathbf{I}}_b(t)] + [\bar{\bar{\mathbf{M}}}_b^s(t)]\frac{d[\bar{\mathbf{I}}_b(t)]}{dt} + [\bar{\bar{\mathbf{A}}}]^T[\bar{\bar{\mathbf{Z}}}_{s_p}^s][\bar{\bar{\mathbf{A}}}][\bar{\mathbf{I}}_b(t)] = -[\bar{\bar{\mathbf{Z}}}_b^q][\bar{\mathbf{I}}_b(t-t_c)]$$

$$-[\bar{\bar{\mathbf{M}}}_b^q(t)]\frac{d[\bar{\mathbf{I}}_b(t-t_A)]}{dt} - [\bar{\bar{\mathbf{A}}}]^T[\bar{\bar{\mathbf{Z}}}_{s_p}][\bar{\mathbf{F}}(t)] - [\bar{\bar{\mathbf{A}}}]^T[\bar{\bar{\mathbf{Z}}}_{s_p}^q][\bar{\mathbf{I}}_{s_p}(t-t_\varphi)] \quad (10.135)$$

For the convenience of explaining various numerical algorithms in the future, this formula is named time domain PI 2 I case.

The time domain spatial 1-d vector hybrid methods based on PEEC basic function

This chapter will review and summarize the relevant content of the frequency domain spatial 1-d vector hybrid algorithm based on PEEC basis functions in Chap. 6, as shown below.

For the frequency domain spatial 1-d vector hybrid algorithm based on PEEC basis function, there are also two solutions to the impedance matrix: (1) considering the impedance matrix of the mutual influence between leakage currents on two conductor segments, and (2) considering the impedance matrix of the mutual influence between leakage currents on two conductor segments connected by two nodes. It can also be abbreviated as the branch matrix approach and the node matrix approach, respectively. The relevant concatenation formulas can be given separately as follows.

- The branch matrix approach for the frequency domain spatial 1-d vector hybrid algorithm based on PEEC basis function

$$[\bar{\bar{\mathbf{Z}}}_b + s\bar{\bar{\mathbf{M}}}_b][\bar{\mathbf{I}}_b] + [\bar{\bar{\mathbf{A}}}]^T[\bar{\mathbf{V}}_p] = 0 \quad (10.136)$$

$$-[\bar{\bar{\mathbf{A}}}][\bar{\mathbf{I}}_b] + [\bar{\bar{\mathbf{K}}}]^T[\bar{\bar{\mathbf{Z}}}_{s_b}]^{-1}[\bar{\bar{\mathbf{K}}}][\bar{\mathbf{V}}_p] = [\bar{\mathbf{F}}] \quad (10.137)$$

where the connotation of these matrices or vectors ($[\bar{\bar{\mathbf{Z}}}_b]$, $[\bar{\bar{\mathbf{M}}}_b]$, $[\bar{\bar{\mathbf{A}}}]$, $[\bar{\mathbf{V}}_p]$, $[\bar{\mathbf{F}}]$ and $[\bar{\mathbf{I}}_b]$) is the same as the corresponding matrices or vectors in the original formula (10.116) (10.117). Except for matrices $[\bar{\bar{\mathbf{Z}}}_{s_b}]$ and $[\bar{\bar{\mathbf{K}}}]$, where $[\bar{\bar{\mathbf{Z}}}_{s_b}]$ is a $2N_b \times 2N_b$ matrix, and $[\bar{\bar{\mathbf{K}}}]$ is a $2N_b \times N_p$ matrix.
For the convenience of explaining various numerical algorithms in the future, there two consecutive formula is named frequency domain PEEC 1 case.

- The node matrix approach for the frequency domain spatial 1-d vector hybrid algorithm based on PEEC basis function

$$[\bar{\bar{\mathbf{Z}}}_b + s\bar{\bar{\mathbf{M}}}_b][\bar{\mathbf{I}}_b] + [\bar{\bar{\mathbf{A}}}]^T[\bar{\mathbf{V}}_p] = 0 \quad (10.138)$$

$$-[\bar{\bar{\mathbf{A}}}][\bar{\mathbf{I}}_b]+[\bar{\bar{\mathbf{Z}}}_{s_p}]^{-1}[\bar{\mathbf{V}}_p]=[\bar{\mathbf{F}}] \tag{10.139}$$

where all matrices and vectors here have exactly the same form and content as the corresponding matrices and vectors of the original Eqs. (10.118) (10.119). However, when calculating the matrix $[\bar{\bar{\mathbf{Z}}}_{s_p}]$, due to the need to further subdivide the leakage current branch in the hybrid algorithm based on PEEC basis functions, compared with the hybrid algorithm based on PI basis functions, the length of the leakage current branch is half smaller.

For the convenience of explaining various numerical algorithms in the future, there two consecutive formula is named frequency domain PEEC 2 case.

In order to reduce the unknown quantity and thus reduce the computational complexity, the two consecutive formulas mentioned above can also be further simplified as shown below.

$$\left[[\bar{\bar{\mathbf{Z}}}_b+s\bar{\bar{\mathbf{M}}}_b]+[\bar{\bar{\mathbf{A}}}]^T[\bar{\bar{\mathbf{Z}}}_{s_p}][\bar{\bar{\mathbf{A}}}]\right][\bar{\mathbf{I}}_b]=-[\bar{\bar{\mathbf{A}}}]^T[\bar{\bar{\mathbf{Z}}}_{s_p}][\bar{\mathbf{F}}] \tag{10.140}$$

For the convenience of explaining various numerical algorithms in the future, this formula is named frequency domain PEEC 2 I case.

The entire process of evolving from a spatial 1-d hybrid algorithm based on PEEC basis functions in the frequency domain to a spatial 1-d hybrid algorithm based on PEEC basis functions in the time domain is exactly the same as the evolution process of the spatial 1-d hybrid algorithm based on PI basis functions introduced in the previous section, and will not be repeated in this section. The three relevant formulas for directly providing the spatial 1-d hybrid algorithm based on PEEC basis functions in the time domain are as follows.

- The branch matrix approach

$$[\bar{\bar{\mathbf{Z}}}_b^s][\bar{\mathbf{I}}_b(t)]+[\bar{\bar{\mathbf{M}}}_b^s]\frac{d[\bar{\mathbf{I}}_b(t)]}{dt}+[\bar{\bar{\mathbf{A}}}]^T[\bar{\mathbf{V}}_p]=-[\bar{\bar{\mathbf{Z}}}_b^q][\bar{\mathbf{I}}_b(t-t_c)]-[\bar{\bar{\mathbf{M}}}_b^q]\frac{d[\bar{\mathbf{I}}_b(t-t_A)]}{dt} \tag{10.141}$$

$$-[\bar{\bar{\mathbf{A}}}][\bar{\mathbf{I}}_b(t)]+[\bar{\bar{\mathbf{K}}}]^T[\bar{\bar{\mathbf{Z}}}_{s_b}^s]^{-1}[\bar{\bar{\mathbf{K}}}][\bar{\mathbf{V}}_p(t)]=[\bar{\mathbf{F}}(t)]-[\bar{\bar{\mathbf{K}}}]^T[\bar{\bar{\mathbf{Z}}}_{s_b}^s]^{-1}[\bar{\bar{\mathbf{Z}}}_{s_b}^q][\bar{\mathbf{I}}_{s_b}(t-t_\varphi)] \tag{10.142}$$

For the convenience of explaining various numerical algorithms in the future, these two consecutive is named time domain PEEC 1 case.

- The node matrix approach

$$[\bar{\bar{\mathbf{Z}}}_b^s][\bar{\mathbf{I}}_b(t)] + [\bar{\bar{\mathbf{M}}}_b^s]\frac{d[\bar{\mathbf{I}}_b(t)]}{dt} + [\bar{\bar{\mathbf{A}}}]^T[\bar{\mathbf{V}}_p] = -[\bar{\bar{\mathbf{Z}}}_b^q][\bar{\mathbf{I}}_b(t - t_c)] - [\bar{\bar{\mathbf{M}}}_b^q]\frac{d[\bar{\mathbf{I}}_b(t - t_A)]}{dt} \tag{10.143}$$

$$-[\bar{\bar{\mathbf{A}}}][\bar{\mathbf{I}}_b(t)] + [\bar{\bar{\mathbf{Z}}}_{s_p}^s]^{-1}[\bar{\mathbf{V}}_p(t)] = [\bar{\mathbf{F}}(t)] - [\bar{\bar{\mathbf{Z}}}_{s_p}^s]^{-1}[\bar{\bar{\mathbf{Z}}}_{s_p}^q][\bar{\mathbf{I}}_{s_p}(t - t_\varphi)] \tag{10.144}$$

For the convenience of explaining various numerical algorithms in the future, these two consecutive is named time domain PEEC 2 case.

$$[\bar{\bar{\mathbf{Z}}}_b^s][\bar{\mathbf{I}}_b(t)] + [\bar{\bar{\mathbf{M}}}_b^s(t)]\frac{d[\bar{\mathbf{I}}_b(t)]}{dt} + [\bar{\bar{\mathbf{A}}}]^T[\bar{\bar{\mathbf{Z}}}_{s_p}^s][\bar{\bar{\mathbf{A}}}][\bar{\mathbf{I}}_b(t)] = -[\bar{\bar{\mathbf{Z}}}_b^q][\bar{\mathbf{I}}_b(t - t_c)]$$
$$- [\bar{\bar{\mathbf{M}}}_b^q(t)]\frac{d[\bar{\mathbf{I}}_b(t - t_A)]}{dt} - [\bar{\bar{\mathbf{A}}}]^T[\bar{\bar{\mathbf{Z}}}_{s_p}][\bar{\mathbf{F}}(t)] - [\bar{\bar{\mathbf{A}}}]^T[\bar{\bar{\mathbf{Z}}}_{s_p}^q][\bar{\mathbf{I}}_{s_p}(t - t_\varphi)] \tag{10.145}$$

For the convenience of explaining various numerical algorithms in the future, this formula is named time domain PEEC 2 I case.

10.5.3 The Time Domain Spatial 1-d Vector Hybrid Methods

The vector of nodal voltage $[\bar{\mathbf{V}}_p(t)]$ and branch current $[\bar{\mathbf{I}}_b(t)]$ may be calculated through solving the Eqs. (10.101) and (10.102) for PI 1 U case, or the Eqs. (10.103) and (10.104) for PI 2 U case, the Eqs. (10.111) and (10.112) for PEEC 1 U case, or the Eqs. (10.113) and (10.114) for PEEC 2 U case.

The vector of branch current $[\bar{\mathbf{I}}_b(t)]$ may be calculated through solving the Eq. (10.105) for PI 2 I case, or the Eq. (10.115) for PEEC 2 I case.

The average voltage $[\bar{\mathbf{U}}_a(t)]$ and leakage currents $[\bar{\mathbf{I}}_s(t)]$, the branch voltage $[\bar{\mathbf{U}}_b](t)$, branch current $[\bar{\mathbf{I}}_b(t)]$ can also be calculated [29–31, 36, 62].

Once branch currents and leakage currents have known, the voltage, vector magnetic potential (VMP) A, electrical field intensity (EFI) and magnetic field intensity (MFI) at any point can be calculated [29–31, 36, 62].

The study of the grounding systems performance for time domain spatial 1-d vector hybrid method based on different basic functions (PI or PEEC) has been reduced to the computation of $[\overline{\overline{\mathbf{Z}}}_b^s]$, $[\overline{\overline{\mathbf{Z}}}_b^q]$, $[\overline{\overline{\mathbf{M}}}_b^s]$, $[\overline{\overline{\mathbf{M}}}_b^q]$, $[\overline{\overline{\mathbf{Z}}}_{s_b}^s]$, $[\overline{\overline{\mathbf{Z}}}_{s_b}^q]$, $[\overline{\overline{\mathbf{Z}}}_{s_p}^s]$ or $[\overline{\overline{\mathbf{Z}}}_{s_p}^q]$ matrices.

Computation of $[\overline{\overline{\mathbf{Z}}}_b^s]$, $[\overline{\overline{\mathbf{Z}}}_b^q]$, $[\overline{\overline{\mathbf{M}}}_b^s]$, $[\overline{\overline{\mathbf{M}}}_b^q]$, $[\overline{\overline{\mathbf{Z}}}_{s_b}^s]$, $[\overline{\overline{\mathbf{Z}}}_{s_b}^q]$, $[\overline{\overline{\mathbf{Z}}}_{s_p}^s]$ or $[\overline{\overline{\mathbf{Z}}}_{s_p}^q]$ matrices

1. Internal impedance matrix $[\bar{\bar{\mathbf{Z}}}_b^s]$
   $[\bar{\bar{\mathbf{Z}}}_b^s]$ is a diagonal matrix, whose elements are the product of the internal resistance and the length of the conductor segment, the formula for internal resistance can be found in [29–31, 36, 62].
2. Internal impedance matrix $[\bar{\bar{\mathbf{Z}}}_b^q]$
   $[\bar{\bar{\mathbf{Z}}}_b^q]$ is also a diagonal matrix, whose elements can be seen the formula below.

$$Z_\theta^q = \sum_{c=1}^{M_c} \alpha_c \tag{10.146}$$

3. Mutual induction matrix $[\bar{\bar{\mathbf{M}}}_b^s]$
   Mutual induction matrix $[\bar{\bar{\mathbf{M}}}_b^s]$ is matrix with zero diagonal, whose elements about mutual inductance coefficient between two thin conductors without delay effect, may be obtained with the formula as below

$$M_{\theta,\vartheta}^s = \int_{l_\theta}\int_{l_\vartheta} \hat{r}_\theta \cdot \bar{\bar{G}}_{\xi\zeta}^{A_s^t}(\bar{r}_\theta, \bar{r}_\vartheta) \cdot \hat{r}_\vartheta^T dr_\vartheta dr_\theta \tag{10.147}$$

For an half infinite homogeneous conductivity medium, one has

$$\bar{\bar{G}}_{11}^{A_s^t}(\bar{r}_\vartheta, \bar{r}_\theta) = \frac{\mu}{4\pi}\begin{bmatrix} \frac{1}{R_{\theta\vartheta}} & 0 & -\hat{k}_{01}^{TM}\frac{\partial\left(\ln\left(-(z_\theta+z_\vartheta)+R'_{\theta\vartheta}\right)\right)}{\partial x} \\ 0 & \frac{1}{R_{\theta\vartheta}} & -\hat{k}_{01}^{TM}\frac{\partial\left(\ln\left(-(z_\theta+z_\vartheta)+R'_{\theta\vartheta}\right)\right)}{\partial y} \\ 0 & 0 & \frac{1}{R_{\theta\vartheta}} - \frac{\hat{k}_{01}^{TM}}{R'_{\theta\vartheta}} \end{bmatrix}, \tag{10.148}$$

where $R_{\theta\vartheta} = |\bar{r}_\theta - \bar{r}_\vartheta|$, $R'_{\theta\vartheta} = |\bar{r}_\theta + \bar{r}_\vartheta|$, $\hat{k}_{01}^{TM} = \frac{\sigma_1-\sigma_0}{\sigma_1+\sigma_0}$.

$$M_{\theta,\vartheta}^s = \frac{\mu}{4\pi}\int_{l_\theta}\int_{l_\vartheta} \hat{r}_\theta \cdot \begin{bmatrix} \frac{1}{R_{\theta\vartheta}} & 0 & -\hat{k}_{01}^{TM}\frac{\partial\left(\ln\left(-(z_\theta+z_\vartheta)+R'_{\theta\vartheta}\right)\right)}{\partial x} \\ 0 & \frac{1}{R_{\theta\vartheta}} & -\hat{k}_{01}^{TM}\frac{\partial\left(\ln\left(-(z_\theta+z_\vartheta)+R'_{\theta\vartheta}\right)\right)}{\partial y} \\ 0 & 0 & \frac{1}{R_{\theta\vartheta}} - \frac{\hat{k}_{01}^{TM}}{R'_{\theta\vartheta}} \end{bmatrix} \cdot \hat{r}_\vartheta^T dr_\vartheta dr_\theta \tag{10.149}$$

4. Mutual induction matrix $[\bar{\bar{\mathbf{M}}}_b^q]$
   Mutual induction matrix $[\bar{\bar{\mathbf{M}}}_b^q]$ is matrix with zero diagonal, whose elements about mutual inductance coefficient between two thin conductors with delay effect, may be obtained with the formula as below

$$M^q_{\theta,\vartheta} = \int_{l_\theta}\int_{l_\vartheta} \hat{r}_\theta \cdot \bar{\bar{G}}^{A^t_q}_{\xi\zeta}(\bar{r}_\theta, \bar{r}_\vartheta) \cdot \hat{r}^T_\vartheta dr_\theta dr_\vartheta \tag{10.150}$$

For an half infinite homogeneous conductivity medium, one has

$$\bar{\bar{G}}^{A^t_q}_{11}(\bar{r}_\vartheta, \bar{r}_\theta) = \frac{\mu}{4\pi}\begin{bmatrix} 0\ 0 - \sum_{q_A=1}^{M_A} \alpha_{q_A} \frac{\partial\left(\ln\left(-(z_\theta+z_\vartheta)+R'_{\theta\vartheta}\right)\right)}{\partial x} \\ 0\ 0 - \sum_{q_A=1}^{M_A} \alpha_{q_A} \frac{\partial\left(\ln\left(-(z_\theta+z_\vartheta)+R'_{\theta\vartheta}\right)\right)}{\partial y} \\ 0\ 0 \qquad - \sum_{q_A=1}^{M_A} \alpha_{q_A} \frac{1}{R'_{\theta\vartheta}} \end{bmatrix}. \tag{10.151}$$

$$M^q_{\theta,\vartheta} = \frac{\mu}{4\pi}\int_{l_\theta}\int_{l_\vartheta} \hat{r}_\theta \cdot \begin{bmatrix} 0\ 0 - \sum_{q_A=1}^{M_A} \alpha_{q_A} \frac{\partial\left(\ln\left(-(z_\theta+z_\vartheta)+R'_{\theta\vartheta}\right)\right)}{\partial x} \\ 0\ 0 - \sum_{q_A=1}^{M_A} \alpha_{q_A} \frac{\partial\left(\ln\left(-(z_\theta+z_\vartheta)+R'_{\theta\vartheta}\right)\right)}{\partial y} \\ 0\ 0 \qquad - \sum_{q_A=1}^{M_A} \alpha_{q_A} \frac{1}{R'_{\theta\vartheta}} \end{bmatrix} \cdot \hat{r}^T_\vartheta dr_\vartheta dr_\theta \tag{10.152}$$

5. Mutual impedance matrix $[\bar{\bar{\mathbf{Z}}}^s_{s_b}]$
   Mutual impedance matrix $[\bar{\bar{\mathbf{Z}}}^s_{s_b}]$, whose core is formula of mutual impedance interaction betweens a pars of currents along conductor segments without delay effects.

$$Z^s_{s_b} = \frac{1}{l_\theta}\int_{l_\theta}\frac{1}{l_\vartheta}\int_{l_\vartheta} G^{\varphi^t_s}_{\xi\zeta}(\bar{r}_\theta, \bar{r}_\vartheta) dr_\vartheta dr_\theta \tag{10.153}$$

For an infinite homogeneous conductivity medium, one has

$$G^{\varphi^t_s}_{11}(\bar{r}_\vartheta, \bar{r}_\theta) = \frac{1}{4\pi\sigma_1}\frac{1}{R_{\theta\vartheta}}, \tag{10.154}$$

so

$$Z^s_{s_b} = \frac{1}{4\pi\sigma_1 l_\theta l_\vartheta}\int_{l_\theta}\int_{l_\vartheta}\frac{1}{R_{\theta\vartheta}} dr_\theta dr_\vartheta \tag{10.155}$$

6. Mutual impedance matrix $[\bar{\bar{\mathbf{Z}}}^q_{s_b}]$
   Mutual impedance matrix $[\bar{\bar{\mathbf{Z}}}^q_{s_b}]$, whose core is formula of mutual impedance interaction betweens a pars of currents along conductor segments with delay effects.

$$Z_{s_b}^{q} = \frac{1}{l_\theta}\int_{l_\theta}\frac{1}{l_\vartheta}\int_{l_\vartheta} G_{\xi\varsigma}^{\varphi_q^t}(\bar{r}_\theta, \bar{r}_\vartheta) dr_\vartheta dr_\theta \tag{10.156}$$

For an half infinite homogeneous conductivity medium, one has

$$G_{11}^{\varphi_q^t}(\bar{r}_\vartheta, \bar{r}_\theta) = \frac{1}{4\pi\sigma_1}\left(\sum_{q_0=1}^{M_0}\alpha_{q_0}\frac{1}{R_{\theta\vartheta}} - \sum_{q_\varphi=1}^{M_\varphi}\alpha_{q_\varphi}\frac{1}{R'_{\theta\vartheta}}\right), \tag{10.157}$$

so

$$Z_{s_b}^{q} = \frac{1}{4\pi\sigma_1 l_\theta l_\vartheta}\int_{l_\theta}\int_{l_\vartheta}\left[\sum_{q_0=1}^{M_0}\alpha_{q_0}\frac{1}{R_{\theta\vartheta}} - \sum_{q_\varphi=1}^{M_\varphi}\alpha_{q_\varphi}\frac{1}{R'_{\theta\vartheta}}\right] dr_\theta dr_\vartheta \tag{10.158}$$

7. Mutual impedance matrix $[\bar{\bar{\mathbf{Z}}}_{s_p}^{s}]$

   Mutual impedance matrix $[\bar{\bar{\mathbf{Z}}}_{s_p}^{s}]$, The core of this matrix element formula is the interaction between leakage currents on two sets of conductor segments connected to two nodes without delay effects.

$$Z_{s_p}^{s} = \frac{1}{N_\theta^\varsigma}\sum_{\theta=1}^{N_\theta^\varsigma}\frac{1}{l_\theta^\varsigma}\int_{l_\theta^\varsigma}\frac{1}{N_\vartheta^\upsilon}\sum_{\vartheta=1}^{N_\vartheta^\upsilon}\frac{1}{l_\vartheta^\upsilon}\int_{l_\vartheta^\upsilon} G_{\xi\varsigma}^{\varphi_s^t}(\bar{r}_\varsigma, \bar{r}_\upsilon) dr_\varsigma dr_\upsilon \tag{10.159}$$

where ς, υ are the node numbers of discrete conductor segments in the grounding grid. N_θ^ς and N_ϑ^υ are the total number of conductor segments connecting nodes ς and υ. l_θ^ς and l_ϑ^υ are the lengths of the θ-th and ϑ-th conductor segments connecting nodes ς and υ, respectively.

For an infinite homogeneous conductivity medium, one has

$$G_{11}^{\varphi_s^t}(\bar{r}_\varsigma, \bar{r}_\upsilon) = \frac{1}{4\pi\sigma_1}\frac{1}{R_{\varsigma\upsilon}}, \tag{10.160}$$

$R_{\varsigma\upsilon} = |\bar{r}_\varsigma - \bar{r}_\upsilon|$, so

$$Z_{s_p}^{s} = \frac{1}{4\pi\sigma_1}\frac{1}{N_\theta^\varsigma}\sum_{\theta=1}^{N_\theta^\varsigma}\frac{1}{l_\theta^\varsigma}\int_{l_\theta^\varsigma}\frac{1}{N_\vartheta^\upsilon}\sum_{\vartheta=1}^{N_\vartheta^\upsilon}\frac{1}{l_\vartheta^\upsilon}\int_{l_\vartheta^\upsilon}\frac{1}{R_{\varsigma\upsilon}} dr_\varsigma dr_\upsilon \tag{10.161}$$

8. Mutual impedance matrix $[\bar{\bar{\mathbf{Z}}}^q_{s_p}]$

   Mutual impedance matrix $[\bar{\bar{\mathbf{Z}}}^q_{s_p}]$, The core of this matrix element formula is the interaction between leakage currents on two sets of conductor segments connected to two nodes with delay effects.

$$Z^q_{s_p} = \frac{1}{N^\varsigma_\theta} \sum_{\theta=1}^{N^\varsigma_\theta} \frac{1}{l^\varsigma_\theta} \int_{l^\varsigma_\theta} \frac{1}{N^\upsilon_\vartheta} \sum_{\vartheta=1}^{N^\upsilon_\vartheta} \frac{1}{l^\upsilon_\vartheta} \int_{l^\upsilon_\vartheta} G^{\varphi^t_q}_{\xi\zeta}(\bar{r}_\varsigma, \bar{r}_\upsilon) dr_\varsigma dr_\upsilon \tag{10.162}$$

For an half infinite homogeneous conductivity medium, one has

$$G^{\varphi^t_q}_{11}(\bar{r}_\varsigma, \bar{r}_\upsilon, t) = \frac{1}{4\pi\sigma_1} \left[\sum_{q_0}^{M_0} \alpha_{q_0} \frac{1}{R_{\varsigma\upsilon}} - \sum_{q_\varphi}^{M_\varphi} \alpha_{q_\varphi} \frac{1}{R'_{\varsigma\upsilon}} \right], \tag{10.163}$$

$R_{\varsigma\upsilon} = |\bar{r}_\varsigma - \bar{r}_\upsilon|$, $R\varsigma\upsilon = |\bar{r}_\varsigma + \bar{r}_\upsilon|$, so

$$Z^s_{s_p} = \frac{1}{4\pi\sigma_1} \frac{1}{N^\varsigma_\theta} \sum_{\theta=1}^{N^\varsigma_\theta} \frac{1}{l^\varsigma_\theta} \int_{l^\varsigma_\theta} \frac{1}{N^\upsilon_\vartheta} \sum_{\vartheta=1}^{N^\upsilon_\vartheta} \frac{1}{l^\upsilon_\vartheta} \int_{l^\upsilon_\vartheta} \left[\sum_{q_0}^{M_0} \alpha_{q_0} \frac{1}{R_{\varsigma\upsilon}} - \sum_{q_\varphi}^{M_\varphi} \alpha_{q_\varphi} \frac{1}{R'_{\varsigma\upsilon}} \right] dr_\varsigma dr_\upsilon \tag{10.164}$$

Note the medium surrounding the point current here was considered as half infinite homogeneous medium. However, in any practical case, the earth is represented as a multilayer earth model , at least as two layer media (one is air, another one is earth). The QSCIM can be used to dealt with the multilayered earth model, this has been fully discussed in Chap. 3.

10.5.4 Time Domain Spatial 1-d Vector Hybrid Methods Based on Different Basic Function (PI, PEEC) for Transient Grounding Problems

By observing formulas (10.101)–(10.105) and (10.111)–(10.115), it can be seen that the three effective matrix formulas obtained from the time-domain spatial 1-d vector hybrid algorithm based on two different basis functions (PI and PEEC) discussed in this section are identical in form, except for individual matrix calculation methods. Therefore, this section will discuss the three matrix formula solving methods in a unified manner, without distinguishing between different basis functions.

The matrix equations of the time domain spatial 1-d vector hybrid method can be unified given as follows, referring to the previous section.

- The branch matrix approach

$$[\bar{\bar{\mathbf{Z}}}_b^s][\bar{\mathbf{I}}_b(t)] + [\bar{\bar{\mathbf{M}}}_b^s]\frac{d[\bar{\mathbf{I}}_b(t)]}{dt} + [\bar{\bar{\mathbf{A}}}]^T[\bar{\mathbf{V}}_p(t)] = -[\bar{\bar{\mathbf{Z}}}_b^q][\bar{\mathbf{I}}_b(t-t_c)] - [\bar{\bar{\mathbf{M}}}_b^q]\frac{d[\bar{\mathbf{I}}_b(t-t_A)]}{dt} \tag{10.165}$$

$$-[\bar{\bar{\mathbf{A}}}][\bar{\mathbf{I}}_b(t)] + [\bar{\bar{\mathbf{K}}}]^T[\bar{\bar{\mathbf{Z}}}_{s_b}^s]^{-1}[\bar{\bar{\mathbf{K}}}][\bar{\mathbf{V}}_p(t)] = [\bar{\mathbf{F}}(t)] - [\bar{\bar{\mathbf{K}}}]^T[\bar{\bar{\mathbf{Z}}}_{s_b}^s]^{-1}[\bar{\bar{\mathbf{Z}}}_{s_b}^q][\bar{\mathbf{I}}_{s_b}(t-t_\varphi)] \tag{10.166}$$

- The node matrix approach

$$[\bar{\bar{\mathbf{Z}}}_b^s][\bar{\mathbf{I}}_b(t)] + [\bar{\bar{\mathbf{M}}}_b^s]\frac{d[\bar{\mathbf{I}}_b(t)]}{dt} + [\bar{\bar{\mathbf{A}}}]^T[\bar{\mathbf{V}}_p(t)] = -[\bar{\bar{\mathbf{Z}}}_b^q][\bar{\mathbf{I}}_b(t-t_c)] - [\bar{\bar{\mathbf{M}}}_b^q]\frac{d[\bar{\mathbf{I}}_b(t-t_A)]}{dt} \tag{10.167}$$

$$-[\bar{\bar{\mathbf{A}}}][\bar{\mathbf{I}}_b(t)] + [\bar{\bar{\mathbf{Z}}}_{s_p}^s]^{-1}[\bar{\mathbf{V}}_p(t)] = [\bar{\mathbf{F}}(t)] - [\bar{\bar{\mathbf{Z}}}_{s_p}^s]^{-1}[\bar{\bar{\mathbf{Z}}}_{s_p}^q][\bar{\mathbf{I}}_{s_p}(t-t_\varphi)] \tag{10.168}$$

Least unknown equation

$$[\bar{\bar{\mathbf{Z}}}_b^s][\bar{\mathbf{I}}_b(t)] + [\bar{\bar{\mathbf{M}}}_b^s(t)]\frac{d[\bar{\mathbf{I}}_b(t)]}{dt} + [\bar{\bar{\mathbf{A}}}]^T[\bar{\bar{\mathbf{Z}}}_{s_p}^s][\bar{\bar{\mathbf{A}}}][\bar{\mathbf{I}}_b(t)] = -[\bar{\bar{\mathbf{Z}}}_b^q][\bar{\mathbf{I}}_b(t-t_c)]$$

$$-[\bar{\bar{\mathbf{M}}}_b^q]\frac{d[\bar{\mathbf{I}}_b(t-t_A)]}{dt} - [\bar{\bar{\mathbf{A}}}]^T[\bar{\bar{\mathbf{Z}}}_{s_p}^s][\bar{\mathbf{F}}(t)] + [\bar{\bar{\mathbf{A}}}]^T[\bar{\bar{\mathbf{Z}}}_{s_p}^q][\bar{\mathbf{I}}_{s_p}(t-t_\varphi)] \tag{10.169}$$

Perform time discretization on the three time-domain matrix equations mentioned above, with a time interval of Δt and $t_k = k\Delta t, k = 0, 1, \ldots, N_t$, N_t is the maximum number of discrete time intervals. Thus, there are three discrete time matrix equations.

- The branch matrix approach

$$[\bar{\bar{\mathbf{Z}}}_b^s][\bar{\mathbf{I}}_b(t_k)] + [\bar{\bar{\mathbf{M}}}_b^s]\frac{[\bar{\mathbf{I}}_b(t_k)] - [\bar{\mathbf{I}}_b(t_{k-1})]}{\Delta t} + [\bar{\bar{\mathbf{A}}}]^T[\bar{\mathbf{V}}_p(t_k)] = -[\bar{\bar{\mathbf{Z}}}_b^q][\bar{\mathbf{I}}_b(t_k - t_c^k)]$$

$$-[\bar{\bar{\mathbf{M}}}_b^q]\frac{[\bar{\mathbf{I}}_b(t_k - t_A^k)] - [\bar{\mathbf{I}}_b(t_{k-1} - t_A^k)]}{\Delta t} \tag{10.170}$$

$$-[\bar{\bar{\mathbf{A}}}][\bar{\mathbf{I}}_b(t_k)]+[\bar{\bar{\mathbf{K}}}]^T[\bar{\bar{\mathbf{Z}}}^s_{s_b}]^{-1}[\bar{\bar{\mathbf{K}}}][\bar{\mathbf{V}}_p(t_k)]=[\bar{\mathbf{F}}(t_k)]-[\bar{\bar{\mathbf{K}}}]^T[\bar{\bar{\mathbf{Z}}}^s_{s_b}]^{-1}[\bar{\bar{\mathbf{Z}}}^q_{s_b}][\bar{\mathbf{I}}_{s_b}(t_k-t^k_{\varphi})] \tag{10.171}$$

- The node matrix approach

$$[\bar{\bar{\mathbf{Z}}}^s_b][\bar{\mathbf{I}}_b(t_k)]+[\bar{\bar{\mathbf{M}}}^s_b]\frac{[\bar{\mathbf{I}}_b(t_k)]-[\bar{\mathbf{I}}_b(t_{k-1})]}{\Delta t}+[\bar{\bar{\mathbf{A}}}]^T[\bar{\mathbf{V}}_p(t_k)]=-[\bar{\bar{\mathbf{Z}}}^q_b][\bar{\mathbf{I}}_b(t_k-t^k_c)]$$

$$-[\bar{\bar{\mathbf{M}}}^q_b]\frac{[\bar{\mathbf{I}}_b(t_k-t^k_A)]-[\bar{\mathbf{I}}_b(t_{k-1}-t^k_A)]}{\Delta t} \tag{10.172}$$

$$-[\bar{\bar{\mathbf{A}}}][\bar{\mathbf{I}}_b(t_k)]+[\bar{\bar{\mathbf{Z}}}^s_{s_p}]^{-1}[\bar{\mathbf{V}}_p(t_k)]=[\bar{\mathbf{F}}(t_k)]-[\bar{\bar{\mathbf{Z}}}^s_{s_p}]^{-1}[\bar{\bar{\mathbf{Z}}}^q_{s_p}][\bar{\mathbf{I}}_{s_p}(t_k-t^k_{\varphi})] \tag{10.173}$$

Least unknown equation

$$[\bar{\bar{\mathbf{Z}}}^s_b][\bar{\mathbf{I}}_b(t_k)]+[\bar{\bar{\mathbf{M}}}^s_b(t_k)]\frac{[\bar{\mathbf{I}}_b(t_k)]-[\bar{\mathbf{I}}_b(t_{k-1})]}{\Delta t}+[\bar{\bar{\mathbf{A}}}]^T[\bar{\bar{\mathbf{Z}}}^s_{s_p}][\bar{\bar{\mathbf{A}}}][\bar{\mathbf{I}}_b(t_k)]=-[\bar{\bar{\mathbf{Z}}}^q_b][\bar{\mathbf{I}}_b(t_k-t^k_c)]$$

$$-[\bar{\bar{\mathbf{M}}}^q_b]\frac{[\bar{\mathbf{I}}_b(t_k-t^k_A)]-[\bar{\mathbf{I}}_b(t_{k-1}-t^k_A)]}{\Delta t}-[\bar{\bar{\mathbf{A}}}]^T[\bar{\bar{\mathbf{Z}}}^s_{s_p}][\bar{\mathbf{F}}(t_k)]+[\bar{\bar{\mathbf{A}}}]^T[\bar{\bar{\mathbf{Z}}}^q_{s_p}][\bar{\mathbf{I}}_{s_p}(t_k-t^k_{\varphi})] \tag{10.174}$$

Finally, three discrete time matrix equations are obtained as follows.

- The branch matrix approach

$$\left[\bar{\bar{\mathbf{Z}}}^s_b+\frac{\bar{\bar{\mathbf{M}}}^s_b}{\Delta t}\right][\bar{\mathbf{I}}_b(t_k)]+[\bar{\bar{\mathbf{A}}}]^T[\bar{\mathbf{V}}_p(t_k)]=[\bar{\bar{\mathbf{M}}}^s_b]\frac{[\bar{\mathbf{I}}_b(t_{k-1})]}{\Delta t}-[\bar{\bar{\mathbf{Z}}}^q_b][\bar{\mathbf{I}}_b(t_k-t^k_c)]$$

$$-[\bar{\bar{\mathbf{M}}}^q_b]\frac{[\bar{\mathbf{I}}_b(t_k-t^k_A)]-[\bar{\mathbf{I}}_b(t_{k-1}-t^k_A)]}{\Delta t} \tag{10.175}$$

$$-[\bar{\bar{\mathbf{A}}}][\bar{\mathbf{I}}_b(t_k)]+[\bar{\bar{\mathbf{K}}}]^T[\bar{\bar{\mathbf{Z}}}^s_{s_b}]^{-1}[\bar{\bar{\mathbf{K}}}][\bar{\mathbf{V}}_p(t_k)]=[\bar{\mathbf{F}}(t_k)]-[\bar{\bar{\mathbf{K}}}]^T[\bar{\bar{\mathbf{Z}}}^s_{s_b}]^{-1}[\bar{\bar{\mathbf{Z}}}^q_{s_b}][\bar{\mathbf{I}}_{s_b}(t_k-t^k_{\varphi})] \tag{10.176}$$

- The node matrix approach

$$\left[\bar{\bar{\mathbf{Z}}}^s_b+\frac{\bar{\bar{\mathbf{M}}}^s_b}{\Delta t}\right][\bar{\mathbf{I}}_b(t_k)]+[\bar{\bar{\mathbf{A}}}]^T[\bar{\mathbf{V}}_p(t_k)]=[\bar{\bar{\mathbf{M}}}^s_b]\frac{[\bar{\mathbf{I}}_b(t_{k-1})]}{\Delta t}-[\bar{\bar{\mathbf{Z}}}^q_b][\bar{\mathbf{I}}_b(t_k-t^k_c)]$$

$$
-[\bar{\bar{\mathbf{M}}}_b^q]\frac{[\bar{\mathbf{I}}_b(t_k-t_A^k)]-[\bar{\mathbf{I}}_b(t_{k-1}-t_A^k)]}{\Delta t} \tag{10.177}
$$

$$
-[\bar{\bar{\mathbf{A}}}][\bar{\mathbf{I}}_b(t_k)]+[\bar{\bar{\mathbf{Z}}}_{s_p}^s]^{-1}[\bar{\mathbf{V}}_p(t_k)]=[\bar{\mathbf{F}}(t_k)]-[\bar{\bar{\mathbf{Z}}}_{s_p}^s]^{-1}[\bar{\bar{\mathbf{Z}}}_{s_p}^q][\bar{\mathbf{I}}_{s_p}(t_k-t_\varphi^k)] \tag{10.178}
$$

Least unknown equation

$$
\left[\bar{\bar{\mathbf{Z}}}_b^s+\frac{\bar{\bar{\mathbf{M}}}_b^s}{\Delta t}+[\bar{\bar{\mathbf{A}}}]^T[\bar{\bar{\mathbf{Z}}}_{s_p}^s][\bar{\bar{\mathbf{A}}}]\right][\bar{\mathbf{I}}_b(t_k)]=[\bar{\bar{\mathbf{M}}}_b^s]\frac{[\bar{\mathbf{I}}_b(t_{k-1})]}{\Delta t}-[\bar{\bar{\mathbf{Z}}}_b^q][\bar{\mathbf{I}}_b(t_k-t_c^k)]
$$

$$
-[\bar{\bar{\mathbf{M}}}_b^q]\frac{[\bar{\mathbf{I}}_b(t_k-t_A^k)]-[\bar{\mathbf{I}}_b(t_{k-1}-t_A^k)]}{\Delta t}-[\bar{\bar{\mathbf{A}}}]^T[\bar{\bar{\mathbf{Z}}}_{s_p}^s][\bar{\mathbf{F}}(t_k)]+[\bar{\bar{\mathbf{A}}}]^T[\bar{\bar{\mathbf{Z}}}_{s_p}^q][\bar{\mathbf{I}}_{s_p}(t_k-t_\varphi^k)] \tag{10.179}
$$

The above three discrete time domain matrix equations represent a marching-on-in-time formulation. Accordingly, they can be quickly solved through iterations.

10.6 Numerical Results and Discussions for the Various Algorithms

Based on the numerical algorithms of various unequal electromagnetic models introduced above, a transient impulse response program for the substation grounding grid has been developed, which can simulate the lightning transient impulse grounding problem of the substation grounding grid in a four layer horizontal layered earth.

10.6.1 Verification of the Various Algorithms

To verify the accuracy and effectiveness of the method proposed in this chapter, some cases solved by other authors are studied.

The first example is from [63]. The grounding grid configuration can be seen in Fig. 10.1a. The grounding grid was made of round copper conductors with a cross-sectional area of 50 mm^2 and was buried at 0.5 m. The injected lightning current parameters were set to $T_1 = 65\mu\text{s}$, $T_2 = 185\mu\text{s}$, and $I_m = 6.86$ A, which was generated by a double exponential function, and the feed point was set at the corner of the grid. For the earth model, the conductivities and permittivities of the two-layer earth model were set to $\frac{\sigma_1=50^{-1}S/m}{\sigma_2=20^{-1}S/m}$, $\frac{\varepsilon_1=10\varepsilon_0}{\varepsilon_2=10\varepsilon_0}$, and $h = 0.6$ m. Figure 10.1a, b shows transient voltage curves based on frequency domain and time domain 1-d spatial scalars and vectors Galerkin MM, vectors Galerkin BEM, vectors hybrid methods based on various basis functions (PI and PEEC), and simulated

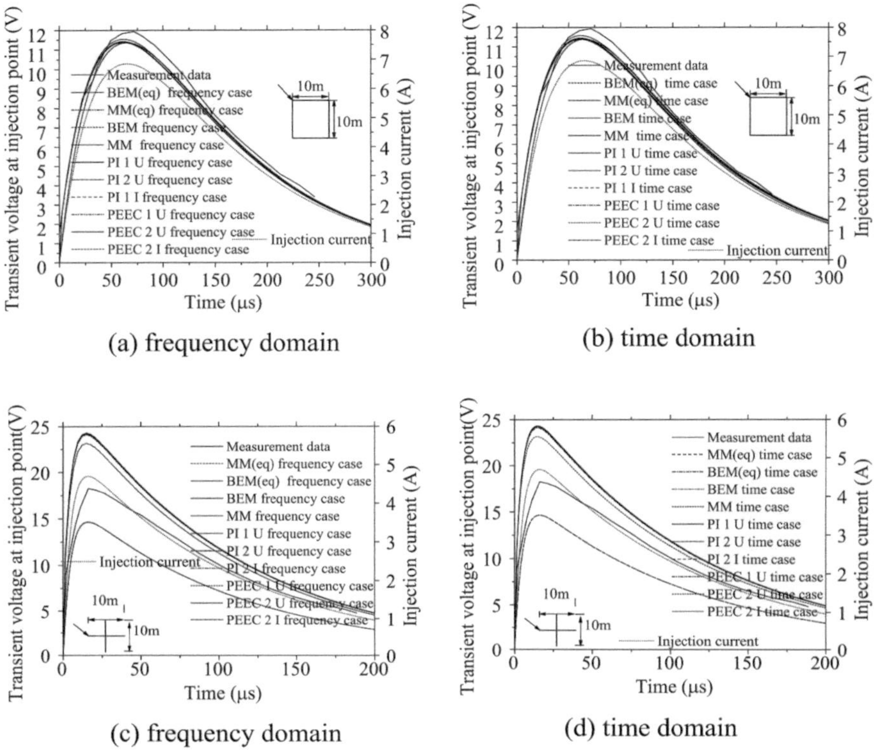

Fig. 10.1 Verification of the impulse response of four simple geometries of grounding grids or rods

measurement data [63], respectively. Among these figures, MM (eq) and BEM (eq) represent 1-d spatial scalars Galerkin MM and Galerkin BEM. The calculated voltage curves were in good agreement with the measured curve in [63]. It should be noted that in this example, the lightning wave is relatively gentle and the size of the grounding grid is relatively small. Therefore, the equipotential model algorithm developed in last chapter is relatively effective and accurate.

The second verification case came from [64]. The grounding grid configuration can be seen in Fig. 10.1c. The grounding grid was also made of round copper conductors with cross-sectional area 50 mm^2 and was buried at 0.5 m. The injected lightning current parameters were set to $T_1 = 16\mu s$, $T_2 = 99\mu s$, and $I_m = 4.7$ A, which was generated by a double exponential function. For the earth model, the conductivities and permittivities of the two-layer earth model were set to $\frac{\sigma_1=50^{-1}S/m}{\sigma_2=20^{-1}S/m}$, $\frac{\varepsilon_1=10\varepsilon_0}{\varepsilon_2=10\varepsilon_0}$, and $h = 0.6$ m. Figure 10.1c, d show the transient voltage curves simulated based on frequency domain and time domain 1-d spatial scalars and vectors Galerkin MM, BEM, vectors hybrid methods based on various basis functions (PI and PEEC), and measurement data [64], respectively. Among these figures, MM (eq) and BEM (eq) represent 1-d spatial scalars Galerkin MM and Galerkin BEM. The calculated voltage curve is very close to the development trend of the measured curve, but there are certain errors. The calculated voltage curve is very close to the development trend of the measured curve, but there are certain errors. This is because the lightning wave waveform in this example is slightly steeper compared to the previous example, and the size of the grounding grid is also relatively small.

Because the steepness of the lightning wave considered in the above tow examples were not high, the third example considered the shock of a lightning wave with high steepness. In the second verification case from [65], the length of horizontal conductor was 15 m, and the conductor was buried at 0.6 m. The soil parameters relating to conductivity and relativity permittivity were 70 Ωm and 15, respectively. The conductor radius was 12 mm. The injected lightning current parameters were set to $T_1 = 0.8\mu s$, $T_2 = 12.5\mu s$, and $I_m = 36.5A$, which was generated by a double exponential function, and the feed point was set at the end of the conductor. Figure 10.2a, b show the transient voltage curves simulated based on frequency domain and time domain 1-d spatial scalars and vectors Galerkin MM, vectors Galerkin BEM, vectors hybrid methods based on various basis functions (PI and PEEC), and measurement data [65], respectively. Among these figures, MM (eq) and BEM (eq) represent 1-d spatial scalars Galerkin MM and Galerkin BEM. From Fig. 10.2a, b, it can be seen that the simulated calculation curves are basically consistent with the measurement curves, while the equipotential model algorithm developed in last chapter yields curves that are completely different. Figure 10.2c, d further provides a longer comparison curve to further illustrate this issue. This example demonstrates that for lightning waves with relatively high steepness, despite the relatively small impact size of the grounding electrode, the calculation results of the equipotential model algorithm developed in this chapter are completely incorrect.

The fourth test case comes from [62, 66]. Three different grounding grid sizes were considered: 10 × 10m, 20 × 20 m, and 30 × 30 m, with $1 \times 1, 2 \times 2$, and 3×3 grids, respectively, as shown in Fig. 10.2e . All of the grids were buried at a depth of 0.5 m in homogeneous earth with resistivity = 1000Ωm and relative permittivity = 9. The injected lightning current was set to $T_1 = 0.36\mu s$, $T_2 = 26.8\mu s$, and $I_0 = 1A$, which was generated by a double exponential function, with the feed point, as shown in Fig. 10.2e. Figure 10.2e shows that the calculated transient voltage curves obtained by using the proposed method and frequency-domain Galerkin BEM [62]. The corresponding measured curves given in Fig. 10.2f by [66]. From Fig. 10.2e, it can be seen that for small-scale grounding grids, the transient voltage curve given in reference [62] is basically consistent with the corresponding curve obtained by the equipotential model algorithm developed in this chapter, and it is also basically consistent with the measurement curve. However, for the other two grounding grids with relatively large sizes, the transient voltage curve given in reference [62] is completely different from the corresponding curve obtained by the equipotential model algorithm developed in this chapter, and there is a significant difference. This is because the lightning wave steepness in this example is relatively large, which poses a great challenge to the correctness of the algorithm in this chapter.

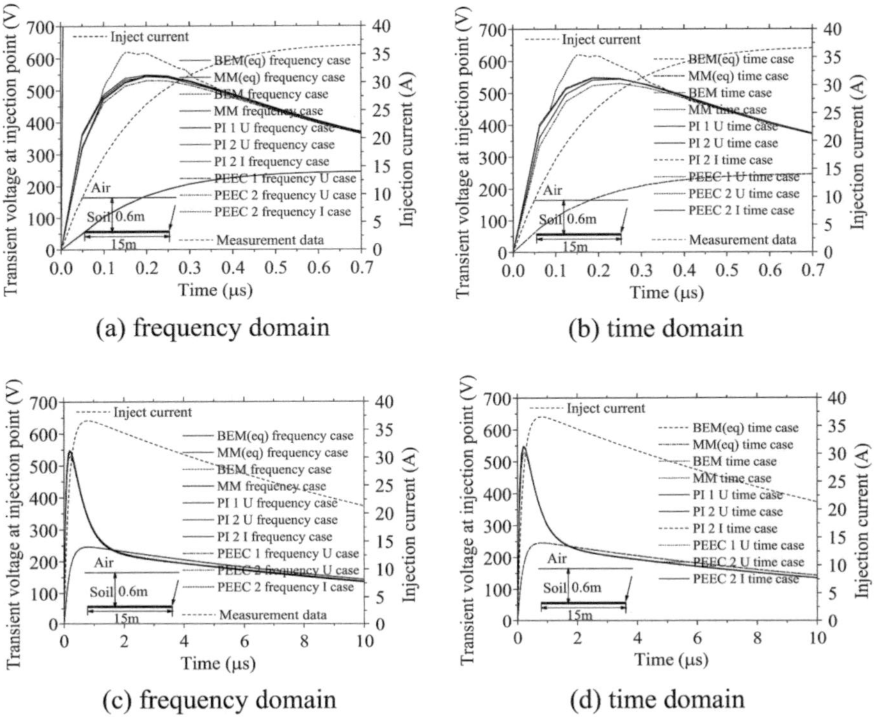

Fig. 10.2 Mutual validation of numerical algorithms in the time and frequency domains

The fifth test case comes from [67]. Figures 10.3 and 10.4a show, respectively, horizontal and vertical grounding electrodes with radius of 12 mm, and the length of horizontal and vertical grounding electrodes is 15 m and 4 m, respectively. The horizontal grounding electrode was buried at a depth of 0.6 m, and both of the electrodes were within in two-layer horizontal earth where the resistivity and relative permittivity ratios for the upper and the lower soil layers were $\frac{\rho_1=300\Omega m}{\rho_2=100\Omega m}$, $\frac{\varepsilon_{r_1}=12}{\varepsilon_{r_2}=14}$, and the upper layer thickness was $h = 2$ m. The injected lightning current was set with parameters of: (a) $T_1 = 0.36\mu$s, $T_2 = 26.8\mu$s, and $I_0 = 3.0$ kA, which was generated by a double exponential function. Figures 10.3 and 10.4a, b show the transient voltage curves of horizontal and vertical grounding rods under lightning current impact, respectively. These curves are simulated based on frequency domain and time domain 1-d spatial scalars and vectors Galerkin MM, vectors Galerkin BEM, vectors hybrid methods based on various basis functions PI and PEEC), and the equipotential model algorithm about scalars Galerkin MM (eq) and Galerkin BEM (eq) were developed in last chapter. Figures 10.3 and 10.4a, b , respectively, show the transient impulse response curves of lightning waves on the horizontal and vertical conductor sections simulated by the 1-d spatial vector Galerkin MM, vector Galerkin BEM, and vector Galerkin hybrid algorithm based on various basis functions (PI and

PEEC). The curves simulated by the transmission line method in the literature [67] are also presented. It can be seen that the curves simulated by the 1-d spatial vector Galerkin MM, vector Galerkin BEM, and hybrid algorithms based on various basis functions are basically consistent with the curves simulated by the transmission line method in the literature. Reference [67]. On the other hands, from Fig. 10.3a, b, it can be seen that for horizontal grounding rods, the curve obtained by the equipotential model algorithm developed in last chapter is significantly different from the curve given in reference [67], with significant errors. As shown in Fig. 10.4a, b, for vertical grounding rods, the curve obtained by the equipotential model algorithm developed in last chapter is basically consistent with the curve given in reference [67], with relatively small errors. This is because the lightning wave waveform in this calculation example is relatively steep, and the correctness of the equipotential model algorithm developed in this chapter is questionable for horizontal grounding rods, while for vertical grounding rods, the correctness of the equipotential model algorithm developed in this chapter is still acceptable. Figures 10.3 and 10.4c, d are enlarged time period images that further illustrate this phenomenon.

From the above four examples, it can be seen that the 1d spatial scalar Galerkin MM and Galerkin BEM developed in previous chapter have certain limitations and a relatively small scope of application for lightning transient impulse grounding problems in substation grounding grids.

In order to further validate the electromagnetic field numerical method in this chapter, the following verification examples will be provided: It should be noted here that: considering the correctness of the EMF numerical hybrid algorithm model based on multiple PI and PEEC basis functions, the following arithmetic example no longer verifies the hybrid algorithm of multiple PI and PEEC basis functions several times, and only the hybrid algorithm of a single PI and PEEC basis function is used to verify it, and it should be pointed out that the hybrid algorithm based on the PI and PEEC basis functions which appears in the following refers to the hybrid algorithm based on PI 2 I and PEEC 2 I basis functions hybrid algorithms.

The sixth verification example : the lightning impulse tests on square grounding grids in two horizontally stratified layers of soil are described in the literature [64] . The source of excitation for all impact tests was a lightning current waveform formed by a double exponential function. A total of 19 tests with transient voltage curves are given in [64]. However, most of the transient voltage curves calculated by the numerical method developed in this book under the same test data simulation do not match with each other, which may be due to the measurement of the inverse performance layered soil model parameters are not accurate enough results. Four test trials were selected here where the transient voltage profiles of the test measurements were in general agreement with the numerical simulation profiles.

In the literature [64], a test grounding grid are given, the size of the grounding grid is $10m \times 10m$. The test grounding grid was of four equal sizes, i.e., $5m \times 5m$ small grids that make up the grounding grid, the grounding grid conductor cross sectional area was 50 square millimeters. The soil was considered as a horizontally layered 2-layer soil model with an upper soil resistivity of 50Ωm and a thickness of 0.6 m and a lower soil resistivity of 20 Ωm. Both layers of soil have a relative dielectric

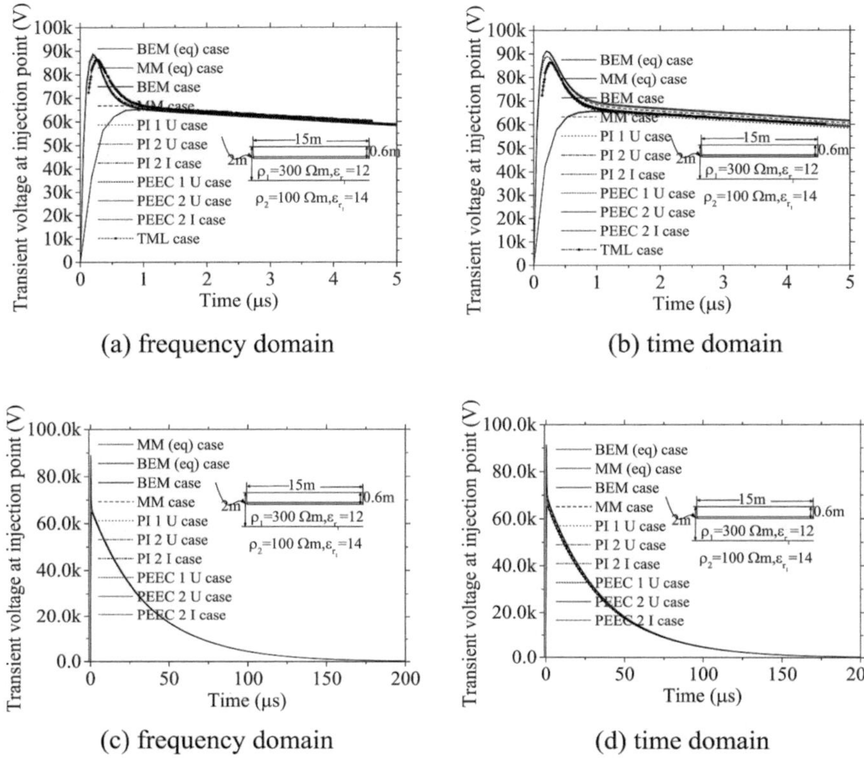

Fig. 10.3 Mutual validation of time-domain and frequency-domain numerical algorithms for horizontal grounding rods in multilayered horizontally stratified soils

Table 10.1 Four choices of experimental biexponential function lightning current parameters

	T_1/T_2 (appointed)	T_1/T_2 (computed)	$\alpha(\mu s^{-1})$	$\beta(\mu s^{-1})$	η	$I_o(A)$
Case 1	7.4/76.0μs	7.0/76.25μs	0.0116947	0.2572883	0.823896	10.4
Case 2	10.5/82.0μs	9.93/82.37μs	0.0118826	0.165206	0.756818	9.7
Case 3	10.0/81.0μs	9.44/81.35μs	0.011854	0.175987	0.767529	9.7

constant of 10. The four selected parameters of the double exponential function of the lightning current for the test can be found in Table 10.1.

The response transient voltage curves at the injection point under three double-exponential-function lightning current shocks numerically simulated by the eight numerical methods developed in this paper, as well as the experimental test transient voltage curves, are presented in the three subfigures (a), (b) and (c) of Fig. 10.5, respectively.

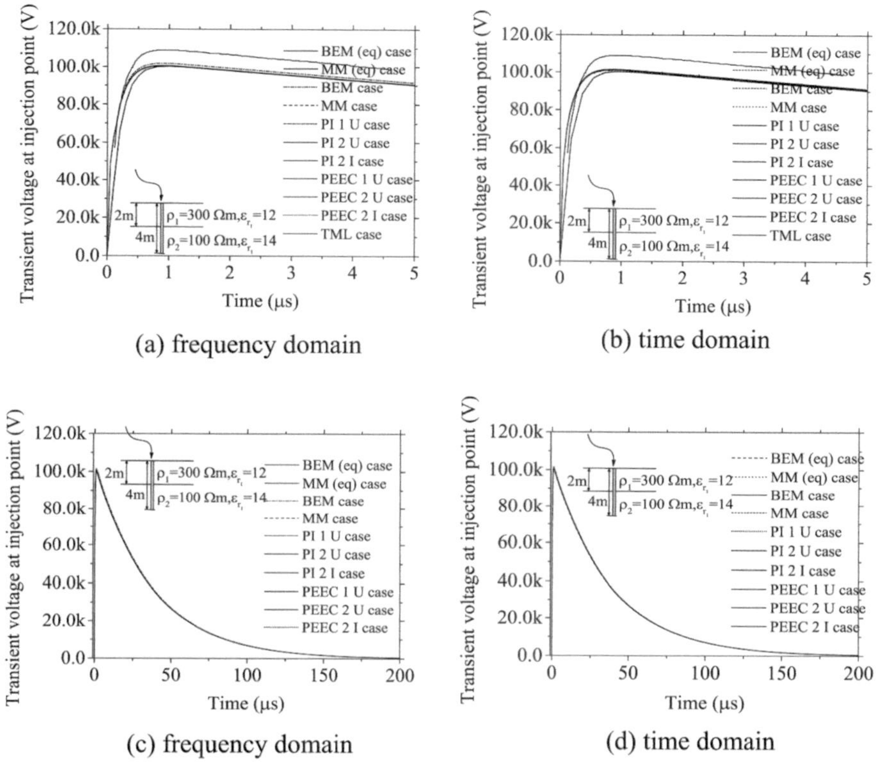

Fig. 10.4 Mutual validation of time-domain and frequency-domain numerical algorithms for vertical grounding rods in multilayered horizontally stratified soils

Observing the three subplots (a) (b) and (c) of Fig. 10.5, it can be seen that the curves of the numerical simulation are basically the same as those of the experimental test.

The seventh verification example comes from the literature [42], a horizontal copper grounding electrode with 30-m-long and 0.01-m radius is considered. This electrode is buried in 1-m depth of soil with a resistivity of 200Ωm and a relative permittivity of 10. Two double-exponential lightning currents are defined as the fast- and slow-fronted currents. The fast-fronted current is represented by $i(t) = 10137.47(e^{-0.02356t} - e^{-12.58497t})A$. This current waveform has 10-kA amplitude, $127.4 - kA/\mu s$ maximum steepness, $0.5 - \mu s$ front time, and $30 - \mu s$ tail time. The slow-fronted current is defined by $i(t) = 25869.92(e^{-0.01900989t} - e^{-0.2922879t})A$. It has 20-kA amplitude, $7 - kA/\mu s$ maximum steepness, $10 - \mu s$ front time, and $50 - \mu s$ tail time. The fast- and slow-fronted currents were applied at one end of the electrode.

The frequency domain and time domain numerical methods developed in this book numerically simulated the transient voltage curves of two types of fast and

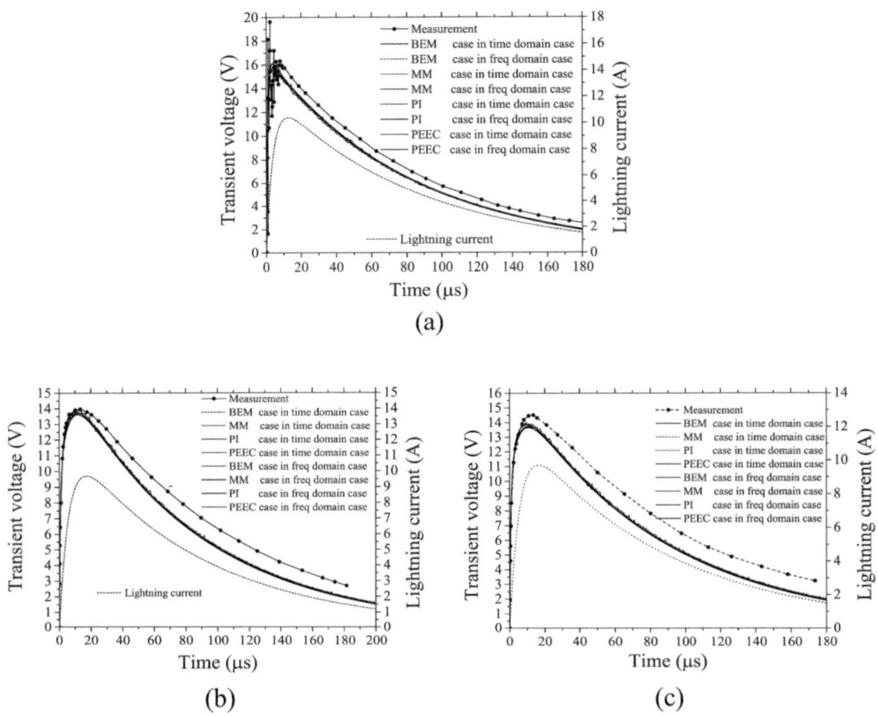

Fig. 10.5 Mutual validation of time and frequency domain numerical algorithms and experiments for simple grounding grids in multilayered horizontally stratified soils

slow lightning current impulse responses of the grounding rod. The obtained curves and the curves in the original literature [42] are presented in the two subplots of Fig. 10.6, respectively.

For comparison, the results using electromagnetic computation were obtained using CDEGS (HIFREQ module). CDEGS numerically computes the voltage response by means of full Maxwell's equations and the MOM EM approach with minimum assumptions. Figure 10.6b shows the comparison of voltage responses from the improved circuit model (IM) and electromagnetic (EM) computation.

All the above examples use a double exponential function to generate lightning current, while the following example uses the Heidler function to generate lightning current.

The example is from reference [43], the following cases are considered in our analysis, based on permutations of the following parameter values:

- Electrode length: 3 m for the case denoted "short electrode", and 30 m for a "long electrode;"
- Resistivity of the earth: 30Ωm for the case denoted "conductive", 300 Ωm for "resistive".

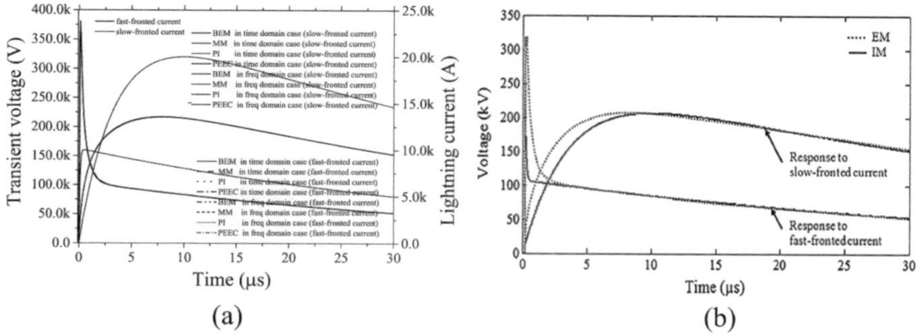

Fig. 10.6 Mutual validation of time and frequency domain numerical algorithms for the transient response of two fast and slow lightning waveforms with results from the open literature

- Lightning current waveforms: typical for the "first" and "subsequent" stroke.

The electrodes are horizontal copper conductors, with a diameter of 1.4 cm, and buried at a depth of 0.8 m in homogeneous earth. The earth has a relative permittivity of ten and a permeability equivalent to that of air.

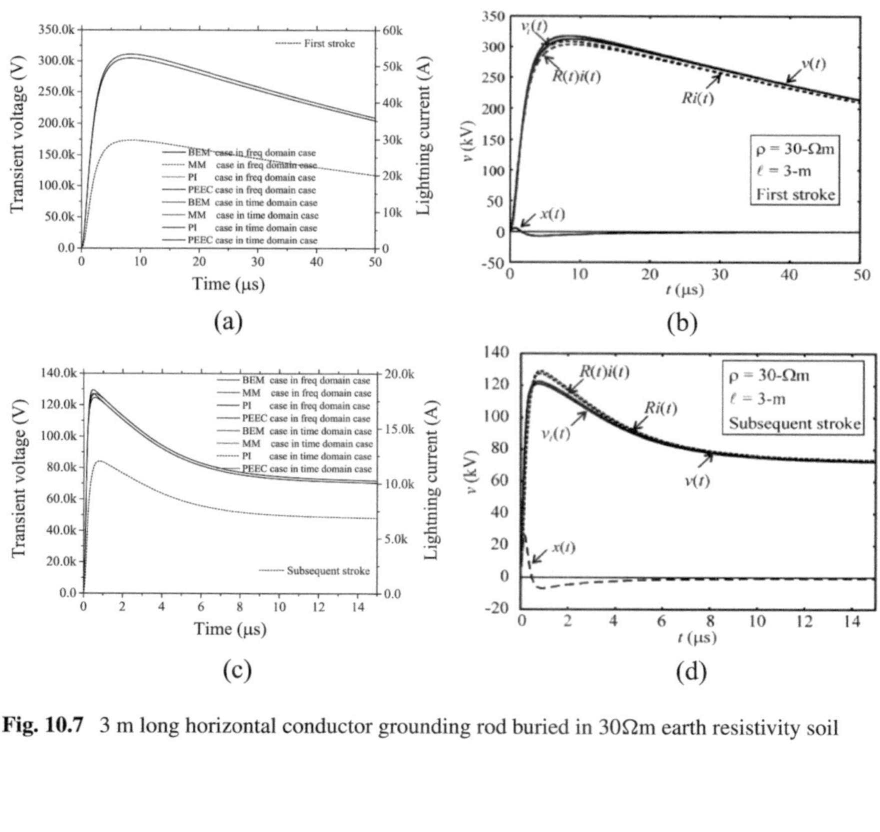

Fig. 10.7 3 m long horizontal conductor grounding rod buried in 30Ωm earth resistivity soil

Two lightning current waveforms corresponding to the typical first and subsequent return strokes, based on the observations of Berger et al. [44], are illustrated in Fig. 10.7a, c.

The current pulses have the following characteristics: (a) the first return stroke current pulse has a peak value of 30 kA, and a zero-to-peak time of about 8μs, (b) the subsequent return stroke current has a peak value of 12 kA, and a zero-to-peak time of about 0.8μs. Therefore, the first stroke is a rather slow rise-time, high intensity pulse, and the subsequent one is a fast rise-time, low intensity pulse. These waveforms appropriately reproduce the observed concave rising portion of the typical recorded lightning current pulses.

Figures 10.7, 10.8, 10.9 and 10.10 show the dependence of the total potential relative to distant ground at the injection point and its resistive and reactive components as a response to the typical injected first and subsequent return stroke current pulses. Figures 10.7 and 10.8 are examples of very conductive soil with a resistivity of 30Ωm. Figure 10.7 is for a short electrode with length 3 m; and Fig. 10.8 is for a longer electrode with length 30 m. Figures 10.7 and 10.8a, b show responses of the typical first return stroke. Figures 10.7 and 10.8c, d show responses of the typical

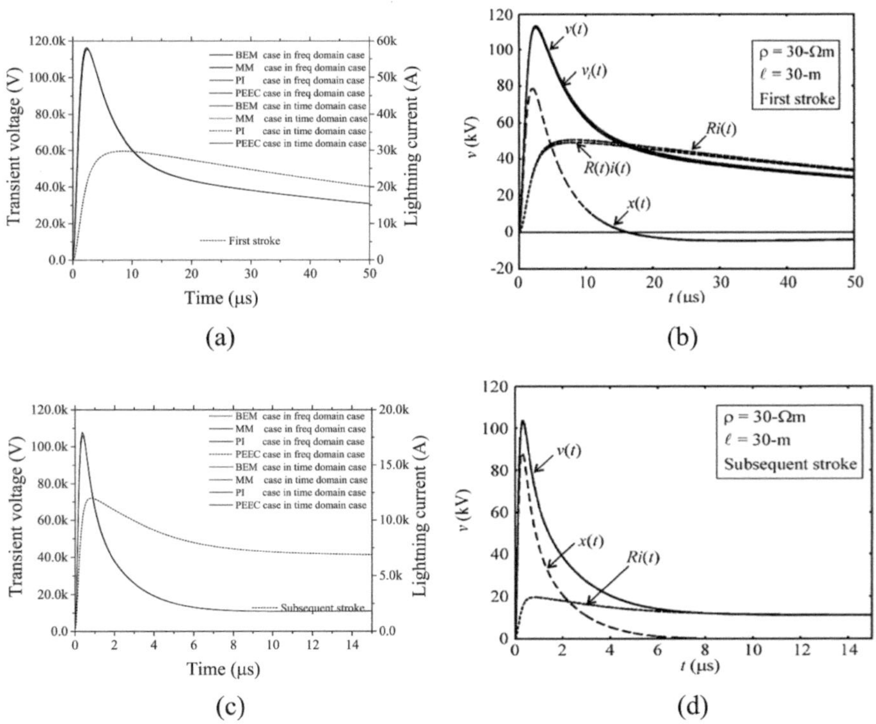

Fig. 10.8 30 m long horizontal conductor grounding rod buried in 30Ωm earth resistivity soil

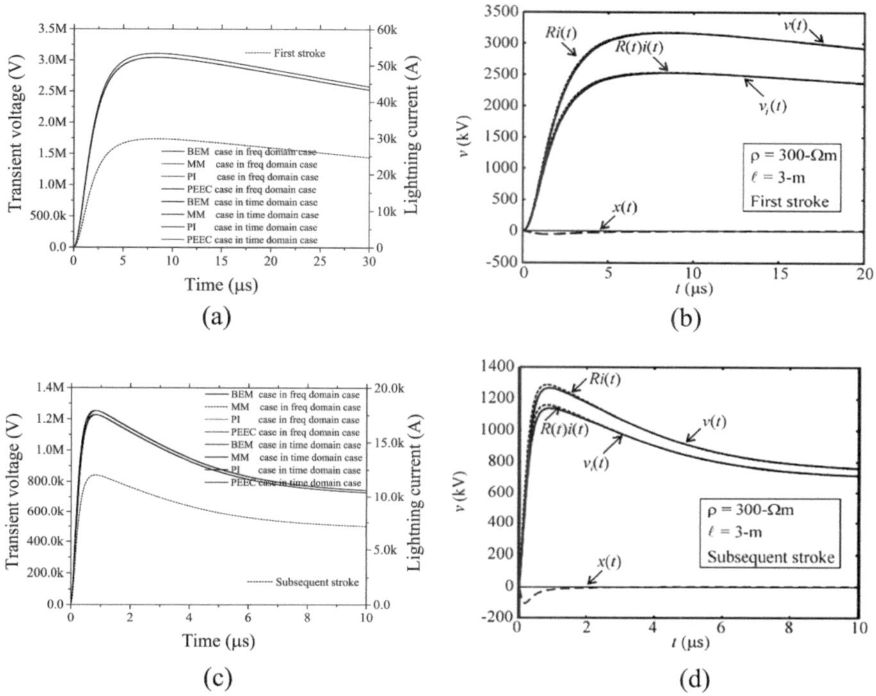

Fig. 10.9 3 m long horizontal conductor grounding rod buried in 300Ωm earth resistivity soil

subsequent return stroke. Figures 10.9 and 10.10 are organized in a similar way, but Figs. 10.9 and 10.10 refer to more resistive soil with a resistivity of 300 m.

Observing Figs. 10.7, 10.8, 10.9 and 10.10, it can be seen that the transient voltage curve simulated by the numerical method developed in this paper is consistent with the transient voltage curve simulated in the original literature [43]. It can be seen that all corresponding curves are basically completely overlapped. This comparison method demonstrates the accuracy and scientificity of the numerical methods developed in this chapter.

It should also be pointed out that the excitation lightning currents in the above verification examples were generated using a double exponential function and a Heidler function, while the following examples were generated using a composite Heidler function.

The arithmetic in this subsection is taken from Literature [45], where a horizontal grounding rod of 30 meters in length is buried in a uniform earth with a resistivity of 300 ohm meters at a depth of 0.5 meters. The earth dielectric constant is taken to be 10 and the radius of the grounding rod is 7 mm.

Representative lightning current waveforms for the first and subsequent strokes used for the excitation of this grounding rod can be seen in Fig. 10.11. Median

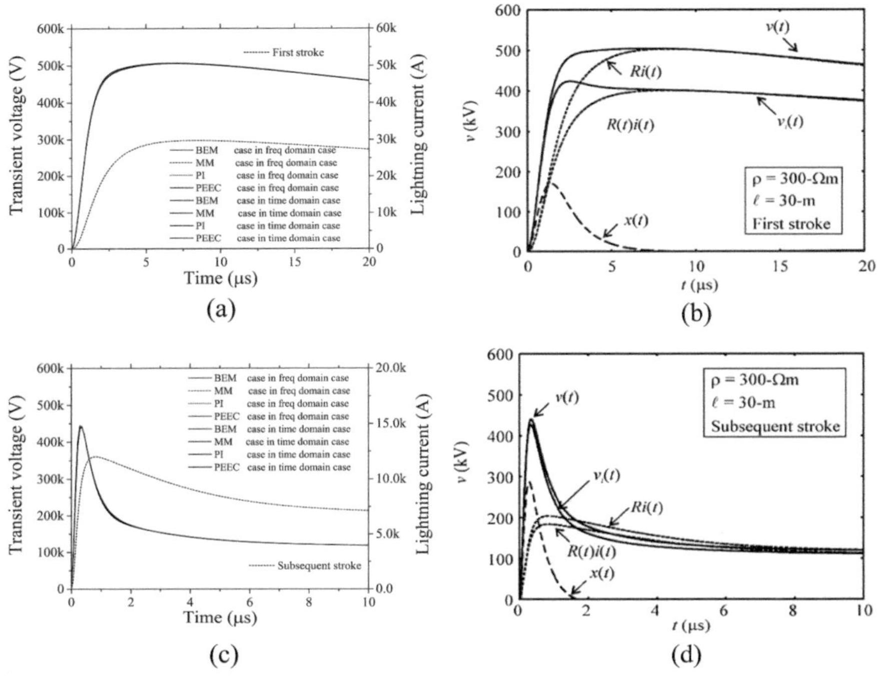

Fig. 10.10 30 m long horizontal conductor grounding rod buried in 300Ωm earth resistivity soil

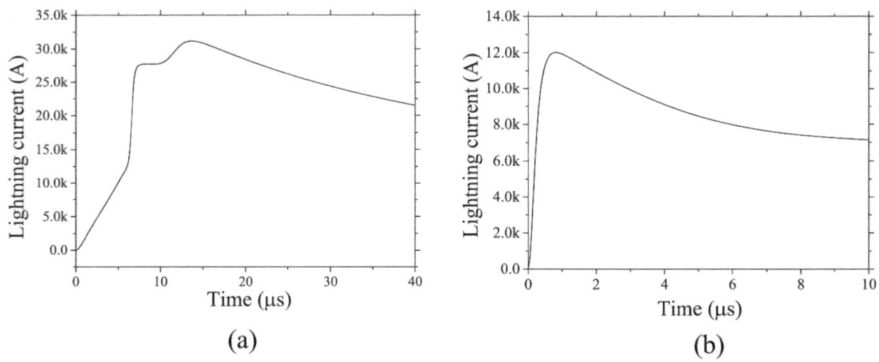

Fig. 10.11 Representative current waveforms of **a** first and **b** subsequent strokes impressed on the electrodes

peak currents and front times: (a) $I_p = 31.1$ kA, $T_{d30} = 3.8\mu s$, $T_{d10} = 5.6\mu s$ and (b) $I_p = 11.8$ kA, $T_{d30} = 0.67\mu s$, $T_{d10} = 0.75\mu s$. Adapted from [46] and [47].

The ground potential rise at the injection point of the horizontal grounding rod under the action of the first and subsequent lightning current strokes can be inverted with reference to Figs. 10.12 and 10.13, respectively.

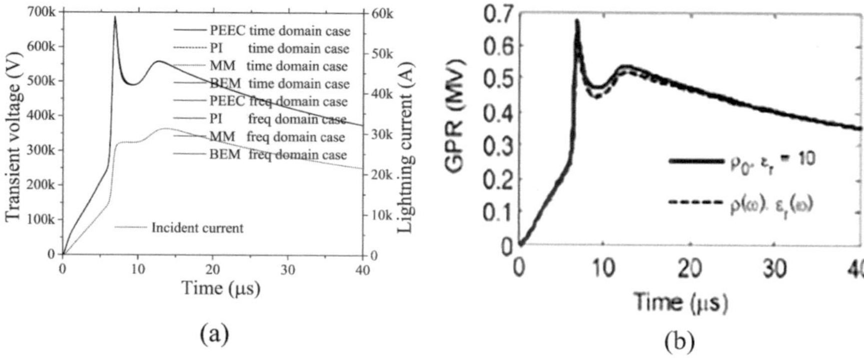

Fig. 10.12 Representative lightning current waveforms for the first strokes

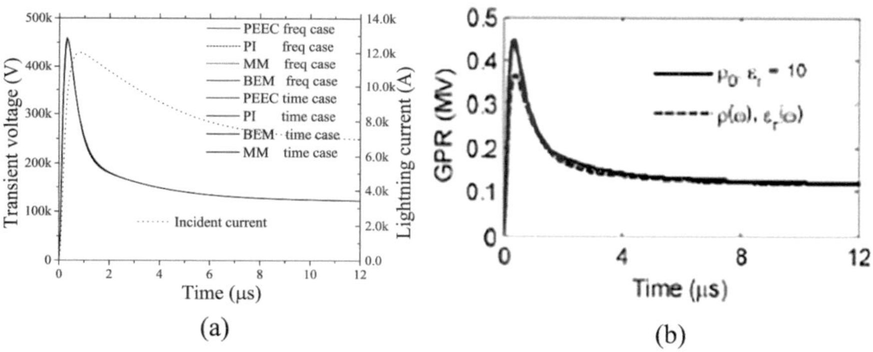

Fig. 10.13 Representative lightning current waveforms for the subsequent strokes

Referring to subfigures (a) and (b) of Fig. 10.12, it is important to point out here that subfigure (a) is the voltage profile of the numerical simulation of the algorithm developed in this book, and subfigure (b) is the voltage profile given in the original reference [45]. Observing and comparing the two subplots, it can be seen that: the two sets of transient voltage curves are basically overlapped and consistent, which further illustrates the correctness and scientificity of the various algorithms developed in this book.

The same can be further observed in the two subfigures (a) and (b) of Fig. 10.13, which differ from the two subfigures above in that the former is the response under the action of the first lightning current stroke, while the latter is the response under the action of the subsequent lightning current stroke. Comparing the transient voltage curves in the latter two subfigures, it can be seen that the two sets of curves are also basically coincident, which likewise further illustrates the correctness and scientific validity of the various frequency-domain and time-domain algorithms developed in this chapter.

10.6.2 Results and Discussions for the Various Algorithms

The numerical examples analyzed and discussed in detail in this section will be divided into two numerical examples based on the lightning currents generated by two different mathematical formulas.

Using a double exponential function to generate excitation source lightning current

A general grounding system can be seen in Fig. 10.14 a, which consists of 4 × 4 square grids and the burial depth of the grounding grid is 0.5 meters. The earth is modelled as horizontal four layers conductive media, whose resistivities and permittivities are $\rho_1 = 100\Omega\ m$ and $\rho_2 = 220\Omega\ m$, $\rho_3 = 310\Omega\ m$, $\rho_4 = 430\Omega\ m$, $\varepsilon_1 = 10\varepsilon_0$, $\varepsilon_2 = 12\varepsilon_0$, $\varepsilon_3 = 13\varepsilon_0$ and $\varepsilon_4 = 14\varepsilon_0$, respectively. The first, second and third layer thickness are $h_1 = 6\,\text{m}$, $h_2 = 12\,\text{m}$ and $h_3 = 8\,\text{m}$. All the conductor radius of grids are 5 mm. The external excited lightning current with $1.29 \times (e^{-0.01910t} - e^{-0.292288t})kA$ is injected from the corner of the grounding system. According to the formula of lightning current, T_1 is 10μs and T_2 is 50μs.

It must be pointed out here that the BEM, MM, and BEM (eq) and MM (eq) in Figures (a) and (b) represent the transient voltage curves at the incident point simulated and calculated by the Galerkin BEM and MM, respectively, representing the

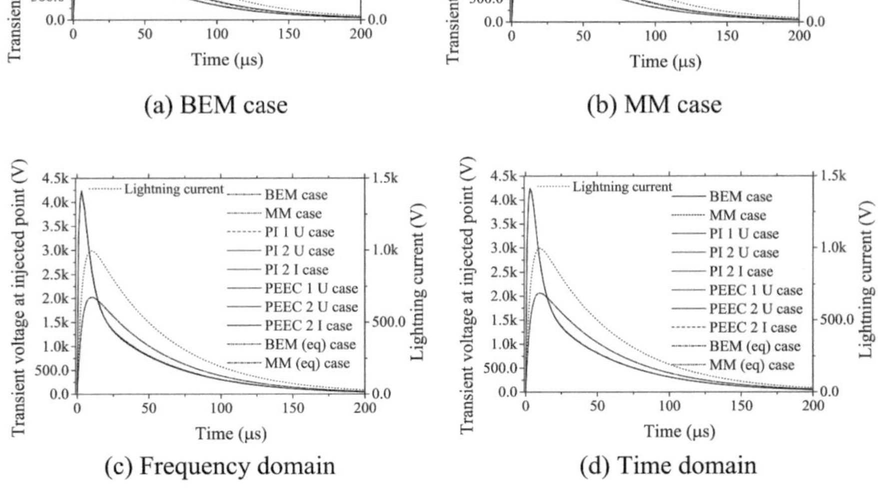

Fig. 10.14 Transient voltage at injection point

unequal and equipotential models. As can be seen from Fig. 10.14a, b, it can be seen that the lightning impulse ground potential calculated by the equipotential electromagnetic field model neglects the mutual inductance effect between the conductor segments of the grounding grid. Compared with the lightning impulse ground potential calculated by the unequal potential electromagnetic field model, its calculation error is very obvious. This conclusion further validates the 1-d spatial scalar Galerkin BEM and MM conclusions of the equipotential electromagnetic field model in the previous section's verification examples. From Fig. 10.14a, b, it can also be seen that the Galerkin BEM and MM for both equipotential and non equipotential electromagnetic field models, based on time-domain and frequency-domain algorithms, simulate and calculate the transient voltage curves at the incident point, which are completely consistent. The time-domain and frequency-domain algorithms related to Galerkin BEM and MM can be mutually verified.

From Fig. 10.14c, d, it can be seen that various frequency and time domain algorithms for the unequal potential electromagnetic field model, such as the Galerkin BEM, MM, and hybrid algorithms based on various basis functions (PI or PEEC), simulate the transient voltage curves at the incident point that completely overlap and are consistent. These two graphs respectively demonstrate the correctness of various frequency domain and time domain algorithms for the unequal potential electromagnetic field model. This conclusion also holds true for the frequency domain and time domain algorithms (scalar Galerkin BEM and MM) of equipotential electromagnetic field models. Meanwhile, comparing the two figures (c) and (d), it can be seen that the transient voltage curves at the incident point simulated by various frequency domain and time domain algorithms are completely consistent, which also proves the correctness of the corresponding frequency domain and time domain algorithms. time-domain and frequency-domain methods. From this perspective, it can be seen that the computational accuracy of various numerical methods in the time domain and frequency domain can be mutually verified.

The Table 10.2 presents the impulse impedance calculated by time-domain and frequency-domain algorithms for numerical models of equal and unequal electromagnetic fields. It can be known that there is a difference in the calculation between the equipotential electromagnetic field numerical model and the equipotential electromagnetic field numerical model.

Table 10.3 shows the impulse impedance of the incident point calculated using a hybrid algorithm based on various basis functions (PI and PEEC) in the frequency

Table 10.2 Calculate impulse impedance

	Impulse impedance			
Methods	BEM(eq)	MM(eq)	BEM	MM
Time domain case	$(2.075, 1.440 \times 10^{-3})$	$(2.070, 1.439 \times 10^{-3})$	$(6.137, 7.371 \times 10^{-4})$	$(6.138, 2.279 \times 10^{-3})$
Frequency domain case	$(2.036, -1.004)$	$(2.032, -1.002)$	$(6.102, -3.157)$	$(6.105, -3.168)$

Table 10.3 Calculate impulse impedance

	Impulse impedance					
Methods	PI 1 U case	PI 2 U case	PI 2 I case	PEEC 1 U case	PEEC 2 U case	PEEC 2 I case
Time domain case	$(6.134, 6.604 \times 10^{-4})$	$(6.135, 6.613 \times 10^{-4})$	$(6.136, 6.615 \times 10^{-4})$	$(6.127, 6.602 \times 10^{-4})$	$(6.135, 6.612 \times 10^{-4})$	$(6.136, 6.615 \times 10^{-4})$
Frequency domain case	(6.137, −3.191)	(6.139, −3.194)	(6.138, −3.194)	(6.132, −3.192)	(6.139, −3.194)	(6.138, −3.194)

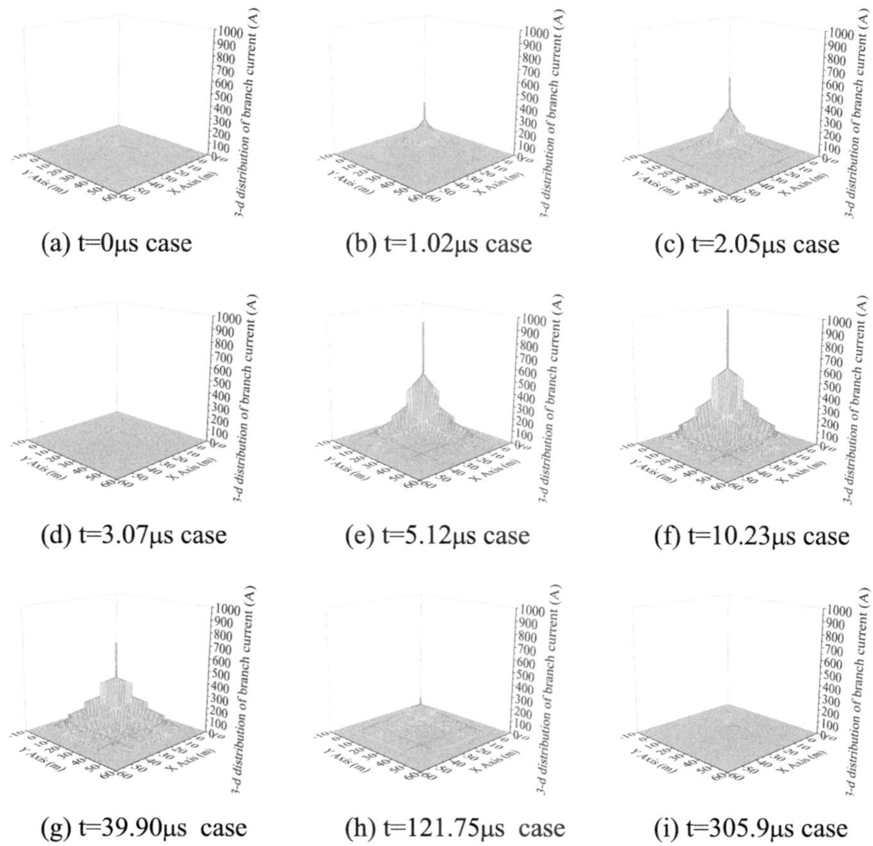

(a) t=0μs case (b) t=1.02μs case (c) t=2.05μs case

(d) t=3.07μs case (e) t=5.12μs case (f) t=10.23μs case

(g) t=39.90μs case (h) t=121.75μs case (i) t=305.9μs case

Fig. 10.15 Each time branch current along the grounding grid (time domain case)

and time domains. It can be seen that the impulse impedance calculated by various algorithms is basically consistent. Combined with the Galerkin BEM and MM impulse impedance calculation values of the unequal potential electromagnetic field model in the table above, it can be seen that the impulse impedance calculated by various algorithms is basically consistent. From this perspective, it once again proves the correctness of various algorithms.

Figures 10.15, 10.16, 10.17, 10.18, 10.19, and 10.20 below show the 3-d spatial distribution of branch current, leakage current, and voltage on the horizontal grounding grid conductor at each time point calculated by a hybrid algorithm based on PI basis functions in both time and frequency domains. Among these pictures, Figs. 10.15 and 10.16 show the 3-d spatial distribution of branch currents on the conductors of the horizontal grounding grid, simulated and calculated using a hybrid algorithm based on PI basis functions in both time and frequency domains. Figures 10.17 and 10.18 show the 3-d spatial distribution of leakage current on the conductors

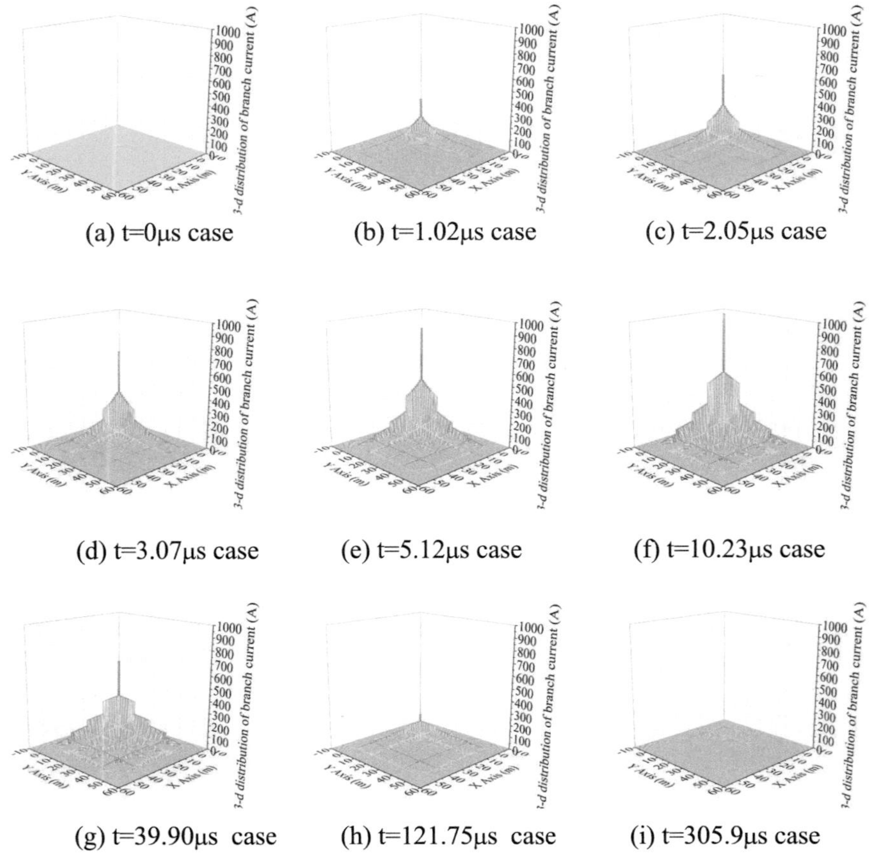

(a) t=0μs case (b) t=1.02μs case (c) t=2.05μs case

(d) t=3.07μs case (e) t=5.12μs case (f) t=10.23μs case

(g) t=39.90μs case (h) t=121.75μs case (i) t=305.9μs case

Fig. 10.16 Each time branch current along the grounding grid (frequency domain case)

of the horizontal grounding grid, simulated and calculated using a hybrid algorithm based on PI basis functions in both time and frequency domains. Figures 10.19 and 10.20 show the 3-d spatial distribution of voltage on the conductors of the horizontal grounding grid, simulated and calculated using a hybrid algorithm based on PI basis functions in both time and frequency domains.

By observing and comparing the 3-d spatial distribution maps of branch current, leakage current, and voltage simulated by time-domain and frequency-domain algorithms at the same time on the conductors of the horizontal grounding grid, it can be seen that the calculation results obtained by the two algorithms (a hybrid algorithm of time-domain and frequency-domain) are basically consistent. Therefore, the correctness of the two algorithms can be verified from this numerical calculation result.

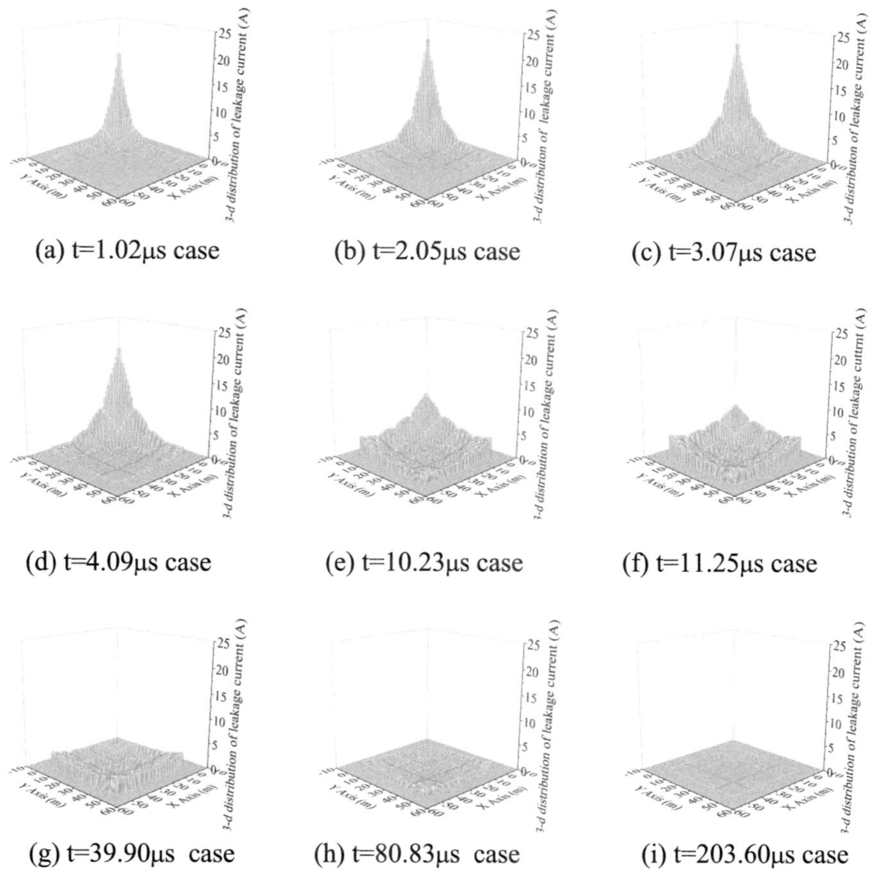

Fig. 10.17 Each time leakage current along the grounding grid (time domain case)

Using the Heidler function to generate excitation source lightning current

The substation grounding grid of this example is buried in five layers of horizontally stratified soil, the grounding grid is buried at a depth of 0.5 m. The substation grounding grid is composed of a square-shaped grounding grid, the entire grid size of 60 meters by 60 meters, consisting of 12 meters by 12 meters of small squares, a total of 5 by 5 a total of 25 small grids, which can be seen in Fig. 10.21a.

The parameters of the five layers of horizontally stratified soil, i.e., resistivity and relative permittivity, are as follows: the first layer of soil has a resistivity of 300Ωm, the first layer of soil has a resistivity of 196.57Ωm, the third layer of soil has a resistivity of 100Ωm, the fourth layer has a resistivity of 400Ωm and the last layer of The resistivity of the soil is 600Ωm; the relative dielectric constant of the first to third and last layers of soil is 10, and only the fourth layer of soil has a relative dielectric constant of 50.

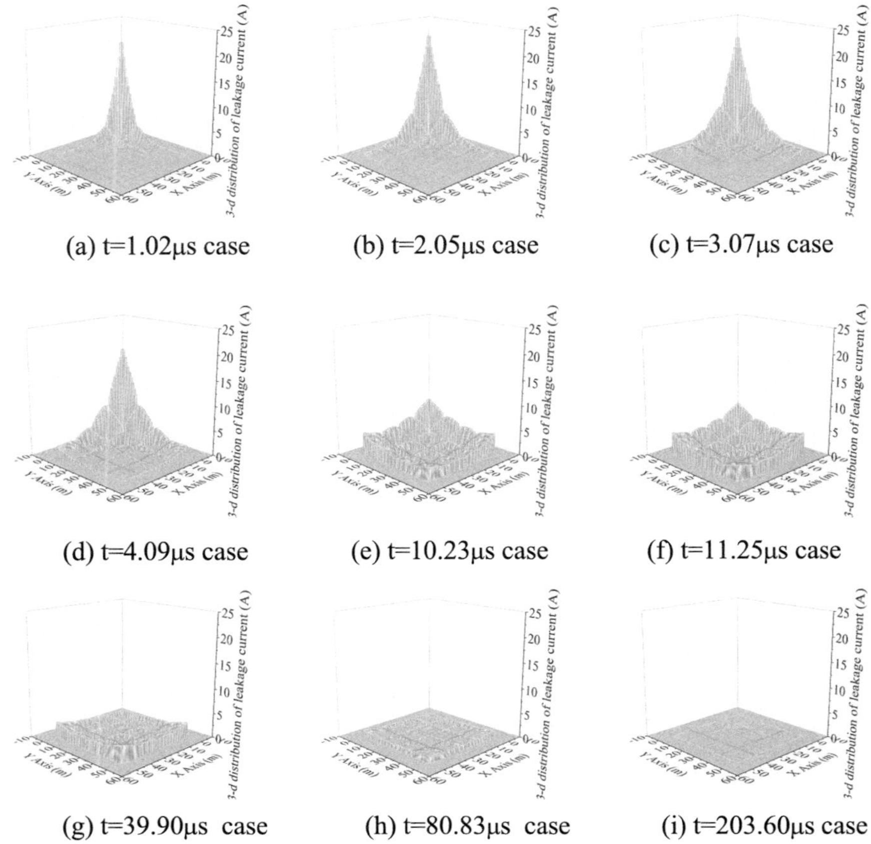

Fig. 10.18 Each time leakage current along the grounding grid (frequency domain case)

The excitation source lightning current is generated by the Heidler function with the following equation:

$$i(t) = \frac{I_0}{\eta} \frac{(\frac{t}{\tau_1})^n}{1 + (\frac{t}{\tau_1})^n} e^{-\frac{t}{\tau_2}} \tag{10.180}$$

where $I_0 = 14.862\ A$, $\eta = 0.7833$, $\tau_1 = 2.4411\mu\text{s}$, $\tau_2 = 18.761\mu\text{s}$ and $n = 5$. The choice of these parameters implies a wavehead time of $T_1 = 3.26\mu\text{s}$ and a half-wave time of $T_2 = 21.34\mu\text{s}$ for the lightning current.

1. Analysis and discussion of transient surge voltages
   Observing the two subfigures (a) and (b) in Fig. 10.21, it can be known that the whole process of the transient voltage curve at the injection point under the lightning impact of the horizontal grounding grid buried in first layer in five horizontally stratified layers of soil simulated by various numerical algorithms

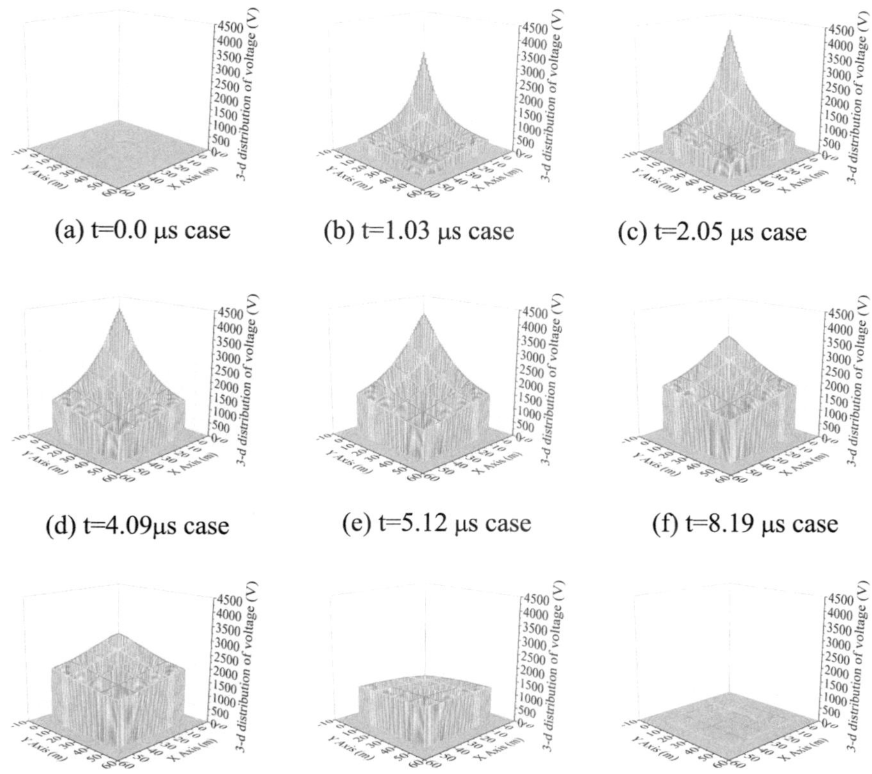

(a) t=0.0 μs case (b) t=1.03 μs case (c) t=2.05 μs case

(d) t=4.09μs case (e) t=5.12 μs case (f) t=8.19 μs case

(g) t=10.23 μs case (h) t=39.90 μs case (i) t=305.91 μs case

Fig. 10.19 Each time voltage along the grounding grid (time domain case)

of frequency-domain and time-domain unequal potential electromagnetic fields appears in subfigure (a), and subfigure (b) is the local zoom-in view near the peak of the numerically simulated transient voltage curve.

Meanwhile, after observing the two subfigures (a) and (b) in Fig. 10.21, it can be seen that the transient voltage profiles simulated and computed by various different unequal potential EMF numerical methods such as vector Galerkin BEM, vector Galerkin MM, as well as vector hybrid algorithms based on various basis functions (PI and PEEC) basically converge and composite with each other. These mutually overlapping transient voltage curves illustrate that the various numerical algorithms for electromagnetic fields with unequal potentials developed in this chapter mutually validate the correctness and accuracy of their own algorithms.

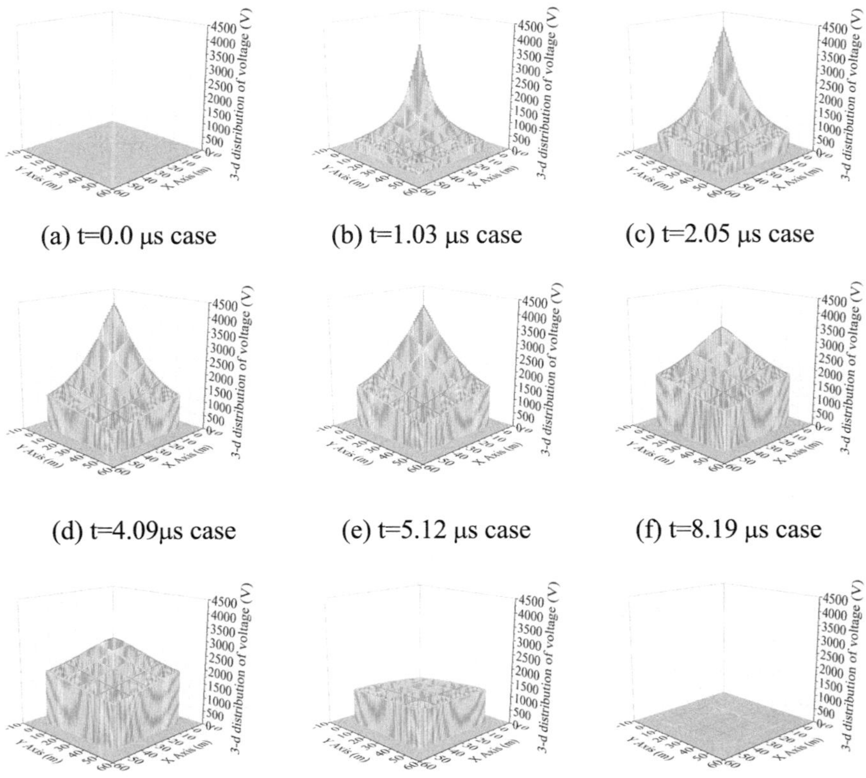

(a) t=0.0 μs case (b) t=1.03 μs case (c) t=2.05 μs case

(d) t=4.09μs case (e) t=5.12 μs case (f) t=8.19 μs case

(g) t=10.23 μs case (h) t=39.90 μs case (i) t=305.91 μs case

Fig. 10.20 Each time voltage along the grounding grid (frequency domain case)

2. Analysis of 3-d distribution plots of voltage, leakage and branch currents in frequency domains
   Each of the following large graphs consists of four large graphs made up of nine smaller graphs showing the 3-d distributions of frequency domain voltages, branch currents, and leakage currents in the grounding grid at key frequencies. The last two large figures show the associated leakage current, both of which further refine the disappearance of that current.
   Observation of Fig. 10.22 can be seen, the voltage distribution of the entire grounding grid is closely related to the frequency, the frequency of zero, the entire grounding grid distribution of voltage amplitude is the largest, with the increase in frequency, the distribution of voltage on the grounding grid as a whole with the increase in frequency and gradually weakened. Obviously, regardless of the frequency, the voltage distribution on the entire grounding grid to meet the law that the voltage amplitude on the injection point is the largest, and and

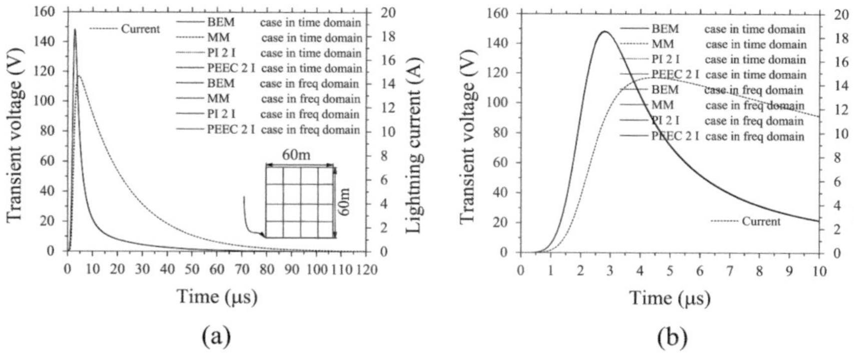

Fig. 10.21 Transient voltage at injection point (Heidler function case)

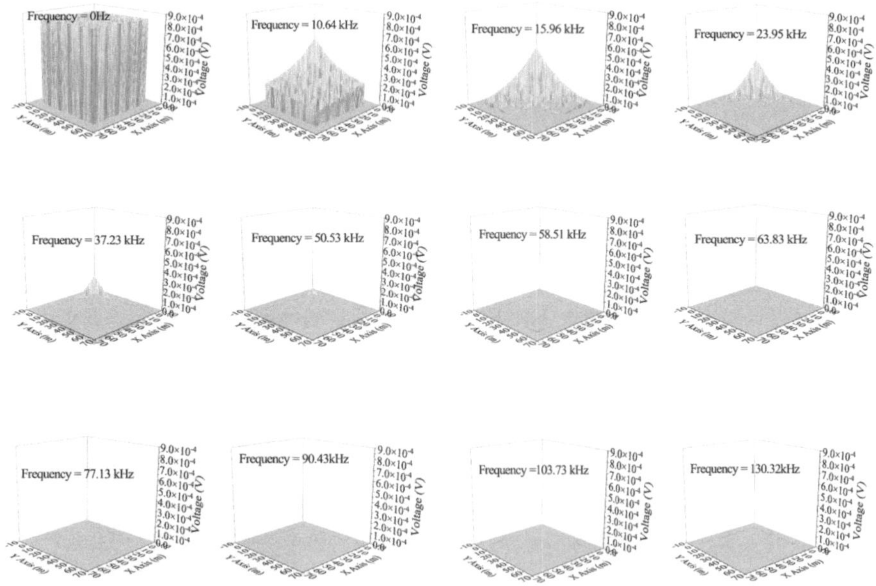

Fig. 10.22 Each frequency voltage along the grounding grid (Heidler function case)

the injection point on the grounding grid and the injection point on the diagonal of the voltage amplitude is the lowest (Figs. 10.23 and 10.24).

Observation of Fig. 10.25 shows that the 3-d distribution of branch currents on the whole grounding grid is basically similar at different frequencies, with the current amplitude at the injection point of the grounding grid being the largest, and the current amplitude at the diagonal point of the injection point of the grounding grid being the smallest. The 3-d distribution of branch currents on the whole grounding grid is gradually weakened with the increase of frequency until it disappears.

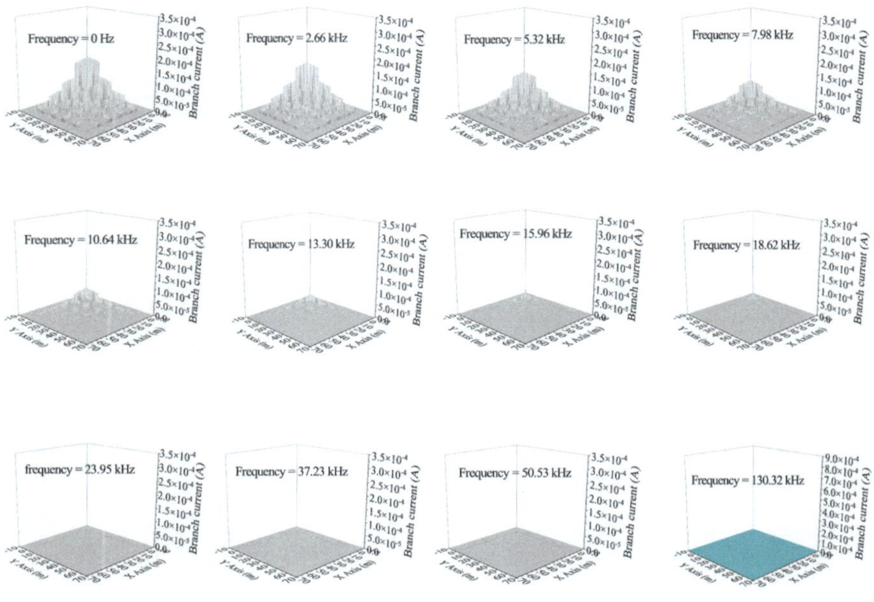

Fig. 10.23 Each frequency branch current along the grounding grid (Heidler function case)

Note that this large graph consists of nine small graphs showing the 3-d distribution of branch currents on the grounding grid at various frequencies, generated in the frequency domain by the lightning current excitation produced by the Heidler function, while the 3-d distribution of branch currents on the grounding grid at various frequencies previously generated in the frequency domain by the double-exponential function generating the lightning current excitation is basically similar, and will not be repeated here.

Similarly, by observing the 3-d distribution of leakage current on the grounding grid as a function of frequency on the nine small graphs in Fig. 10.26, it can be concluded that the leakage current on the grounding grid is slightly different from the 3-d distribution of leakage current in the frequency domain generated by the lightning current excitation generated by the double exponential function. Although they are basically consistent with the 3-d distribution of voltage and branch current on the grounding grid, that is, the leakage current amplitude at the injection point on the grounding grid is the largest, and the leakage current amplitude at the diagonal point of the injection point on the grounding grid is the smallest; But as the frequency continues to increase, the 3-d distribution of leakage current on the grounding grid here weakens and disappears extremely slowly, to the point where a large graph consisting of nine small graphs is added to show the trend of leakage current on the grounding grid as the frequency increases.

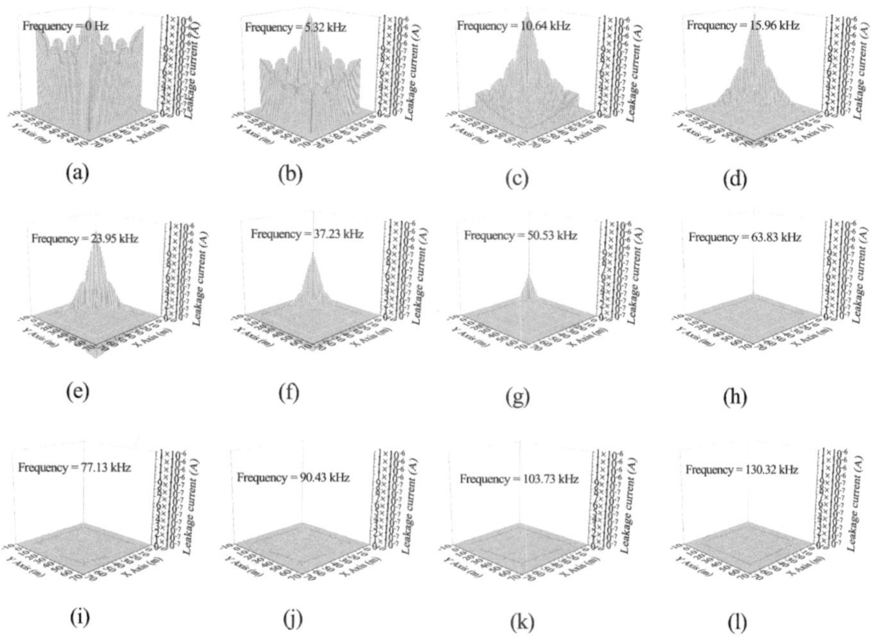

Fig. 10.24 Each frequency leakage current along the grounding grid (Heidler function case)

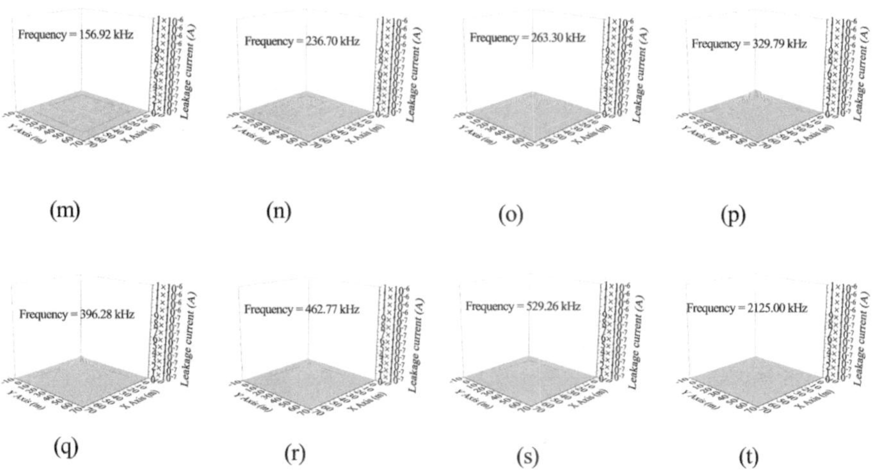

Fig. 10.25 Each frequency leakage current along the grounding grid (Heidler function case)

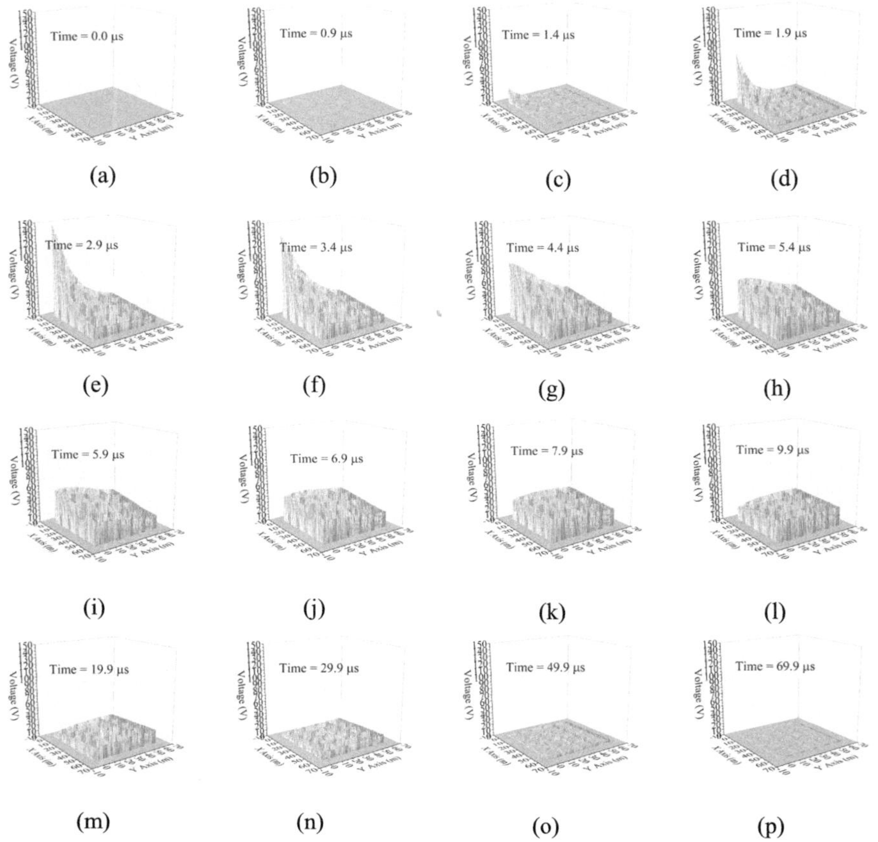

Fig. 10.26 Each time voltage along the grounding grid from time domain methods (Heidler function case)

3. Analysis of 3-d distribution plots of voltage, leakage and branch currents in time domains
   In this subchapter, the 3-d distributions of electromagnetic field quantities (voltages, leakage currents, and branch currents) at various critical moments on the substation grounding grid numerically simulated by the numerical method in the time and frequency domains are given, respectively. The first three 3-d distribution plots (10.26)–(10.28) of EMF quantities (voltage, leakage current and branch current) are obtained from the numerical simulation of time-domain EMF numerical method, while the last three 3-d distribution plots (10.29)–(10.31) of EMF quantities (voltage, leakage current and branch current) are obtained from the numerical simulation of frequency-domain EMF numerical method.
   First of all, observing the 3-d distribution diagrams of voltage, leakage current and branch current at each critical moment on the substation grounding grid simulated and calculated by the time-domain numerical algorithm, it is obvious

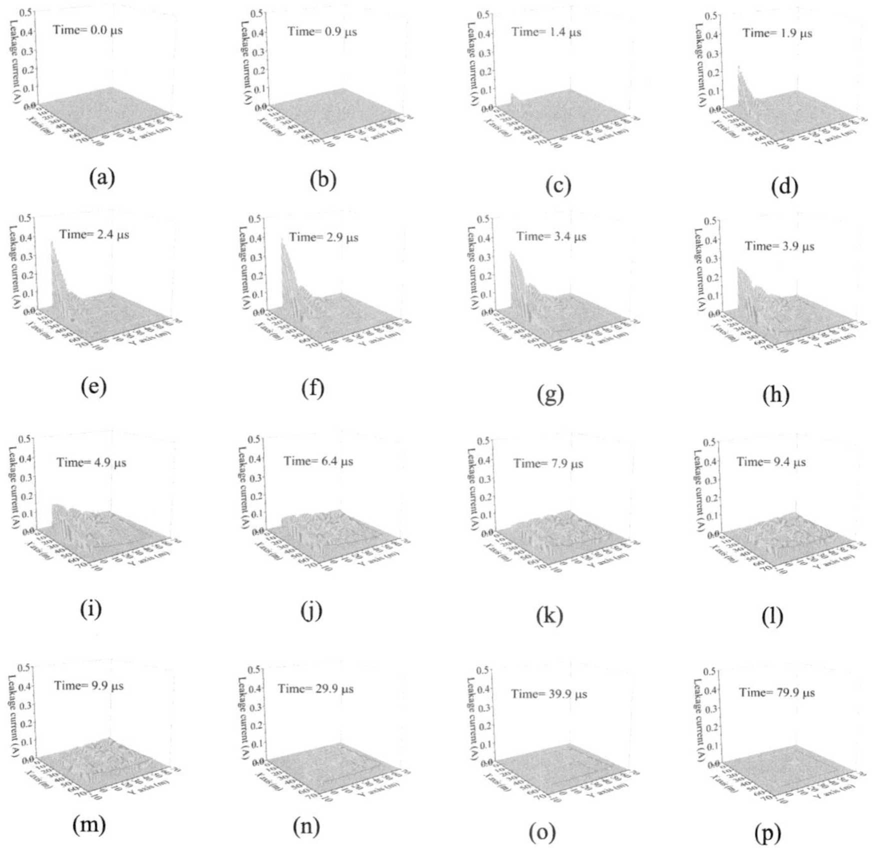

Fig. 10.27 Each time leakage current along the grounding grid from time domain methods (Heidler function case)

that, observing and analyzing each sub-diagram Fig. 10.26a–p of the 3-d spatial-temporal distribution diagrams of the voltage at each critical moment, it can be known that: with the lightning current going from none to all, from small to large, the transient voltage on the substation grounding grid begins to be generated from the injected point from none to all, the From small to large, and, the overall transient voltage distribution on the grounding grid is also from the injection point, with time gradually in the whole grounding grid gradually overall gradually increase, the overall distribution reaches the maximum, and with time to increase gradually weaken.

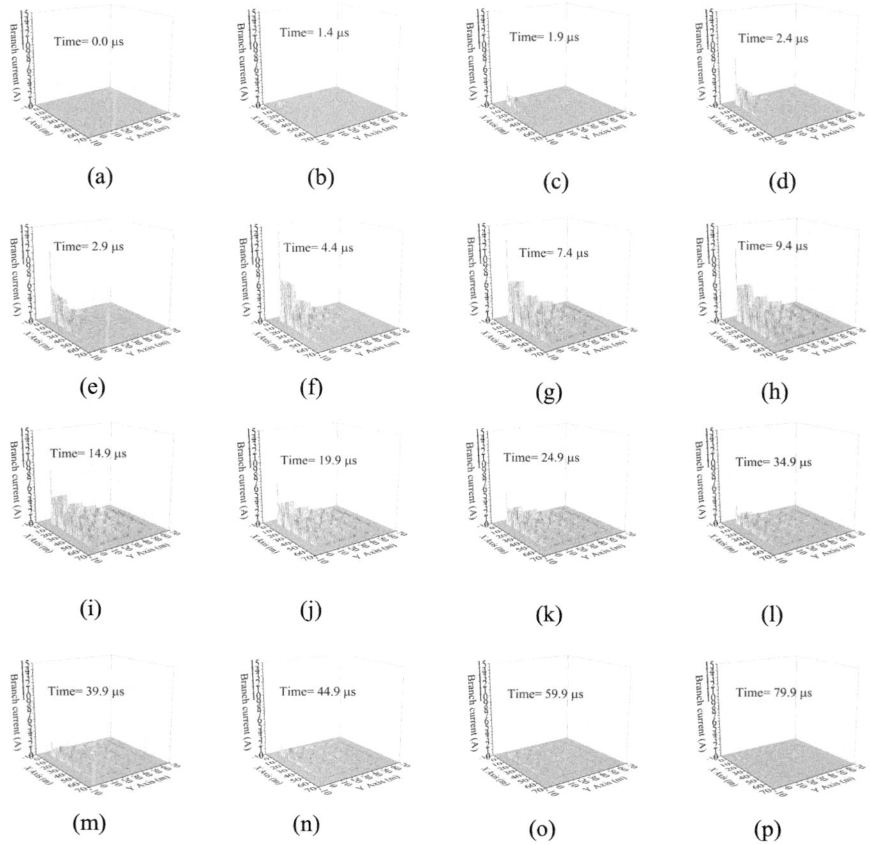

Fig. 10.28 Each time branch current along the grounding grid from time domain methods (Heidler function case)

Then, observing and analyzing each Fig. 10.27a–p of the 3-d spatio-temporal distribution map of leakage current at each critical moment, it can be known that: with the lightning current from none to some, from small to large, the transient leakage current on the grounding grid of the substation begins to be injected from the injection point from none to some, from small to large, and the overall distribution of the transient leakage current in the grounding grid is also starting from the injection point, and with the time gradually gradually in the whole grounding grid gradually. The overall distribution gradually increases, and after the overall distribution reaches the maximum, it gradually decreases with the increase of time. Furthermore, observing and analyzing each Fig. 10.28a–p of the 3-d spatio-temporal distribution diagram of the branch currents at each critical moment, it can be known that: with the lightning current from none to some, from small to large, the transient branch currents on the grounding

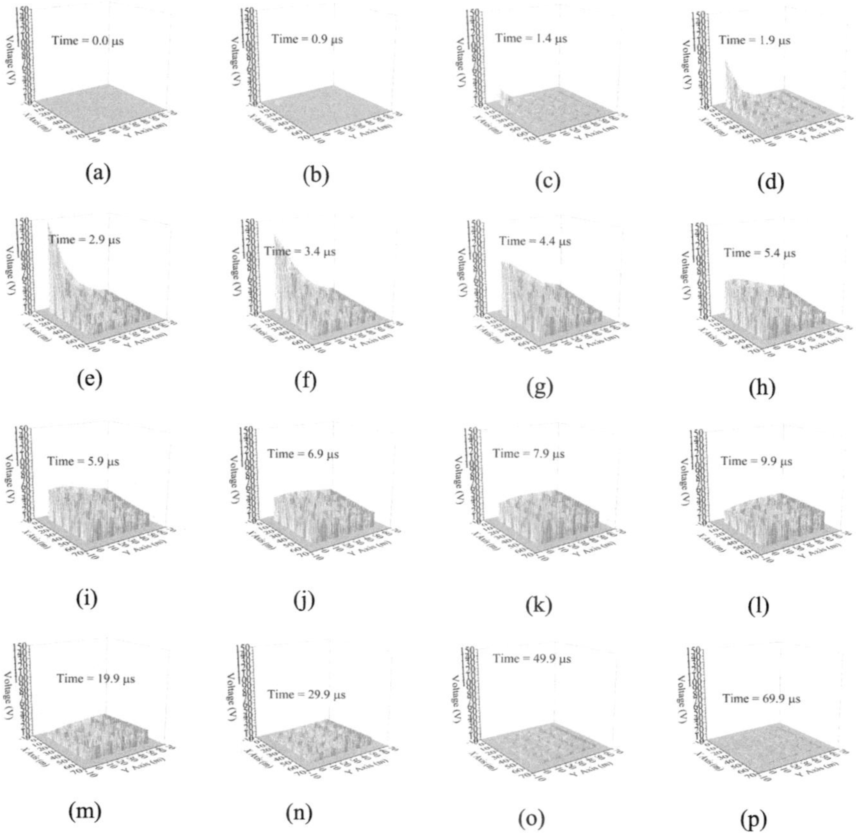

Fig. 10.29 Each time voltage along the grounding grid from frequency domain methods (Heidler function case)

grid of the substation begin to be injected from the injection point from none, from small to large, and the overall distribution of the transient branch currents in the grounding grid is also starting from the injection point, and with the time gradually gradually in the whole grounding grid. The overall distribution gradually increases, and after the overall distribution reaches the maximum, it gradually decreases with the increase of time.

The 3-d spatio-temporal distribution map of electromagnetic field quantities (transient voltage, transient leakage current and transient branch current) on the substation grounding grid numerically simulated by the numerical method of time-domain electromagnetic field is observed and analyzed above; and below, the 3-d spatial and temporal distribution diagrams of electromagnetic field quantities (transient voltage, transient leakage current and transient branch current) on the substation grounding grid numerically simulated by the frequency domain electromagnetic field numerical method can also be observed and analyzed.

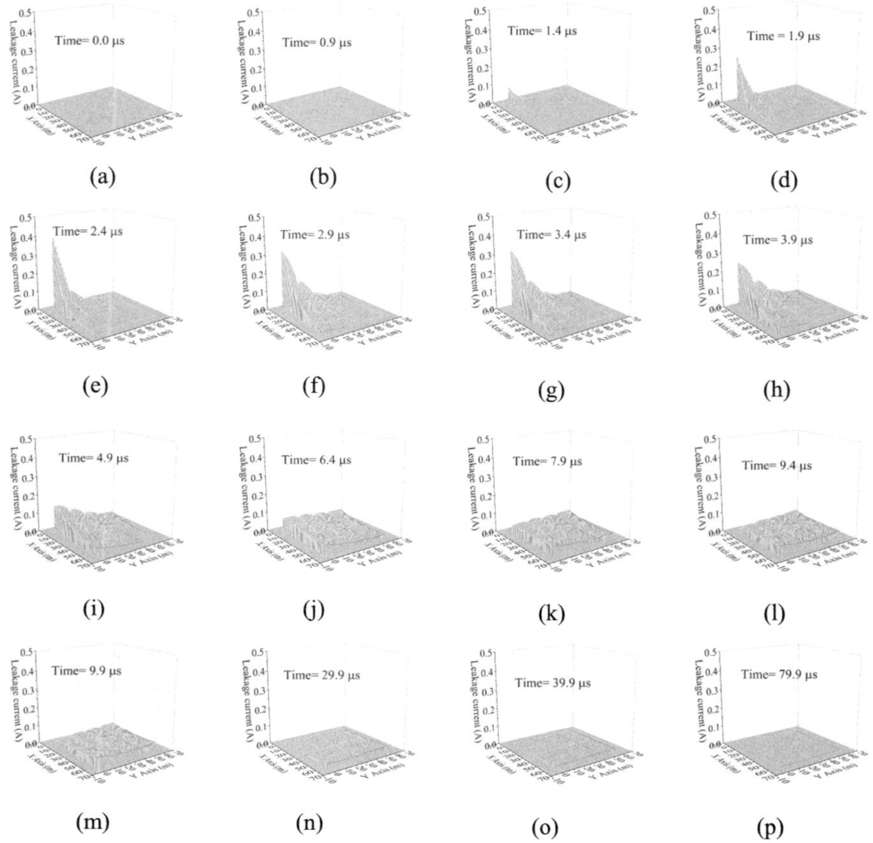

Fig. 10.30 Each time leakage current along the grounding grid from frequency domain methods (Heidler function case)

Obviously, the 3-d spatial and temporal distributions of transient voltages, leakage currents and branch currents associated with the substation grounding grid numerically simulated by the frequency domain and time domain EMF numerical methods are basically similar. It should be emphasized here that the frequency-domain and time-domain EMF numerical algorithms can be validated against each other by relevant comparisons, illustrating the correctness and accuracy of the two types of algorithms (Figs. 10.29, 10.30 and 10.31).

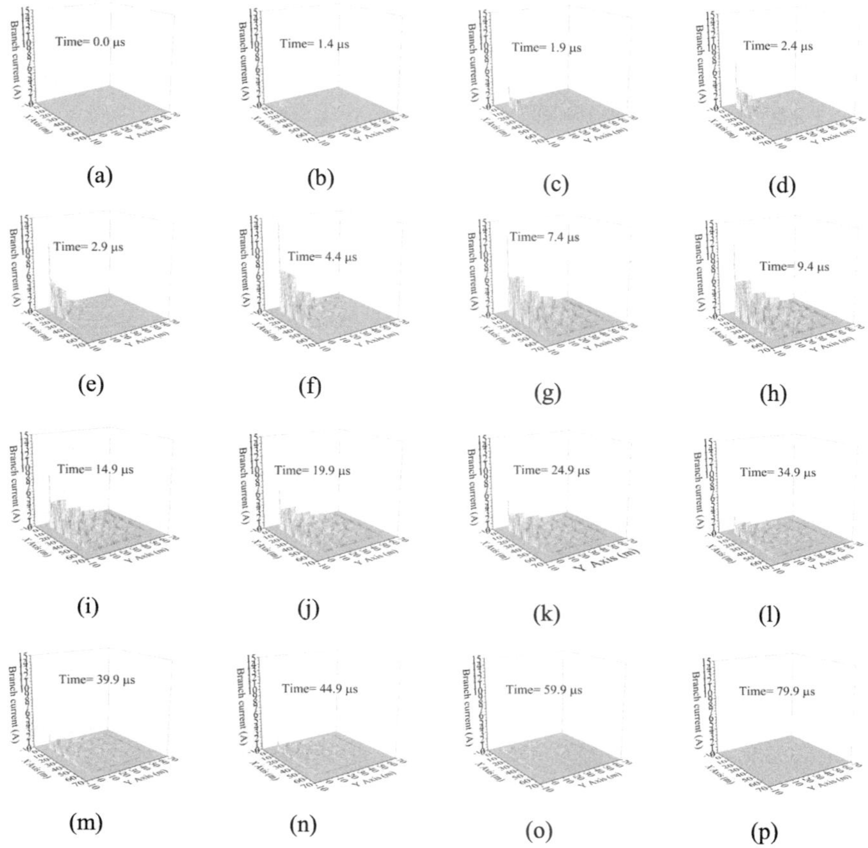

Fig. 10.31 Each time branch current along the grounding grid from frequency domain methods (Heidler function case)

10.7 Summary

This chapter delves into various time-domain numerical calculation methods based on the unequal potential electromagnetic field model, and compares them with various time-domain numerical calculation methods of the equal potential electromagnetic field model. At the same time, it compares them with the frequency domain numerical methods of the corresponding equal potential and unequal potential electromagnetic field models based on Fourier transform. The numerical results indicate that the time-domain and frequency-domain numerical algorithms of the equal and unequal position electromagnetic field models can mutually verify the correctness and accuracy of the calculations. Meanwhile, due to the neglect of mutual inductance between conductors in the substation grounding grid by the equipotential electromagnetic field model, there is a significant error between the time-domain and frequency-domain numerical algorithms of the equipotential and non equipotential electromagnetic field models, indicating the unscientific nature of using the equipotential electromagnetic field numerical model to simulate lightning impulse grounding problems in the substation grounding grid.

References

1. G. Ala, E. Francomano, E. Toscano, F. Viola, Finite difference time domain simulation of soil ionization in grounding systems under lightning surge conditions. Appl. Numer. Anal. Comput. Mathem. **1**(1), 90–103 (2004)
2. R.L. Stoll, G. Chen, N. Pilling, Comparison of two simple high-frequency earthing electrodes. IEE Proc. Gen. Transmission Distrib. **47**(1), 219–224 (2004)
3. Y. Baba, N. Nagaoka, A. Ametani, Modeling of thin wires in a lossy medium for fdtd simulations. IEEE Trans. Electromagn. Compat. **47**(1), 54–60 (2005)
4. G. Ala, P.L. Buccheri, P. Romano, F. Viola, Finite difference time domain simulation of earth electrodes soil ionisation under lightning surge condition. Iet Sci. Measur. Technol. **2**(3), 134–145 (2008)
5. G. Ala, M.L.D. Silvestre, F. Viola, Soil ionization due to high pulse transient currents leaked by earth electrodes. Progress Electromagn. Res. B **14**(14), 1–21 (2009)
6. R. Xiong, B. Chen, Y.-F. Mao, W. Deng, Q. Wu, Y.-Y. Qiu, Fdtd modeling of the earthing conductor in the transient grounding resistance analysis. IEEE Antennas Wirel. Propag. Lett. **11**, 957–960 (2012)
7. A. Ametani, T. Chikaraa, H. Morii, T. Kubo, Impedance characteristics of grounding electrodes on earth surface. Electr. Power Syst. Res. **38**, 38–43 (2012)
8. R. Xiong, B. Chen, C. Gao, Y. Yi, W. Yang, Fdtd calculation model for the transient analyses of grounding systems. IEEE Trans. Electromagn. Compat. **56**(5), 1155–1162 (2014)
9. M.E.M. Rizk, E.A. Badran, M.H. Abdel-Rahman, Fdtd calculation of transient electromagnetic fields in the grounding system of wind towers due to direct lightning strikes. J. Lightning Res. **6**, 1–13 (2014)
10. K. Yamamoto, K. Yoshioka, S. Sumi, S. Yanagawa, S. Sekioka, Mutual influence of a deeply buried grounding electrode and the surrounding grounding mesh. IEEE Trans. Indus. Appl. **51**(6), 4900–4906 (2015)
11. Y. Du, B. Li, M. Chen, The extended thin-wire model of lossy roun wire structures for fdtd simulations. IEEE Trans. Power Delivery **32**(6), 2472–2480 (2017)
12. R. Alipioa, M.T.C. de Barrosb, M.A.O. Schroederc, K. Yamamotod, Analysis of the lightning impulse and low-frequency performance of wind farm grounding systems. Electric Power Syst. Res. **180**, 106068 (2020)
13. C. Li, S. Wang, W. Wang, C. Zhang, L. Jiang, X. Wen, H. Lu, Equivalent circuit model of tower based on full-wave analysis. Energy Reports **8**, 1026–1033 (2022)
14. J. Cao, Y. Du, Y. Ding, B. Li, Y. Zhang R. Qi, and Z. Li. Lightning surge analysis of transmission line towers with a hybrid fdtd-peec method. IEEE Trans. Power Delivery **37**(2), 1275–1284 (2022)
15. A. Boutadjine, B. Nekhoul, B. Harrat, M. Melit, Realistic modeling of the grounding system transient behavior with frequency dependence. Electric Power Syst. Res. **203**, 107644 (2022)
16. B. Nekhoul, B. Harrat, A. Boutadjine, M. Melit, A simplified numerical modeling of the transient behavior of grounding systems considering soil ionization. Electric Power Syst. Res. **211**, 108182 (2022)
17. K.A. Carvalho, F.H. Silveira, Evaluation of the lightning performance of grounding electrodes using fdtd-based computational model: Influence of absorbing boundary conditions and representation of the frequency-dependence of soil parameters. Electric Power Syst. Res. **213**, 108728 (2022)

18. L. Jiang, X. Dong, X. Zhou, J. Wang, J. Song, Q. Ma, Propagation characteristics of lightning radiation field on three-layer vertical layered ground based on cpml absorption boundary. IEEE Trans. Electromagn. Compatibility **65**(4), 1191–1201 (2023)
19. K. de A. Carvalho, F.H. Silveira, Fdtd-based computational model applied to assess the influence of the frequency dependence of soil parameters on the lightning response of grounding electrodes for typical parameters of first and subsequent return stroke currents. Electric Power Syst. Res. **229**, 110170 (2024)
20. J. Li, T. Yuan, Q. Yang, W. Sima, C. Sun, M. Zahn, Numerical and experimental investigation of grounding electrode impulse-current dispersal regularity considering the transient ionization phenomenon. IEEE Trans. Power Delivery **26**(4), 2647–2658 (2011)
21. A. Habjanic, M. Trlep, The simulation of the soil ionization phenomenon around the grounding system by the finite element method. IEEE Trans. Magn. **42**(4), 867–870 (2006)
22. R. Syafruddin, Hetty Fadriani, Salamatul Afiyah, Givy Devira Ramady, and Rahmad Hidayat. Voltages transient analysis in electric grounding systems. J. Phys. Conf. Series **1783**, 012059 (2021)
23. S. Sesnic, D. Poljak, S. Tkachenko, Time domain analytical modeling of a straight tin wire buried in loaay mdium. Progress Electromagn. Res. **121**, 485–504 (2011)
24. M.R. Alemi, S.H.H. Sadeghi, H. Askarian-Abyaneh, A marching-on-in-time method of moments for transient modeling of a vertical electrode buried in a lossy medium. IEEE Trans. Electromagn. Compat. **64**(6), 2170–2178 (2022)
25. M.R. Alemi, S.H.H. Sadeghi, H. Askarian-Abyaneh, A marching-on-in-time method of moments for computation of transient potential rise of grounding grids exposed to lightning strikes. IEEE Trans. Electromagn. Compat. **65**(5), 1484–1491 (2023)
26. M.R. Alemi, S.H.H. Sadeghi, H. Askarian-Abyaneh, A marching-on-in-time method of moments for transient modeling of a vertical electrode buried in a lossy medium. IEEE Trans. Electromagn. Compat. **64**(6), 2170–2178 (2022)
27. Z.-X. Li, P. Li, X. Zhao, Galerkin boundary element method for simulating lightning response of grounding grid in horizontal multilayered soil model considering soil ionization effect. Int. J. Numer. Modelling Electron. Network Dev. fields **37**(5), e3285 (2024)
28. M. Ramamoorty, M.M.B. Narayanan, S. Parameswaran, D. Mukhedkar, Transient performance of grounding grids. IEEE Trans. Power Delivery **4**(4), 2053–2059 (1989)
29. Z.-X. Li, S.-W. Rao, Transient lightning response of grounding grids in a horizontal multi-layered soil model that considers soil ionization effects with time-domain quasi-static complex images. Int. J. Numer. Modelling Electron. Network Dev. Fields **32**(1), e2463 (2019)
30. Z.-X. Li, P. Li, K.-C. Wang, Transient lightning response to grounding grids buried in horizontal multilayered earth model considering time domain quasi-static complex image method and soil ionization effect. COMPEL Int. J. Comput. Mathem. Electr. Electron. Eng. **40**(3), 516–534 (2021)
31. Z.-X. Li, W.-J. Chen, K.-C. Wang, Frequency domain and time domain analysis of the transient behavior of buried grounding grids in horizontal multilayered earth model. Electr. Eng. **104**(4), 2515–2529 (2022)
32. M. Moradi, An electrostatic approach for simulation of grounding systems lightning response in time domain. IEEE Trans. Power Delivery **37**(6), 5289–5298 (2022)
33. B.P. Silva, S. Visacro, F.H. Silveira, Hem-td: new time-domain electromagnetic model for calculating the lightning response of electric systems and their components. IEEE Trans. Power Delivery **37**(6), 4848–4857 (2022)
34. B.P. Silva, S. Visacro, F.H. Silveira, New approaches to represent the frequency dependence of soil parameters on the time-domain hybrid electromagnetic model (hem-td). IEEE Trans. Electromagn. Compat. **65**(6), 1793–1800 (2023)
35. P. Yutthagowith, A. Ametani, F. Rachidi, N. Nagaoka, Y. Baba, Application of a partial element equivalent circuit method to lightning surge analyses. Electr. Power Syst. Res. **99**, 30–37 (2013)
36. Z.-X. Li, Lightning response transient characteristics of substation grounding grid in multilayer horizontal stratified earth model considering soil ionization effect. IEEE Trans. Electromagn. Compat. **65**(6), 1665–1673 (2023)

37. M.I. Lorentzou, N.D. Hatziargyriou, B.C. Papadias, Time domain analysis of grounding electrodes impulse response. IEEE Trans. Power Delivery **18**(2), 517–524 (2003)
38. A.I. Ramirez, A. Semlyen, R. Iravani, Modeling nonuniform transmission lines for time domain simulation of electromagnetic transients. IEEE Trans. Power Delivery **18**(3), 968–974 (2003)
39. N. Theethayi, Y. Baba, F. Rachidi, R. Thottappillil, On the choice between transmission line equations and full-wave maxwell's equations for transient analysis of buried wires. IEEE Trans. Electromagn. Compat. **50**(2), 347–357 (2008)
40. D. Poljak, S. Antonijevic, K.E.K. Drissi, K. Kerroum, Transient response of straight thin wires located at different heights above a ground plane using antenna theory and transmission line approach. IEEE Trans. Electromagn. Compat. **52**(1), 108–116 (2010)
41. D.S. Gazzana, A.S. Bretas, G.A.D. Dias, M. Tello, D.W.P. Thomas, C. Christopoulos, The transmission line modeling method to represent the soil ionization phenomenon in grounding systems. IEEE Trans. Magn. **50**(2), 505–508 (2014)
42. M. Mokhtari, Z. Abdul-Malek, Z. Salam, An improved circuit-based model of a grounding electrode by considering the current rate of rise and soil ionization factors. IEEE Trans. Power Delivery **30**(1), 211–219 (2015)
43. L. Grcev, Time- and frequency-dependent lightning surge characteristics of grounding electrodes. IEEE Trans. Power Delivery **24**(4), 2186–2196 (2009)
44. K. Berger, R.B. Anderson, H. Kroninger, Parameters of lightning flashes. Electra **41**, 23–37 (1975)
45. R. Alipio, S. Visacro, Impulse efficiency of grounding electrodes: effect of frequency-dependent soil parameters. IEEE Trans. Electromagn. Compat. **29**(2), 716–723 (2014)
46. S. Visacro, A representative curve for lightning current waveshape of first negative stroke. Geophys. Res. Lett. **31**, L07112 (2004)
47. A. De Conti, S. Visacro, Analytical representation of single- and double-peaked lightning current waveforms. IEEE Trans. Electromagn. Compat. **49**(2), 448–451 (2007)
48. S. Gazzana Daniel, A.D. Dias Guilherme, C. Leborgne Roberto, S. Bretas Arturo, T. Marcos, W.P. Thomas Dave, C. Christos, Novel formulation to determine the potential on the soil surface generated by a lightning surge. IEEE Trans. Magn. **52**(3), 1–4 (2016)
49. S. Chiheb, O. Kherif, M. Teguar, A. Mekhaldi, N. Harid, Transient behaviour of grounding electrodes in uniform and in vertically stratified soil using state space representation. IET Sci. Measur. Technol. **12**(4), 427–435 (2018)
50. O. Kherif, S. Chiheb, M. Teguar, A. Mekhaldi, N. Harid, Time-domain modeling of grounding systems?impulse response incorporating nonlinear and frequency-dependent aspects. IEEE Trans. Electromagn. Compat. **60**(4), 907–916 (2018)
51. S. Gazzana Daniel, B. Tronchoni Alex, C. Leborgne Roberto, S. Bretas Arturo, W.P. Thomas Dave, C. Christos, An improved soil ionization representation to numerical simulation of impulsive grounding systems. IEEE Trans. Magn. **54**(3), 1–4 (2018)
52. V. Steinsland, L.H. Sivertsen, E. Cimpan, S. Zhang, A new approach to include complex grounding system in lightning transient studies and emi evaluations. Energies **12**, 3142 (2019)
53. M. Moradi, Analysis of transient performance of grounding system considering frequency-dependent soil parameters and ionization. IEEE Trans. Electromagn. Compat. **62**(3), 785–797 (2020)
54. S.S. Sajadi, S.H.H. Sadeghi, Parametric dependence of lightning impulse behavior of grounding electrodes buried in a dispersive and ionized lossy soil under first- and subsequent-stroke currents. COMPEL Int. J. Comput. Mathema. Electr. Electron. Eng. **39**(4), 757–773 (2020)
55. B. Salarieh, H.M.J.D. Silva, B. Kordi, Electromagnetic transient modeling of grounding electrodes buried in frequency dependent soil with variable water content. Electr. Power Syst. Res. **189**, 106595 (2020)
56. K. Sheshyekani, M. Akbari, B. Tabei, R. Kazemi, Wideband modeling of large grounding systems to interface with electromagnetic transient solvers. IEEE Trans. Power Delivery **29**(4), 1868–1876 (2014)
57. J. Gholinezhad, R. Shariatinasab, Interfacing electromagnetic model of tower-footing impedance with the emtp software package. Electr. Power Energy Syst. **105**, 394–403 (2019)

58. A.R.J.D. Arao, J.S.L. Colqui, C.M.D. Seixas, S. Kurokawa, B. Salarieh, J.P. Filho, B. Kordi, Computation of ground potential rise and grounding impedance of simple arrangement of electrodes buried in frequency-dependent stratified soil. Electr. Power Syst. Res. **198**, 107364 (2021)
59. A. Mehri, R. Ghanizadeh, M. Beiraghi, Transient analysis of grounding electrodes in a two-layer soil considering frequency dependence of soil. Electr. Eng. **104**, 1019–1028 (2022)
60. H. Chen, Y. Du, Lightning grounding grid model considering both the frequency-dependent behavior and ionization phenomenon. IEEE Trans. Electromagn. Compat. **61**(1), 157–165 (2019)
61. R. Alipio, D. Conceioa, A.D. Conti, K. Yamamoto, R.N. Dias, S. Visacro, A comprehensive analysis of the effect of frequency-dependent soil electrical parameters on the lightning response of wind-turbine grounding systems. Electr. Power Syst. Res. **175**, 105927 (2019)
62. Z.-X. Li, Y. Yu, K.-C. Wang, Lightning response of a grounding system buried in multilayered earth based on the galerkin's boundary element and quasi-static complex image methods. IEEE Trans. Electromagn. Compat. **63**(4), 1118–1127 (2021)
63. Z. Stojkovic et al., Experimental investigation of grounding grid impulse characteristic. Europ. Trans. Electr. Power **8**(6), 417–425 (1998)
64. Z. Stojkovic et al., Sensitivity analysis of experimentally determined grounding grid impulse characteristic. IEEE Trans. Power Delivery **13**(3), 1136–1142 (1998)
65. L. Grcev, Computer analysis of transient voltages in large grounding systems. IEEE Trans. Power Delivery **11**(2), 815–823 (1996)
66. L. Grcev, M. Heimbach, Frequency dependent and transient characteristics of substation grounding systems. IEEE Trans. Power Delivery **12**(1), 172–178 (1997)
67. M. Jamali, M. Niasati, M. Jazaeri, A two-layer soil model for the calculation of electrical parameters of grounding systems under lightning strikes. Electr. Power Components Syst. **47**(12), 181–191 (2019)

Chapter 11
Numerical Methods for Grounding Problem at Transient State V

Various frequency and time domain numerical methods for numerical simulation of grounding problems in the case of lightning transient shock response considering the effect of frequency characteristics of soil parameters.

11.1 Preface

The previous chapters have all studied the grounding problem of substation grounding grids without considering the frequency characteristics of soil electrical parameters (resistivity and dielectric constant). This chapter will further investigate the lightning transient impulse grounding problem of substation grounding grids in multi-layer horizontal layered soil, taking into account the frequency characteristics of soil electrical parameters (resistivity and dielectric constant).

11.2 Introduction

This chapter mainly discusses various frequency-domain and time-domain vector Galerkin numerical methods used for numerical simulation of lightning transient impulse grounding problems in substation grounding grids in multi-layer horizontal layered soil, taking into account the frequency characteristics of soil electrical parameters (resistivity and dielectric constant). The question of the dependence of soil parameters on frequency characteristics has been intensively studied in the literature [1] specifically with the help of experimental methods. It should be pointed out that many literature [2–9] has developed a hybrid algorithm based on PI basis functions that takes into account the frequency characteristics of soil electrical parameters and is used to numerically simulate the lightning transient impulse grounding problem of substation grounding grids in uniform soil. There are also literature [10] introducing

Z.-X. Li, *Numerical Simulation Algorithm of Electromagnetic Field for Grounding Problems in Power System Substation Grounding Grids*,
https://doi.org/10.1007/978-981-96-0904-8_11

a hybrid algorithm based on PEEC basis functions that considers the frequency characteristics of soil electrical parameters for numerical simulation of lightning transient impulse grounding problems in substation grounding grids in multi-layer horizontal layered soil. Literature [11], on the other hand, is based on the frequency-domain transmission line method to simulate the lightning impact response of substation grounding grids in a homogeneous earth considering the frequency-dependent characteristics of soil parameters. In addition, literature has also introduced the vector Galerkin BEM [12] and MM [13], which consider the frequency characteristics of soil electrical parameters and are used for numerical simulation of lightning transient impulse grounding problems in substation grounding grids in multi-layer horizontal layered soil. Literature [14–16] is based on the finite element method (FEM) to numerically simulate the lightning impact response of substation grounding grids in a homogeneous earth with frequency-dependent soil parameters. There is also some literatures on the problem of lightning transient shock response of substation grounding grids in a homogeneous earth considering the frequency dependence of soil parameters based on the simulation of the commercial software HFSS [17] and FEKO [18], the problem of lightning transient shock response of substation grounding grids in a homogeneous earth considering the frequency dependence of soil parameters has also been numerically simulated in the literature [19] using EMTP software. On the other hand, there are also literature [20–22] introducing a time-domain hybrid algorithm based on PI basis functions that considers the frequency characteristics of soil electrical parameters and is used for numerical simulation of lightning transient impulse grounding problems in a substation grounding grids in uniform soil. The time-domain finite-difference method of solving the transmission line equation has also been used in the literature [23–25] to numerically simulate the lightning transient shock response of a grounding rod in a homogeneous geodetic soil taking into account the frequency characteristics of the soil parameters. In addition, there is also literature on the transformation of the frequency-domain FEM [16], MM [27, 28] and the hybrid method [29] into a time-domain method based on a vector fitting approach for numerically modeling the lightning transient shock response of grounding rods in a homogeneous earth.

11.3 How to Consider the Frequency Characteristics of Soil Electrical Parameters

The following chapters will provide a detailed and systematic introduction to the study of how to consider the frequency characteristics of soil electrical parameters in the lightning transient impulse grounding problem of substation grounding grids in multi-layer horizontal layered soil.

11.3.1 The Frequency Characteristics of Soil Parameters

There has been significant development in the literature on the frequency characteristics of electrical parameters (resistivity and dielectric constant) of the earth's soil at present. The following chapter will introduce the formulas related to the frequency characteristics of soil electrical parameters given in mainstream literature as follows.

Frequency-dependent soil parameters of Equation I: Cole-Cole equation

According to [30], the form of complex conductivity can be expressed as follows:

$$\tilde{\sigma} = \sigma_0 \left[1 + \frac{m}{1-m}(1 - \frac{1}{1+(j\omega\tau)^c(1-m)}) \right] \tag{11.1}$$

where σ_0 is dc conductivity, $\omega = 2\pi f$ is angular frequency, f is frequency, m is chargeability, c is frequency dependence, τ is central relaxation time, and j is $\sqrt{-1}$. The ranges of m and c are (0, 1.0).

Equation (11.1), as the expression of complex conductivity, is used here. If the model parameters have been determined, complex conductivity at different frequencies can be obtained based on (11.1). As the relevant parameters of the soil are usually conductivity and permittivity, complex conductivity can be expressed by using them as $\tilde{\sigma} = \sigma(f) + j\omega\varepsilon_r(f)\varepsilon_0$, where $\varepsilon_r(f)$ is relative permittivity, ε_0 is the permittivity of vacuum, and $\sigma(f)$ is the conductivity of soil.

Equation II for the frequency dependence of soil parameters

The soil resistivity and dielectric constant are considered to be variable in frequency domain, that is, for the same model, they have different values at different frequencies. Reference [31] compared the theoretical calculation results with actual measurement data, which were obtained by different formulas put forward by researchers. The following expression proposed in Ref. [32] is based on the experiment results of Ref. [33], which can well describe the frequency characteristics of soil parameters:

$$\sigma(\omega) = \sigma_0 \left[1 + \sqrt{\frac{2\omega\varepsilon_\infty}{\sigma_0}} \right] \tag{11.2}$$

$$\varepsilon_r(\omega) = \frac{\varepsilon_\infty}{\varepsilon_0} \left[1 + \sqrt{\frac{2\sigma_0}{\omega\varepsilon_\infty}} \right] \tag{11.3}$$

where σ_0 is supposed to be the conductivity measured under DC field, ε_0 is vacuum dielectric constant. $\omega = 2\pi f$ is angular frequency, f is frequency, ε_∞ is the limit of dielectric constant in high frequency and is set as $8\varepsilon_0$ according to the literature.

Equation III for the frequency dependence of soil parameters

Referring to Literature [4], the equation for the frequency characteristics dependent on the soil parameters can be given directly as the following equation

$$\sigma(\omega) = 1 + [1.2 \times 10^{-6} \times \sigma_0^{-0.73}] \times (f - 100.0)^{0.65} \tag{11.4}$$

$$\varepsilon_r(\omega) = 7.6 \times 10^3 \times f^{-0.4} + 1.3 \tag{11.5}$$

Only homogeneous conductive medium is considered here. This chapter will further discuss the parameter-dependent frequency domain characteristics of multi-layer horizontal layered conductive medium.

11.3.2 The Vector and Scalar Green's Function of Point Sources in Multilayered Horizontal Layered Soil Contain the Frequency Characteristics of Soil Parameters

The previous section mainly introduced formulas that include frequency characteristics for soil electrical parameters, such as resistivity and dielectric constant. However, in fact, in a multi-layer horizontal layered soil model, all soil electrical parameters, such as resistivity and dielectric constant, have formulas that include frequency characteristics.

In the previous Chap. 3, the frequency and time domain dyadic Green's function of vector dipoles and the frequency and time domain scalar Green's function of scalar monopoles in the multi-layer horizontal layered soil model had been mainly introduced, as well as the corresponding QSCIM. These quasi-static dyadic Green's function and scalar Green's function in frequency domain and time domain are obtained based on the soil parameters, that is, resistivity and dielectric constant are constants. In fact, if the soil parameters, namely resistivity and dielectric constant, are no longer constants but change with frequency, then the quasi-static dyadic Green's function and scalar Green's function in frequency domain and time domain, including frequency characteristics, can also be obtained. This means that the dyadic and scalar Green's function in frequency domain and time domain of point sources in multilayered horizontal layered soil obtained based on the frequency domain and time domain QSCIM include the frequency characteristics of soil parameters in each layer, namely resistivity and dielectric constant.

The various frequency domain and time domain vector numerical methods introduced in this chapter include mutual inductance matrix and mutual impedance matrix, and the mutual inductance matrix and mutual impedance matrix respectively contain the dyadic Green's function and scalar Green's function obtained by the QSCIM. The following will specifically discuss how to consider the frequency domain characteristics of soil parameters for each vector numerical method.

11.4 The Frequency Domain Spatial 1-d Vector Galerkin BEM Considering the Frequency Characteristics of Soil Parameters

11.4.1 The Frequency Domain Spatial 1-d Vector Galerkin BEM Considering the Frequency Characteristics of Soil Parameters

This chapter will review and summarize the relevant content of the spatial 1-d vector Galerkin BEM for grounding grid buried in horizontal N_e-layer soil in Chap. 6, as follows.

Referring to the formula (6.7) related to the spatial 1-d vector Galerkin BEM in Chap. 6, this formula can be rewritten as follows.

$$[\bar{\bar{\mathbf{Z}}}_p][\bar{\mathbf{I}}_p] = 0 \tag{11.6}$$

where $[\bar{\bar{\mathbf{Z}}}_p]$ is $N_p \times N_p$ matrix, who includes the mutual impedance and induction interaction betweens a parts of conductor segments currents. $[\bar{\mathbf{I}}_p]$ is $N_p \times 1$ vector of unknown currents.

Matrix equation (11.6) is assembled from the following equations.

$$\sum_{\theta=1}^{N_b} Z_{l_\theta} \int_{l_\theta} [\bar{q}(\bar{r}_\theta)][\bar{p}(\bar{r}_\theta)]dr_\theta[\bar{I}_\theta^\chi] + \sum_{\vartheta=1}^{N_b} \int_{l_\theta}\int_{l_\vartheta} \hat{r}_\theta \cdot j\omega \bar{\bar{G}}_{\xi\zeta}^{A^f}(\bar{r}_\theta, \bar{r}_\vartheta, \omega) \cdot \hat{r}_\vartheta^T [\bar{q}(\bar{r}_\theta)][\bar{p}(\bar{r}_\vartheta)]dr_\vartheta dr_\theta [\bar{I}_\vartheta^\chi]$$

$$+\sum_{\vartheta=1}^{N_b} \int_{l_\theta}\int_{l_\vartheta} G_{\xi\zeta}^{\varphi^f}(\bar{r}_\theta, \bar{r}_\vartheta, \omega) \frac{\partial[\bar{q}(\bar{r}_\theta)]}{\partial r_\theta} \frac{\partial[\bar{p}(\bar{r}_\vartheta)]}{\partial r_\vartheta} dr_\vartheta dr_\theta [\bar{I}_\vartheta^\chi] = 0 \tag{11.7}$$

where $\xi = 0, 1, \ldots, N_e$; $\zeta = 0, 1, \ldots, N_e$. $\theta = 1, 2, \ldots, N_b$; $\chi = 1, 2, \ldots, N_p$; $\bar{r}_\theta$ and $\bar{r}_\vartheta$ is location vector along θ-th and ϑ-th conductors segment, respectively; r_θ and r_ϑ, respectively, parameter variable along the θ-th and ϑ-th thin conductor axis. Z_{l_θ} is the internal impedance of the θ-th thin conductor segment. $\hat{r}_\theta$ and $\hat{r}_\vartheta$ are, respectively, the 1×3 dimensional unit director vector along the axis of the θ-th and ϑ-th cylindrical conductor segments. The symbol T represents matrix transposition. j is a imaginary unit, ω is angular frequency. $[\bar{p}(\bar{r}_\vartheta)]$ and $[\bar{q}(\bar{r}_\theta)]$ are 1×2 dimensional 1 order basis functions and 2×1 dimensional 1 order test functions, respectively. $[\bar{p}(\bar{r}_\vartheta)] = \begin{bmatrix} \frac{l_\vartheta - r_\vartheta}{l_\vartheta} & \frac{r_\vartheta}{l_\vartheta} \end{bmatrix}$ and $[\bar{q}(\bar{r}_\theta)] = \begin{bmatrix} \frac{l_\theta - r_\theta}{l_\theta} \\ \frac{r_\theta}{l_\theta} \end{bmatrix}$. $G_{\xi\zeta}^{\varphi^f}(\bar{r}_\theta, \bar{r}_\vartheta, \omega)$ and $\bar{\bar{G}}_{\xi\zeta}^{A^f}(\bar{r}_\theta, \bar{r}_\vartheta, \omega)$ are, respectively, frequency domain scalar monopole Green's function and vector dipole dyadic Green's function within horizontal multilayered earth model, which had been fully discussed in front Chap. 3. It must be pointed out that the dyadic and scalar Green's function of point sources in multilayered horizontal layered soil discussed in this section contain the frequency characteristics of soil parameters.

11.4.2 The Frequency Domain Spatial 1-d Vector Galerkin BEM Considering the Frequency Characteristics of Soil Parameters for Simulating Transient Grounding Problems

Traditionally, the lightning transient impulse grounding problem of substation grounding grid can be simulated and calculated through Fourier transform, transforming the time-domain lightning transient impulse grounding grid problem into the frequency-domain steady-state grounding grid problem.

The main steps include:

Firstly, the continuous lightning wave time domain function is reasonably evolved into a discrete lightning wave time domain function.

Then, the discrete lightning wave time domain function is transformed into a discrete lightning wave frequency domain function through Fourier transform

Subsequently, the spatial 1-d vector Galerkin BEM was used to obtain the steady-state response of lightning current at different discrete frequencies, in order to obtain the electromagnetic field response of the grounding grid at each frequency point.

Finally, the frequency domain response of the electromagnetic field at each frequency point on the grounding grid is converted into the time domain response of the grounding grid field at each time point through the inverse Fourier transform.

The above content is a solution to the entire time domain lightning wave current response of the substation grounding grid. Specifically, the spatial 1-d vector Galerkin BEM used in this chapter mainly solves the transient voltage response of lightning flowing into the grounding grid considering the frequency characteristics of soil parameters.

11.5 The Frequency Domain Spatial 1-d Vector Galerkin MM Considering the Frequency Characteristics of Soil Parameters

11.5.1 The Frequency Domain Spatial 1-d Vector Galerkin MM Considering the Frequency Characteristics of Soil Parameters

This chapter will review and summarize the relevant content of the frequency domain spatial 1-d vector Galerkin MM in Chap. 6. The core content is as follows.

Referring to the formula (6.24) related to the spatial 1-d vector Galerkin MM in Chap. 6, the formula can be rewritten as follows.

$$\begin{bmatrix} \overset{=00}{\mathbf{Z}_b} & \overset{=01}{\mathbf{Z}_b} \\ \overset{=10}{\mathbf{Z}_b} & \overset{=11}{\mathbf{Z}_b} \end{bmatrix} \begin{bmatrix} \overset{-0}{\mathbf{I}_b} \\ \overset{-1}{\mathbf{I}_b} \end{bmatrix} = 0 \tag{11.8}$$

where the $[\overset{-0}{\mathbf{I}_b}]$ is $N_b \times 1$ vector of unknown branch currents and the $[\overset{-1}{\mathbf{I}_b}]$ is $N_b \times 1$ vector of unknown leakage currents; $[\overset{=00}{\mathbf{Z}_b}]$ is $N_b \times N_b$ matrix, who presents the mutual induction interaction betweens a parts of conductor segments branch currents, $[\overset{=01}{\mathbf{Z}_b}]$ is $N_b \times N_b$ matrix, who presents the mutual impedance and induction interaction betweens a parts of conductor segments branch and leakage currents, $[\overset{=10}{\mathbf{Z}_b}]$ is $N_b \times N_b$ matrix, who presents the mutual impedance and induction interaction betweens a parts of conductor segments leakage and branch currents, $[\overset{=11}{\mathbf{Z}_b}]$ is $N_b \times N_b$ matrix, who presents the mutual impedance and induction interaction betweens a parts of conductor segments leakage currents.

Matrix equation (11.8) is assembled from the following equations.

$$\sum_{\theta=1}^{N_b} Z_{l_\theta} \int_{l_\theta} [\bar{q}(\bar{r}_\theta)][\bar{p}(\bar{r}_\theta)] dr_\theta [\bar{I}_\theta^\chi] + \sum_{\vartheta=1}^{N_b} \int_{l_\theta} \int_{l_\vartheta} \hat{r}_\theta \cdot j\omega \bar{\bar{G}}_{\xi\varsigma}^{Af}(\bar{r}_\theta, \bar{r}_\vartheta, \omega) \cdot \hat{r}_\vartheta^T [\bar{q}(\bar{r}_\theta)][\bar{p}(\bar{r}_\vartheta)] dr_\vartheta dr_\theta [\bar{I}_\vartheta^\chi]$$

$$+ \sum_{\vartheta=1}^{N_b} \int_{l_\theta} \int_{l_\vartheta} G_{\xi\varsigma}^{\varphi f}(\bar{r}_\theta, \bar{r}_\vartheta, \omega) \frac{\partial[\bar{q}(\bar{r}_\theta)]}{\partial r_\theta} \frac{\partial[\bar{p}(\bar{r}_\vartheta)]}{\partial r_\vartheta} dr_\vartheta dr_\theta [\bar{I}_\vartheta^\chi] = 0 \tag{11.9}$$

where $\theta = 1, 2, \ldots, N_b$; $\chi = 0, 1$; $\theta = 1, \ldots, N_b$. $\bar{r}_\theta$ and $\bar{r}_\vartheta$ is location vector along θ-th and ϑ-th conductors segment, respectively; r_θ and r_ϑ, respectively, parameter variable along the θ-th and ϑ-th thin conductor axis. Z_{l_θ} is the internal impedance of the θ-th thin conductor segment. $\hat{r}_\theta$ and $\hat{r}_\vartheta$ are, respectively, the 1×3 dimensional unit director vector along the axis of the θ-th and ϑ-th cylindrical conductor segments. j is a imaginary unit, ω is angular frequency. $[\bar{p}(\bar{r}_\vartheta)]$ and $[\bar{q}(\bar{r}_\theta)]$ are 1×2 dimensional 1 order basis functions and 2×1 dimensional 1 order test functions, respectively. $[\bar{p}(\bar{r}_\vartheta)] = \begin{bmatrix} 1 & -\frac{r_\vartheta}{l_\vartheta} \end{bmatrix}$ and $[\bar{q}(\bar{r}_\theta)] = \begin{bmatrix} 1 \\ -\frac{r_\theta}{l_\theta} \end{bmatrix}$. $G_{\xi\varsigma}^{\varphi f}(\bar{r}_\theta, \bar{r}_\vartheta, \omega)$ and $\bar{\bar{G}}_{\xi\varsigma}^{Af}(\bar{r}_\theta, \bar{r}_\vartheta, \omega)$ are, respectively, scalar monopole Green's function and vector dipole dyadic Green's function within horizontal multilayered earth model, which had been fully discussed in front Chap. 3. It must be pointed out that the dyadic and scalar Green's function of point sources in multilayered horizontal layered soil discussed in this section contain the frequency characteristics of soil parameters.

11.5.2 The Frequency Domain Spatial 1-d Vector Galerkin MM Considering the Frequency Characteristics of Soil Parameters for Simulating Transient Grounding Problems

It is completely the same as using the 1-d spatial vector Galerkin BEM in the above chapter to solve the transient impulse grounding problem of lightning waves on the substation grounding grid. Traditionally, the lightning transient impulse grounding problem of substation grounding grid is simulated and calculated through Fourier transform, transforming the time-domain lightning transient impulse grounding grid problem into the frequency-domain steady-state grounding grid grounding problem. The main steps are also roughly the same and will not be repeated. It should be noted that the steady-state frequency domain response of the grounding grid is solved using the spatial 1-d vector Galerkin MM.

Therefore, by combining forward and backward Fourier transform methods, the spatial 1-d Galerkin MM developed in this chapter can effectively simulate the lightning transient impulse grounding problem of substation grounding grid considering the frequency characteristics of soil parameters.

11.6 The Frequency Domain Spatial 1-d Vector Hybrid Methods Based on Different Basic Function (PI, PEEC, T and TPEEC) Considering the Frequency Characteristics of Soil Parameters

11.6.1 The Frequency Domain Spatial 1-d Vector Hybrid Methods Based on Different Basic Function (PI and PEEC) Considering the Frequency Characteristics of Soil Parameters

This chapter will review and summarize the relevant content of the spatial 1-d vector hybrid algorithm based on PI or PEEC basis functions in Chap. 6, as shown below.

For the spatial 1-d vector hybrid algorithm based on PI or PEEC basis function, the relevant concatenation formulas can be given separately as follows.

- The branch matrix approach

$$[\bar{\bar{\mathbf{Z}}}_b + j\omega\bar{\bar{\mathbf{M}}}_b][\bar{\mathbf{I}}_b] + [\bar{\bar{\mathbf{A}}}]^T[\bar{\mathbf{V}}_p] = 0 \tag{11.10}$$

$$-[\bar{\bar{\mathbf{A}}}][\bar{\mathbf{I}}_b]+[\bar{\bar{\mathbf{K}}}]^T[\bar{\bar{\mathbf{Z}}}_{s_b}]^{-1}[\bar{\bar{\mathbf{K}}}][\bar{\mathbf{V}}_p]=[\bar{\mathbf{F}}] \tag{11.11}$$

where the connotation of these matrices or vectors ($[\bar{\bar{\mathbf{Z}}}_b]$, $[\bar{\bar{\mathbf{M}}}_b]$, $[\bar{\bar{\mathbf{A}}}]$, $[\bar{\mathbf{V}}_p]$, $[\bar{\mathbf{F}}]$ and $[\bar{\mathbf{I}}_b]$) here have exactly the same form and content for both PI and PEEC basic function. Except for matrices $[\bar{\bar{\mathbf{Z}}}_{s_b}]$ and $[\bar{\bar{\mathbf{K}}}]$, where $[\bar{\bar{\mathbf{Z}}}_{s_b}]$ is a $N_b \times N_b$ matrix for PI basic function and a $2N_b \times 2N_b$ matrix for PEEC basic function, and $[\bar{\bar{\mathbf{K}}}]$ is a $N_b \times N_p$ matrix for PI basic function and a $2N_b \times N_p$ matrix for PEEC basic function.

- The node matrix approach

$$[\bar{\bar{\mathbf{Z}}}_b+j\omega\bar{\bar{\mathbf{M}}}_b][\bar{\mathbf{I}}_b]+[\bar{\bar{\mathbf{A}}}]^T[\bar{\mathbf{V}}_p]=0 \tag{11.12}$$

$$-[\bar{\bar{\mathbf{A}}}][\bar{\mathbf{I}}_b]+[\bar{\bar{\mathbf{Z}}}_{s_p}]^{-1}[\bar{\mathbf{V}}_p]=[\bar{\mathbf{F}}] \tag{11.13}$$

where all matrices and vectors here have exactly the same form and content for both PI and PEEC basic function. However, when calculating the matrix $[\bar{\bar{\mathbf{Z}}}_{s_p}]$, due to the need to further subdivide the leakage current branch in the hybrid algorithm based on PEEC basis functions, compared with the hybrid algorithm based on PI basis functions, the length of the leakage current branch is half smaller.

In order to reduce the unknown quantity and thus reduce the computational complexity, the two consecutive formulas mentioned above can be further simplified as shown below.

- The branch matrix approach

$$\left[[\bar{\bar{\mathbf{K}}}]^T[\bar{\bar{\mathbf{Z}}}_{s_b}]^{-1}[\bar{\bar{\mathbf{K}}}]+[\bar{\bar{\mathbf{A}}}]\left[[\bar{\bar{\mathbf{Z}}}_b]+j\omega[\bar{\bar{\mathbf{M}}}_b]\right]^{-1}[\bar{\bar{\mathbf{A}}}]^T\right][\bar{\mathbf{V}}_p]=[\bar{\mathbf{F}}] \tag{11.14}$$

This equation is called PI 1 case in this book.

- The node matrix approach

$$\left[[\bar{\bar{\mathbf{Z}}}_{s_p}]^{-1}+[\bar{\bar{\mathbf{A}}}]\left[[\bar{\bar{\mathbf{Z}}}_b]+j\omega[\bar{\bar{\mathbf{M}}}_b]\right]^{-1}[\bar{\bar{\mathbf{A}}}]^T\right][\bar{\mathbf{V}}_p]=[\bar{\mathbf{F}}] \tag{11.15}$$

This equation is called PI 2 case in this book.

$$\left[[\bar{\bar{\mathbf{Z}}}_b+j\omega\bar{\bar{\mathbf{M}}}_b]+[\bar{\bar{\mathbf{A}}}]^T[\bar{\bar{\mathbf{Z}}}_{s_p}][\bar{\bar{\mathbf{A}}}]\right][\bar{\mathbf{I}}_b]=-[\bar{\bar{\mathbf{A}}}]^T[\bar{\bar{\mathbf{Z}}}_{s_p}][\bar{\mathbf{F}}] \tag{11.16}$$

This equation is called PI 2 I case in this book.

The above five matrix formulas are all assembled from the following three formulas as shown in the following equation

- The continuation matrix equation system based on branch matrix approach

$$[\bar{\bar{\mathbf{Z}}}_b + j\omega \sum_{\theta=1}^{N_b} \sum_{\vartheta=1}^{N_b} \int_{l_\theta} \int_{l_\vartheta} \hat{r}_\theta \cdot \bar{\bar{G}}_{\xi\zeta}^{Af}(\bar{r}_\theta, \bar{r}_\vartheta, \omega) \cdot \hat{r}_\vartheta^T dr_\vartheta dr_\theta][\bar{\mathbf{I}}_b] + [\bar{\bar{\mathbf{A}}}]^T[\bar{\mathbf{V}}_p] = 0 \tag{11.17}$$

$$-[\bar{\bar{\mathbf{A}}}][\bar{\mathbf{I}}_b] + [\bar{\bar{\mathbf{K}}}]^T[\sum_{\theta=1}^{N_b} \sum_{\vartheta=1}^{N_b} \int_{l_\theta} \int_{l_\vartheta} G_{\xi\zeta}^{\varphi f}(\bar{r}_\theta, \bar{r}_\vartheta, \omega) \frac{1}{l_\theta} \frac{1}{l_\vartheta} dr_\vartheta dr_\theta]^{-1}[\bar{\bar{\mathbf{K}}}][\bar{\mathbf{V}}_p] = [\bar{\mathbf{F}}] \tag{11.18}$$

- The continuation matrix equation system based on node matrix approach

$$[\bar{\bar{\mathbf{Z}}}_b + j\omega \sum_{\theta=1}^{N_b} \sum_{\vartheta=1}^{N_b} \int_{l_\theta} \int_{l_\vartheta} \hat{r}_\theta \cdot \bar{\bar{G}}_{\xi\zeta}^{Af}(\bar{r}_\theta, \bar{r}_\vartheta, \omega) \cdot \hat{r}_\vartheta^T dr_\vartheta dr_\theta][\bar{\mathbf{I}}_b] + [\bar{\bar{\mathbf{A}}}]^T[\bar{\mathbf{V}}_p] = 0 \tag{11.19}$$

$$-[\bar{\bar{\mathbf{A}}}][\bar{\mathbf{I}}_b] + [\frac{1}{N_\theta^\varsigma} \sum_{\theta=1}^{N_\theta^\varsigma} \frac{1}{l_\theta^\varsigma} \int_{l_\theta^\varsigma} \frac{1}{N_\vartheta^\upsilon} \sum_{\vartheta=1}^{N_\vartheta^\upsilon} \frac{1}{l_\vartheta^\upsilon} \int_{l_\vartheta^\upsilon} G_{\xi\zeta}^{\varphi f}(\bar{r}_\varsigma, \bar{r}_\upsilon, \omega) dr_\varsigma dr_\upsilon]^{-1}[\bar{\mathbf{V}}_p] = [\bar{\mathbf{F}}] \tag{11.20}$$

where ς, υ are the node numbers of discrete conductor segments in the grounding grid. N_θ^ς and N_ϑ^υ are the total number of conductor segments connecting nodes ς and υ. l_θ^ς and l_ϑ^υ are the lengths of the θ-th and ϑ-th conductor segments connecting nodes ς and υ, respectively.

- The single matrix equation based on branch matrix method

$$[[\bar{\bar{\mathbf{K}}}]^T[\sum_{\theta=1}^{N_b} \sum_{\vartheta=1}^{N_b} \int_{l_\theta} \int_{l_\vartheta} G_{\xi\zeta}^{\varphi f}(\bar{r}_\theta, \bar{r}_\vartheta, \omega) \frac{1}{l_\theta} \frac{1}{l_\vartheta} dr_\vartheta dr_\theta]^{-1}[\bar{\bar{\mathbf{K}}}]$$

$$+ [\bar{\bar{\mathbf{A}}}] \left[\bar{\bar{\mathbf{Z}}}_b + j\omega \sum_{\theta=1}^{N_b} \sum_{\vartheta=1}^{N_b} \int_{l_\theta} \int_{l_\vartheta} \hat{r}_\theta \cdot \bar{\bar{G}}_{\xi\zeta}^{Af}(\bar{r}_\theta, \bar{r}_\vartheta, \omega) \cdot \hat{r}_\vartheta^T dr_\vartheta dr_\theta \right]^{-1} [\bar{\bar{\mathbf{A}}}]^T][\bar{\mathbf{V}}_p] = [\bar{\mathbf{F}}] \tag{11.21}$$

- The single matrix equation based on node matrix approach

$$[[\frac{1}{N_\theta^\varsigma}\sum_{\theta=1}^{N_\theta^\varsigma}\frac{1}{l_\theta^\varsigma}\int_{l_\theta^\varsigma}\frac{1}{N_\vartheta^\upsilon}\sum_{\vartheta=1}^{N_\vartheta^\upsilon}\frac{1}{l_\vartheta^\upsilon}\int_{l_\vartheta^\upsilon}G_{\xi\zeta}^{\varphi f}(\bar{r}_\varsigma,\bar{r}_\upsilon,\omega)dr_\varsigma dr_\upsilon]^{-1}$$

$$+[\bar{\bar{\mathbf{A}}}]\left[\bar{\bar{\mathbf{Z}}}_b+j\omega\sum_{\theta=1}^{N_b}\sum_{\vartheta=1}^{N_b}\int_{l_\theta}\int_{l_\vartheta}\hat{r}_\theta\cdot\bar{\bar{G}}_{\xi\zeta}^{Af}(\bar{r}_\theta,\bar{r}_\vartheta,\omega)\cdot\hat{r}_\vartheta^T dr_\vartheta dr_\theta\right]^{-1}[\bar{\bar{\mathbf{A}}}]^T][\bar{\mathbf{V}}_p]=[\bar{\mathbf{F}}] \tag{11.22}$$

$$[[\bar{\bar{\mathbf{Z}}}_b+j\omega\sum_{\theta=1}^{N_b}\sum_{\vartheta=1}^{N_b}\int_{l_\theta}\int_{l_\vartheta}\hat{r}_\theta\cdot\bar{\bar{G}}_{\xi\zeta}^{Af}(\bar{r}_\theta,\bar{r}_\vartheta,\omega)\cdot\hat{r}_\vartheta^T dr_\vartheta dr_\theta]$$

$$+[\bar{\bar{\mathbf{A}}}]^T[\frac{1}{N_\theta^\varsigma}\sum_{\theta=1}^{N_\theta^\varsigma}\frac{1}{l_\theta^\varsigma}\int_{l_\theta^\varsigma}\frac{1}{N_\vartheta^\upsilon}\sum_{\vartheta=1}^{N_\vartheta^\upsilon}\frac{1}{l_\vartheta^\upsilon}\int_{l_\vartheta^\upsilon}G_{\xi\zeta}^{\varphi f}(\bar{r}_\varsigma,\bar{r}_\upsilon,\omega)dr_\varsigma dr_\upsilon][\bar{\bar{\mathbf{A}}}]][\bar{\mathbf{I}}_b]$$

$$=-[\bar{\bar{\mathbf{A}}}]^T[\frac{1}{N_\theta^\varsigma}\sum_{\theta=1}^{N_\theta^\varsigma}\frac{1}{l_\theta^\varsigma}\int_{l_\theta^\varsigma}\frac{1}{N_\vartheta^\upsilon}\sum_{\vartheta=1}^{N_\vartheta^\upsilon}\frac{1}{l_\vartheta^\upsilon}\int_{l_\vartheta^\upsilon}G_{\xi\zeta}^{\varphi f}(\bar{r}_\varsigma,\bar{r}_\upsilon,\omega)dr_\varsigma dr_\upsilon][\bar{\mathbf{F}}] \tag{11.23}$$

11.6.2 The Frequency Domain Spatial 1-d Vector Hybrid Methods Based on Different Basic Function (T and TPEEC) Considering the Frequency Characteristics of Soil Parameters

This chapter will review and summarize the relevant content of the spatial 1-d vector hybrid algorithm based on T and TPEEC basis function in Chap. 6, as shown below.

There is only one solution for the T based spatial 1-d vector hybrid algorithm, as shown below.

$$\left[\left[\bar{\bar{\mathbf{A}}}_{pb}\bar{\bar{\mathbf{Y}}}_{bb}\bar{\bar{\mathbf{A}}}_{bp}^T\right]-\left[\bar{\bar{\mathbf{A}}}_{pb}\bar{\bar{\mathbf{Y}}}_{bb}\bar{\bar{\mathbf{A}}}_{bb}^T\left[\bar{\bar{\mathbf{A}}}_{bb}\bar{\bar{\mathbf{Y}}}_{bb}\bar{\bar{\mathbf{A}}}_{bb}^T+\bar{\bar{\mathbf{Z}}}_{s_b}^{-1}\right]^{-1}\bar{\bar{\mathbf{A}}}_{bb}\bar{\bar{\mathbf{Y}}}_{bb}\bar{\bar{\mathbf{A}}}_{bp}^T\right]\right][\bar{\mathbf{V}}_p]=[\bar{\mathbf{F}}] \tag{11.24}$$

The matrix formula (11.24) is assembled from the following equation.

$$[[\bar{\bar{\mathbf{A}}}_{pb}\bar{\bar{\mathbf{Y}}}_{bb}\bar{\bar{\mathbf{A}}}_{bp}^{T}] - [\bar{\bar{\mathbf{A}}}_{pb}\bar{\bar{\mathbf{Y}}}_{bb}\bar{\bar{\mathbf{A}}}_{bb}^{T}[\bar{\bar{\mathbf{A}}}_{bb}\bar{\bar{\mathbf{Y}}}_{bb}\bar{\bar{\mathbf{A}}}_{bb}^{T}+$$

$$[\sum_{\theta=1}^{N_b}\sum_{\vartheta=1}^{N_b}\int_{l_\theta}\int_{l_\vartheta} G_{\xi\zeta}^{\varphi f}(\bar{r}_\theta, \bar{r}_\vartheta, \omega)\frac{1}{l_\theta}\frac{1}{l_\vartheta}dr_\vartheta dr_\theta]^{-1}]^{-1}\bar{\bar{\mathbf{A}}}_{bb}\bar{\bar{\mathbf{Y}}}_{bb}\bar{\bar{\mathbf{A}}}_{bp}^{T}]][\bar{\mathbf{V}}_p] = [\bar{\mathbf{F}}] \quad (11.25)$$

where $[\bar{\bar{\mathbf{Y}}}_b] = \begin{bmatrix} \bar{\bar{\mathbf{Y}}}_{bb} & \bar{\bar{\mathbf{Y}}}_{bb} \\ \bar{\bar{\mathbf{Y}}}_{bb} & \bar{\bar{\mathbf{Y}}}_{bb} \end{bmatrix}$, and $[\bar{\bar{\mathbf{Y}}}_b] = -[\bar{\bar{\mathbf{Z}}}_b + j\omega\bar{\bar{\mathbf{M}}}_b]^{-1}$, therefore, further reference can be made to the matrix formula below.

$$[\bar{\bar{\mathbf{Y}}}_b] = -[\bar{\bar{\mathbf{Z}}}_b + j\omega\sum_{\theta=1}^{N_b}\sum_{\vartheta=1}^{N_b}\int_{l_\theta}\int_{l_\vartheta} \hat{r}_\theta \cdot \bar{\bar{G}}_{\xi\zeta}^{Af}(\bar{r}_\theta, \bar{r}_\vartheta, \omega) \cdot \hat{r}_\vartheta^T dr_\vartheta dr_\theta]^{-1} \quad (11.26)$$

There is only one solution for the TPEEC based spatial 1-d vector hybrid algorithm, as shown below.

$$\left[[\bar{\bar{\mathbf{D}}} - \bar{\bar{\mathbf{C}}}]\frac{1}{2}[\bar{\bar{\mathbf{Z}}}]_{s_b}^{-1}[\bar{\bar{\mathbf{K}}}] - [\bar{\bar{\mathbf{D}}} + \bar{\bar{\mathbf{C}}}][\bar{\bar{\mathbf{Z}}}_b + j\omega\bar{\bar{\mathbf{M}}}_b]^{-1}[\bar{\bar{\mathbf{A}}}]^T\right][\bar{\mathbf{V}}_p] = [\bar{\mathbf{F}}_p] \quad (11.27)$$

The matrix formula (11.27) is assembled from the following equation.

$$[[\bar{\bar{\mathbf{D}}} - \bar{\bar{\mathbf{C}}}]\frac{1}{2}[\sum_{\theta=1}^{N_b}\sum_{\vartheta=1}^{N_b}\int_{l_\theta}\int_{l_\vartheta} G_{\xi\zeta}^{\varphi f}(\bar{r}_\theta, \bar{r}_\vartheta, \omega)\frac{1}{l_\theta}\frac{1}{l_\vartheta}dr_\vartheta dr_\theta]_{s_b}^{-1}[\bar{\bar{\mathbf{K}}}]$$

$$- [\bar{\bar{\mathbf{D}}} + \bar{\bar{\mathbf{C}}}][\bar{\bar{\mathbf{Z}}}_b + j\omega\sum_{\theta=1}^{N_b}\sum_{\vartheta=1}^{N_b}\int_{l_\theta}\int_{l_\vartheta} \hat{r}_\theta \cdot \bar{\bar{G}}_{\xi\zeta}^{Af}(\bar{r}_\theta, \bar{r}_\vartheta, \omega) \cdot \hat{r}_\vartheta^T dr_\vartheta dr_\theta]^{-1}[\bar{\bar{\mathbf{A}}}]^T][\bar{\mathbf{V}}_p] = [\bar{\mathbf{F}}_p]$$

$$(11.28)$$

From formulas (11.17), (11.18), (11.19), (11.20), (11.21), (11.22), (11.23), (11.25), (11.26), and (11.28) above, it can be seen that: $G_\varphi^f(\bar{r}_\theta, \bar{r}_\vartheta, \omega)$ and $\bar{\bar{G}}_A^f(\bar{r}_\theta, \bar{r}_\vartheta, \omega)$ are, respectively, scalar monopole Green's function and vector dipole dyadic Green's function within horizontal multilayered earth model, which had been fully discussed in front Chap. 3. It must be pointed out that the dyadic and scalar Green's function of point sources in multilayered horizontal layered soil discussed in this section contain the frequency characteristics of soil parameters.

11.6.3 *The Frequency Domain Spatial 1-d Vector Hybrid Method Based on Different Basic Functions (PI, PEEC, T, and TPEEC) Considering the Frequency Characteristics of Soil Parameters for Simulating Transient Grounding Problems*

It is also completely the same as using the 1-d spatial vector Galerkin BEM in the above chapter to solve the transient impulse grounding problem of lightning waves on the substation grounding grid.

Traditionally, the lightning transient impulse grounding problem of substation grounding grid is simulated and calculated through Fourier transform, transforming the time-domain lightning transient impulse grounding grid problem into the frequency-domain steady-state grounding grid grounding problem. The main steps are also roughly the same and will not be repeated. It should be noted that the steady-state frequency domain response of the grounding grid is solved using the spatial 1-d vector hybrid methods based on different basic function (PI, PEEC, T and TPEEC).

Therefore, by combining forward and backward Fourier transform methods, the spatial 1-d vector hybrid methods based on different basic function (PI, PEEC, T and TPEEC) developed in this chapter can effectively simulate the lightning transient impulse grounding problem of substation grounding grid considering the frequency characteristics of soil parameters.

11.7 The Tme Domain Spatial 1-d Vector Galerkin BEM Considering the Frequency Characteristics of Soil Parameters

11.7.1 *The Time Domain Spatial 1-d Vector Galerkin BEM Considering the Frequency Characteristics of Soil Parameters*

This chapter will review and summarize the relevant content of the time domain spatial 1-d vector Galerkin BEM in Chap. 10, as follows.

$$[\bar{\bar{\mathbf{Z}}}_b^s][I_p(t)] + [\bar{\bar{\mathbf{M}}}_b^s]\frac{d[I_b(t)]}{dt} + [\bar{\bar{\mathbf{Z}}}_s^s][I_p(t)] = -[\bar{\bar{\mathbf{Z}}}_b^q][I_p(t - t_c)] - [\bar{\bar{\mathbf{M}}}_b^q]\frac{d[I_p(t - t_A)]}{dt} - [\bar{\bar{\mathbf{Z}}}_s^q][I_p(t - t_\varphi)] \tag{11.29}$$

where $[\bar{\bar{\mathbf{Z}}}_b^s]$ and $[\bar{\bar{\mathbf{Z}}}_b^q]$ are the self impedance matrices of grounding grid conductors without and with delay effects in the $N_p \times N_p$ dimension, respectively. $[\bar{\bar{\mathbf{M}}}_b^s]$ and $[\bar{\bar{\mathbf{M}}}_b^q]$ are the mutual induction matrices of grounding grid conductors without and with delay effects in the $N_p \times N_p$ dimension, respectively. $[\bar{\bar{\mathbf{Z}}}_s^s]$ and $[\bar{\bar{\mathbf{Z}}}_s^q]$ are the mutual impedance matrices of grounding grid conductors without and with delay effects in the $N_p \times N_p$ dimension, respectively.

Matrix equation (11.29) is assembled from the following equations.

$$
\begin{aligned}
&[\sum_{\theta=1}^{N_b}\left(Z_{l_0}\int_{l_\theta}[\bar{q}(\bar{r}_\theta)][\bar{p}(\bar{r}_\theta)]dr_\theta[\bar{I}_\theta^\chi(t)] + Z_l\int_{l_\theta}[\bar{q}(\bar{r}_\theta)][\bar{p}(\bar{r}_\theta)]dr_\theta[\bar{I}_\theta^\chi(t-t_c)]\right)\\
&+\sum_{\vartheta=1}^{N_b}[\int_{l_\theta}\int_{l_\vartheta}\hat{r}_\theta\cdot\bar{\bar{G}}_{\xi\zeta}^{A_s^t}(\bar{r}_\theta,\bar{r}_\vartheta)\cdot\hat{r}_\vartheta^T[\bar{q}(\bar{r}_\theta)][\bar{p}(\bar{r}_\vartheta)]dr_\vartheta dr_\theta\frac{d[\bar{I}_\vartheta^\chi(t)]}{dt}\\
&+\int_{l_\theta}\int_{l_\vartheta}\hat{r}_\theta\cdot\bar{\bar{G}}_{\xi\zeta}^{A_q^t}(\bar{r}_\theta,\bar{r}_\vartheta)\cdot\hat{r}_\vartheta^T[\bar{q}(\bar{r}_\theta)][\bar{p}(\bar{r}_\vartheta)]dr_\vartheta dr_\theta\frac{d[\bar{I}_\vartheta^\chi(t-t_A)]}{dt}]\\
&+\sum_{\vartheta=1}^{N_b}[\int_{l_\theta}\int_{l_\vartheta}G_{\xi\zeta}^{\varphi_s^t}(\bar{r}_\theta,\bar{r}_\vartheta)\frac{\partial[\bar{q}(\bar{r}_\theta)]}{\partial r_\theta}\frac{d\bar{p}(\bar{r}_\vartheta)}{dr_\vartheta}dr_\vartheta dr_\theta[\bar{I}_\vartheta^\chi(t)]\\
&+\int_{l_\theta}\int_{l_\vartheta}G_{\xi\zeta}^{\varphi_q^t}(\bar{r}_\theta,\bar{r}_\vartheta)\frac{\partial[\bar{q}(\bar{r}_\theta)]}{\partial r_\theta}\frac{\partial\bar{p}(\bar{r}_\vartheta)}{\partial r_\vartheta}dr_\vartheta dr_\theta[\bar{I}_\vartheta^\chi(t-t_\varphi)]] = 0
\end{aligned}
\tag{11.30}
$$

where $\theta = 1, 2, \ldots, N_b$. $G_\varphi^{s_t}(\bar{r}_\theta, \bar{r}_\vartheta)$ and $\bar{\bar{G}}_A^{s_t}(\bar{r}_\theta, \bar{r}_\vartheta)$ are, respectively, time domain scalar monopole Green's function and vector dipole dyadic Green's function within horizontal multilayered earth model without delay effects; $G_\varphi^{q_t}(\bar{r}_\theta, \bar{r}_\vartheta)$ and $\bar{\bar{G}}_A^{q_t}(\bar{r}_\theta, \bar{r}_\vartheta)$ are, respectively, time domain scalar monopole Green's function and vector dipole dyadic Green's function within horizontal multilayered earth model with delay effects, all these had been fully discussed in front Chap. 3. It must be pointed out that the dyadic and scalar Green's function of point sources in multilayered horizontal layered soil discussed in this section contain the frequency characteristics of soil parameters.

11.7.2 The Time Domain Spatial 1-d Vector Galerkin BEM Considering the Frequency Characteristics of Soil Parameters for Simulating Transient Grounding Problems

$$\left[\bar{\bar{\mathbf{Z}}}_b^s + \frac{\bar{\bar{\mathbf{M}}}_b^s}{\Delta t} + \bar{\bar{\mathbf{Z}}}_s^s\right][I_p(t_k)] = \frac{[I_b(t_{k-1})]}{\Delta t} - [\bar{\bar{\mathbf{Z}}}_b^q][I_p(t_k - t_c^k)]$$
$$- [\bar{\bar{\mathbf{M}}}_b^q]\frac{[I_p(t_k - t_A^k)] - [I_p(t_{k-1} - t_A^k)]}{\Delta t} - [\bar{\bar{\mathbf{Z}}}_s^q][I_p(t_k - t_\varphi^k)] \quad (11.31)$$

The excitation is incorporated into the formulation through the boundary condition:

$$I_{p_1}(t_k) = I_g(t_k) \quad (11.32)$$

where $I_g(t_k)$ denotes the lightning current generator and $I_{p_1}(t_k)$ current in the injection point at t_k moment.

At a junction consisting of two or more segments the continuity properties of the electric field must be satisfied, which is ensured by applying the Kirchhoff current law

$$\sum_{m=1}^{n} I_{p_m}(t_k) = 0 \quad (11.33)$$

And continuity equation

$$[\frac{\partial I_{p_1}(t_k)}{\partial r_1}]_{at\ junction} = [\frac{\partial I_{p_2}(t_k)}{\partial r_2}]_{at\ junction} = \cdots = [\frac{\partial I_{p_n}(t_k)}{\partial r_n}]_{at\ junction} \quad (11.34)$$

The condition (11.34) ensures the discontinuities in charge per unit length to be ruled out in passing from one conductor to another across the junction.

At the conductor free ends, the total current vanishes,i.e.is forced to be zero.

Equation (11.31) represents a marching-on-in-time formulation. Accordingly, it can be quickly solved through iterations. Therefore, the time domain spatial 1-d vector Galerkin BEM developed in this chapter can effectively simulate the lightning transient impulse grounding problem of substation grounding grid considering the frequency characteristics of soil parameters.

11.8 The Time Domain Spatial 1-d Vector Galerkin MM Considering the Frequency Characteristics of Soil Parameters

11.8.1 The Time Domain Spatial 1-d Vector Galerkin MM Considering the Frequency Characteristics of Soil Parameters

This chapter will review and summarize the relevant content of the time domain spatial 1-d vector Galerkin MM in Chap. 10, as follows.

$$\begin{bmatrix} \bar{\bar{\mathbf{Z}}}_b^{s_{00}} & \bar{\bar{\mathbf{Z}}}_b^{s_{01}} \\ \bar{\bar{\mathbf{Z}}}_b^{s_{10}} & \bar{\bar{\mathbf{Z}}}_b^{s_{11}} + \bar{\bar{\mathbf{Z}}}_s^{s_{11}} \end{bmatrix} \begin{bmatrix} \bar{\mathbf{I}}_b^0(t) \\ \bar{\mathbf{I}}_b^1(t) \end{bmatrix} + \begin{bmatrix} \bar{\bar{\mathbf{M}}}_b^{s_{00}} & \bar{\bar{\mathbf{M}}}_b^{s_{01}} \\ \bar{\bar{\mathbf{M}}}_b^{s_{10}} & \bar{\bar{\mathbf{M}}}_b^{s_{11}} \end{bmatrix} \frac{d}{dt} \begin{bmatrix} \bar{\mathbf{I}}_b^0(t) \\ \bar{\mathbf{I}}_b^1(t) \end{bmatrix}$$

$$= -\begin{bmatrix} \bar{\bar{\mathbf{Z}}}_b^{q_{00}} & \bar{\bar{\mathbf{Z}}}_b^{q_{01}} \\ \bar{\bar{\mathbf{Z}}}_b^{q_{10}} & \bar{\bar{\mathbf{Z}}}_b^{q_{11}} \end{bmatrix} \begin{bmatrix} \bar{\mathbf{I}}_b^0(t-t_c) \\ \bar{\mathbf{I}}_b^1(t-t_c) \end{bmatrix} - \begin{bmatrix} \bar{\bar{\mathbf{M}}}_b^{q_{00}} & \bar{\bar{\mathbf{M}}}_b^{q_{01}} \\ \bar{\bar{\mathbf{M}}}_b^{q_{10}} & \bar{\bar{\mathbf{M}}}_b^{q_{11}} \end{bmatrix} \frac{d}{dt} \begin{bmatrix} \bar{\mathbf{I}}_b^0(t-t_A) \\ \bar{\mathbf{I}}_b^1(t-t_A) \end{bmatrix}$$

$$- \begin{bmatrix} 0 & 0 \\ 0 & \bar{\bar{\mathbf{Z}}}_s^{q_{11}} \end{bmatrix} \begin{bmatrix} \bar{\mathbf{I}}_b^0(t-t_\varphi) \\ \bar{\mathbf{I}}_b^1(t-t_\varphi) \end{bmatrix} \tag{11.35}$$

The matrix formula (11.35) is assembled from the following equation.

$$\sum_{\theta=1}^{N_b}\left(Z_{l_0}\int_{l_\theta}[\bar{q}(\bar{r}_\theta)][\bar{p}(\bar{r}_\theta)]dr_\theta[\bar{I}_\theta^\chi(t)] + Z_l\int_{l_\theta}[\bar{q}(\bar{r}_\theta)][\bar{p}(\bar{r}_\theta)]dr_\theta[\bar{I}_\theta^\chi(t-t_c)]\right)$$

$$+\sum_{\vartheta=1}^{N_b}[\int_{l_\theta}\int_{l_\vartheta}\hat{r}_\theta\cdot\bar{\bar{G}}_{\xi\zeta}^{A_s^t}(\bar{r}_\theta,\bar{r}_\vartheta)\cdot\hat{r}_\vartheta^T[\bar{q}(\bar{r}_\theta)][\bar{p}(\bar{r}_\vartheta)]dr_\vartheta dr_\theta\frac{d[\bar{I}_\vartheta^\chi(t)]}{dt}$$

$$+\int_{l_\theta}\int_{l_\vartheta}\hat{r}_\theta\cdot\bar{\bar{G}}_{\xi\zeta}^{A_q^t}(\bar{r}_\theta,\bar{r}_\vartheta)\cdot\hat{r}_\vartheta^T[\bar{q}(\bar{r}_\theta)][\bar{p}(\bar{r}_\vartheta)]dr_\vartheta dr_\theta\frac{d[\bar{I}_\vartheta^\chi(t-t_A)]}{dt}]$$

$$+\sum_{\vartheta=1}^{N_b}[\int_{l_\theta}\int_{l_\vartheta}G_{\xi\zeta}^{\varphi_s^t}(\bar{r}_\theta,\bar{r}_\vartheta)\frac{\partial[\bar{q}(\bar{r}_\theta)]}{\partial r_\theta}\frac{d[\bar{p}(\bar{r}_\vartheta)]}{dr_\vartheta}dr_\vartheta dr_\theta[\bar{I}_\vartheta^\chi(t)]$$

$$+\int_{l_\theta}\int_{l_\vartheta}G_{\xi\zeta}^{\varphi_q^t}(\bar{r}_\theta,\bar{r}_\vartheta)\frac{\partial[\bar{q}(\bar{r}_\theta)]}{\partial r_\theta}\frac{\partial[\bar{p}(\bar{r}_\vartheta)]}{\partial r_\vartheta}dr_\vartheta dr_\theta[\bar{\mathbf{I}}_\vartheta^\chi(t-t_\varphi)]]=0 \tag{11.36}$$

where $\theta = 1, 2, \ldots, N_b$; $\chi = 0, 1$; $G_{\xi\zeta}^{\varphi_s^t}(\bar{r}_\theta, \bar{r}_\vartheta)$ and $\bar{\bar{G}}_{\xi\zeta}^{A_s^t}(\bar{r}_\theta, \bar{r}_\vartheta)$ are, respectively, time domain scalar monopole Green's function and vector dipole dyadic Green's function within horizontal multilayered earth model without delay effects; $G_{\xi\zeta}^{\varphi_q^t}(\bar{r}_\theta, \bar{r}_\vartheta)$ and $\bar{\bar{G}}_{\xi\zeta}^{A_q^t}(\bar{r}_\theta, \bar{r}_\vartheta)$ are, respectively, time domain scalar monopole Green's function and vector dipole dyadic Green's function within horizontal multilayered earth model with delay effects, all these had been fully discussed in front Chap. 3. It must be pointed out that the dyadic and scalar Green's function of point sources in multilayered horizontal layered soil discussed in this section contain the frequency characteristics of soil parameters.

11.8.2 The Time Domain Spatial 1-d Vector Galerkin MM Considering the Frequency Characteristics of Soil Parameters for Simulating Transient Grounding Problems

$$\begin{bmatrix} \bar{\bar{\mathbf{Z}}}_b^{s00} + \frac{\bar{\bar{\mathbf{M}}}_b^{s00}}{\Delta t} & \bar{\bar{\mathbf{Z}}}_b^{s01} + \frac{\bar{\bar{\mathbf{M}}}_b^{s01}}{\Delta t} \\ \bar{\bar{\mathbf{Z}}}_b^{s10} + \frac{\bar{\bar{\mathbf{M}}}_b^{s10}}{\Delta t} & \bar{\bar{\mathbf{Z}}}_b^{s11} + \bar{\bar{\mathbf{Z}}}_s^{s11} + \frac{\bar{\bar{\mathbf{M}}}_b^{s11}}{\Delta t} \end{bmatrix} \begin{bmatrix} \bar{\mathbf{I}}_b^0(t_k) \\ \bar{\mathbf{I}}_b^1(t_k) \end{bmatrix} = \frac{1}{\Delta t} \begin{bmatrix} \bar{\bar{\mathbf{M}}}_b^{s00} & \bar{\bar{\mathbf{M}}}_b^{s01} \\ \bar{\bar{\mathbf{M}}}_b^{s10} & \bar{\bar{\mathbf{M}}}_b^{s11} \end{bmatrix} \begin{bmatrix} \bar{\mathbf{I}}_b^0(t_{k-1}) \\ \bar{\mathbf{I}}_b^1(t_{k-1}) \end{bmatrix}$$
$$- \begin{bmatrix} \bar{\bar{\mathbf{Z}}}_b^{q00} & \bar{\bar{\mathbf{Z}}}_b^{q01} \\ \bar{\bar{\mathbf{Z}}}_b^{q10} & \bar{\bar{\mathbf{Z}}}_b^{q11} \end{bmatrix} \begin{bmatrix} \bar{\mathbf{I}}_b^0(t_k - t_c^k) \\ \bar{\mathbf{I}}_b^1(t_k - t_c^k) \end{bmatrix} - \begin{bmatrix} \bar{\bar{\mathbf{M}}}_b^{q00} & \bar{\bar{\mathbf{M}}}_b^{q01} \\ \bar{\bar{\mathbf{M}}}_b^{q10} & \bar{\bar{\mathbf{M}}}_b^{q11} \end{bmatrix} \begin{bmatrix} \frac{\bar{\mathbf{I}}_b^0(t_k - t_A^k) - \bar{\mathbf{I}}_b^0(t_{k-1} - t_A^k)}{\Delta t} \\ \frac{\bar{\mathbf{I}}_b^1(t_k - t_A^k) - \bar{\mathbf{I}}_b^1(t_{k-1} - t_A^k)}{\Delta t} \end{bmatrix}$$
$$- \begin{bmatrix} 0 & 0 \\ 0 & \bar{\bar{\mathbf{Z}}}_s^{q11} \end{bmatrix} \begin{bmatrix} \bar{\mathbf{I}}_b^0(t_k - t_\varphi^k) \\ \bar{\mathbf{I}}_b^1(t_k - t_\varphi^k) \end{bmatrix} \tag{11.37}$$

The excitation is incorporated into the formulation through the boundary condition:

$$I_{b_1}^0(t_k) = I_g(t_k) \tag{11.38}$$

where $I_g(t_k)$ denotes the lightning current generator and $I_{b_1}^0(t_k)$ current in the injection point at t_k moment.

At a junction consisting of two or more segments the continuity properties of the electric field must be satisfied, which is ensured by applying the Kirchhoff current law

$$\sum_{m=1}^{n} I_{b_m}^0(t_k) = 0 \tag{11.39}$$

where n is the total number of conductor segments connected to the node.

and continuity equation

$$[\frac{\partial I_{b_1}^0(t_k)}{\partial r_1}]_{at\ junction} = [\frac{\partial I_{b_2}^0(t_k)}{\partial r_2}]_{at\ junction} = \cdots = [\frac{\partial I_{b_n}^0(t_k)}{\partial r_n}]_{at\ junction} \quad (11.40)$$

The condition (11.40) ensures the discontinuities in charge per unit length to be ruled out in passing from one conductor to another across the junction.

At the conductor free ends, the total current vanishes,i.e.is forced to be zero.

Equation (11.37) represents a marching-on-in-time formulation. Accordingly, it can be quickly solved through iterations. Therefore, the time domain spatial 1-d vector Galerkin MM developed in this chapter can effectively simulate the lightning transient impulse grounding problem of substation grounding grid considering the frequency characteristics of soil parameters.

11.9 The Time Domain Spatial 1-d Vector Hybrid Methods Based on Different Basic Function (PI and PEEC) Considering the Frequency Characteristics of Soil Parameters

11.9.1 The Time Domain Spatial 1-d Vector Hybrid Methods Based on Different Basic Function (PI and PEEC) Considering the Frequency Characteristics of Soil Parameters

This chapter will review and summarize the relevant content of the time domain spatial 1-d vector hybrid methods based on different basic function (PI and PEEC) in Chap. 10, as follows.

- The branch matrix approach

$$[\bar{\bar{\mathbf{Z}}}_b^s][\bar{\mathbf{I}}_b(t)] + [\bar{\bar{\mathbf{M}}}_b^s]\frac{d[\bar{\mathbf{I}}_b(t)]}{dt} + [\bar{\bar{\mathbf{A}}}]^T[\bar{\mathbf{V}}_p(t)] = -[\bar{\bar{\mathbf{Z}}}_b^q][\bar{\mathbf{I}}_b(t-t_c)] - [\bar{\bar{\mathbf{M}}}_b^q]\frac{d[\bar{\mathbf{I}}_b(t-t_A)]}{dt} \quad (11.41)$$

$$-[\bar{\bar{\mathbf{A}}}][\bar{\mathbf{I}}_b(t)] + [\bar{\bar{\mathbf{K}}}]^T[\bar{\bar{\mathbf{Z}}}_{s_b}^s]^{-1}[\bar{\bar{\mathbf{K}}}][\bar{\mathbf{V}}_p(t)] = [\bar{\mathbf{F}}(t)] - [\bar{\bar{\mathbf{K}}}]^T[\bar{\bar{\mathbf{Z}}}_{s_b}^s]^{-1}[\bar{\bar{\mathbf{Z}}}_{s_b}^q][\bar{\mathbf{I}}_{s_b}(t-t_\varphi)] \quad (11.42)$$

where the connotation of these matrices or vectors ($[\bar{\bar{\mathbf{Z}}}_b]$, $[\bar{\bar{\mathbf{M}}}_b]$, $[\bar{\bar{\mathbf{A}}}]$, $[\bar{\mathbf{V}}_p]$, $[\bar{\mathbf{F}}]$ and $[\bar{\mathbf{I}}_b]$) here have exactly the same form and content for both PI and PEEC basic function. Except for matrices $[\bar{\bar{\mathbf{Z}}}_{s_b}]$ and $[\bar{\bar{\mathbf{K}}}]$, where $[\bar{\bar{\mathbf{Z}}}_{s_b}]$ is a $N_b \times N_b$ matrix for PI basic function and a $2N_b \times 2N_b$ matrix for PEEC basic function, and $[\bar{\bar{\mathbf{K}}}]$ is a $N_b \times N_p$ matrix for PI basic function and a $2N_b \times N_p$ matrix for PEEC basic function.

- The node matrix approach

$$[\bar{\bar{\mathbf{Z}}}_b^s][\bar{\mathbf{I}}_b(t)] + [\bar{\bar{\mathbf{M}}}_b^s]\frac{d[\bar{\mathbf{I}}_b(t)]}{dt} + [\bar{\bar{\mathbf{A}}}]^T[\bar{\mathbf{V}}_p(t)] = -[\bar{\bar{\mathbf{Z}}}_b^q][\bar{\mathbf{I}}_b(t-t_c)] - [\bar{\bar{\mathbf{M}}}_b^q]\frac{d[\bar{\mathbf{I}}_b(t-t_A)]}{dt} \tag{11.43}$$

$$-[\bar{\bar{\mathbf{A}}}][\bar{\mathbf{I}}_b(t)] + [\bar{\bar{\mathbf{Z}}}_{s_p}^s]^{-1}[\bar{\mathbf{V}}_p(t)] = [\bar{\mathbf{F}}(t)] - [\bar{\bar{\mathbf{Z}}}_{s_p}^s]^{-1}[\bar{\bar{\mathbf{Z}}}_{s_p}^q][\bar{\mathbf{I}}_{s_p}(t-t_\varphi)] \tag{11.44}$$

where all matrices and vectors here have exactly the same form and content for both PI and PEEC basic function. However, when calculating the matrix $[\bar{\bar{\mathbf{Z}}}_{s_p}]$, due to the need to further subdivide the leakage current branch in the hybrid algorithm based on PEEC basis functions, compared with the hybrid algorithm based on PI basis functions, the length of the leakage current branch is half smaller.

$$[\bar{\bar{\mathbf{Z}}}_b^s][\bar{\mathbf{I}}_b(t)] + [\bar{\bar{\mathbf{M}}}_b^s]\frac{d[\bar{\mathbf{I}}_b(t)]}{dt} + [\bar{\bar{\mathbf{A}}}]^T[\bar{\bar{\mathbf{Z}}}_{s_p}^s][\bar{\bar{\mathbf{A}}}][\bar{\mathbf{I}}_b(t)] = -[\bar{\bar{\mathbf{Z}}}_b^q][\bar{\mathbf{I}}_b(t-t_c)]$$

$$-[\bar{\bar{\mathbf{M}}}_b^q]\frac{d[\bar{\mathbf{I}}_b(t-t_A)]}{dt} - [\bar{\bar{\mathbf{A}}}]^T[\bar{\bar{\mathbf{Z}}}_{s_p}^s][\bar{\mathbf{F}}(t)] + [\bar{\bar{\mathbf{A}}}]^T[\bar{\bar{\mathbf{Z}}}_{s_p}^q][\bar{\mathbf{I}}_{s_p}(t-t_\varphi)] \tag{11.45}$$

Matrix equations (11.41), (11.42), (11.43), (11.44) and (11.45) are assembled from the following equations.

- The branch matrix approach

$$[\bar{\bar{\mathbf{Z}}}_b^s][\bar{\mathbf{I}}_b(t)] + [\bar{\bar{\mathbf{M}}}_b^s]\frac{d[\bar{\mathbf{I}}_b(t)]}{dt} + [\bar{\bar{\mathbf{A}}}]^T[\bar{\mathbf{V}}_p(t)] = -[\bar{\bar{\mathbf{Z}}}_b^q][\bar{\mathbf{I}}_b(t-t_c)]$$

$$-[\sum_{\theta=1}^{N_b}\sum_{\vartheta=1}^{N_b}\int_{l_\theta}\int_{l_\vartheta}\hat{r}_\theta \cdot \bar{\bar{G}}_{\xi\zeta}^{A_q'}(\bar{r}_\theta,\bar{r}_\vartheta)\cdot\hat{r}_\vartheta^T dr_\vartheta dr_\theta]\frac{d[\bar{\mathbf{I}}_b(t-t_A)]}{dt} \tag{11.46}$$

$$-[\bar{\bar{\mathbf{A}}}][\bar{\mathbf{I}}_b(t)] + [\bar{\bar{\mathbf{K}}}]^T[\bar{\bar{\mathbf{Z}}}_{s_b}^s]^{-1}[\bar{\bar{\mathbf{K}}}][\bar{\mathbf{V}}_p(t)] = [\bar{\mathbf{F}}(t)]$$

$$-[\bar{\bar{\mathbf{K}}}]^T[\bar{\bar{\mathbf{Z}}}^s_{s_b}]^{-1}[\sum_{\theta=1}^{N_b}\sum_{\vartheta=1}^{N_b}\int_{l_\theta}\int_{l_\vartheta} G^{\varphi^t_q}_{\xi\varsigma}(\bar{r}_\theta,\bar{r}_\vartheta)\frac{1}{l_\theta}\frac{1}{l_\vartheta}dr_\vartheta dr_\theta][\bar{\mathbf{I}}_{s_b}(t-t_\varphi)] \quad (11.47)$$

- The node matrix approach

$$[\bar{\bar{\mathbf{Z}}}^s_b][\bar{\mathbf{I}}_b(t)] + [\bar{\bar{\mathbf{M}}}^s_b]\frac{d[\bar{\mathbf{I}}_b(t)]}{dt} + [\bar{\bar{\mathbf{A}}}]^T[\bar{\mathbf{V}}_p(t)] = -[\bar{\bar{\mathbf{Z}}}^q_b][\bar{\mathbf{I}}_b(t-t_c)]$$

$$-[\sum_{\theta=1}^{N_b}\sum_{\vartheta=1}^{N_b}\int_{l_\theta}\int_{l_\vartheta}\hat{r}_\theta\cdot\bar{\bar{G}}^{A^t_q}_{\xi\varsigma}(\bar{r}_\theta,\bar{r}_\vartheta)\cdot\hat{r}^T_\vartheta dr_\vartheta dr_\theta]\frac{d[\bar{\mathbf{I}}_b(t-t_A)]}{dt} \quad (11.48)$$

$$-[\bar{\bar{\mathbf{A}}}][\bar{\mathbf{I}}_b(t)] + [\bar{\bar{\mathbf{Z}}}^s_{s_p}]^{-1}[\bar{\mathbf{V}}_p(t)] = [\bar{\mathbf{F}}(t)]$$

$$-[\bar{\bar{\mathbf{Z}}}^s_{s_p}]^{-1}[\frac{1}{N^\varsigma_\theta}\sum_{\theta=1}^{N^\varsigma_\theta}\frac{1}{l^\varsigma_\theta}\int_{l^\varsigma_\theta}\frac{1}{N^\upsilon_\vartheta}\sum_{\vartheta=1}^{N^\upsilon_\vartheta}\frac{1}{l^\upsilon_\vartheta}\int_{l^\upsilon_\vartheta}G^{\varphi^t_q}_{\xi\varsigma}(\bar{r}_\varsigma,\bar{r}_\upsilon)dr_\varsigma dr_\upsilon][\bar{\mathbf{I}}_{s_p}(t-t_\varphi)] \quad (11.49)$$

Least unknown matrix equation

$$[\bar{\bar{\mathbf{Z}}}^s_b][\bar{\mathbf{I}}_b(t)] + [\bar{\bar{\mathbf{M}}}^s_b(t)]\frac{d[\bar{\mathbf{I}}_b(t)]}{dt} + [\bar{\bar{\mathbf{A}}}]^T[\bar{\bar{\mathbf{Z}}}^s_{s_p}][\bar{\bar{\mathbf{A}}}][\bar{\mathbf{I}}_b(t)] = -[\bar{\bar{\mathbf{Z}}}^q_b][\bar{\mathbf{I}}_b(t-t_c)]$$

$$-[\sum_{\theta=1}^{N_b}\sum_{\vartheta=1}^{N_b}\int_{l_\theta}\int_{l_\vartheta}\hat{r}_\theta\cdot\bar{\bar{G}}^{A^t_q}_{\xi\varsigma}(\bar{r}_\theta,\bar{r}_\vartheta)\cdot\hat{r}^T_\vartheta dr_\vartheta dr_\theta]\frac{d[\bar{\mathbf{I}}_b(t-t_A)]}{dt} - [\bar{\bar{\mathbf{A}}}]^T[\bar{\bar{\mathbf{Z}}}^s_{s_p}][\bar{\mathbf{F}}(t)]$$

$$+[\bar{\bar{\mathbf{A}}}]^T[\frac{1}{N^\varsigma_\theta}\sum_{\theta=1}^{N^\varsigma_\theta}\frac{1}{l^\varsigma_\theta}\int_{l^\varsigma_\theta}\frac{1}{N^\upsilon_\vartheta}\sum_{\vartheta=1}^{N^\upsilon_\vartheta}\frac{1}{l^\upsilon_\vartheta}\int_{l^\upsilon_\vartheta}G^{\varphi^t_q}_{\xi\varsigma}(\bar{r}_\varsigma,\bar{r}_\upsilon)dr_\varsigma dr_\upsilon][\bar{\mathbf{I}}_{s_p}(t-t_\varphi)] \quad (11.50)$$

where $G^{s_t}_\varphi(\bar{r}_\theta,\bar{r}_\vartheta)$ and $\bar{\bar{G}}^{s_t}_A(\bar{r}_\theta,\bar{r}_\vartheta)$ are, respectively, time domain scalar monopole Green's function and vector dipole dyadic Green's function within horizontal multilayered earth model without delay effects; $G^{q_t}_\varphi(\bar{r}_\theta,\bar{r}_\vartheta)$ and $\bar{\bar{G}}^{q_t}_A(\bar{r}_\theta,\bar{r}_\vartheta)$ are, respectively, time domain scalar monopole Green's function and vector dipole dyadic Green's function within horizontal multilayered earth model with delay effects, all these had been fully discussed in front Chap. 3. It must be pointed out that the dyadic and scalar Green's function of point sources in multilayered horizontal layered soil discussed in this section contain the frequency characteristics of soil parameters.

11.9.2 The Time Domain Spatial 1-d Vector Hybrid Methods Based on Different Basic Function (PI and PEEC) Considering the Frequency Characteristics of Soil Parameters for Simulating Transient Grounding Problems

- The branch matrix approach

$$\left[\bar{\bar{\mathbf{Z}}}_b^s + \frac{\bar{\bar{\mathbf{M}}}_b^s}{\Delta t}\right][\bar{\mathbf{I}}_b(t_k)] + [\bar{\bar{\mathbf{A}}}]^T[\bar{\mathbf{V}}_p(t_k)] = [\bar{\bar{\mathbf{M}}}_b^s]\frac{[\bar{\mathbf{I}}_b(t_{k-1})]}{\Delta t} - [\bar{\bar{\mathbf{Z}}}_b^q][\bar{\mathbf{I}}_b(t_k - t_c^k)]$$

$$- [\bar{\bar{\mathbf{M}}}_b^q]\frac{[\bar{\mathbf{I}}_b(t_k - t_A^k)] - [\bar{\mathbf{I}}_b(t_{k-1} - t_A^k)]}{\Delta t} \tag{11.51}$$

$$- [\bar{\bar{\mathbf{A}}}][\bar{\mathbf{I}}_b(t_k)] + [\bar{\bar{\mathbf{K}}}]^T[\bar{\bar{\mathbf{Z}}}_{s_b}^s]^{-1}[\bar{\bar{\mathbf{K}}}][\bar{\mathbf{V}}_p(t_k)] = [\bar{\mathbf{F}}(t_k)] - [\bar{\bar{\mathbf{K}}}]^T[\bar{\bar{\mathbf{Z}}}_{s_b}^s]^{-1}[\bar{\bar{\mathbf{Z}}}_{s_b}^q][\bar{\mathbf{I}}_{s_b}(t_k - t_\varphi^k)] \tag{11.52}$$

This equation is called PI 1 case in this book.

- The node matrix approach

$$\left[\bar{\bar{\mathbf{Z}}}_b^s + \frac{\bar{\bar{\mathbf{M}}}_b^s}{\Delta t}\right][\bar{\mathbf{I}}_b(t_k)] + [\bar{\bar{\mathbf{A}}}]^T[\bar{\mathbf{V}}_p(t_k)] = [\bar{\bar{\mathbf{M}}}_b^s]\frac{[\bar{\mathbf{I}}_b(t_{k-1})]}{\Delta t} - [\bar{\bar{\mathbf{Z}}}_b^q][\bar{\mathbf{I}}_b(t_k - t_c^k)]$$

$$- [\bar{\bar{\mathbf{M}}}_b^q]\frac{[\bar{\mathbf{I}}_b(t_k - t_A^k)] - [\bar{\mathbf{I}}_b(t_{k-1} - t_A^k)]}{\Delta t} \tag{11.53}$$

$$- [\bar{\bar{\mathbf{A}}}][\bar{\mathbf{I}}_b(t_k)] + [\bar{\bar{\mathbf{Z}}}_{s_p}^s]^{-1}[\bar{\mathbf{V}}_p(t_k)] = [\bar{\mathbf{F}}(t_k)] - [\bar{\bar{\mathbf{Z}}}_{s_p}^s]^{-1}[\bar{\bar{\mathbf{Z}}}_{s_p}^q][\bar{\mathbf{I}}_{s_p}(t_k - t_\varphi^k)] \tag{11.54}$$

This equation is called PI 2 case in this book.
Least unknown equation

$$\left[\bar{\bar{\mathbf{Z}}}_b^s + \frac{\bar{\bar{\mathbf{M}}}_b^s}{\Delta t} + [\bar{\bar{\mathbf{A}}}]^T[\bar{\bar{\mathbf{Z}}}_{s_p}^s][\bar{\bar{\mathbf{A}}}]\right][\bar{\mathbf{I}}_b(t_k)] = [\bar{\bar{\mathbf{M}}}_b^s]\frac{[\bar{\mathbf{I}}_b(t_{k-1})]}{\Delta t} - [\bar{\bar{\mathbf{Z}}}_b^q][\bar{\mathbf{I}}_b(t_k - t_c^k)]$$
$$- [\bar{\bar{\mathbf{M}}}_b^q]\frac{[\bar{\mathbf{I}}_b(t_k - t_A^k)] - [\bar{\mathbf{I}}_b(t_{k-1} - t_A^k)]}{\Delta t} - [\bar{\bar{\mathbf{A}}}]^T[\bar{\bar{\mathbf{Z}}}_{s_p}^s][\bar{\mathbf{F}}(t_k)] + [\bar{\bar{\mathbf{A}}}]^T[\bar{\bar{\mathbf{Z}}}_{s_p}^q][\bar{\mathbf{I}}_{s_p}(t_k - t_\varphi^k)] \tag{11.55}$$

This equation is called PI 2 I case in this book.

The above three discrete time domain matrix equations represent a marching-on-in-time formulation. Accordingly, they can be quickly solved through iterations.

11.10 Numerical Example Analysis and Discussion

In this section, numerical examples will be used to analyze the effect of considering the frequency characteristics of soil parameters on the lightning impact response of substation grounding grids.

11.10.1 Validation Examples

First of all, it should be noted that the acronyms "freq case" and "time case" imply various algorithms in the frequency and time domains.

The first arithmetic example is to validate the algorithms in this chapter by comparing them with measurements in the literature [4]. The first and subsequent lightning current strokes are given in (a) and (b) in Fig. 11.1, respectively. Three lengths of horizontal ground rods are given in this example, 3 m, 10 m and 30 m respectively. They were buried in soil with a uniform resistivity of 100 ohm-m, a relative soil dielectric constant of 20, and satisfy equations (11.2) and (11.3) for the frequency characteristics of the soil parameters. These horizontal ground rods are buried in the soil at a depth of 0.5 m and have a radius of 7 mm. Comparative plots of the GPR at the injection point of the horizontal ground rod, numerically simulated by various algorithms in the frequency and time domains, and measured, are shown in subplots (a), (b), (c) and (b) of Figs. 11.2, 11.3, and 11.4, respectively. Further, the GPR under the action of first lightning current stroke simulated by the frequency and time domain algorithms, respectively, is shown in subfigures (a) and (b). The GPR under subsequent lightning current strokes simulated by the frequency domain and time domain algorithms, respectively, is shown in subfigures (c) and (d).

Observing all sub-figures (a), (b), (c) and (d) of Figs. 11.2, 11.3, and 11.4, it can be known that: the transient voltage curves simulated by numerical simulation of frequency and time domain algorithms are basically the same as the measured curves for the responses under the impacts of two kinds of lightning currents, irrespective of the length of the horizontal grounding rods, but at the same time it can be seen that there is a certain acceptable error in the results of the calculations of the frequency and time domain.

The second example is from Literature [34], where the lightning current waveform is the same as in the previous example (see Fig. 11.1), the horizontal ground rod is 30 m, the burial depth is 0.5 m and have a radius of 7 mm, the resistivity of the uniform earth is 300 ohm-meters, and the relative dielectric constant is 10.

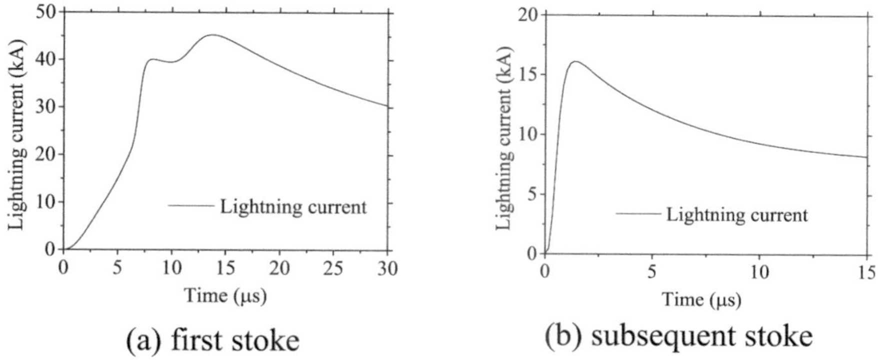

Fig. 11.1 Representative lightning current waves of: **a** first stroke and **b** subsequent lightning stroke. Adapted from [4]

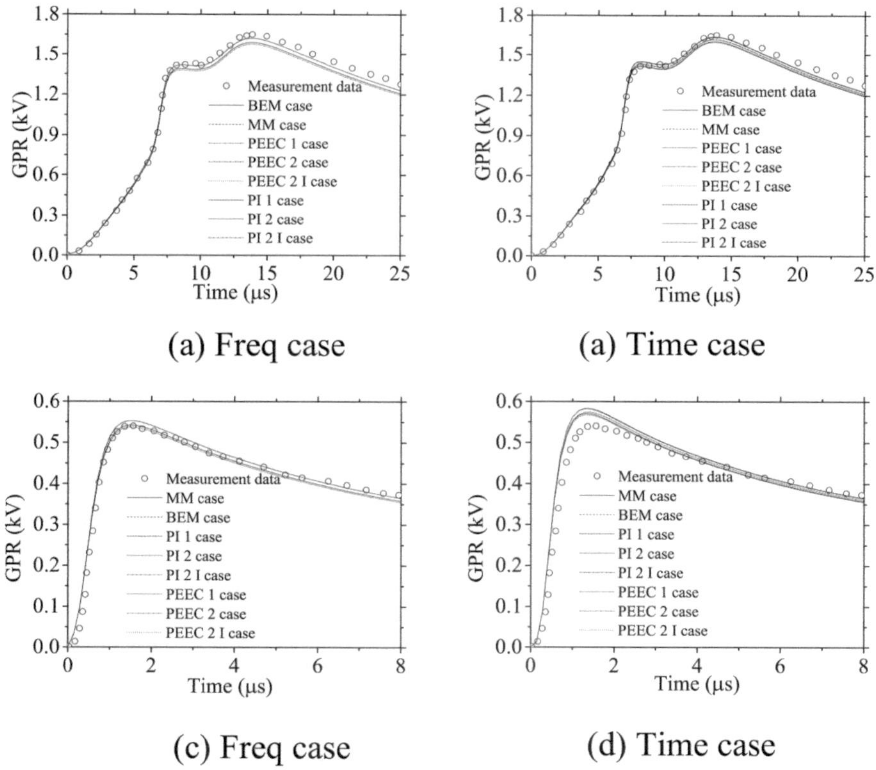

Fig. 11.2 GPR at the injection point of a horizontal grounding rod with a length of 3 m

First of all, it should be noted that the icon "reference case" in all subfigures of Fig. 11.5 refers to the results of numerical simulations in Literature [34] and is used as a standard of comparison.

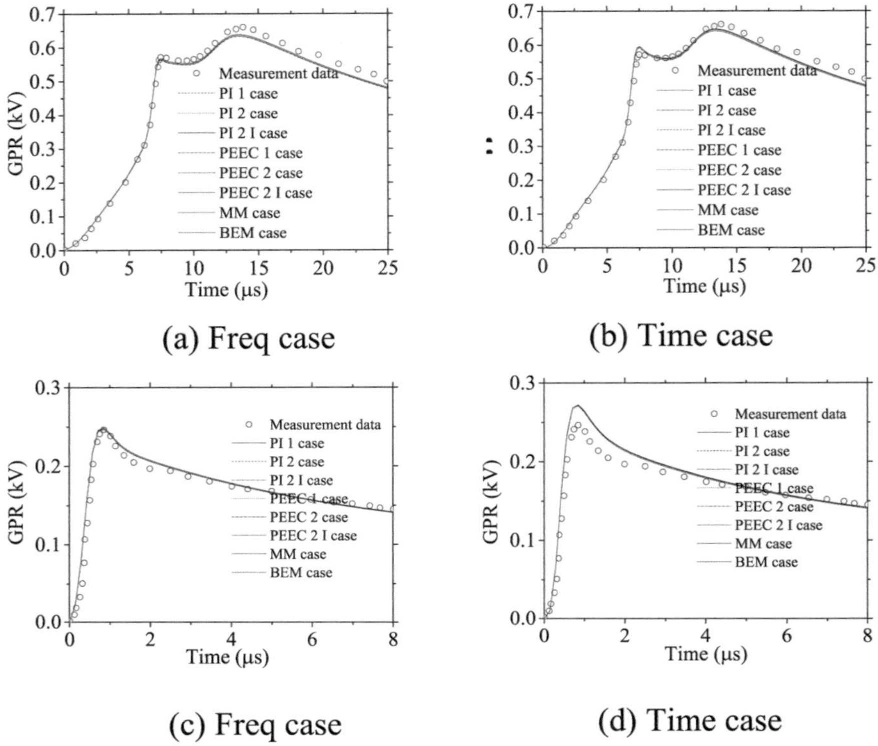

Fig. 11.3 GPR at the injection point of a horizontal grounding rod with a length of 10 m

Observing all the sub-diagrams in Fig. 11.3, it can be known that: no matter the time domain algorithm or the frequency domain algorithm, all the algorithms simulate the transient voltage curves and the reference curves are basically the same, even though the time domain algorithms have a certain amount of computational error, but it can still be in the numerical error within the acceptable range.

11.10.2 Numerical Results and Discussions

As in the previous chapter, the general grounding system used is shown in Fig. 11.7a, consisting of three 3-square grids and a grounding depth of 0.5 m. The conductor radius of all grids is 7mm. The earth is simulated as a horizontal four layer conductive medium, with a resistivity and permittivity of $\rho_1 = 100\Omega\text{m}$ and $\rho_2 = 500\ \Omega\text{m}$, $\rho_3 = 400\ \Omega\text{m}$, $\rho_4 = 300\ \Omega\text{m}$, $\varepsilon_1 = 10\varepsilon_0$, $\varepsilon_2 = 12\varepsilon_0$, $\varepsilon_3 = 13\varepsilon_0$ and $\varepsilon_4 = 12\varepsilon_0$, respectively. The thickness of the first, second, and third layers is $h_1 = 1m$, $h_2 = 12m$ and

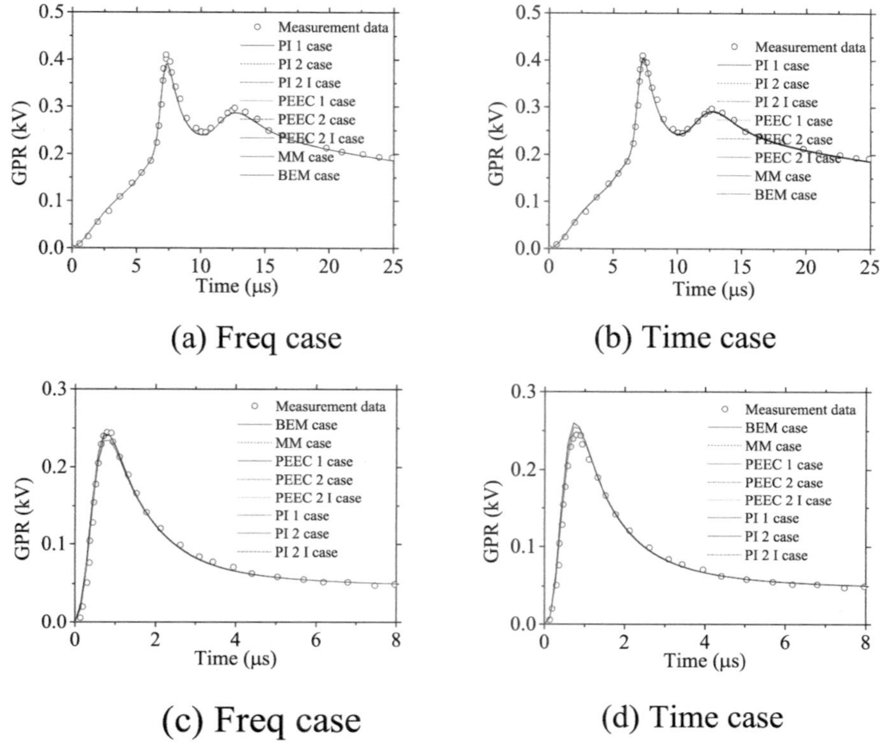

Fig. 11.4 GPR at the injection point of a horizontal grounding rod with a length of 30 m

$h_3 = 8m$, respectively. In order to study the relationship between the impact characteristics of grounding grids and the frequency characteristics of soil conductive media, this calculation example adopts the cole-cole equations (11.1) model, those soil parameters in (11.1) are selected as follows: 1) The parameters of soil in the first layer are $\sigma_0 = 100^{-1}S/m, m = 0.9, c = 0.59$, and $\tau = 4.2 \times 10^{-8}$. 2) Those of the second layer are $\sigma_0 = 500^{-1}S/m, m = 0.95, c = 0.63$, and $\tau = 1.8 \times 10^{-7}$. 3) The third layer has parameters $\sigma_0 = 400^{-1}S/m, m = 0.8, c = 0.68$, and $\tau = 1.0 \times 10^{?}$. 4) The fourth layer has parameters $\sigma_0 = 300^{-1}S/m$, $m = 0.85$, $c = 0.72$, and $\tau = 3.2 \times 10^{-7}$, which can be seen in Fig. 11.6.

The external excited lightning current with $i(t) = 22.9 \times (e^{-0.01910t} - e^{-0.292288t})kA$ is injected from the corner of the grounding system. According to the formula of lightning current, the lightning parameters are T_1 is $10\mu s$, T_2 is $50\mu s$ and $I_m = 17.7kA$.

The impulse impedance results and discussions

The impulse impedance calculated by the hybrid algorithm simulation of various basis functions in the time and frequency domains can be found in Tables (11.1), (11.2), and (11.3). The words "T" and "F" in these tables represent the corresponding

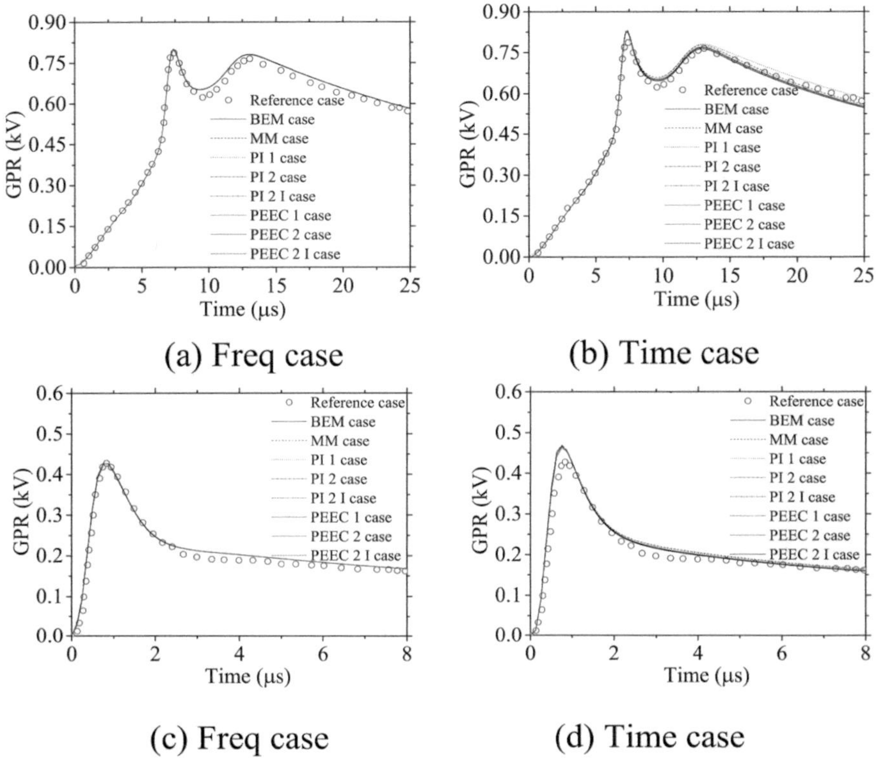

Fig. 11.5 GPR at the injection point of a horizontal grounding rod with a length of 30 m

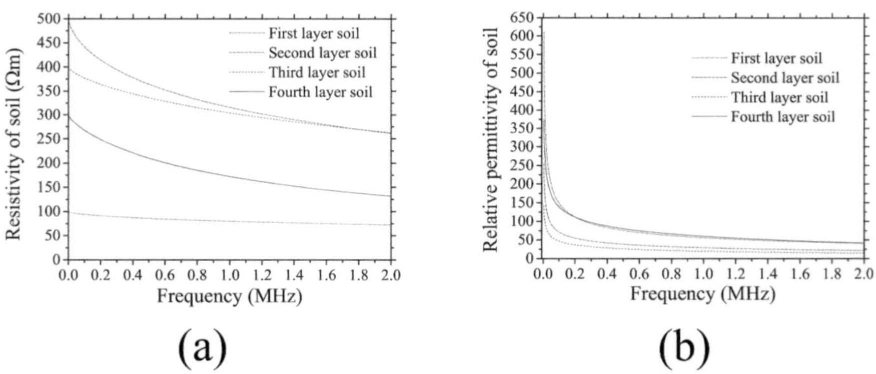

Fig. 11.6 Relationship between resistivity and relative dielectric constant with frequency of four layered soil layers

Table 11.1 Calculated impulse impedance in both frequency and time domains (a)

	Impulse impedance calculated without considering soil frequency characteristics							
	BEM	MM	PI 1	PI 1 I	PI 2	PEEC 1	PEEC 1I	PEEC 2
T	(7.84, −9.2 × 10^{-2})	(7.84, −9.2 × 10^{-2})	(7.84, −9.2 × 10^{-2})	(7.84, −9.2 × 10^{-2})	(7.84, −9.2 × 10^{-2})	(7.84, −9.2 × 10^{-2})	(7.84, −9.2 × 10^{-2})	(7.82, −9.2 × 10^{-2})
F	(7.80, −5.04)	(7.81, −5.06)	(7.82, −5.05)	(7.82, −5.05)	(7.82, −5.05)	(7.82, −5.05)	(7.82, −5.05)	(7.83, −5.06)

Table 11.2 Calculated impulse impedance in both frequency and time domains (b)

	Impulse impedance calculated considering soil frequency characteristics							
	BEM	MM	PI 1	PI 1 I	PI 2	PEEC 1	PEEC 1 I	PEEC 2
T	(7.48, −0.53)	(7.48, −0.53)	(7.48, −0.52)	(7.48, −0.52)	(7.64, −0.34)	(7.48, −0.52)	(7.48, −0.52)	(7.55, −0.43)
F	(7.53, −5.08)	(7.53, −5.08)	(7.54, −5.09)	(7.54, −5.09)	(7.54, −5.09)	(7.54, −5.09)	(7.54, −5.08)	(7.53, −5.09)

Table 11.3 Calculated impulse impedance in frequency domain (c)

	Impulse impedance	
	T case	TPEEC case
Without considering soil frequency characteristics	(7.81, −5.05)	(7.79, −5.06)
Considering soil frequency characteristics	(7.57, −5.08)	(7.52, −5.11)

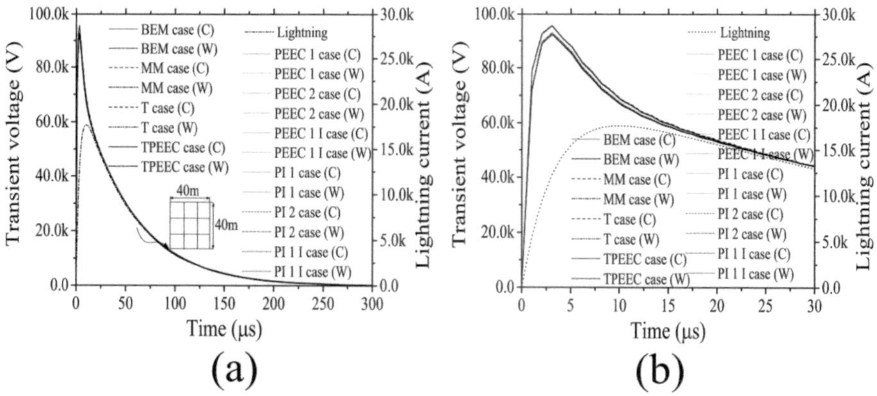

Fig. 11.7 Transient voltage at injection point (Frequency domain case)

algorithms in the time and frequency domains, respectively. Observing the impulse impedance data in the three tables, it can be seen that the impulse impedance data calculated by various time-domain and frequency-domain algorithms are close, and after considering the frequency characteristics of soil parameters, the calculated impulse impedance data is relatively reduced (Tables 11.1, 11.2 and 11.3).

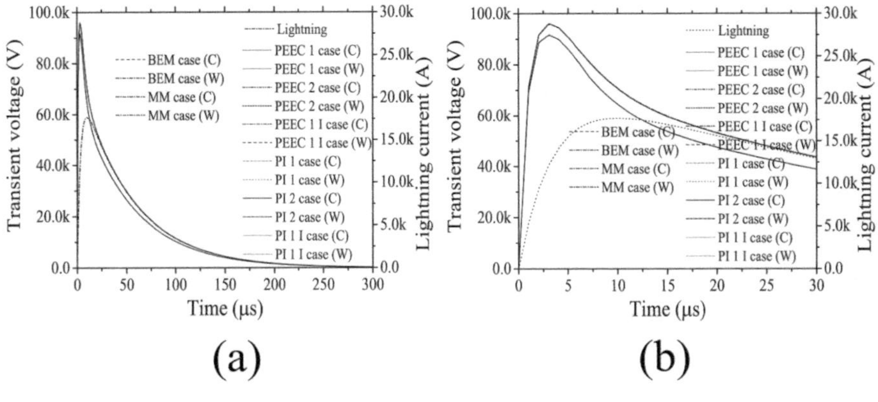

Fig. 11.8 Transient voltage at injection point (Time domain case)

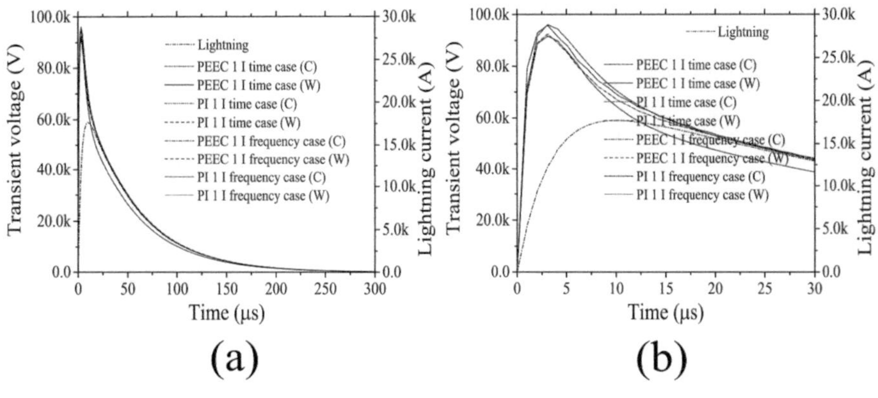

Fig. 11.9 Transient voltage at injection point (Both frequency and time domains)

Analysis and discussion of transient surge voltages

Various time-domain and frequency-domain algorithms are used to simulate and calculate the transient voltage curves of lightning current at the injection point, as shown in Figs. 11.7, 11.8, and 11.9. Figure 11.7 shows the simulation results of the frequency-domain algorithm, Fig. 11.8 shows the simulation results of the frequency-domain algorithm, and Fig. 11.9 shows the numerical simulation results of the corresponding time-domain and frequency-domain hybrid algorithms selected. The values in the three graphs are the results of considering and not considering the frequency domain characteristics of soil parameters, respectively. Moreover, in these three figures, the words "C" and "F" in parentheses respectively indicate that the influence of soil parameter frequency characteristics is not taken into account in the corresponding simulation calculations. In these three graphs, there are two small pictures (a) and (b) on the left and right sides of each graph. Here, picture (b) is an enlarged view of picture (a) at the peak time.

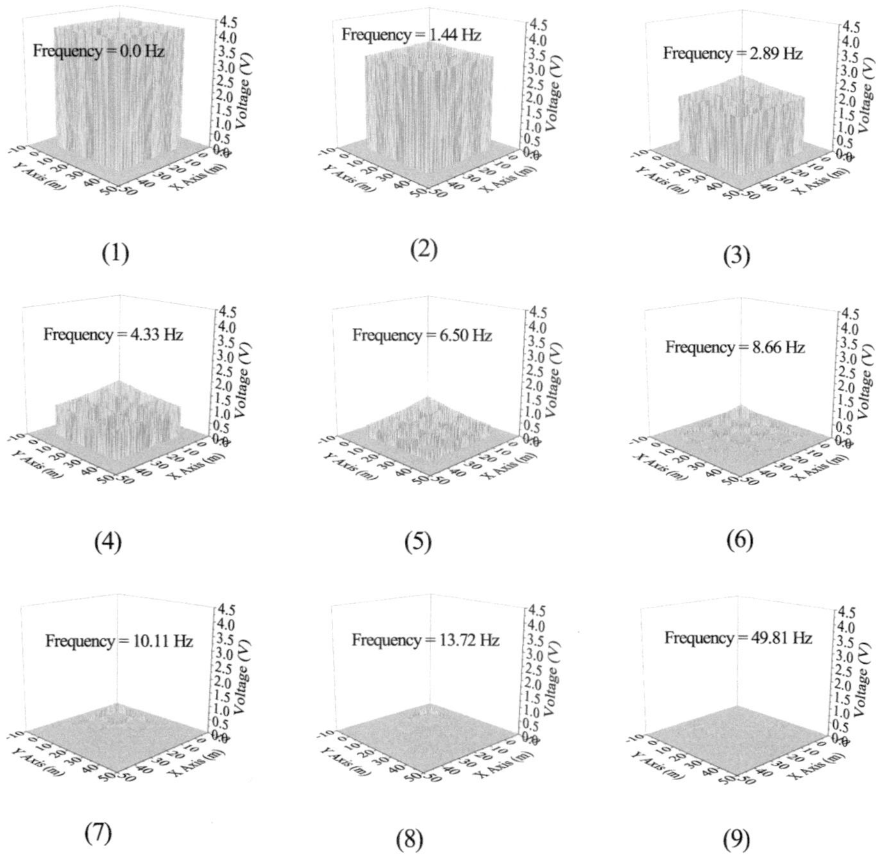

Fig. 11.10 Harmonic voltage along the grounding grid (frequency domain case)

Observing the three pictures, it can be seen that the transient voltage curves simulated by various frequency domain and time domain algorithms are basically consistent. Moreover, considering the influence of soil parameter frequency characteristics, the simulated curves correspondingly decrease. This conclusion is consistent with the conclusion of the impulse impedance data in the three tables.

By analyzing the data from the three pictures and tables above, it can be seen that the algorithms in the frequency domain and time domain can be mutually validated, and the algorithms in the two domains can also be mutually validated. These all demonstrate the correctness of the algorithms in the frequency domain and time domain.

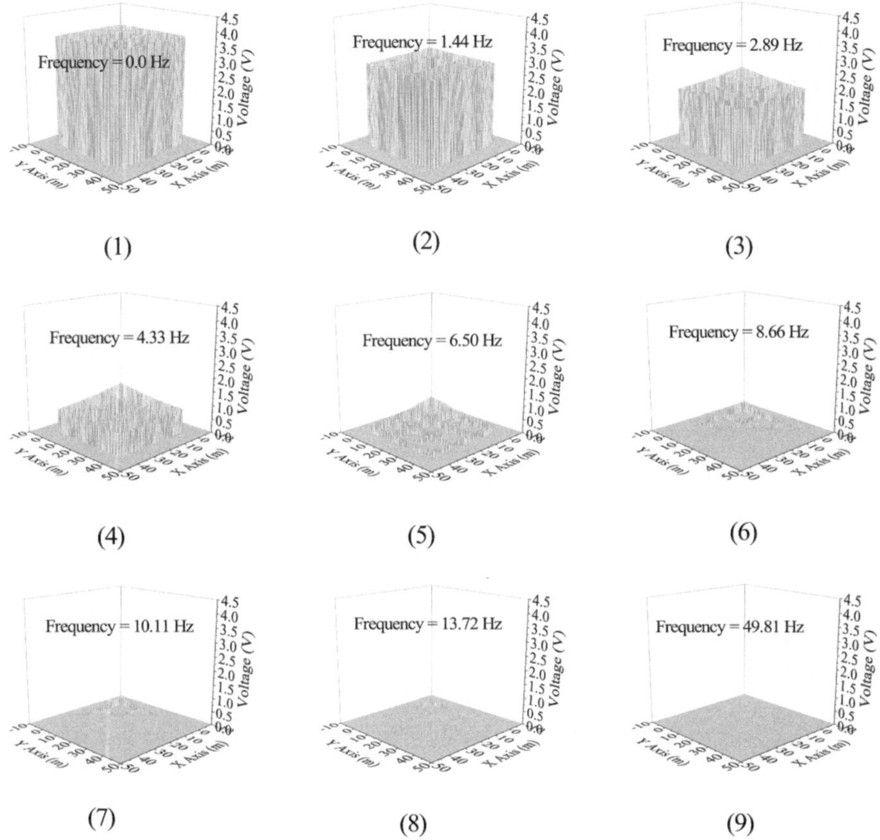

Fig. 11.11 Harmonic voltage along the grounding grid (time domain case)

Analysis of 3-d distribution plots of voltage, branch currents and leakage currents in frequency and time domains

The frequency and time domain 3-d distribution plots of voltage, branch currents and leakage currents on the conductors of the substation grounding grid are shown below, respectively. Among them, Figs. 11.10, 11.11, 11.12, 11.13, 11.14 and 11.15 are the frequency domain 3-d distribution plots while Figs. 11.16, 11.17, 11.18, 11.19, 11.20 and 11.21 are the time domain 3-d distribution plots. It is important to note that the voltages, branching and leakage currents in the frequency and time domains on the conductors of the substation grounding grid here are obtained from numerical simulations taking into account the effect of the frequency characteristics of the soil.

Analysis of 3-d distribution plots of voltage, branch currents and leakage currents in frequency domains

It should be further noted that Figs. 11.10, 11.11, 11.12, 11.13, 11.14 and 11.15 show the 3-d distributions of each key harmonic of voltage, branch current, and leakage current, respectively, in the frequency domain. Only the 3-d distribution

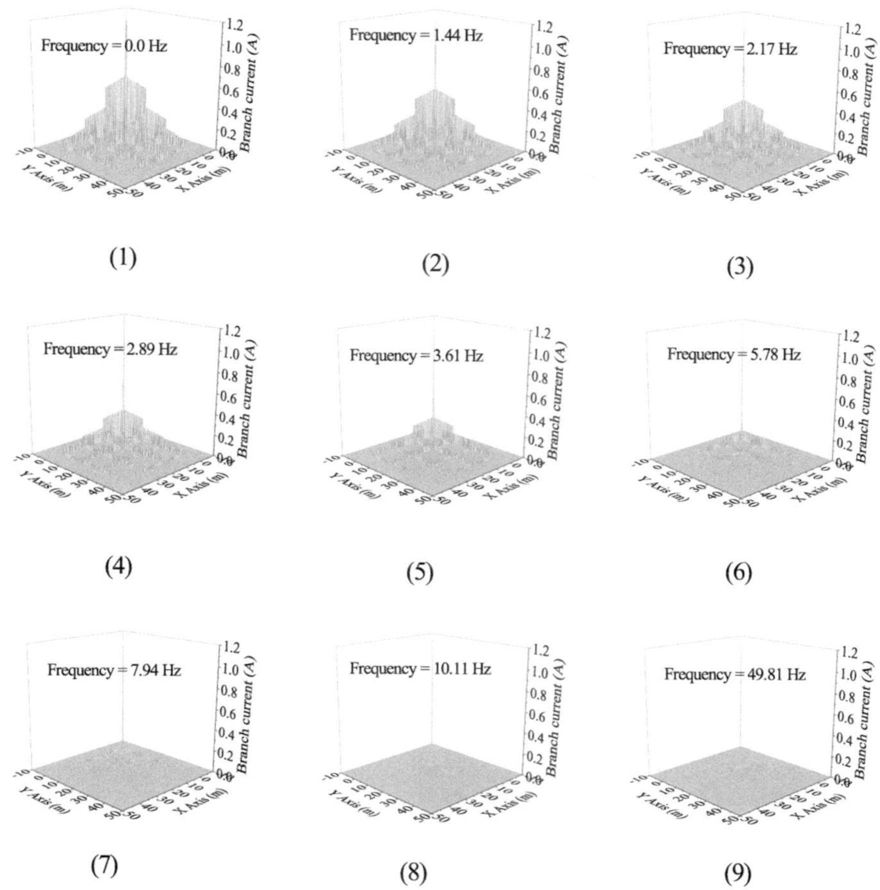

Fig. 11.12 Harmonic branch current along the grounding grid (frequency domain case)

of the positive part of each critical harmonic of the voltage, branch current and leakage current in the frequency domain is of interest here. Of the three sets of 3-d distribution plots of voltage, branch currents and leakage currents here, the former, Figures 11.10, 11.12, 11.13 and 11.14, are obtained from numerical simulations based on the PI 2 I basis function hybrid algorithm in the frequency domain. The latter Figs. 11.11, 11.13, 11.14 and 11.15 are obtained by numerical simulation of time-domain voltages, branch currents and leakage currents based on the time-domain PI 2 I basis function hybrid algorithm, and then the corresponding frequency-domain voltages, branch currents and leakage currents are obtained by the Fourier transform method.

Observing here the two 3-d voltage distributions Figs. 11.10 and 11.11, first note that the voltage amplitude of the 3-d distribution is maximum at frequency zero, after which it gradually diminishes with increasing frequency until it disappears. It should

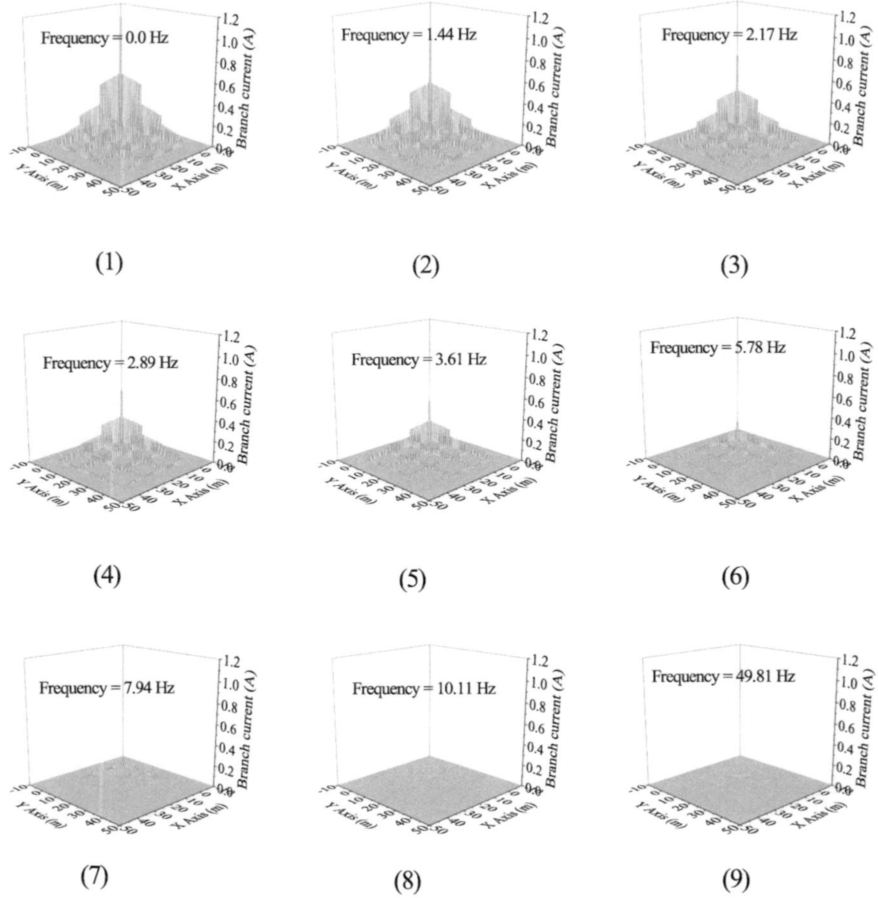

Fig. 11.13 Harmonic branch current along the grounding grid (time domain case)

be noted that the maximum value of the 3-d voltage distribution at each frequency is at the point of incidence, and then as the frequency increases, the maximum point of the 3-d voltage distribution at each frequency remains at the point of incidence, even though the voltage magnitude in the 3-d voltage distribution decreases accordingly. Secondly, by comparing the two plots of the same frequency in the voltage distribution graph, it can be seen that the two 3-d voltage distribution graphs of the same frequency are basically the same.

Having previously focused on the 3-d voltage distribution plots, the following is a further look at the 3-d branch current distribution plots Figs. 11.12 and 11.13. First, similar to the 3-d voltage distribution plots, the overall amplitude of the branch currents in the 3-d branch current distribution plots is the largest at frequency zero, after which the amplitude of the branch currents in the 3-d plots diminishes until it disappears as the frequency increases; Moreover, in the 3-d distribution plots for

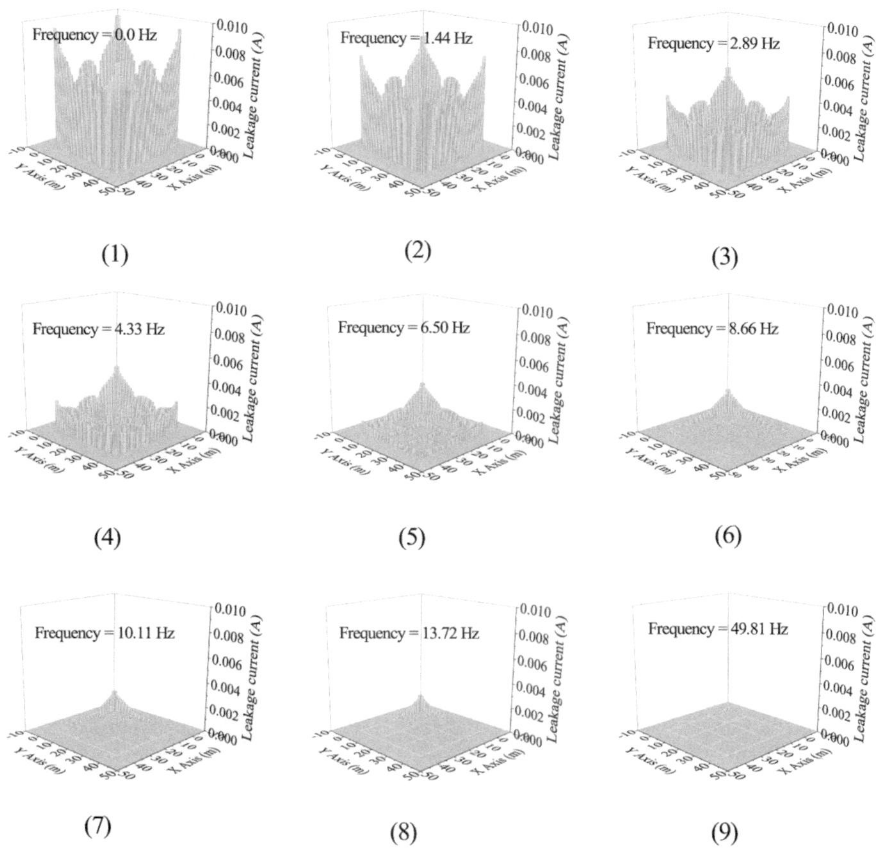

Fig. 11.14 Harmonic leakage current along the grounding grid (frequency domain case)

each frequency, it can be observed that the largest values of branch currents across the grounding grid occur at the point of incidence. Secondly, it is also consistent with the 3-d voltage distribution pattern, respectively, two consecutive observations of the same frequency of the two plots, it can be seen that the two algorithms in the frequency domain and time domain to obtain the 3-d branch voltage distribution map is basically the same.

The 3-d voltage and branch current distribution plots were observed above, respectively, and further attention is given to the 3-d leakage current distribution plot below. Consistent with the observation of the 3-d voltage and branch current distribution graphs, the maximum amplitude of the overall leakage current occurs at zero frequency, after which the amplitude of the overall distribution of leakage current gradually decreases with the increase of frequency until it disappears. Also in the single frequency case, the 3-d leakage current profile shows that the largest magnitude

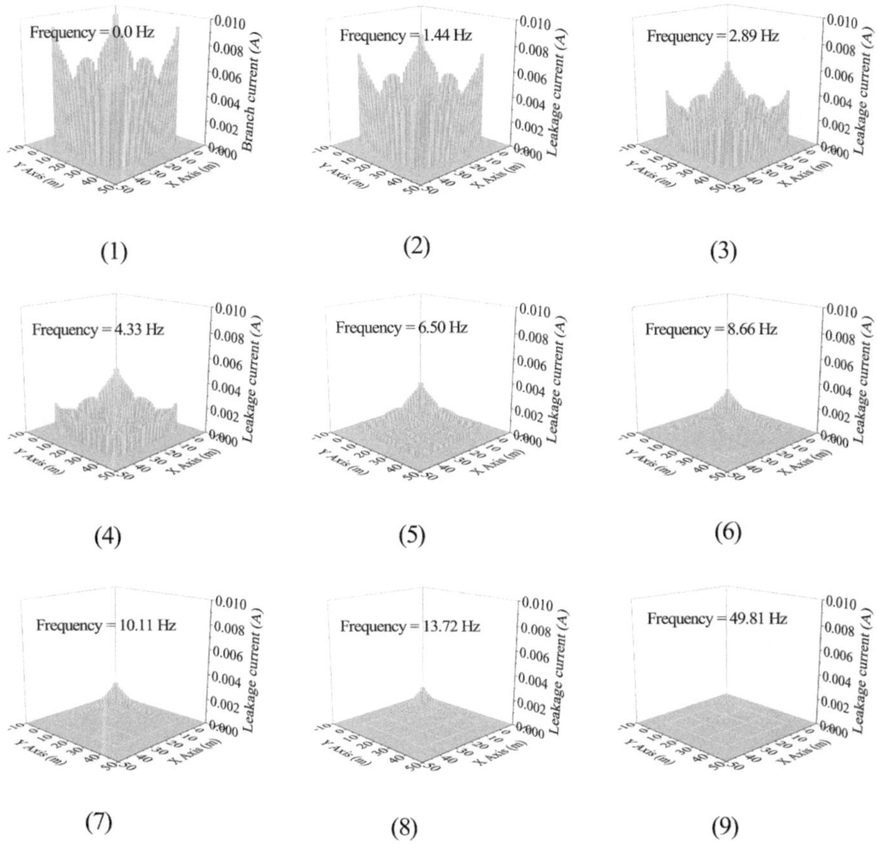

Fig. 11.15 Harmonic leakage current along the grounding grid (time domain case)

occurs at the point of incidence, and this is consistently maintained until it disappears with increasing frequency. Second, as in the case of 3-d voltage and branch current distributions, the frequency and time domain algorithms obtain essentially identical 3-d distributions of leakage currents by observing the two 3-d leakage current distributions at the same frequencies in sequence, respectively.

Analysis of 3-d distribution plots of voltage, branch currents and leakage currents in time domains

The 3-d distribution diagrams of voltage, branch currents and leakage currents on conductors on substation grounding grids in the frequency domain have been analyzed and discussed above, and the 3-d distribution diagrams of voltage, branch currents and leakage currents on conductors on substation grounding grids in the time domain are then analyzed and discussed in the following.

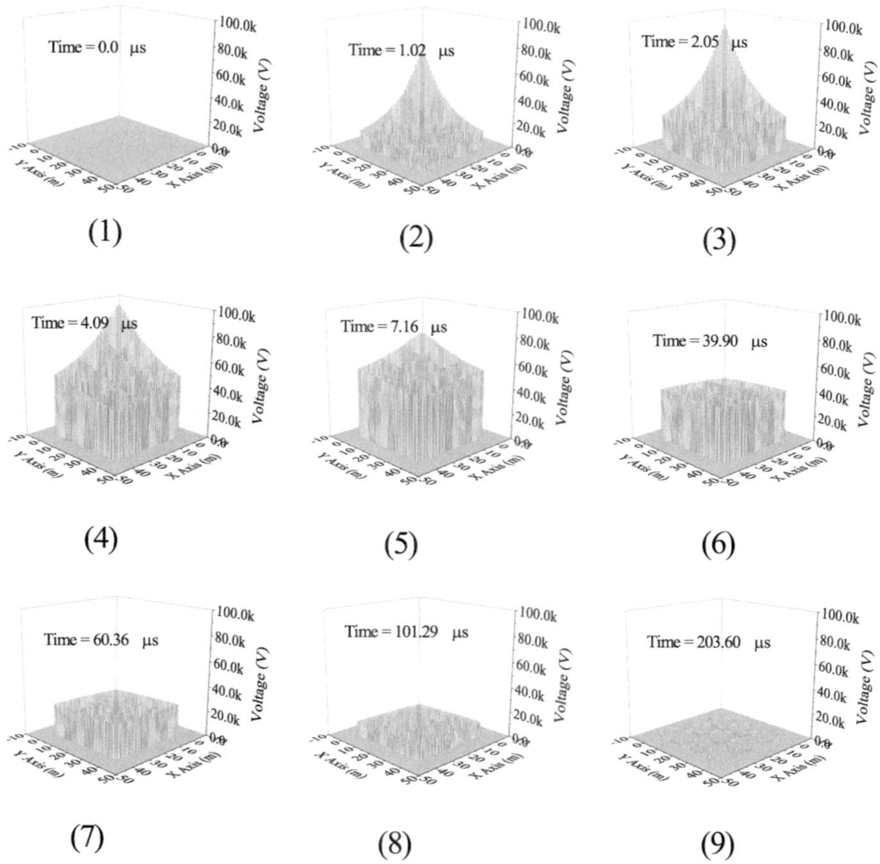

Fig. 11.16 Each time voltage along the grounding grid (time domain case)

Observing Figs. 11.16, 11.17, 11.18, 11.19, 11.20 and 11.21, it is known that Figs. 11.16, 11.17, 11.18, 11.19, 11.20 and 11.21, are 3-d distributions of voltages, branch currents, and leakage currents in time domain obtained by numerical simulation of hybrid algorithms based on PI 2 I basis functions in time and frequency domains, respectively. It is important to point out that the hybrid algorithm based on PI 2 I basis functions in the frequency domain numerically simulates the voltage, branch current and leakage current in the frequency domain of the grounding grid, and then the voltage, branch current and leakage current in the time domain are obtained by Fourier transform.

Here, observing Figs. 11.16, 11.17, 11.18, 11.19, 11.20 and 11.21, one can see the 3-d distribution of voltages, branch currents, and leakage currents on the conductors of the grounding grid in the time domain at critical moments. It is also important to realize that only the positive portion of the voltage, branch and leakage currents in

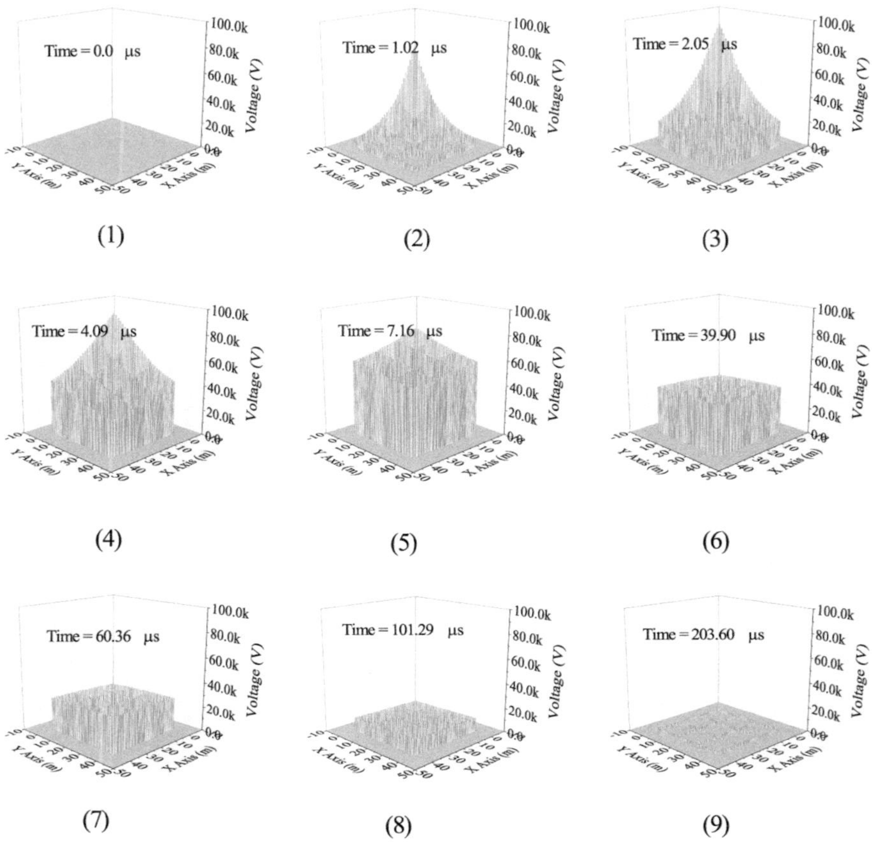

Fig. 11.17 Each time voltage along the grounding grid (frequency domain case)

the time domain are of interest here. Unlike the 3-d distribution of voltage, branch and leakage current in the frequency domain above, the 3-d distribution pattern of voltage, branch and leakage current in the time domain here has some correlation with the incident lightning current. All of them start from zero, gradually increase to the maximum, and then further gradually weaken to disappear.

Observing Figs. 11.16 and 11.17, it can be seen that the transient voltage at the lightning current injection point of the substation grounding grid starts from zero, gradually increases until the maximum value, and then gradually decreases until it disappears. The 3-d distributions of the critical moments are given in the figures, respectively. Unlike the frequency domain voltage 3-d distribution case above, the largest magnitude voltage on the time domain transient voltage 3-d distribution plot starts to occur at the lightning current injection point on the grounding grid, but after the overall reduction of the transient voltage, the largest magnitude voltage occurs

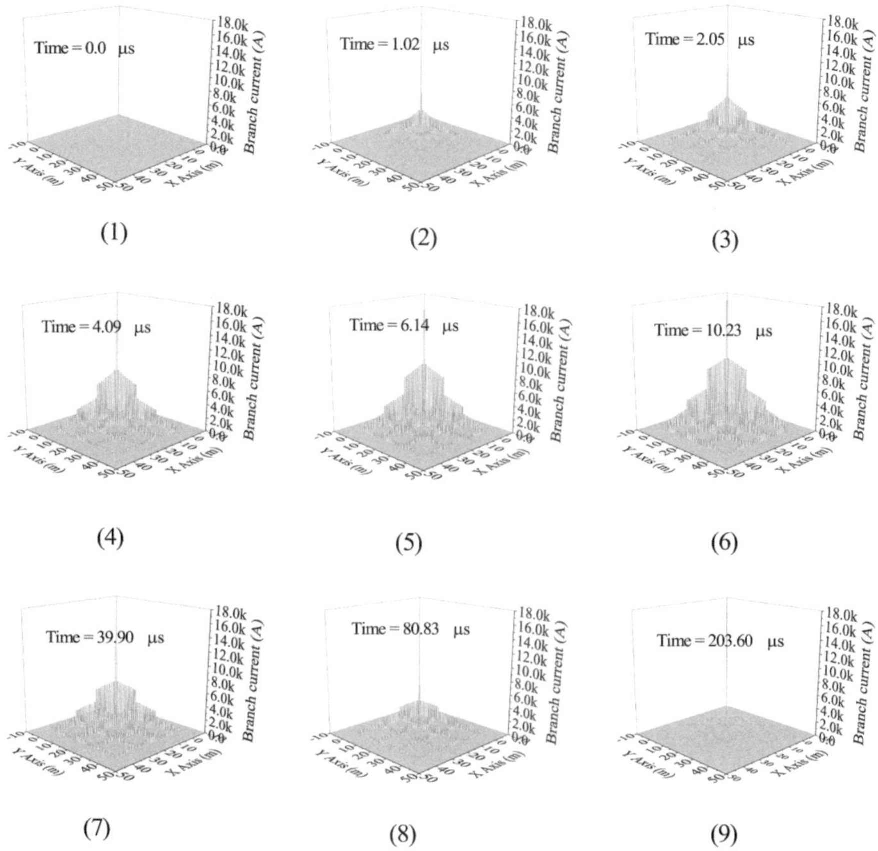

Fig. 11.18 Each time branch current along the grounding grid (time domain case)

at the diagonal point of the lightning current injection point on the grounding grid, while the voltage at the injection point morphs to the lowest magnitude. At the same time, by comparing the 3-d transient voltage distribution graphs of the same moments obtained from the simulation of time-domain and frequency-domain algorithms in sequence, respectively, it can be known that the 3-d transient voltage distribution graphs of the same moments are basically the same.

Observing Figs. 11.18 and 11.19, it can be seen that unlike the 3-d distribution of frequency domain branch currents, the 3-d distribution of time domain branch currents at different times starts from zero amplitude as a whole, gradually increases with time, reaches the maximum amplitude as a whole, and gradually weakens with time until it disappears. Throughout the entire time period, the maximum amplitude of the branch current at each moment occurred at the injection point of the grounding grid lightning current, without any displacement phenomenon. At the same time, it can also be seen that the frequency domain and time domain algorithms simulate the

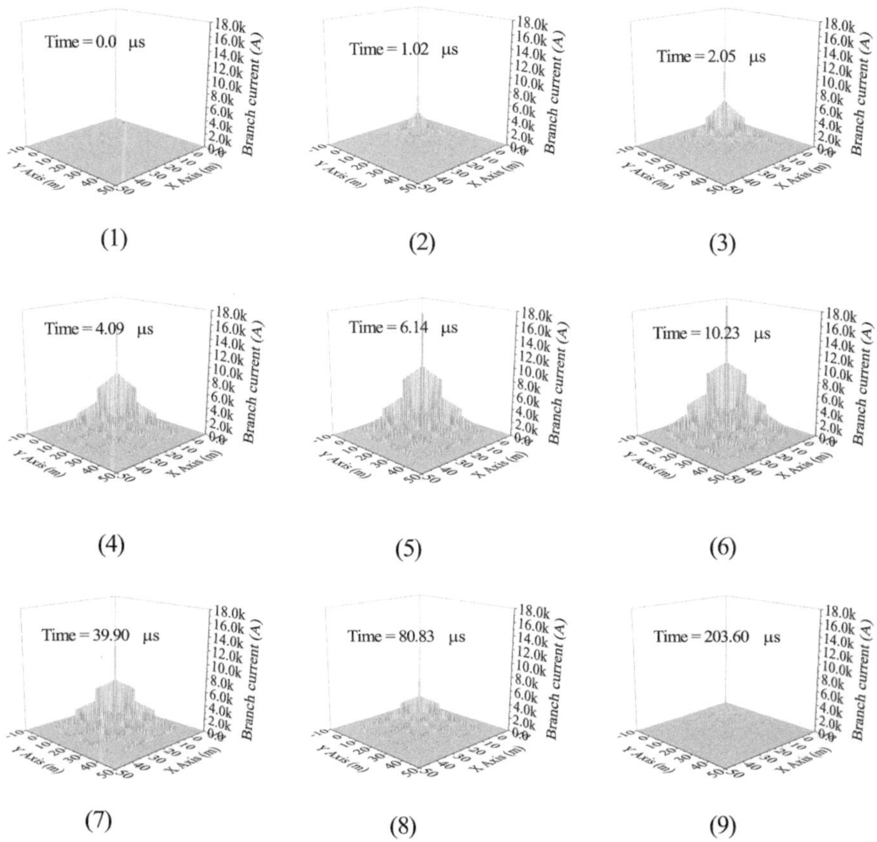

Fig. 11.19 Each time branch current along the grounding grid (frequency domain case)

same time 3-d branch current distribution map. By comparing the two 3-d distribution maps at the same time in sequence, it can be seen that the two are basically consistent.

Finally, observing Figs. 11.20 and 11.21, it can also be seen that throughout the entire time period, the leakage current on the grounding grid conductor of the substation also starts from zero amplitude, gradually increases with time, reaches the maximum amplitude, and then gradually weakens with time until it disappears to zero. Similarly, at each moment, the maximum amplitude of the leakage current on the 3-d leakage current distribution map of the entire grounding grid occurs at the lightning current injection point. Similarly, as time goes on, there is no displacement phenomenon. At the same time, by observing the frequency domain and time domain algorithms sequentially, the 3-d leakage current distribution map at the same time can be obtained. It can be known that the two are basically consistent.

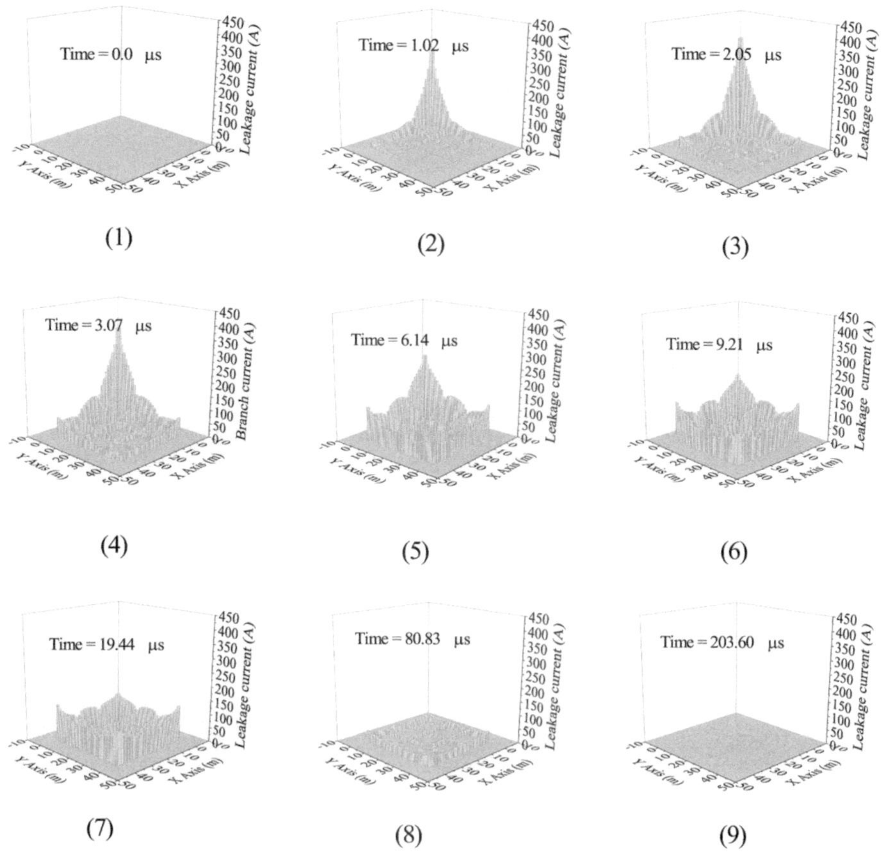

Fig. 11.20 Each time leakage current along the grounding grid (time domain case)

11.11 Summary

In this chapter, the correctness and validity of the simulation results obtained by various frequency domain and time domain algorithms are discussed whether the frequency effect of soil parameters is considered or not. Through the analysis and discussion of numerical examples, the following conclusions are obtained: under the premise of not considering the influence of soil parameter frequency effect, the lightning transient response of substation grounding grid simulated by various frequency and time domain algorithms is higher than that simulated by various frequency and time domain algorithms under the consideration of the influence of soil parameter frequency effect.

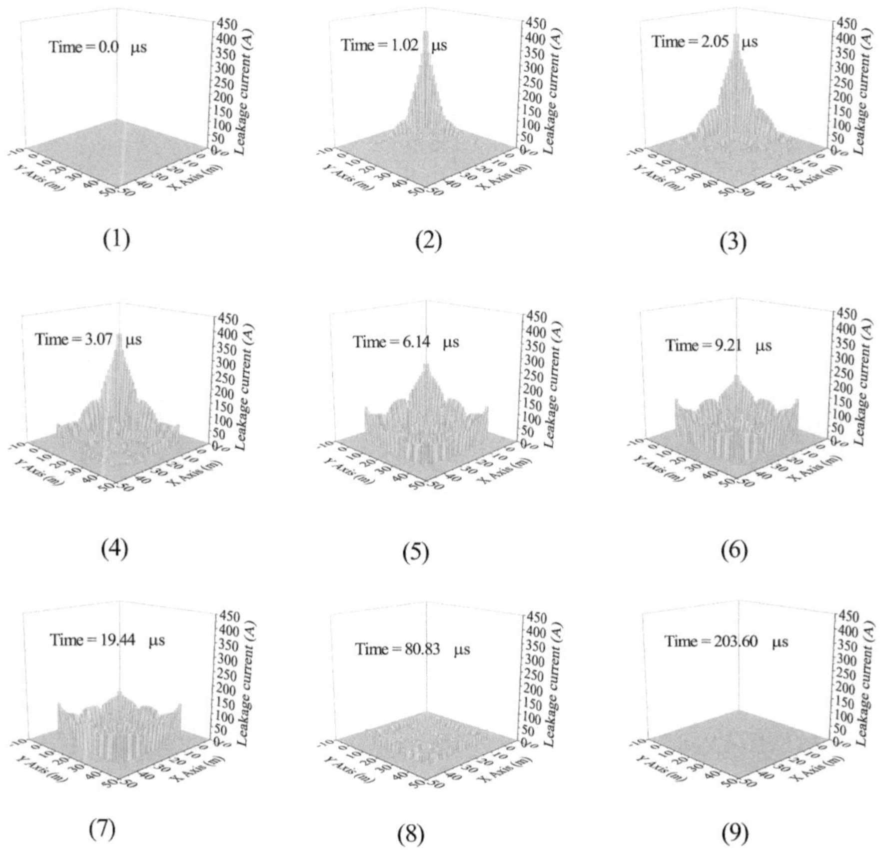

Fig. 11.21 Each time leakage current along the grounding grid (frequency domain case)

References

1. A.D.B. Manjunath, F. Khan, N. Noyanbayev, etc., Investigation into variation of resistivity and permittivity of aqueous solutions and soils with frequency and current density. IEEE Trans. Electromagn. Compatib. **62**(4), 443–455 (2022)
2. S. Visacro, R. Alipio, M.H.M. Vale, C. Pereira, The response of grounding electrodes to lightning currents: The effect of frequency-dependent soil resistivity and permittivity. IEEE Trans. Electromagn. Compat. **53**(2), 401–406 (2011)
3. R.S. Alipio, M.A.O. Schroeder, M.M. Afonso, T.A.S. Oliveira, S.C. Assis, Electric fields of grounding electrodes with frequency dependent soil parameters. Electr. Power Syst. Res. **83**, 220–226 (2012)
4. S. Visacro, R. Alipio, Impulse efficiency of grounding electrodes: Effect of frequency-dependent soil parameters. IEEE Trans. Power Delivery **27**(2), 927–935 (2012)
5. R. Alipio, S. Visacro, Frequency dependence of soil parameters: Experimental results, predicting formula and influence on the lightning response of grounding electrodes. IEEE Trans. Power Delivery **29**(2), 716–723 (2014)

6. F.H. Silveira, S. Visacro, R. Alipio, A. De Conti, Lightning-induced voltages over lossy ground: The effect of frequency dependence of electrical parameters of soil. IEEE Trans. Electromagn. Compat. **56**(5), 1129–1136 (2014)
7. R. Alipio, S. Visacro, Modeling the frequency dependence of electrical parameters of soil. IEEE Trans. Electromagn. Compat. **56**(5), 1163–1171 (2014)
8. S. Visacro, R. Alipio, C. Pereira, M. Guimarèaes, M.A.O. Schroeder, Lightning response of grounding grids: Simulated and experimental results. IEEE Trans. Power Delivery **57**(1), 121–127 (2015)
9. M.A.O. Schroeder, R.A.R. Moura, F.A. Assis, J.O.S. Paulino, Evaluation of the transmission tower and frequency-dependent soil parameters influence on the grounding potential rise waveforms. Electr. Power Syst. Res. **222**, 109495 (2023)
10. Z.-X. Li, K.-L. Gao, S.-W. Rao, Lightning response of a grounding system buried in multiple layers in earth using the peec method based on the quasi-static complex image method. Electr. Power Compon. Syst. **48**(3), 291–303 (2020)
11. A.C.S. de Lima, C. Portela, Inclusion of frequency-dependent soil parameters in transmission-line modeling. Electr. Power Components Syst. **22**(1), 492–499 (2007)
12. Z.-X. Li, Y. Yin, K.-C. Wang, Lightning response of a grounding system buried in multilayered earth based on the galerkin's boundary element and quasi-static complex image methods. IEEE Trans. Electromagn. Compat. **63**(4), 1118–1127 (2021)
13. M. Nazari, R. Moini, S. Fortin, F.P. Dawalibi, F. Rachidi, Impact of frequency-dependent soil models on grounding system performance for direct and indirect lightning strikes. IEEE Trans. Electromagn. Compat. **63**(1), 134–144 (2021)
14. M. Akbari, K. Sheshyekani, M.R. Alemi, The effect of frequency dependence of soil electrical parameters on the lightning performance of grounding systems. IEEE Trans. Electromagn. Compat. **55**(4), 739–746 (2013)
15. G.V.N. Bezerra, F.A. Moreira, T.V. Ferreira, R. Alipio, Concrete encased grounding: Lightning response analysis considering the frequency dependence of soil. IEEE Trans. Electromagn. Compat. **66**(3), 879–889 (2024)
16. B. Salarieh, J. De Silva, B. Kordi, High frequency response of grounding electrodes: effect of soil dielectric constant. IET Gen. Transmission Distrib. **14**(15), 2915–2921 (2020)
17. R. Batista, R. Alipio, R.A.R. Moura, M.A.O. Schroeder, Assessment of frequency-dependent parameters of stratified soils for lightning response of earth electrodes. IEEE Trans. Electromagn. Compat. **64**(6), 2292–2295 (2022)
18. W.L.M. de Azevedo, W.C. da Silva, A.R.J. de Araéo, J.P. Filho, Comparison of soil modeling concerning physical factors: Application to transient analysis in wind turbines. Electr. Power Energy Syst. **155**, 109505 (2024)
19. S. Piliškic, I. Uglešic, B. Jurišic, Evaluating the overvoltage performance of an overhead line taking into account the frequency-dependence of its towers grounding electrodes with high soil resistivity. Int. J. Electr. Power Energy Syst. **116**, 105547 (2020)
20. R. Alipio, S. Visacro, Time-domain analysis of frequency-dependent electrical parameters of soil. IEEE Trans. Electromagn. Compat. **59**(3), 873–878 (2017)
21. M. Moradi, An electrostatic approach for simulation of grounding systems lightning response in time domain. IEEE Trans. Power Delivery **7**(6), 5289–5298 (2022)
22. M. Moradi, Comparison of different approaches for incorporating frequency dependence of soil parameters in time domain simulation of grounding system. IEEE Trans. Power Delivery **39**(4), 2028–2040 (2024)
23. R. Alipio, M.T.C. de Barros, M.A.O. Schroeder, K. Yamamoto, Analysis of the lightning impulse and low-frequency performance of wind farm grounding systems. Electr. Power Syst. Res. **180**, 106068 (2020)
24. A. Boutadjine, B. Nekhoul, B. Harrat, M. Melit, Realistic modeling of the grounding system transient behavior with frequency dependence. Electr. Power Syst. Res. **203**, 107644 (2022)
25. K.A. Carvalho, F.H. Silveira, Evaluation of the lightning performance of grounding electrodes using fdtd-based computational model: Influence of absorbing boundary conditions and representation of the frequency-dependence of soil parameters. Electr. Power Syst. Res. **213**, 108728 (2022)

26. B. Salarieh, H.M.J. De Silvab, B. Kordi, Electromagnetic transient modeling of grounding electrodes buried in frequency dependent soil with variable water content. Electr. Power Syst. Res. **189**, 106595 (2020)
27. A.R.J. de AraéÂ·o, J.S.L. Colqui, C.M. de Seixas, S. Kurokawa, B. Salarieh, J.P. Filho, B. Kordi, Computation of ground potential rise and grounding impedance of simple arrangement of electrodes buried in frequency-dependent stratified soil. Electr. Power Syst. Res. **198**, 107364 (2021)
28. A. Mehri, R. Ghanizadeh, M. Beiraghi, Transient analysis of grounding electrodes in a two-layer soil considering frequency dependence of soil. Electr. Eng. **104**, 1019–1028 (2022)
29. R. Alipio, D. Conceição, A. De Conti, K. Yamamoto, R.N. Dias, S. Visacro, A comprehensive analysis of the effect of frequency-dependent soil electrical parameters on the lightning response of wind-turbine grounding systems. Electr. Power Syst. Res. **175**, 105927 (2019)
30. A. Tarasov, K. Titov, On the use of the cole-cole equations in spectral induced polarization. Geophysical J. Int. **195**(1), 352–356 (2013)
31. D. Cavka, N. Mora, F. Rachidi, A comparison of frequency-dependent soil models: application to the analysis of grounding systems. IEEE Trans. Electromagn. Compat. **56**(1), 177–187 (2014)
32. M.A. Messier, The propagation of an electromagnetic impulse through soil: influence of frequency dependent parameters (technical report no. mrc-n-415). Mission Res. Corporation (1980)
33. J.H. Scott, Electrical and magnetic properties of rock and soil (open-file report no. 83–915). Department of Interior, U.S. Geological Survey (1966)
34. R. Alipio, S. Visacro, Frequency dependence of soil parameters: Effect on the lightning response of grounding electrodes. IEEE Trans. Electromagn. Compat. **55**(1), 132–139 (2013)

Chapter 12
Numerical Methods for Grounding Problem at Transient State VI

Various time-domain numerical methods for grounding problems in the case of numerical simulation of lightning transient shock response when considering the effect of soil ionization.

12.1 Preface

The lightning transient impulse grounding problem of substation grounding grids in multi-layer horizontal soil discussed and studied in the previous chapters did not consider soil ionization phenomenon. This chapter will further scientifically and systematically study and discuss how to consider the lightning transient impulse grounding problem of substation grounding grids in multi-layer horizontal soil considering soil ionization phenomenon.

12.2 Introduction

This chapter mainly discusses the study of soil ionization phenomenon and its application in numerical simulation of transient lightning impulse grounding of substation grounding grids in multi-layer horizontal layered soil. At present, literature [1–6] has been published on the consideration of soil ionization phenomena based on hybrid algorithms for electromagnetic field numerical calculations and their application in numerical simulation of lightning transient pulse grounding problems in substation grounding grids. Among them, Refs. [1, 2] are about time-domain hybrid algorithms based on PEEC basis functions; Refs. [3–6] are related to time-domain hybrid algorithms based on PI basis functions. At the same time, it can be known that Refs. [1, 5] are related to the lightning transient grounding problem of substation grounding grids

Z.-X. Li, *Numerical Simulation Algorithm of Electromagnetic Field for Grounding Problems in Power System Substation Grounding Grids*,
https://doi.org/10.1007/978-981-96-0904-8_12

buried in uniform earth; Refs. [2, 3, 6] are related to the lightning transient grounding problem of substation grounding grids buried in multi-layer horizontal layered soil. It should be noted that the authors of this book published an article [7] describing the numerical simulation of lightning transient shock response considering soil ionization effects for substation grounding grids in multilayer horizontally layered soils using a time-domain Galerkin BEM. In contrast, the literature published by the authors of this book [2–4, 6] addresses the problem of numerical simulation of lightning transient shocks for calculating substation grounding grids in multilayered horizontally stratified geodesic substations considering soil ionization effects, and all of them are based on hybrid algorithms with various basis functions in the time domain.

12.3 Soil Ionization Effects

Considering that when a large value lightning current impacts the grounding grids, the soil near the conductor surface of the grounding grids undergoes a drastic change in conductivity due to ionization, which causes a significant change in the electromagnetic field around the conductor of the grounding grids. So soil ionization is usually unnegligible in the numerical simulation of lightning grounding. When the leakage current along grounding grid conductor's surface exceeds the critical value, the soil ionization will occur.

When the soil ionization occurs, the grounding conductor will be surrounded by the ionized soil, and the voltage drop can be negligible. The ionized soil that rapidly decreases the resistivity around the alternative conductor can be effectively simulated by increasing the virtual equivalent radius of the grounding grid conductor.

A mathematical method that iteratively calculates the radius of the virtual equivalent conductor due to the occurrence of soil ionization phenomenon at each time step will scientifically simulate the soil ionization phenomenon, which has been fully discussed in next section.

12.4 The Time Domain Spatial 1-d Vector Galerkin BEM, MM and Hybrid Method for Solving First Order Linear Differential Equation

In this section, an in-depth discussion is presented on how the time-domain spatial 1-d vector Galerkin BEM, MM, and a hybrid algorithm that takes into account the effects of soil ionization can be used to numerically simulate the lightning transient shock response of a substation grounding grid in a multilayered, horizontally stratified soil.

12.4.1 Time Domain Spatial 1-d Vector Galerkin BEM for Solving First Order Linear Differential Equation

This chapter will review and summarize the relevant content of the time domain spatial 1-d vector Galerkin BEM in Chap. 10, the time domain spatial 1-d vector Galerkin BEM is mainly used to solve the first order linear differential equation as shown in the following formula as follows.

$$[\bar{\bar{\mathbf{Z}}}_b^s][I_p(t)] + [\bar{\bar{\mathbf{M}}}_b^s]\frac{d[I_b(t)]}{dt} + [\bar{\bar{\mathbf{Z}}}_s^s][I_p(t)] = -[\bar{\bar{\mathbf{Z}}}_b^q][I_p(t-t_c)] - [\bar{\bar{\mathbf{M}}}_b^q]\frac{d[I_p(t-t_A)]}{dt} - [\bar{\bar{\mathbf{Z}}}_s^q][I_p(t-t_\varphi)] \tag{12.1}$$

where $[\bar{\bar{\mathbf{Z}}}_b^s]$ and $[\bar{\bar{\mathbf{Z}}}_b^q]$ are the self impedance matrices of grounding grid conductors without and with delay effects in the $N_p \times N_p$ dimension, respectively. $[\bar{\bar{\mathbf{M}}}_b^s]$ and $[\bar{\bar{\mathbf{M}}}_b^q]$ are the mutual induction matrices of grounding grid conductors without and with delay effects in the $N_p \times N_p$ dimension, respectively. $[\bar{\bar{\mathbf{Z}}}_s^s]$ and $[\bar{\bar{\mathbf{Z}}}_s^q]$ are the mutual impedance matrices of grounding grid conductors without and with delay effects in the $N_p \times N_p$ dimension, respectively.

By discretizing the first order linear differential equation in time, the time domain spatial 1-d vector Galerkin BEM can solve the first order linear differential equation by iterative method. The derivation process of this method is fully discussed in Chap. 10. This section directly provides the calculation formula for solving this method as follows.

$$\left[\bar{\bar{\mathbf{Z}}}_b^s + \frac{\bar{\bar{\mathbf{M}}}_b^s}{\Delta t} + \bar{\bar{\mathbf{Z}}}_s^s\right][I_p(t_k)] = \frac{[I_b(t_{k-1})]}{\Delta t} - [\bar{\bar{\mathbf{Z}}}_b^q][I_p(t_k - t_c^k)] - [\bar{\bar{\mathbf{M}}}_b^q]\frac{[I_p(t_k - t_A^k)] - [I_p(t_{k-1} - t_A^k)]}{\Delta t}$$
$$- [\bar{\bar{\mathbf{Z}}}_s^q][I_p(t_k - t_\varphi^k)] \tag{12.2}$$

The excitation is incorporated into the formulation through the boundary condition:

$$I_{p_1}(t_k) = I_g(t_k) \tag{12.3}$$

where $I_g(t_k)$ denotes the lightning current generator and $I_{p_1}(t_k)$ current in the injection point at t_k moment.

At a junction consisting of two or more segments the continuity properties of the electric field must be satisfied, which is ensured by applying the Kirchhoff current law

$$\sum_{m=1}^{n} I_{p_m}(t_k) = 0 \tag{12.4}$$

And continuity equation

$$[\frac{\partial I_{p_1}(t_k)}{\partial r_1}]_{at\ junction} = [\frac{\partial I_{p_2}(t_k)}{\partial r_2}]_{at\ junction} = \cdots = [\frac{\partial I_{p_n}(t_k)}{\partial r_n}]_{at\ junction} \tag{12.5}$$

The condition (12.5) ensures the discontinuities in charge per unit length to be ruled out in passing from one conductor to another across the junction.

At the conductor free ends, the total current vanishes,i.e.is forced to be zero.

Equation (12.2) represents a marching-on-in-time formulation. Accordingly, it can be quickly solved through iterations. Therefore, the time domain spatial 1-d vector Galerkin BEM developed in this chapter can effectively simulate the lightning transient impulse grounding problem of substation grounding grid.

12.4.2 *Time Domain Spatial 1-d Vector Galerkin MM for Solving First Order Linear Differential Equation*

This chapter will review and summarize the relevant content of the time domain spatial 1-d vector Galerkin MM in Chap. 10, the time domain spatial 1-d vector Galerkin MM is mainly used to solve the first order linear differential equation as shown in the following formula as follows.

$$\begin{bmatrix} \bar{\bar{\mathbf{Z}}}_b^{s00} & \bar{\bar{\mathbf{Z}}}_b^{s01} \\ \bar{\bar{\mathbf{Z}}}_b^{s10} & \bar{\bar{\mathbf{Z}}}_b^{s11} + \bar{\bar{\mathbf{Z}}}_s^{s11} \end{bmatrix} \begin{bmatrix} \bar{\mathbf{I}}_b^0(t) \\ \bar{\mathbf{I}}_b^1(t) \end{bmatrix} + \begin{bmatrix} \bar{\bar{\mathbf{M}}}_b^{s00} & \bar{\bar{\mathbf{M}}}_b^{s01} \\ \bar{\bar{\mathbf{M}}}_b^{s10} & \bar{\bar{\mathbf{M}}}_b^{s11} \end{bmatrix} \frac{d}{dt} \begin{bmatrix} \bar{\mathbf{I}}_b^0(t) \\ \bar{\mathbf{I}}_b^1(t) \end{bmatrix}$$

$$= -\begin{bmatrix} \bar{\bar{\mathbf{Z}}}_b^{q00} & \bar{\bar{\mathbf{Z}}}_b^{q01} \\ \bar{\bar{\mathbf{Z}}}_b^{q10} & \bar{\bar{\mathbf{Z}}}_b^{q11} \end{bmatrix} \begin{bmatrix} \bar{\mathbf{I}}_b^0(t-t_c) \\ \bar{\mathbf{I}}_b^1(t-t_c) \end{bmatrix} - \begin{bmatrix} \bar{\bar{\mathbf{M}}}_b^{q00} & \bar{\bar{\mathbf{M}}}_b^{q01} \\ \bar{\bar{\mathbf{M}}}_b^{q10} & \bar{\bar{\mathbf{M}}}_b^{q11} \end{bmatrix} \frac{d}{dt} \begin{bmatrix} \bar{\mathbf{I}}_b^0(t-t_A) \\ \bar{\mathbf{I}}_b^1(t-t_A) \end{bmatrix}$$

$$- \begin{bmatrix} 0 & 0 \\ 0 & \bar{\bar{\mathbf{Z}}}_s^{q11} \end{bmatrix} \begin{bmatrix} \bar{\mathbf{I}}_b^0(t-t_\varphi) \\ \bar{\mathbf{I}}_b^1(t-t_\varphi) \end{bmatrix} \tag{12.6}$$

By discretizing the first order linear differential equation in time, the time domain spatial 1-d vector Galerkin BEM can solve the first order linear differential equation by iterative method. The derivation process of this method is fully discussed in Chap. 10. This section directly provides the calculation formula for solving this method as follows.

$$\begin{bmatrix} \bar{\bar{\mathbf{Z}}}_b^{s00} + \frac{\bar{\bar{\mathbf{M}}}_b^{s00}}{\Delta t} & \bar{\bar{\mathbf{Z}}}_b^{s01} + \frac{\bar{\bar{\mathbf{M}}}_b^{s01}}{\Delta t} \\ \bar{\bar{\mathbf{Z}}}_b^{s10} + \frac{\bar{\bar{\mathbf{M}}}_b^{s10}}{\Delta t} & \bar{\bar{\mathbf{Z}}}_b^{s11} + \bar{\bar{\mathbf{Z}}}_s^{s11} + \frac{\bar{\bar{\mathbf{M}}}_b^{s11}}{\Delta t} \end{bmatrix} \begin{bmatrix} \bar{\mathbf{I}}_b^0(t_k) \\ \bar{\mathbf{I}}_b^1(t_k) \end{bmatrix}$$

$$= \frac{1}{\Delta t} \begin{bmatrix} \bar{\bar{\mathbf{M}}}_b^{s00} & \bar{\bar{\mathbf{M}}}_b^{s01} \\ \bar{\bar{\mathbf{M}}}_b^{s10} & \bar{\bar{\mathbf{M}}}_b^{s11} \end{bmatrix} \begin{bmatrix} \bar{\mathbf{I}}_b^0(t_{k-1}) \\ \bar{\mathbf{I}}_b^1(t_{k-1}) \end{bmatrix} - \begin{bmatrix} \bar{\bar{\mathbf{Z}}}_b^{q00} & \bar{\bar{\mathbf{Z}}}_b^{q01} \\ \bar{\bar{\mathbf{Z}}}_b^{q10} & \bar{\bar{\mathbf{Z}}}_b^{q11} \end{bmatrix} \begin{bmatrix} \bar{\mathbf{I}}_b^0(t_k - t_c^k) \\ \bar{\mathbf{I}}_b^1(t_k - t_c^k) \end{bmatrix}$$

$$- \begin{bmatrix} \bar{\bar{\mathbf{M}}}_b^{q00} & \bar{\bar{\mathbf{M}}}_b^{q01} \\ \bar{\bar{\mathbf{M}}}_b^{q10} & \bar{\bar{\mathbf{M}}}_b^{q11} \end{bmatrix} \begin{bmatrix} \frac{\bar{\mathbf{I}}_b^0(t_k-t_A^k)-\bar{\mathbf{I}}_b^0(t_{k-1}-t_A^k)}{\Delta t} \\ \frac{\bar{\mathbf{I}}_b^1(t_k-t_A^k)-\bar{\mathbf{I}}_b^1(t_{k-1}-t_A^k)}{\Delta t} \end{bmatrix} - \begin{bmatrix} 0 & 0 \\ 0 & \bar{\bar{\mathbf{Z}}}_s^{q11} \end{bmatrix} \begin{bmatrix} \bar{\mathbf{I}}_b^0(t_k - t_\varphi^k) \\ \bar{\mathbf{I}}_b^1(t_k - t_\varphi^k) \end{bmatrix} \tag{12.7}$$

The excitation is incorporated into the formulation through the boundary condition:

$$I^0_{b_1}(t_k) = I_g(t_k) \tag{12.8}$$

where $I_g(t_k)$ denotes the lightning current generator and $I^0_{b_1}(t_k)$ current in the injection point at t_k moment.

At a junction consisting of two or more segments the continuity properties of the electric field must be satisfied, which is ensured by applying the Kirchhoff current law

$$\sum_{m=1}^{n} I^0_{b_m}(t_k) = 0 \tag{12.9}$$

where n is the total number of conductor segments connected to the node.

And continuity equation

$$[\frac{\partial I^0_{b_1}(t_k)}{\partial r_1}]_{at\ junction} = [\frac{\partial I^0_{b_2}(t_k)}{\partial r_2}]_{at\ junction} = \cdots = [\frac{\partial I^0_{b_n}(t_k)}{\partial r_n}]_{at\ junction} \tag{12.10}$$

The condition (12.10) ensures the discontinuities in charge per unit length to be ruled out in passing from one conductor to another across the junction.

At the conductor free ends, the total current vanishes,i.e.is forced to be zero.

Equation (12.7) represents a marching-on-in-time formulation. Therefore, it can be quickly solved through iteration. Therefore, the time-domain spatial 1-d vector Galerkin MM developed in this chapter can effectively simulate the lightning transient pulse grounding problem of the substation grounding grids.

12.4.3 The Time Domain Spatial 1-d Vector Hybrid Method for Solving First Order Linear Differential Equation

This chapter will review and summarize the relevant content of the time domain spatial 1-d vector hybrid methods based on different basic function (PI and PEEC) in Chap. 10, the time domain spatial 1-d vector hybrid methods based on different basic function (PI and PEEC) is mainly used to solve the first order linear differential equations as shown in the following formula as follows.

- The branch matrix approach

$$[\bar{\bar{\mathbf{Z}}}^s_b][\bar{\mathbf{I}}_b(t)] + [\bar{\bar{\mathbf{M}}}^s_b]\frac{d[\bar{\mathbf{I}}_b(t)]}{dt} + [\bar{\bar{\mathbf{A}}}]^T[\bar{\mathbf{V}}_p(t)] = -[\bar{\bar{\mathbf{Z}}}^q_b][\bar{\mathbf{I}}_b(t-t_c)] - [\bar{\bar{\mathbf{M}}}^q_b]\frac{d[\bar{\mathbf{I}}_b(t-t_A)]}{dt} \tag{12.11}$$

$$-[\bar{\bar{\mathbf{A}}}][\bar{\mathbf{I}}_b(t)] + [\bar{\bar{\mathbf{K}}}]^T[\bar{\bar{\mathbf{Z}}}^s_{s_b}]^{-1}[\bar{\bar{\mathbf{K}}}][\bar{\mathbf{V}}_p(t)] = [\bar{\mathbf{F}}(t)] - [\bar{\bar{\mathbf{K}}}]^T[\bar{\bar{\mathbf{Z}}}^s_{s_b}]^{-1}[\bar{\bar{\mathbf{Z}}}^q_{s_b}][\bar{\mathbf{I}}_{s_b}(t-t_\varphi)] \tag{12.12}$$

where the connotation of these matrices or vectors ($[\bar{\bar{\mathbf{Z}}}_b]$, $[\bar{\bar{\mathbf{M}}}_b]$, $[\bar{\bar{\mathbf{A}}}]$, $[\bar{\mathbf{V}}_p]$, $[\bar{\mathbf{F}}]$ and $[\bar{\mathbf{I}}_b]$) here have exactly the same form and content for both PI and PEEC basic function. Except for matrices $[\bar{\bar{\mathbf{Z}}}_{s_b}]$ and $[\bar{\bar{\mathbf{K}}}]$, where $[\bar{\bar{\mathbf{Z}}}_{s_b}]$ is a $N_b \times N_b$ matrix for PI basic function and a $2N_b \times 2N_b$ matrix for PEEC basic function, and $[\bar{\bar{\mathbf{K}}}]$ is a $N_b \times N_p$ matrix for PI basic function and a $2N_b \times N_p$ matrix for PEEC basic function.

- The node matrix approach

$$[\bar{\bar{\mathbf{Z}}}_b^s][\bar{\mathbf{I}}_b(t)] + [\bar{\bar{\mathbf{M}}}_b^s]\frac{d[\bar{\mathbf{I}}_b(t)]}{dt} + [\bar{\bar{\mathbf{A}}}]^T[\bar{\mathbf{V}}_p(t)] = -[\bar{\bar{\mathbf{Z}}}_b^q][\bar{\mathbf{I}}_b(t-t_c)] - [\bar{\bar{\mathbf{M}}}_b^q]\frac{d[\bar{\mathbf{I}}_b(t-t_A)]}{dt} \tag{12.13}$$

$$-[\bar{\bar{\mathbf{A}}}][\bar{\mathbf{I}}_b(t)] + [\bar{\bar{\mathbf{Z}}}_{s_p}^s]^{-1}[\bar{\mathbf{V}}_p(t)] = [\bar{\mathbf{F}}(t)] - [\bar{\bar{\mathbf{Z}}}_{s_p}^s]^{-1}[\bar{\bar{\mathbf{Z}}}_{s_p}^q][\bar{\mathbf{I}}_{s_p}(t-t_\varphi)] \tag{12.14}$$

where all matrices and vectors here have exactly the same form and content for both PI and PEEC basic function. However, when calculating the matrix $[\bar{\bar{\mathbf{Z}}}_{s_p}]$, due to the need to further subdivide the leakage current branch in the hybrid algorithm based on PEEC basis functions, compared with the hybrid algorithm based on PI basis functions, the length of the leakage current branch is half smaller.
The two consecutive equations (12.13) and (12.14) based on the node matrix equation method can be further transformed into one equation form, which is formed by the minimum unknown quantity.

$$\begin{aligned}[\bar{\bar{\mathbf{Z}}}_b^s][\bar{\mathbf{I}}_b(t)] + [\bar{\bar{\mathbf{M}}}_b^s]\frac{d[\bar{\mathbf{I}}_b(t)]}{dt} + [\bar{\bar{\mathbf{A}}}]^T[\bar{\bar{\mathbf{Z}}}_{s_p}^s][\bar{\bar{\mathbf{A}}}][\bar{\mathbf{I}}_b(t)] = -[\bar{\bar{\mathbf{Z}}}_b^q][\bar{\mathbf{I}}_b(t-t_c)] \\ -[\bar{\bar{\mathbf{M}}}_b^q]\frac{d[\bar{\mathbf{I}}_b(t-t_A)]}{dt} - [\bar{\bar{\mathbf{A}}}]^T[\bar{\bar{\mathbf{Z}}}_{s_p}^s][\bar{\mathbf{F}}(t)] + [\bar{\bar{\mathbf{A}}}]^T[\bar{\bar{\mathbf{Z}}}_{s_p}^q][\bar{\mathbf{I}}_{s_p}(t-t_\varphi)]\end{aligned} \tag{12.15}$$

By discretizing the first order linear differential equation in time, the time domain spatial 1-d vector hybrid methods based on different basic function (PI and PEEC) can solve the first order linear differential equations by iterative method. The derivation process of this method is fully discussed in Chap. 10. This section directly provides the calculation formulas for solving this method as follows.

- The branch matrix approach

$$\begin{aligned}\left[\bar{\bar{\mathbf{Z}}}_b^s + \frac{\bar{\bar{\mathbf{M}}}_b^s}{\Delta t}\right][\bar{\mathbf{I}}_b(t_k)] + [\bar{\bar{\mathbf{A}}}]^T[\bar{\mathbf{V}}_p(t_k)] = [\bar{\bar{\mathbf{M}}}_b^s]\frac{[\bar{\mathbf{I}}_b(t_{k-1})]}{\Delta t} - [\bar{\bar{\mathbf{Z}}}_b^q][\bar{\mathbf{I}}_b(t_k - t_c^k)] \\ -[\bar{\bar{\mathbf{M}}}_b^q]\frac{[\bar{\mathbf{I}}_b(t_k - t_A^k)] - [\bar{\mathbf{I}}_b(t_{k-1} - t_A^k)]}{\Delta t}\end{aligned} \tag{12.16}$$

$$-[\bar{\bar{\mathbf{A}}}][\bar{\mathbf{I}}_b(t_k)] + [\bar{\bar{\mathbf{K}}}]^T[\bar{\bar{\mathbf{Z}}}_{s_b}^s]^{-1}[\bar{\bar{\mathbf{K}}}][\bar{\mathbf{V}}_p(t_k)] = [\bar{\mathbf{F}}(t_k)] - [\bar{\bar{\mathbf{K}}}]^T[\bar{\bar{\mathbf{Z}}}_{s_b}^s]^{-1}[\bar{\bar{\mathbf{Z}}}_{sb}^q][\bar{\mathbf{I}}_{s_b}(t_k - t_\varphi^k)] \tag{12.17}$$

This system of Eqs. (12.16) (12.17) is called PEEC or PI 1 case.

- The node matrix approach

$$\left[\bar{\bar{\mathbf{Z}}}_b^s + \frac{\bar{\bar{\mathbf{M}}}_b^s}{\Delta t}\right][\bar{\mathbf{I}}_b(t_k)] + [\bar{\bar{\mathbf{A}}}]^T[\bar{\mathbf{V}}_p(t_k)] = [\bar{\bar{\mathbf{M}}}_b^s]\frac{[\bar{\mathbf{I}}_b(t_{k-1})]}{\Delta t} - [\bar{\bar{\mathbf{Z}}}_b^q][\bar{\mathbf{I}}_b(t_k - t_c^k)] - [\bar{\bar{\mathbf{M}}}_b^q]\frac{[\bar{\mathbf{I}}_b(t_k - t_A^k)] - [\bar{\mathbf{I}}_b(t_{k-1} - t_A^k)]}{\Delta t} \tag{12.18}$$

$$-[\bar{\bar{\mathbf{A}}}][\bar{\mathbf{I}}_b(t_k)] + [\bar{\bar{\mathbf{Z}}}_{s_p}^s]^{-1}[\bar{\mathbf{V}}_p(t_k)] = [\bar{\mathbf{F}}(t_k)] - [\bar{\bar{\mathbf{Z}}}_{s_p}^s]^{-1}[\bar{\bar{\mathbf{Z}}}_{s_p}^q][\bar{\mathbf{I}}_{s_p}(t_k - t_\varphi^k)] \tag{12.19}$$

This system of Eqs. (12.18) (12.19) is called PEEC or PI 2 case.

Least unknown equation

$$\left[\bar{\bar{\mathbf{Z}}}_b^s + \frac{\bar{\bar{\mathbf{M}}}_b^s}{\Delta t} + [\bar{\bar{\mathbf{A}}}]^T[\bar{\bar{\mathbf{Z}}}_{s_p}^s][\bar{\bar{\mathbf{A}}}]\right][\bar{\mathbf{I}}_b(t_k)] = [\bar{\bar{\mathbf{M}}}_b^s]\frac{[\bar{\mathbf{I}}_b(t_{k-1})]}{\Delta t} - [\bar{\bar{\mathbf{Z}}}_b^q][\bar{\mathbf{I}}_b(t_k - t_c^k)] - [\bar{\bar{\mathbf{M}}}_b^q]\frac{[\bar{\mathbf{I}}_b(t_k - t_A^k)] - [\bar{\mathbf{I}}_b(t_{k-1} - t_A^k)]}{\Delta t} - [\bar{\bar{\mathbf{A}}}]^T[\bar{\bar{\mathbf{Z}}}_{s_p}^s][\bar{\mathbf{F}}(t_k)] + [\bar{\bar{\mathbf{A}}}]^T[\bar{\bar{\mathbf{Z}}}_{s_p}^q][\bar{\mathbf{I}}_{s_p}(t_k - t_\varphi^k)] \tag{12.20}$$

This equations (12.20) is called PEEC or PI 2 I case.

The above three discrete time domain matrix equations represent a marching-on-in-time formulation. Accordingly, they can be quickly solved through iterations.

12.5 The Time Domain Spatial 1-d Vector Galerkin BEM, MM and Hybrid Method for Solving First Order Nonlinear Differential Equation Considering Soil Ionization Effect

As lightning currents can be huge, soil ionization usually cannot be ignored in numerical simulations of lightning protection rounding. Soil ionization will occur when the leaked current in the earth exceeds a critical value. When soil ionization occurs, the grounding conductor will be surrounded by ionized soil (corona type discharge), and voltage drops can be neglected in this region. Increasing the equivalent radius of the grounding conductors can be used to deal with this situation. The ionization effect depends on electrical and geometric parameters, and the ionization effect can be approximately simulated using the nonlinear resistance generated by the virtual electrode with varying current radius.

An iterative scheme is listed at each time step to accurately reflect the effect of soil ionization:

- First step, the time domain spatial 1-d vector Galerkin BEM, MM or hybrid method can be chosen for solving first order linear differential equation, which has been fully discussed above. When calculating the current distribution along the grounding grid in a time step, the radius of each segment is set as the radius from the previous time step.
- Then, second step, according to the leakage currents, the electric field on the surface of each segment can be calculated using Eq. (12.1):

$$E(\bar{\mathbf{r}}, t) = \frac{I_s(\bar{\mathbf{r}}, t)}{2\pi r_c \sigma \Delta l} \tag{12.21}$$

  where r_c is the original radius of the segment, Δl is the length of the segment, and σ is the conductivity of the soil around the segment. The critical value to determine whether soil ionization has occurred or not is defined as E_0. Soil ionization occurs when the value of the electric field on the surface of a segment is higher than E_0. The equivalent radius of the corresponding segment during the ionization is calculated using Eq. (12.2):

$$r_e = r_c \frac{E(\bar{\mathbf{r}}, t)}{E_0} \tag{12.22}$$

- Third step, after the radii are updated, the time domain spatial 1-d vector Galerkin BEM, MM or hybrid method can be chosen again for solving the first order linear differential equation, which has been fully discussed above. The current distribution along the grounding grid has been calculated again. As such, another group of new equivalent radii of the ionized segments will be obtained.
- Fourth step, The radius of the virtual electrode increases until the electric field around the electrode does not exceed the breakdown strength of the soil. International council on large electric systems (CIGRE) [8] recommended an empirical formula for the variation of nonlinear resistance with injection current pulse, which can be expressed as

$$R_g(I) = \frac{R_0}{1 + \frac{I(t)}{I_g}} \tag{12.23}$$

  where $I_g = \frac{E0}{2\pi \sigma_g R_g^2}$, σ_g is the conductivity of the earth around the grounding grid conductors, and $I(t)$ is the current injected into the electrode. R_0 is the ground resistance with low current at low frequency. This empirical formula is used to further correct the effect of virtual electrode radius in the time-domain spatial 1-d vector Galerkin BEM, MM or hybrid method developed in this section.
- Last step, if the difference between the two groups of equivalent radii can be neglected (such as the relative errors are smaller than 3% for all the segments), the final results at this time are obtain. Otherwise, go back to step 3.

In summary, the nonlinear and dynamic soil ionization effects are considered in our time domain, mathematical model.

12.6 Results and Discussions

12.6.1 *Validation of Various Time-Domain Algorithms When Considering Soil Ionization Effects*

The previous sub-chapters of this chapter described how time-domain algorithms such as the spatial 1-d vector Galerkin BEM, MM, and hybrid algorithms based on various basis functions can simulate by iterative methods to simulate the lightning transient shock response of a substation grounding grid considering soil ionization effects. In this subchapter, the correctness of the algorithm will be verified by a numerical simulation example comparing with the measured data. It should be noted that the words "Y" and "N" in the figure indicate that the time-domain algorithm takes into account and does not take into account soil ionization effects in the numerical simulations, respectively.

The first case comes from Geri [9]. Here, the length of the horizontal conductors was 5 m, which are buried 0.6 m below earth. The conductivity and relativity permittivity of the half-homogenous earth are $43^{-1}\frac{S}{m}$ and 10, respectively. The radii of the horizontal conducting bars were 4 mm. In Fig. 12.1a, an injected current whose amplitude 25 kA provoked soil ionization. The simulation results are also presented in Fig. 12.1a, where we compare the model with the measured values of Geri [9]. It is clear that, basically, agreement is obtained by the expedited methodology herein presented with the observational data. It should be remarked that for this analysis, a constant value was used for the computed time window and for all electrodes, which was equal to $350\frac{kV}{m}$.

The second case for verification is also from Ref. [9]. Here, the length of the vertical bars was 1 m, and they were on the surface of the ground. The conductivity

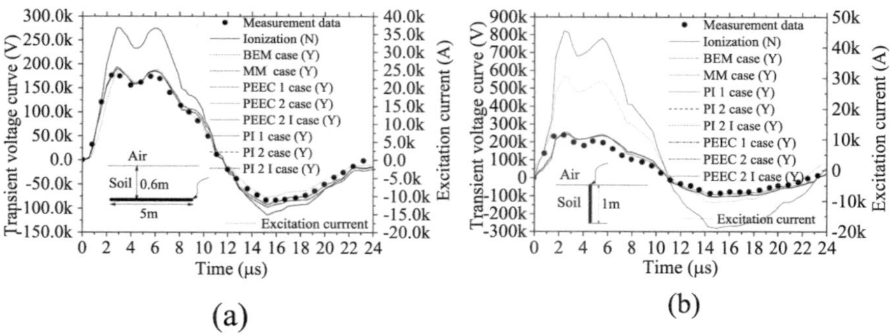

Fig. 12.1 The ionization phenomenon for a grounding rod

and relative of the half-homogenous soil were 43 Ωm and 10, respectively, and the radius of the vertical conducting bar was 25 mm. Figure 12.1b shows an injected current with an amplitude of 30 kA to provoke the ionization of soil. The results of the simulation are also presented in Fig. 12.1b, where we compare the model with the measured values from Ref. [9]. It is clear that a basic agreement was obtained between the results of the proposed method and observational data. In this example, the criterion of soil ionization was 350 kV/m [9].

12.6.2 Analysis and Discussion of Numerical Simulation Examples When Considering Soil Ionization Effects

As in the previous chapter, the general grounding system used is shown in Fig. 12.2a, consisting of three 3-square grids and a grounding depth of 0.5 m. The conductor radius of all grids is 7 mm. The earth is simulated as a horizontal four layer conductive medium, with a resistivity and permittivity of $\rho_1 = 100\,\Omega\text{m}$ and $\rho_2 = 500\,\Omega\text{m}$, $\rho_3 = 400\,\Omega\text{m}$, $\rho_4 = 300\,\Omega\text{m}$, $\varepsilon_1 = 10\varepsilon_0$, $\varepsilon_2 = 12\varepsilon_0$, $\varepsilon_3 = 13\varepsilon_0$ and $\varepsilon_4 = 12\varepsilon_0$, respectively. The thickness of the first, second, and third layers is $h_1 = 1$ m, $h_2 = 12$ m and $h_3 = 8$ m, respectively.

The external excited lightning current with $i(t) = 22.9 \times (e^{-0.01910t} - e^{-0.292288t})$ kA is injected from the corner of the grounding system. According to the formula of lightning current, the lightning parameters are T_1 is 10 μs, T_2 is 50μs and $I_m = 17.7$ kA. The reason for using high amplitude lightning current is to ensure the occurrence of soil ionization phenomenon.

The impulse impedance results and discussions

The impulse impedance calculated by the algorithm simulation in the time and frequency domains can be found in Tables (12.1). The words "T" and "F" in the table represent the corresponding algorithms in the time and frequency domains, respectively. The abbreviations "TN" and "TI" are used here to denote the results of numerical

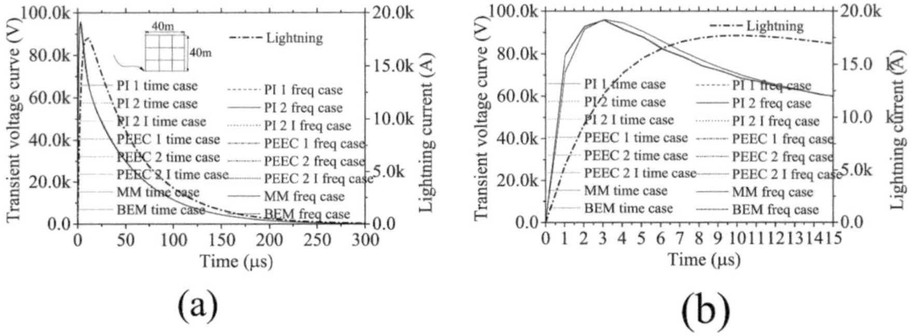

Fig. 12.2 Transient voltage at injection point (Frequency and time domain case)

Table 12.1 Calculated impulse impedance in both frequency and time domains (a)

	BEM	MM	PI 1	PI 1 I	PI 2	PEEC 1	PEEC 1 I	PEEC 2
F	(7.80, −5.04)	(7.81, −5.06)	(7.82, −5.05)	(7.82, −5.05)	(7.82, −5.05)	(7.82, −5.05)	(7.82, −5.05)	(7.83, −5.06)
TN	$(7.84, -9.2 \times 10^{-2})$	$(7.84, -9.2 \times 10^{-2})$	$(7.84, -9.2 \times 10^{-2})$	$(7.84, -9.2 \times 10^{-2})$	$(7.84, -9.2 \times 10^{-2})$	$(7.84, -9.2 \times 10^{-2})$	$(7.84, -9.2 \times 10^{-2})$	$(7.82, -9.2 \times 10^{-2})$
TN	$(7.33, 3.9 \times 10^{-2})$	$(7.44, -6.1 \times 10^{-2})$	$(7.35, -9.2 \times 10^{-2})$	$(7.35, -9.2 \times 10^{-2})$	$(7.34, 1.5 \times 10^{-2})$	$(7.45, -6.1 \times 10^{-2})$	$(7.35, -9.2 \times 10^{-2})$	$(7.52, -3.7 \times 10^{-2})$

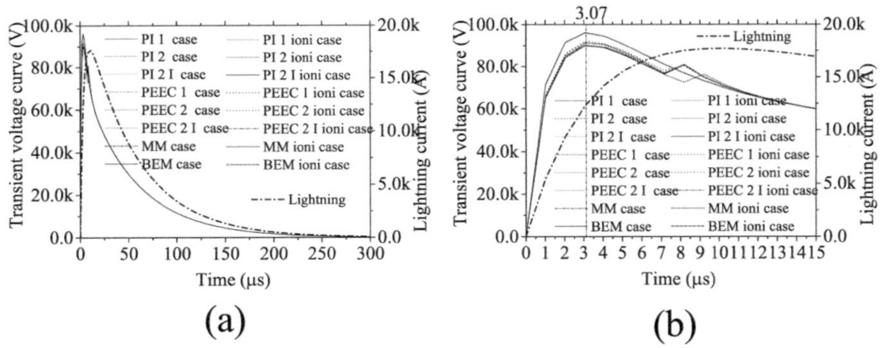

Fig. 12.3 Transient voltage at injection point (Time domain case)

simulations of the time-domain numerical method without and with soil ionization phenomena, respectively.

Observing the impulse impedance data in the the table, it can be seen that the impulse impedance data calculated by various time-domain and frequency-domain algorithms are close, and after accounting for soil ionization effects, the calculated impulse impedance data are relatively reduced (Table 12.1).

Analysis and discussion of transient surge voltages

Figures 12.2 and 12.3 give the plots of transient impulse voltages at the lightning current injection point of the substation grounding grid numerically simulated by different numerical frequency-domain and time-domain methods with and without taking into account the soil ionization phenomenon, respectively. It is important to note that the abbreviations "time" and "freq" in the figure represent the time and frequency domains, respectively. The abbreviation "ioni" denotes the phenomenon of soil ionization.

Figure 12.2a, b show the full-time transient voltage profiles and the local time amplification profiles calculated by frequency and time domain algorithms, respectively, at the lightning current injection point of the substation grounding grid. Observation of Fig. 12.2 shows that the curves simulated by various mathematical algorithms in the frequency and time domains, respectively, are basically consistent. This situation is in perfect agreement with the analysis of the impulse impedance values above.

The full-time transient voltage curves and the local-time amplification curves at the lightning current injection point of the substation grounding grid simulated by the time-domain algorithm with or without considering the soil ionization phenomenon are given in Fig. 12.3a, b, respectively. Observing and analyzing the two subplots (a) and (b) in this figure, it can be known that the transient impulse voltage profile obtained from the simulation with considering the effect of soil ionization becomes relatively lower in comparison with the transient impulse voltage profile obtained without considering the effect of soil ionization. This phenomenon is also consistent with the results of the analysis of the impulse voltage values above.

Analysis of 3-d distribution plots of voltage, leakage and branch currents and virtual conductor radius with soil ionization effects in time domains

In this section, the 3-d distributions of transient voltages, leakage and branch currents, and virtual conductor radii at critical moments obtained from numerical simulations based on the PI 2 I case time-domain hybrid algorithm under large-value lightning current impacts are given.

Observing the nine 3-d transient voltage distribution subplots in Fig. 12.4, it can be seen that the temporal and spatial distribution of the transient voltage on the whole grounding grid is basically closely related to the excitation source, i.e., the lightning current. First of all, the transient voltage induced on the whole grounding grid starts from the lightning current injection and gradually increases from near zero value, and the overall transient voltage reaches the maximum at the moment of 3.07 μs. Referring to the 2-d Fig. 12.3b in the previous section, it can be seen that the transient voltage curve at the lightning current injection point of the grounding grid reaches the maximum value at this time. In the 4th (moment of 3.07 μs) 3-d transient voltage distribution subplot, it also shows the maximum value of transient voltage at the injection point of the grounding grid, while the transient voltage at

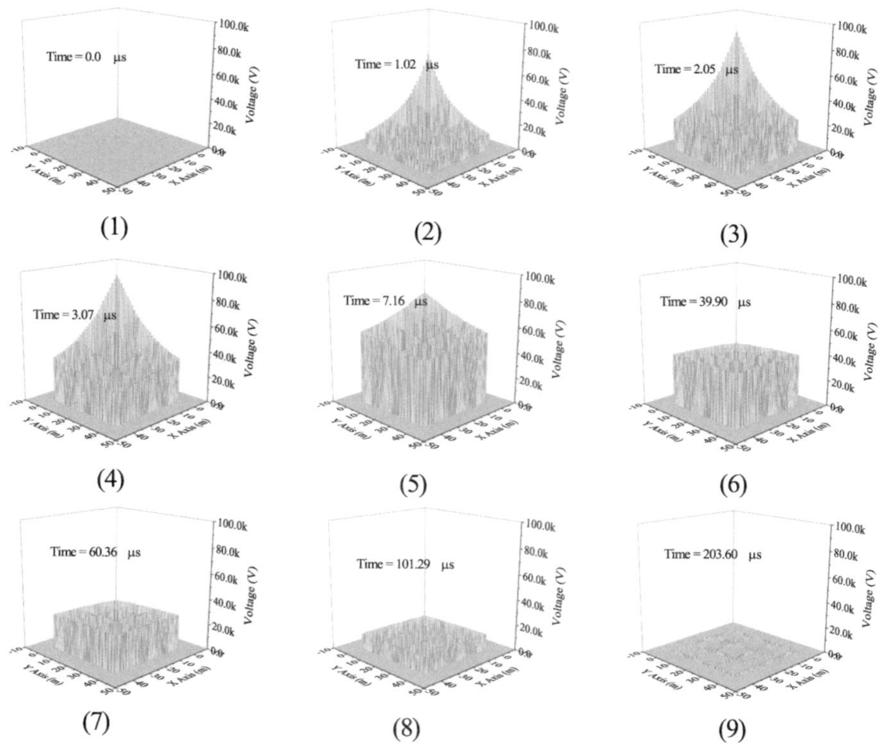

Fig. 12.4 3-d plot for voltage along the grid (time domain with soil ionization effect case)

the diagonal point of the injection point is minimum at that moment. After 3.07 μs, the overall transient voltage induced on the grounding grid gradually decreases from its peak value until it disappears. It is basically the same as that of the lightning current, which starts from zero and increases gradually, reaches the peak and then gradually decreases until it disappears, except that the peak occurs at a different moment. The peak of the lightning current occurs at 10.23 μs, while the peak of the induced transient voltage occurs at 3.07 μs.

Here observe the nine small diagrams in Fig. 12.5, the leakage current on the grounding grid and the temporal and spatial distribution of the transient voltage induced on the grounding grid are basically the same, and the change rule of the lightning current of the excitation source is the same. At each critical moment, the largest value of leakage current occurs at the injection point of lightning current on the grounding grid; at the same time, in the whole time period, as in the case of transient voltage induced on the grounding grid, the largest value of leakage current also occurs at the moment of 3.07 μs.

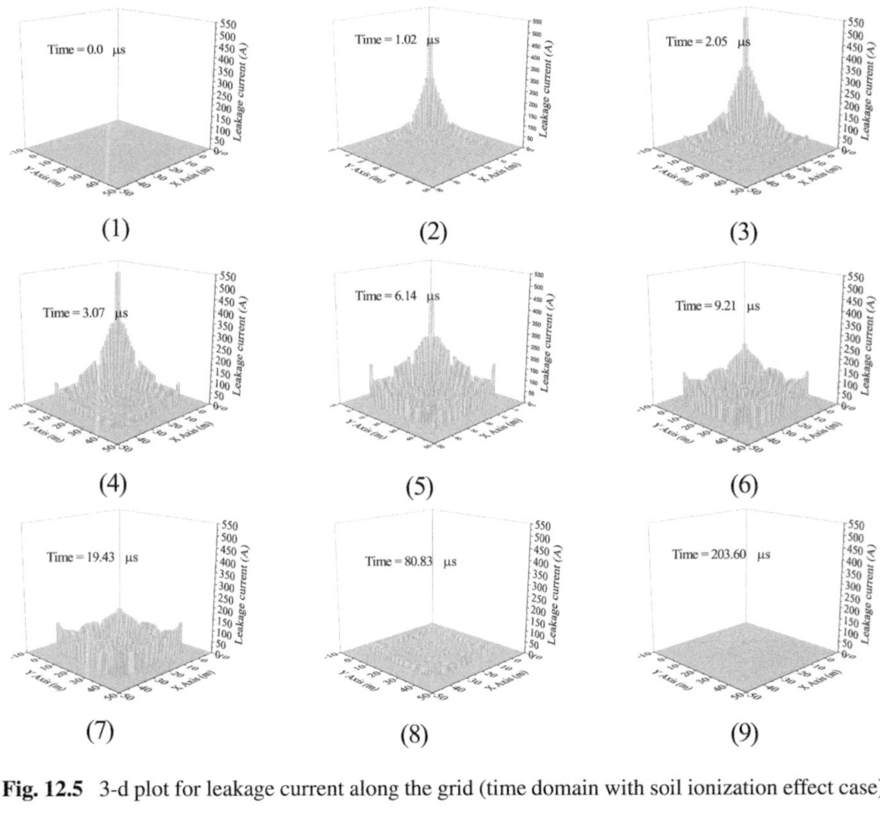

Fig. 12.5 3-d plot for leakage current along the grid (time domain with soil ionization effect case)

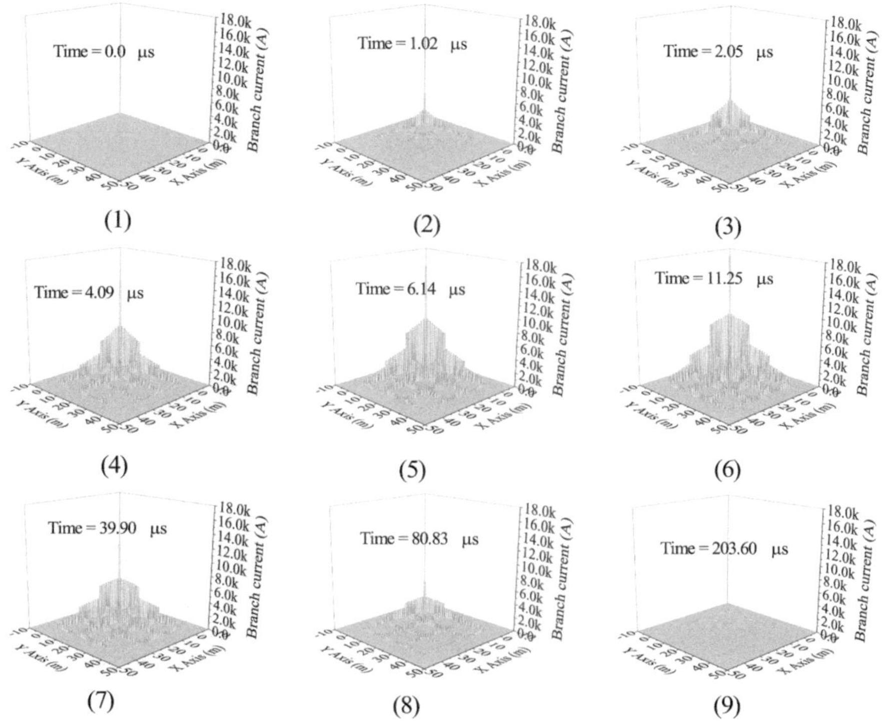

Fig. 12.6 3-d plot for branch current along the grid (time domain with soil ionization effect case)

Next, observing the nine subplots in Fig. 12.6, it can be known that the overall temporal and spatial distribution of the transient branch currents on the grounding grid is basically the same as that of the transient voltages and leakage currents. It is also the same as the development of the excitation source lightning current, and the largest value of the branch current in each moment occurs at the point of lightning current injection on the grounding grid, only that the moment of the largest value of the branch current occurs in the whole time period is slightly delayed to 4.09 μs.

The time-domain algorithm for the soil ionization phenomenon occurring under the action of large-value lightning currents is realized by numerical simulation of the increasing virtual conductor radius approach. In the absence of soil ionization, the virtual conductor half disappears and recedes to the actual radius of the conductor. The nine subplots in Fig. 12.7 reflect the process of development of the virtual radius of the conductor of the grounding grid at the critical moment. It also starts from nothing and gradually increases until the maximum virtual radius appears, and then gradually decreases until it disappears. The maximum virtual radius occurs at the moment of 3.07 μs during the whole period of soil ionization phenomenon. This moment is the same moment when the maximum values of transient voltage and leakage current on the grounding grid occur.

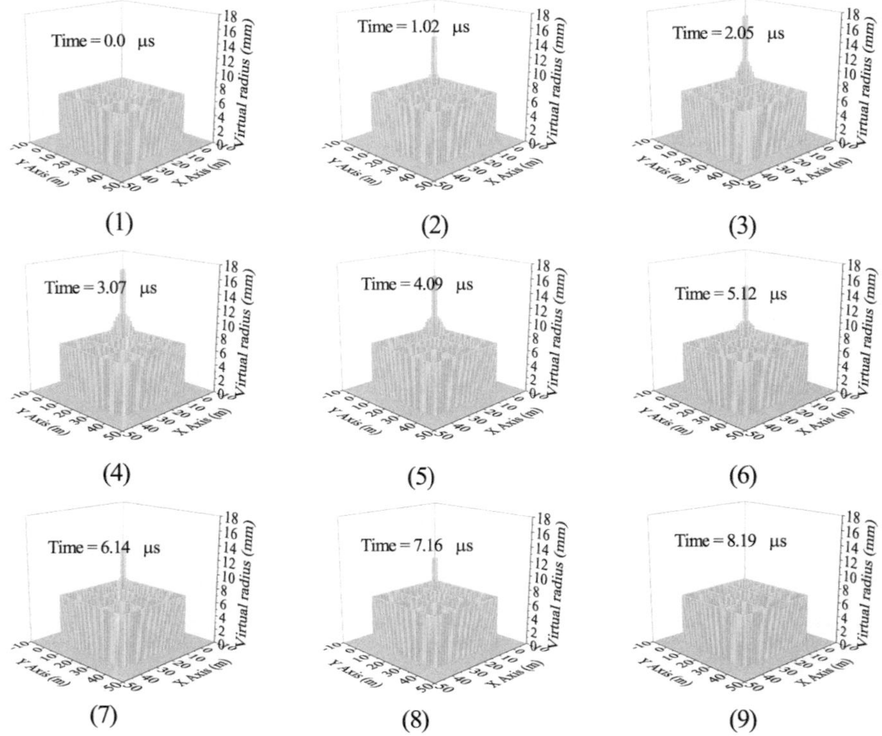

Fig. 12.7 3-d plot for virtual radius of the grid (time domain with soil ionization effect case)

12.7 Summary

In this chapter, the correctness and validity of the simulation results obtained by various time-domain algorithms with or without soil ionization effects are discussed. Through the analysis and discussion of numerical examples, the following conclusion is obtained: the lightning transient response of substation grounding grid simulated by the time domain algorithm without considering the influence of soil ionization effect is higher than that simulated by the time domain algorithm considering the influence of soil ionization effect.

References

1. H. Chen, Y. Du, Lightning grounding grid model considering both the frequency-dependent behavior and ionization phenomenon. IEEE Trans. Electromagn. Compat. **61**(1), 157–165 (2019)
2. Z.-X. Li, Lightning response transient characteristics of substation grounding grid in multilayer horizontal stratified earth model considering soil ionization effect. IEEE Trans. Electromagn. Compat. **65**(6), 1665–1673 (2023)

3. Z.-X. Li, S.-W. Rao, Transient lightning response of grounding grids in a horizontal multi-layered soil model that considers soil ionization effects with time-domain quasi-static complex images. Int. J. Numer. Model. Electron. Network Dev. Fields **32**(1), e2463 (2019)
4. Z.-X. Li, P. Li, K.-C. Wang, Transient lightning response to grounding grids buried in horizontal multilayered earth model considering time domain quasi-static complex image method and soil ionization effect. COMPEL Int. J. Computation Mathem. Electr. Electron. Eng. **40**(3), 516–534 (2021)
5. B.P. Silva, S. Visacro, F.H. Silveira, Hem-td: New time-domain electromagnetic model for calculating the lightning response of electric systems and their components. IEEE Trans. Power Delivery **37**(6), 4848–4857 (2022)
6. Z.-X. Li, W.-J. Chen, K.-C. Wang, Frequency domain and time domain analysis of the transient behavior of buried grounding grids in horizontal multilayered earth model. Electr. Eng. **104**(4), 2515–2529 (2022)
7. Z.-X. Li, P. Li, X. Zhao, Galerkin boundary element method for simulating lightning response of grounding grid in horizontal multilayered soil model considering soil ionization effect. Int. J. Numer. Model. Electron. Network Dev. Fields **37**(5), e3285 (2024)
8. L. Grcev, Modeling of grounding electrodes under lightning currents. IEEE Trans. Electromagn. Compat. **51**(3), 559–571 (2009)
9. S. Visacro, R. Alipio, C. Pereira, M. Guimarèaes, M.A.O., Schroeder. Behaviour of grounding systems excited by high impulse currents: the model and its validation. IEEE Trans. Power Delivery **14**(3), 1008–1017 (1999)

Chapter 13
Numerical Methods for Grounding Problem at Transient State VII

Various time-domain numerical methods for grounding problems in the case of numerical simulation of lightning transient shock response when considering the effects of soil ionization and parametric frequency characteristics.

13.1 Preface

The lightning transient impulse grounding problem of substation grounding grids in multi-layer horizontal soil discussed and studied in the previous chapters has respectively considered the frequency characteristics of soil electrical parameters and soil ionization phenomenon. This chapter will further scientifically and systematically study and discuss how to comprehensively consider the frequency characteristics of soil electrical parameters and soil ionization phenomena in the lightning transient impulse grounding problem of substation grounding grids in multi-layer horizontal soil.

13.2 Introduction

In recent years, literature [1–3] has been published describing the lightning impact response of grounding rods for soil ionization phenomena considering the frequency characteristics of soil parameters. It is important to note, however, that the numerical methods in this literature have been discussed only for the problem of lightning impulse response to soil ionization phenomena occurring in ground rods in uniformly soils. On the other hand, the authors of this book published the time-domain Galerkin BEM for numerical simulation of lightning transient shock grounding problems in substation grounding grid in multilayer horizontally stratified soils taking into account soil ionization and frequency effects [4].

Z.-X. Li, *Numerical Simulation Algorithm of Electromagnetic Field for Grounding Problems in Power System Substation Grounding Grids*,
https://doi.org/10.1007/978-981-96-0904-8_13

In this chapter, the lightning transient impulse response of substation grounding grids in multilayer horizontally stratified soils combining soil ionization phenomena and frequency-dependent properties of soil parameters will be discussed in depth.

13.3 The Soil Ionization Effects Considering the Frequency Characteristics of Soil Parameters

How the phenomenon of soil ionization is reflected in the process of solving the first order differential equation was fully discussed in the previous chapter. This chapter will further discuss how to simulate and calculate soil ionization phenomena when soil parameters include frequency characteristics.

When a large amount of lightning current is discharged through the grounding electrode, although the soil parameters around the electrode contain frequency characteristics, the electric field around the electrode may still exceed the breakdown strength of the soil. This leads to a low resistance area near the electrode. The ionization effect depends on electrical and geometric parameters, and can also be approximated by using the nonlinear resistance generated by virtual electrodes with varying current radii. The radius of the virtual electrode increases until the electric field around the electrode does not exceed the breakdown strength of the soil.

13.4 The Time Domain Spatial 1-d Vector Galerkin BEM, MM and Hybrid Method for Solving First Order Linear Differential Equation Considering the Frequency Characteristics of Soil Parameters

This chapter will review and summarize the relevant content of the time domain spatial 1-d vector Galerkin BEM, MM and hybrid method for grounding grid buried in horizontal N_e-layer soil in Chaps. 10 and 12, the time domain spatial 1-d vector Galerkin BEM, MM and hybrid method are mainly used to solve the first order linear differential equation as shown in the following formula as follows.

- The time domain spatial 1-d vector Galerkin BEM

$$[\bar{\bar{\mathbf{Z}}}_b^s][\bar{\mathbf{I}}_p(t)] + [\bar{\bar{\mathbf{M}}}_b^s]\frac{d[\bar{\mathbf{I}}_p(t)]}{dt} + [\bar{\bar{\mathbf{Z}}}_s^s][\bar{\mathbf{I}}_p(t)] = -[\bar{\bar{\mathbf{Z}}}_b^q][\bar{\mathbf{I}}_p(t - t_c)] - [\bar{\bar{\mathbf{M}}}_b^q]\frac{d[\bar{\mathbf{I}}_p(t - t_A)]}{dt} - [\bar{\bar{\mathbf{Z}}}_s^q][\bar{\mathbf{I}}_p(t - t_\varphi)] \tag{13.1}$$

where $\bar{\mathbf{I}}_p(t)$ is $N_b \times 1$ vector of the unknown current quantity, which includes the zero order branch current and the first order leakage current; $[\bar{\bar{\mathbf{Z}}}_b^s]$ and $[\bar{\bar{\mathbf{Z}}}_b^q]$

are the self impedance matrices of grounding grid conductors without and with delay effects in the $N_p \times N_p$ dimension, respectively. $[\bar{\bar{\mathbf{M}}}_b^s]$ and $[\bar{\bar{\mathbf{M}}}_b^q]$ are the mutual induction matrices of grounding grid conductors without and with delay effects in the $N_p \times N_p$ dimension, respectively. $[\bar{\bar{\mathbf{Z}}}_s^s]$ and $[\bar{\bar{\mathbf{Z}}}_s^q]$ are the mutual impedance matrices of grounding grid conductors without and with delay effects in the $N_p \times N_p$ dimension, respectively.

- The time domain spatial 1-d vector Galerkin MM

$$\begin{bmatrix} \bar{\bar{\mathbf{Z}}}_b^{s00} & \bar{\bar{\mathbf{Z}}}_b^{s01} \\ \bar{\bar{\mathbf{Z}}}_b^{s10} & \bar{\bar{\mathbf{Z}}}_b^{s11} + \bar{\bar{\mathbf{Z}}}_s^{s11} \end{bmatrix} \begin{bmatrix} \bar{\mathbf{I}}_b^0(t) \\ \bar{\mathbf{I}}_b^1(t) \end{bmatrix} + \begin{bmatrix} \bar{\bar{\mathbf{M}}}_b^{s00} & \bar{\bar{\mathbf{M}}}_b^{s01} \\ \bar{\bar{\mathbf{M}}}_b^{s10} & \bar{\bar{\mathbf{M}}}_b^{s11} \end{bmatrix} \frac{d}{dt} \begin{bmatrix} \bar{\mathbf{I}}_b^0(t) \\ \bar{\mathbf{I}}_b^1(t) \end{bmatrix}$$

$$= - \begin{bmatrix} \bar{\bar{\mathbf{Z}}}_b^{q00} & \bar{\bar{\mathbf{Z}}}_b^{q01} \\ \bar{\bar{\mathbf{Z}}}_b^{q10} & \bar{\bar{\mathbf{Z}}}_b^{q11} \end{bmatrix} \begin{bmatrix} \bar{\mathbf{I}}_b^0(t - t_c) \\ \bar{\mathbf{I}}_b^1(t - t_c) \end{bmatrix} - \begin{bmatrix} \bar{\bar{\mathbf{M}}}_b^{q00} & \bar{\bar{\mathbf{M}}}_b^{q01} \\ \bar{\bar{\mathbf{M}}}_b^{q10} & \bar{\bar{\mathbf{M}}}_b^{q11} \end{bmatrix} \frac{d}{dt} \begin{bmatrix} \bar{\mathbf{I}}_b^0(t - t_A) \\ \bar{\mathbf{I}}_b^1(t - t_A) \end{bmatrix}$$

$$- \begin{bmatrix} 0 & 0 \\ 0 & \bar{\bar{\mathbf{Z}}}_s^{q11} \end{bmatrix} \begin{bmatrix} \bar{\mathbf{I}}_b^0(t - t_\varphi) \\ \bar{\mathbf{I}}_b^1(t - t_\varphi) \end{bmatrix} \quad (13.2)$$

where the $[\bar{\mathbf{I}}_b^0(t)]$ is $N_b \times 1$ vector of unknown branch currents and the $[\bar{\mathbf{I}}_b^1(t)]$ is $N_b \times 1$ vector of unknown leakage currents; $[\bar{\bar{\mathbf{Z}}}_b^{00}]$ is $N_b \times N_b$ matrix, who presents the mutual induction interaction betweens a parts of conductor segments branch currents, $[\bar{\bar{\mathbf{Z}}}_b^{01}]$ is $N_b \times N_b$ matrix, who presents the mutual impedance and induction interaction betweens a parts of conductor segments branch and leakage currents, $[\bar{\bar{\mathbf{Z}}}_b^{10}]$ is $N_b \times N_b$ matrix, who presents the mutual impedance and induction interaction betweens a parts of conductor segments branch and leakage currents, $[\bar{\bar{\mathbf{Z}}}_b^{11}]$ is $N_b \times N_b$ matrix, who presents the mutual impedance and induction interaction betweens a parts of conductor segments leakage currents.

- The time domain spatial 1-d vector hybrid methods

$$[\bar{\bar{\mathbf{Z}}}_b^s][\bar{\mathbf{I}}_b(t)] + [\bar{\bar{\mathbf{M}}}_b^s]\frac{d[\bar{\mathbf{I}}_p(t)]}{dt} + [\bar{\bar{\mathbf{A}}}]^T[\bar{\bar{\mathbf{Z}}}_{s_p}^s][\bar{\bar{\mathbf{A}}}][\bar{\mathbf{I}}_b(t)] = -[\bar{\bar{\mathbf{Z}}}_b^q][\bar{\mathbf{I}}_b(t - t_c)]$$

$$- [\bar{\bar{\mathbf{M}}}_b^q]\frac{d[\bar{\mathbf{I}}_b(t - t_A)]}{dt} - [\bar{\bar{\mathbf{A}}}]^T[\bar{\bar{\mathbf{Z}}}_{s_p}^s][\bar{\mathbf{F}}_p(t)] + [\bar{\bar{\mathbf{A}}}]^T[\bar{\bar{\mathbf{Z}}}_{s_p}^q][\bar{\mathbf{I}}_{s_p}(t - t_\varphi)] \quad (13.3)$$

where the connotation of these matrices or vectors ($[\bar{\bar{\mathbf{Z}}}_b^s]$, $[\bar{\bar{\mathbf{Z}}}_b^q]$, $[\bar{\bar{\mathbf{M}}}_b^s]$, $[\bar{\bar{\mathbf{M}}}_b^q]$, $[\bar{\bar{\mathbf{A}}}]$, $[\bar{\mathbf{F}}_p(t)]$, $[\bar{\mathbf{I}}_b(t)]$) and $[\bar{\mathbf{I}}_{s_p}(t)]$) here have exactly the same form and content for both PI and PEEC basic function, which means that all $[\bar{\bar{\mathbf{Z}}}_b^s]$, $[\bar{\bar{\mathbf{Z}}}_b^q]$, $[\bar{\bar{\mathbf{M}}}_b^s]$ and $[\bar{\bar{\mathbf{M}}}_b^q]$ are $N_b \times N_b$ matrices for both PI and PEEC basic function; the $[\bar{\bar{\mathbf{A}}}]$ is $N_p \times N_b$

matrix for both PI and PEEC basic function; the $\bar{\mathbf{F}}_p$ is $N_p \times 1$ vector for both PI and PEEC basic function; $\bar{\mathbf{I}}_b(t)$ and $\bar{\mathbf{I}}_{s_p}(t)$ are the $N_b \times 1$ vector of unknown branch current and $N_p \times 1$ vector of the unknown leakage current on the conductor segment connected to the node for both PI and PEEC basic function, respectively. Except for matrices $[\bar{\bar{\mathbf{Z}}}^s_{s_p}]$, $[\bar{\bar{\mathbf{Z}}}^q_{s_p}]$ and $[\bar{\bar{\mathbf{K}}}]$, where $[\bar{\bar{\mathbf{K}}}]$ is a $N_b \times N_p$ matrix for PI basic function and a $2N_b \times N_p$ matrix for PEEC basic function. When calculating the matrix $[\bar{\bar{\mathbf{Z}}}^s_{s_p}]$ and $[\bar{\bar{\mathbf{Z}}}^p_{s_p}]$, due to the need to further subdivide the leakage current branch in the hybrid algorithm based on PEEC basis functions, compared with the hybrid algorithm based on PI basis functions, the length of the leakage current branch is half smaller.

Observing the first order linear differential equation (13.1), (13.2) and (13.3), it can be seen that there are frequency related factors in the element formula of the equation matrix containing delay effect in the three first order differential equations. See the following formulas for details.

- The time domain spatial 1-d vector Galerkin BEM

$$[\bar{\bar{\mathbf{M}}}^q_b] = \sum_{\theta=1}^{N_b}\sum_{\vartheta=1}^{N_b}\int_{l_\theta}\int_{l_\vartheta} \hat{r}_\theta \cdot \bar{\bar{G}}^{A^t_q}_{\xi\zeta}(\bar{r}_\theta, \bar{r}_\vartheta) \cdot \hat{r}^T_\vartheta [\bar{q}(\bar{r}_\theta)][\bar{p}(\bar{r}_\vartheta)] dr_\vartheta dr_\theta \tag{13.4}$$

$$[\bar{\bar{\mathbf{Z}}}^q_b] = \sum_{\theta=1}^{N_b}\sum_{\vartheta=1}^{N_b}\int_{l_\theta}\int_{l_\vartheta} G^{\varphi^t_q}_{\xi\zeta}(\bar{r}_\theta, \bar{r}_\vartheta) \frac{\partial[\bar{q}(\bar{r}_\theta)]}{\partial r_\theta}\frac{\partial \bar{p}(\bar{r}_\vartheta)}{\partial r_\vartheta} dr_\vartheta dr_\theta \tag{13.5}$$

where $\xi = 0, 1, \ldots, N_e$; $\zeta = 0, 1, \ldots, N_e$. $\bar{r}_\theta$ and $\bar{r}_\vartheta$ is location vector along θ-th and ϑ-th conductors segment, respectively; r_θ and r_ϑ, respectively, parameter variable along the θ-th and ϑ-th thin conductor axis. $\hat{r}_\theta$ and $\hat{r}_\vartheta$ are, respectively, the 1×3 dimensional unit director vector along the axis of the θ-th and ϑ-th cylindrical conductor segments. The symbol T represents matrix transposition. $[\bar{p}(\bar{r}_\vartheta)]$ and $[\bar{q}(\bar{r}_\theta)]$ are 1×2 dimensional 1 order basis functions and 2×1 dimensional 1 order test functions, respectively. $[\bar{p}(\bar{r}_\vartheta)] = \begin{bmatrix} \frac{l_\vartheta - r_\vartheta}{l_\vartheta} & \frac{r_\vartheta}{l_\vartheta} \end{bmatrix}$ and $[\bar{q}(\bar{r}_\theta)] = \begin{bmatrix} \frac{l_\theta - r_\theta}{l_\theta} \\ \frac{r_\theta}{l_\theta} \end{bmatrix}$.

- The time domain spatial 1-d vector Galerkin MM

$$[\bar{\bar{\mathbf{M}}}^q_b] = \begin{bmatrix} \bar{\bar{\mathbf{M}}}^{q00}_b & \bar{\bar{\mathbf{M}}}^{q01}_b \\ \bar{\bar{\mathbf{M}}}^{q10}_b & \bar{\bar{\mathbf{M}}}^{q11}_b \end{bmatrix} = \sum_{\theta=1}^{N_b}\sum_{\vartheta=1}^{N_b}\int_{l_\theta}\int_{l_\vartheta} \hat{r}_\theta \cdot \bar{\bar{G}}^{A^t_q}_{\xi\zeta}(\bar{r}_\theta, \bar{r}_\vartheta) \cdot \hat{r}^T_\vartheta [\bar{q}(\bar{r}_\theta)][\bar{p}(\bar{r}_\vartheta)] dr_\vartheta dr_\theta \tag{13.6}$$

$$[\bar{\bar{\mathbf{Z}}}^{q11}_b] = \sum_{\theta=1}^{N_b}\sum_{\vartheta=1}^{N_b}\int_{l_\theta}\int_{l_\vartheta} G^{\varphi^t_q}_{\xi\zeta}(\bar{r}_\theta, \bar{r}_\vartheta) \frac{\partial[\bar{q}(\bar{r}_\theta)]}{\partial r_\theta}\frac{\partial \bar{p}(\bar{r}_\vartheta)}{\partial r_\vartheta} dr_\vartheta dr_\theta \tag{13.7}$$

where $[\bar{p}(\bar{r}_\vartheta)] = \begin{bmatrix} 1 & -\frac{r_\vartheta}{l_\vartheta} \end{bmatrix}$ and $[\bar{q}(\bar{r}_\theta)] = \begin{bmatrix} 1 \\ -\frac{r_\theta}{l_\theta} \end{bmatrix}$.

- The time domain spatial 1-d vector hybrid method

$$[\bar{\bar{\mathbf{M}}}_b^q] = \sum_{\theta=1}^{N_b}\sum_{\vartheta=1}^{N_b}\int_{l_\theta}\int_{l_\vartheta}\hat{r}_\theta \cdot \bar{\bar{G}}_{\xi\zeta}^{A_q^t}(\bar{r}_\theta, \bar{r}_\vartheta)\cdot \hat{r}_\vartheta^T dr_\vartheta dr_\theta \tag{13.8}$$

$$[\bar{\bar{\mathbf{Z}}}_p^q] = \frac{1}{N_\theta^\varsigma}\sum_{\theta=1}^{N_\theta^\varsigma}\frac{1}{l_\theta^\varsigma}\int_{l_\theta^\varsigma}\frac{1}{N_\vartheta^\upsilon}\sum_{\vartheta=1}^{N_\vartheta^\upsilon}\frac{1}{l_\vartheta^\upsilon}\int_{l_\vartheta^\upsilon} G_{\xi\zeta}^{\varphi_q^t}(\bar{r}_\varsigma, \bar{r}_\upsilon)dr_\varsigma dr_\upsilon \tag{13.9}$$

where ς, υ are the node numbers of discrete conductor segments in the grounding grid. N_θ^ς and N_ϑ^υ are the total number of conductor segments connecting nodes ς and υ. l_θ^ς and l_ϑ^υ are the lengths of the θ-th and ϑ-th conductor segments connecting nodes ς and υ, respectively.

where $G_{\xi\zeta}^{\varphi_q^t}(\bar{r}_\theta, \bar{r}_\vartheta)$ or $G_\varphi^{q_t}(\bar{r}_\varsigma, \bar{r}_\upsilon)$ and $\bar{\bar{G}}_{\xi\zeta}^{A_q^t}(\bar{r}_\theta, \bar{r}_\vartheta)$ are, respectively, time domain scalar monopole Green's function and vector dipole dyadic Green's function within horizontal multilayered earth model with delay effects, all these had been fully discussed in front Chap. 3. It must be pointed out that the dyadic and scalar Green's function of point sources in multilayered horizontal layered soil discussed in this section contain the frequency characteristics of soil parameters. This means that these three first order differential equations (13.1), (13.2) and (13.3) all consider the frequency characteristics of soil parameters.

By discretizing the first order linear differential equation in time, the time domain spatial 1-d vector Galerkin BEM, Galerkin MM and Galerkin hybrid method can solve the first order linear differential equation by iterative method. The derivation process of this method is fully discussed in Chap. 10. This section directly provides the calculation formula for solving this method as follows.

- The time domain spatial 1-d vector Galerkin BEM

$$\left[\bar{\bar{\mathbf{Z}}}_b^s + \frac{\bar{\bar{\mathbf{M}}}_b^s}{\Delta t} + \bar{\bar{\mathbf{Z}}}_s^s\right][I_p(t_k)] = \frac{[I_b(t_{k-1})]}{\Delta t} - [\bar{\bar{\mathbf{Z}}}_b^q][I_p(t_k - t_c^k)]$$

$$- [\bar{\bar{\mathbf{M}}}_b^q]\frac{[I_p(t_k - t_A^k)] - [I_p(t_{k-1} - t_A^k)]}{\Delta t} - [\bar{\bar{\mathbf{Z}}}_s^q][I_p(t_k - t_\varphi^k)] \tag{13.10}$$

where $t_c^k = \frac{t_c}{\Delta t}$, $t_A^k = \frac{t_A}{\Delta t}$ and $t_\varphi^k = \frac{t_\varphi}{\Delta t}$.
The excitation is incorporated into the formulation through the boundary condition:

$$I_{p_1}(t_k) = I_g(t_k) \tag{13.11}$$

where $I_g(t_k)$ denotes the lightning current generator and $I_{p_1}(t_k)$ current in the injection point at t_k moment.

At a junction consisting of two or more segments the continuity properties of the electric field must be satisfied, which is ensured by applying the Kirchhoff current law

$$\sum_{m=1}^{n} I_{p_m}(t_k) = 0 \tag{13.12}$$

And continuity equation

$$[\frac{\partial I_{p_1}(t_k)}{\partial r_1}]_{at\ junction} = [\frac{\partial I_{p_2}(t_k)}{\partial r_2}]_{at\ junction} = \cdots = [\frac{\partial I_{p_n}(t_k)}{\partial r_n}]_{at\ junction} \tag{13.13}$$

The condition (13.13) ensures the discontinuities in charge per unit length to be ruled out in passing from one conductor to another across the junction.
At the conductor free ends, the total current vanishes,i.e.is forced to be zero.

- The time domain spatial 1-d vector Galerkin MM

$$\begin{bmatrix} \bar{\bar{\mathbf{Z}}}_b^{s00} + \frac{\bar{\bar{\mathbf{M}}}_b^{s00}}{\Delta t} & \bar{\bar{\mathbf{Z}}}_b^{s01} + \frac{\bar{\bar{\mathbf{M}}}_b^{s01}}{\Delta t} \\ \bar{\bar{\mathbf{Z}}}_b^{s10} + \frac{\bar{\bar{\mathbf{M}}}_b^{s10}}{\Delta t} & \bar{\bar{\mathbf{Z}}}_b^{s11} + \bar{\bar{\mathbf{Z}}}_s^{s11} + \frac{\bar{\bar{\mathbf{M}}}_b^{s11}}{\Delta t} \end{bmatrix} \begin{bmatrix} \bar{\mathbf{I}}_b^{0}(t_k) \\ \bar{\mathbf{I}}_b^{1}(t_k) \end{bmatrix}$$
$$= \frac{1}{\Delta t} \begin{bmatrix} \bar{\bar{\mathbf{M}}}_b^{s00} & \bar{\bar{\mathbf{M}}}_b^{s01} \\ \bar{\bar{\mathbf{M}}}_b^{s10} & \bar{\bar{\mathbf{M}}}_b^{s11} \end{bmatrix} \begin{bmatrix} \bar{\mathbf{I}}_b^{0}(t_{k-1}) \\ \bar{\mathbf{I}}_b^{1}(t_{k-1}) \end{bmatrix} - \begin{bmatrix} \bar{\bar{\mathbf{Z}}}_b^{q00} & \bar{\bar{\mathbf{Z}}}_b^{q01} \\ \bar{\bar{\mathbf{Z}}}_b^{q10} & \bar{\bar{\mathbf{Z}}}_b^{q11} \end{bmatrix} \begin{bmatrix} \bar{\mathbf{I}}_b^{0}(t_k - t_c^k) \\ \bar{\mathbf{I}}_b^{1}(t_k - t_c^k) \end{bmatrix}$$
$$- \begin{bmatrix} \bar{\bar{\mathbf{M}}}_b^{q00} & \bar{\bar{\mathbf{M}}}_b^{q01} \\ \bar{\bar{\mathbf{M}}}_b^{q10} & \bar{\bar{\mathbf{M}}}_b^{q11} \end{bmatrix} \begin{bmatrix} \frac{\bar{\mathbf{I}}_b^{0}(t_k - t_A^k) - \bar{\mathbf{I}}_b^{0}(t_{k-1} - t_A^k)}{\Delta t} \\ \frac{\bar{\mathbf{I}}_b^{1}(t_k - t_A^k) - \bar{\mathbf{I}}_b^{1}(t_{k-1} - t_A^k)}{\Delta t} \end{bmatrix} - \begin{bmatrix} 0 & 0 \\ 0 & \bar{\bar{\mathbf{Z}}}_s^{q11} \end{bmatrix} \begin{bmatrix} \bar{\mathbf{I}}_b^{0}(t_k - t_\varphi^k) \\ \bar{\mathbf{I}}_b^{1}(t_k - t_\varphi^k) \end{bmatrix} \tag{13.14}$$

The excitation is incorporated into the formulation through the boundary condition:

$$I_{b_1}^0(t_k) = I_g(t_k) \tag{13.15}$$

where $I_g(t_k)$ denotes the lightning current generator and $I_{b_1}^0(t_k)$ current in the injection point at t_k moment.
At a junction consisting of two or more segments the continuity properties of the electric field must be satisfied, which is ensured by applying the Kirchhoff current law

$$\sum_{m=1}^{n} I_{b_m}^0(t_k) = 0 \tag{13.16}$$

And continuity equation

$$[\frac{\partial I_{b_1}^0(t_k)}{\partial r_1}]_{at\ junction} = [\frac{\partial I_{b_2}^0(t_k)}{\partial r_2}]_{at\ junction} = \cdots = [\frac{\partial I_{b_n}^0(t_k)}{\partial r_n}]_{at\ junction} \quad (13.17)$$

The condition (13.17) ensures the discontinuities in charge per unit length to be ruled out in passing from one conductor to another across the junction.
At the conductor free ends, the total current vanishes,i.e.is forced to be zero.

- The time domain spatial 1-d vector Galerkin hybrid method

$$\left[\bar{\bar{\mathbf{Z}}}_b^s + \frac{\bar{\bar{\mathbf{M}}}_b^s}{\Delta t} + [\bar{\bar{\mathbf{A}}}]^T[\bar{\bar{\mathbf{Z}}}_{s_p}^s][\bar{\bar{\mathbf{A}}}]\right][\bar{\mathbf{I}}_b(t_k)] = [\bar{\bar{\mathbf{M}}}_b^s]\frac{[\bar{\mathbf{I}}_b(t_{k-1})]}{\Delta t} - [\bar{\bar{\mathbf{Z}}}_b^q][\bar{\mathbf{I}}_b(t_k - t_c^k)]$$

$$-[\bar{\bar{\mathbf{M}}}_b^q]\frac{[\bar{\mathbf{I}}_b(t_k - t_A^k)] - [\bar{\mathbf{I}}_b(t_{k-1} - t_A^k)]}{\Delta t} - [\bar{\bar{\mathbf{A}}}]^T[\bar{\bar{\mathbf{Z}}}_{s_p}^s][\bar{\mathbf{F}}(t_k)] + [\bar{\bar{\mathbf{A}}}]^T[\bar{\bar{\mathbf{Z}}}_{s_p}^q][\bar{\mathbf{I}}_{s_p}(t_k - t_\varphi^k)] \quad (13.18)$$

Equations (13.16), (13.17) and (13.18) represents a marching-on-in-time formulation. Accordingly, it can be quickly solved through iterations. Therefore, the time domain spatial 1-d vector Galerkin BEM, MM and hybrid method developed in this chapter can effectively simulate the lightning transient impulse grounding problem of substation grounding grid considering frequency characteristics of soil parameters.

13.5 The Time Domain Spatial 1-d Vector Galerkin BEM, MM and Hybrid Method for Solving First Order Nonlinear Differential Equation Consider Soil Ionization Effects with the Frequency Characteristics of Soil Parameters

The entire idea scheme for solving the nonlinear first order differential equation of soil ionization was discussed in depth in the previous chapter. In order to ensure consistency in writing, this chapter once again delves into the overall scheme for solving soil ionization phenomena, and more importantly, emphasizes the frequency characteristics of soil electrical parameters.

Due to the high lightning current, soil ionization phenomenon cannot be ignored in numerical simulation of lightning protection grounding. When the leakage current in the soil exceeds the critical value, the electric field strength at that location will be higher than the critical field strength, that is, the electric field strength at that location will exceed the soil insulation strength, and thus soil ionization will occur. When the soil undergoes ionization, the grounding conductor will be surrounded by the ionized soil (corona discharge), and the voltage drop in this area can be ignored. Adding the equivalent radius of the grounding conductor in numerical simulation can be used to effectively handle this situation. The ionization effect depends on the electrical parameters containing frequency characteristics and the geometric parameters of the metal conductor of the grounding grid, and the nonlinear resistance generated by virtual electrodes with varying current radii can be used to approximate the soil ionization effect.

At each time step, an iterative scheme can be listed to accurately reflect the impact of nonlinear factors in soil ionization with frequency characteristic electrical parameters:

- In the first step, the time domain spatial 1-d vector Galerkin BEM, MM or hybrid method can be selected to solve the first-order linear differential equation taking into account the frequency characteristics of soil electrical parameters. The entire method has been fully discussed above and will not be repeated. When calculating the current distribution along the grounding grid in a time step, the radius of each segment is set to the radius of the previous time step.
- Then, in the second step, based on the calculated leakage current on the conductor section of the grounding grid, the formula (13.19) can be used to calculate the electric field on the surface of each conductor section.

$$E(\bar{\mathbf{r}}, t) = \frac{I_s(\bar{\mathbf{r}}, t)}{2\pi r_c \sigma \Delta l} \tag{13.19}$$

where r_c is the original radius of the segment, Δl is the length of the segment, and σ is the conductivity of the soil around the segment. The critical value to determine whether soil ionization has occurred or not is defined as E_0. Soil ionization occurs when the value of the electric field on the surface of a segment is higher than E_0. The equivalent radius of the corresponding segment during the ionization is calculated using Equation (13.20):

$$r_e = r_c \frac{E(\bar{\mathbf{r}}, t)}{E_0} \tag{13.20}$$

- Third, after updating the radius, the time-domain spatial 1-d vector Galerkin BEM, MM or hybrid method can be again chosen to solve the first-order linear differential equation considering frequency characteristic of soil parameters, which has been fully discussed above. The current distribution along the grounding grid was calculated again. In this way, another new set of equivalent radii for the soil ionization around the grounding grid conductors will be obtained.

- Fourth step, The radius of the virtual electrode increases until the electric field around the electrode does not exceed the breakdown strength of the soil. International council on large electric systems (CIGRE) [5] recommended an empirical formula for the variation of nonlinear resistance with injection current pulse, which can be expressed as

$$R_g(I) = \frac{R_0}{1 + \frac{I(t)}{I_g}} \tag{13.21}$$

where $I_g = \frac{E0}{2\pi\sigma_g R_g^2}$, σ_g is the conductivity of the earth around the grounding grid conductors, and $I(t)$ is the current injected into the electrode. R_0 is the ground resistance with low current at low frequency. This empirical formula is used to further correct the effect of virtual electrode radius in the time-domain spatial 1-d vector Galerkin BEM, MM or hybrid method developed in this section.
- In the final step, if the difference between the two sets of equivalent radii is negligible (for example, the relative error of all line segments is less than 3%), the final result is obtained at this time. Otherwise, go back to Step 3.

In conclusion, the frequency characteristics of soil electrical parameters as well as nonlinear and dynamic soil ionization effects are considered in our time-domain mathematical model.

13.6 Results and Discussions

As in the previous chapter, the general grounding system used is shown in Fig. 13.1a, consisting of four 4-square grids and a grounding depth of 0.5 m. The conductor radius of all grids is 7 mm. The earth is simulated as a horizontal four layer conductive medium, with a resistivity and permittivity of $\rho_1 = 100\,\Omega$m and $\rho_2 = 600\,\Omega$m, $\rho_3 = 500\,\Omega$m, $\rho_4 = 300\,\Omega$m, $\varepsilon_1 = 10\varepsilon_0$, $\varepsilon_2 = 12\varepsilon_0$, $\varepsilon_3 = 13\varepsilon_0$ and $\varepsilon_4 = 12\varepsilon_0$, respectively. The thickness of the first, second, and third layers is $h_1 = 1$ m, $h_2 = 12$ m and $h_3 = 8$ m, respectively. In order to study the relationship between the impact characteristics of grounding grids and the frequency characteristics of soil conductive media, this calculation example adopts the cole-cole equations model, those soil parameters are selected as follows: 1) The parameters of soil in the first layer are $\sigma_0 = 100^{-1} S/m, m = 0.9, c = 0.59$, and $\tau = 4.2 \times 10^{-8}$. 2) Those of the second layer are $\sigma_0 = 600^{-1} S/m, m = 0.95, c = 0.63$, and $\tau = 1.8 \times 10^{-7}$. 3) The third layer has parameters $\sigma_0 = 500^{-1} S/m, m = 0.8, c = 0.68$, and $\tau = 1.0 \times 10^{7}$. 4) The fourth layer has parameters $\sigma_0 = 300^{-1} S/m,\ m = 0.85,\ c = 0.72$, and $\tau = 3.2 \times 10^{-7}$.

The external excited lightning current with $i(t) = 22.9 \times (e^{-0.01910t} - e^{-0.292288t})kA$ is injected from the corner of the grounding system. According to the formula of lightning current, the lightning parameters are T_1 is 10μs, T_2 is 50μs and $I_m = 17.7kA$. The reason for using high amplitude lightning current is to ensure the occurrence of soil ionization phenomenon.

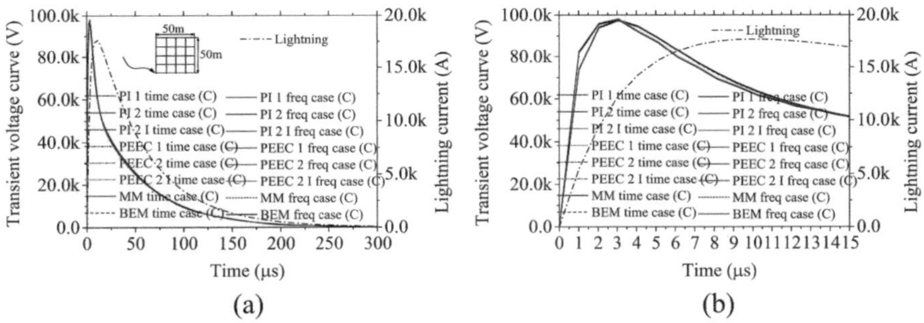

Fig. 13.1 Transient voltage at injection point (Frequency and time domain without frequency characteristic case)

13.6.1 The Impulse Impedance Results and Discussions

The impulse impedance calculated by the algorithm simulation in the time and frequency domains can be found in Tables 13.1 and 13.2. The words "T" and "F" in the table represent the corresponding algorithms in the time and frequency domains, respectively. The abbreviations "TN" and "TI" are used here to denote the results of numerical simulations of the time-domain numerical method without and with soil ionization phenomena, respectively.

Observation of the impulse impedance data in Tables 13.1 and 13.2 shows that the impulse impedance data calculated by the various time and frequency domain algorithms are relatively close to each other whether or not the frequency characteristics of the soil parameters are taken into account, and that the calculated impulse impedance data are relatively reduced after taking into account the ionization effect of the soil. At the same time, it can be seen that the pulse impedance data calculated by the various algorithms are relatively reduced after considering the frequency characteristics of the soil parameters.

13.6.2 Analysis and Discussion of Transient Surge Voltages

The following seven figures give the transient voltage curves at the lightning current injection point of the substation grounding grid for various calculation premises. Among them, Figs. 13.1 and 13.2 give the transient voltage curves simulated by various algorithms in the frequency and time domains, respectively, under the influence of whether or not to consider the frequency properties of soil parameters, Figs. 13.3 and 13.4 give the transient voltage curves simulated by various algorithms in the time domain, respectively, under the influence of whether or not to consider the frequency properties of soil parameters and the effect of ionization of the soil, whereas Figs. 13.5 and 13.6 give the transient voltage curves simulated by various algorithms

Table 13.1 Calculated impulse impedance in both frequency and time domains (a)

	Impulse impedance calculated without considering soil frequency characteristics							
	BEM	MM	PI 1	PI 2	PI 2 I	PEEC 1	PEEC 2	PEEC 2 I
F	(7.99, −4.36)	(7.99, −4.37)	(8.01, −4.37)	(8.01, −4.38)	(8.01, −4.38)	(7.97, −4.39)	(8.01, −4.38)	(8.01, −4.38)
TN	$(7.94, -6.0 \times 10^{-2})$	$(7.94, -6.0 \times 10^{-2})$	$(7.94, -5.9 \times 10^{-2})$	$(7.98, -8.3 \times 10^{-2})$	$(7.98, -8.3 \times 10^{-2})$	$(7.93, -5.9 \times 10^{-2})$	$(7.94, -5.9 \times 10^{-2})$	$(7.94, -5.9 \times 10^{-2})$
TI	$(7.54, 5.2 \times 10^{-3})$	$(7.65, -4.3 \times 10^{-2})$	$(7.53, -2.1 \times 10^{-3})$	$(7.59, -8.3 \times 10^{-2})$	$(7.59, -8.3 \times 10^{-2})$	$(7.55, -6.1 \times 10^{-2})$	$(7.56, -6.0 \times 10^{-2})$	$(7.56, -6.0 \times 10^{-2})$

Table 13.2 Calculated impulse impedance in both frequency and time domains (a)

	Impulse impedance calculated with considering soil frequency characteristics							
	BEM	MM	PI 1	PI 2	PI 2 I	PEEC 1	PEEC 2	PEEC 2 I
F	(7.74, −4.45)	(7.74, −4.46)	(7.76, −4.47)	(7.76, −4.47)	(7.76, −4.47)	(7.75, −4.47)	(7.76, −4.47)	(7.76, −4.47)
TN	(7.54, −0.61)	(7.56, −0.63)	(7.69, , −0.44)	(7.79, −0.57)	(7.79, −0.57)	(7.61, −0.52)	(7.54, −0.61)	(7.54, −0.61)
TI	(7.11, −0.58)	(7.12, −0.59)	(7.29, −0.40)	(7.41, −0.58)	(7.41, −0.58)	(7.28, −0.51)	(7.16, −0.61)	(7.16, −0.61)

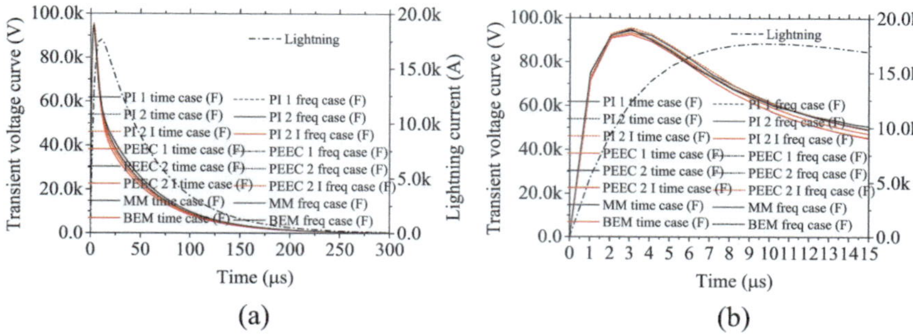

Fig. 13.2 Transient voltage at injection point (Frequency and time domain with frequency characteristic case)

in the frequency domain or the time domain, respectively, under the influence of whether or not to consider the frequency properties of soil parameters. transient voltage curves obtained from various algorithmic simulations in the time domain with or without considering the influence of the frequency characteristics of the soil parameter, while Fig. 13.7 gives the transient voltage curves obtained from various algorithmic simulations in the time domain with or without considering the influence of the frequency characteristics of the soil parameter and with or without considering the influence of the ionization effect of the soil. It is important to note that the seven plots consist of two subplots (a) and (b), respectively, where subplot (a) is the transient voltage profile for the full time period and subplot (b) is a localized zoomed-in plot near the peak of this transient voltage profile to facilitate detailed observation. It is also important to note that in the following seven figures, the abbreviations "time" and "freq" denote the time and frequency domains, respectively, while the capital letters "F" and "C" denote the frequency characteristics with and without considering the soil parameters, respectively, while the capital letters "I" and "N" denote the effects of soil ionization with and without considering the effects of soil ionization, respectively, and various combinations of these four capital letters denote different meanings.

Firstly, observing and analyzing Fig. 13.1, it can be known that: under the premise of not considering the influence of frequency characteristics of soil parameters, the transient voltage curves at the lightning current injection point of the substation grounding grid simulated by various algorithms in frequency and time domains basically conform to the consistency, which shows the correctness of the algorithms in frequency and time domains.

Then observe and analyze Fig. 13.2, it can be known that: under the premise of considering the influence of frequency characteristics of soil parameters, the transient voltage curves at the lightning current injection point of the substation grounding grid simulated by various algorithms in frequency and time domains basically conform to the consistency, which also explains the correctness of the algorithms in frequency and time domains.

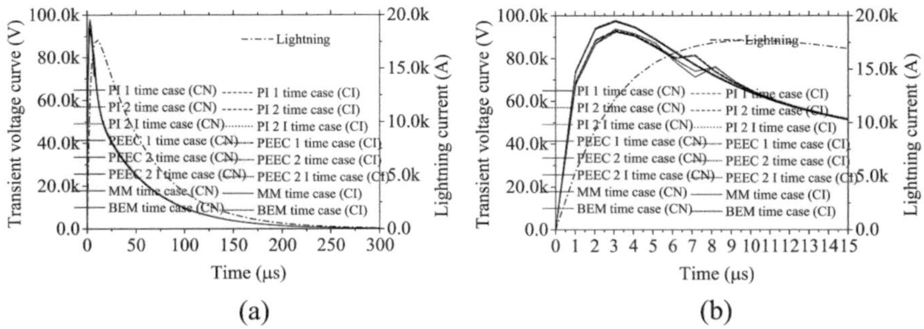

Fig. 13.3 Transient voltage at injection point (Time domain without frequency characteristic & with and without soil ionization effect case)

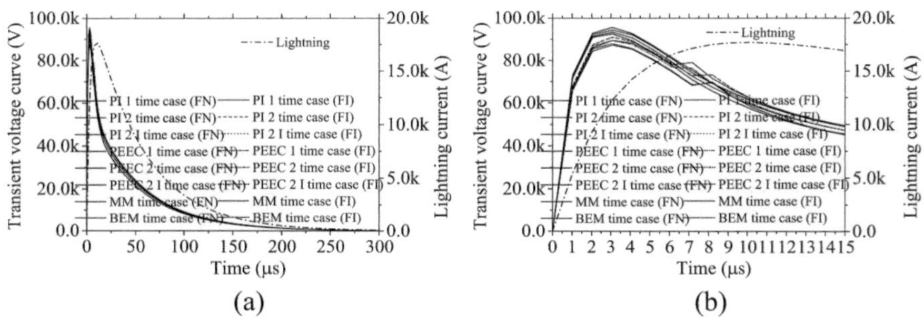

Fig. 13.4 Transient voltage at injection point (Time domain with frequency characteristic & with and without soil ionization effect case)

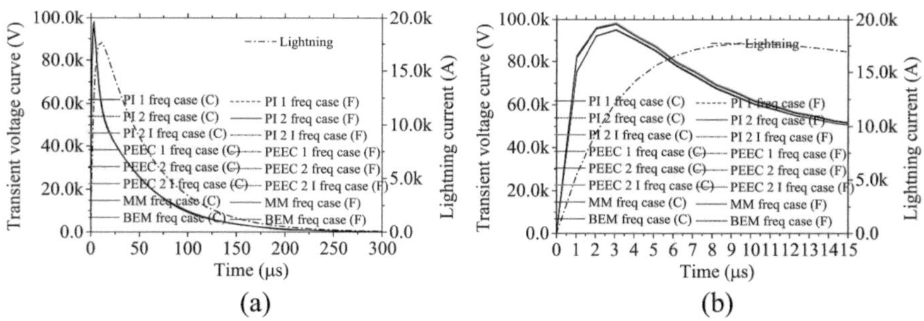

Fig. 13.5 Transient voltage at injection point (freq domain with and without soil frequency characteristic case)

Then observe and analyze Fig. 13.3, it can be known that: under the premise of the frequency characteristics of soil parameters are not considered, respectively, without considering and considering the influence of soil ionization effect, the transient voltage curve of the substation grounding grid lightning current injection point simulated by various algorithms in the time domain basically conforms to the same, and at the

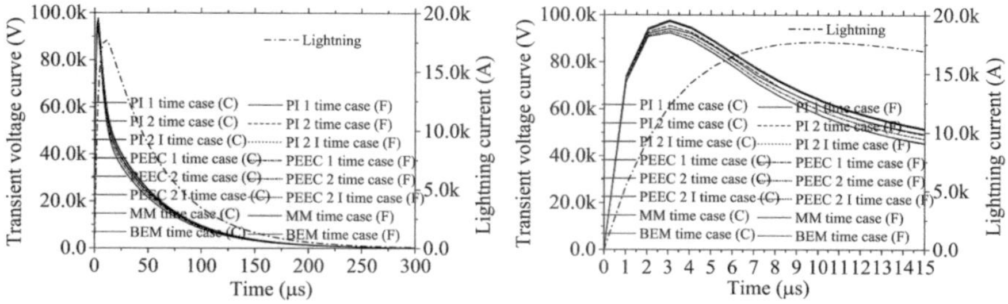

(a) (b)

Fig. 13.6 Transient voltage at injection point (Time domain with and without soil frequency characteristic case)

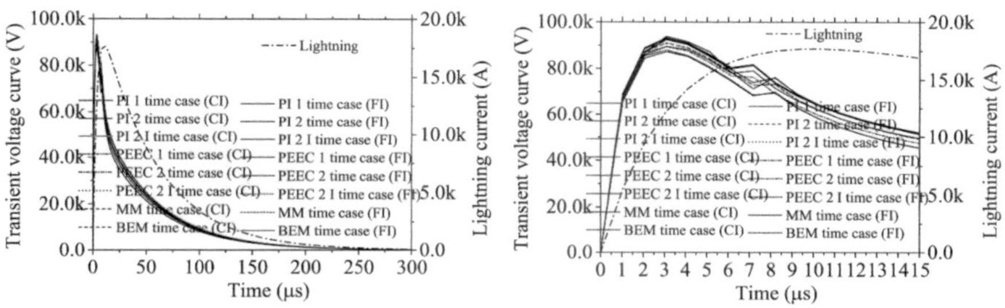

(a) (b)

Fig. 13.7 Transient voltage at injection point (Time domain with soil ionization effect & with and without soil frequency characteristic case)

same time, it can be seen that, without taking into account the soil ionization effect of the transient voltage curve obtained by simulation is higher than that obtained by the simulation of the ionization effect of the soil is taken into account.

Then observe and analyze Fig. 13.4, it can be known that: under the premise of considering the influence of soil parameter frequency characteristics, respectively, without considering and considering the influence of soil ionization effect, the transient voltage curve of the substation grounding grid lightning current injection point simulated by various algorithms in the time domain basically conforms to the same, and at the same time, it can be seen that the transient voltage curve obtained by simulating the effect of ionization of the soil without considering the effect of ionization of the soil is higher than that of the simulation of ionization effect obtained by considering the effect of the soil transient voltage curve.

And then continue to observe and analyze Fig. 13.5, it can be known that: respectively, without considering and considering the effect of soil soil parameter frequency characteristics, the frequency domain of the various algorithms simulated by the substation grounding grid lightning current injection point of the transient voltage curve basically in line with the same, and at the same time it can be seen, do not take into account the effect of the soil parameter frequency characteristics of the transient voltage curve obtained by simulation is higher than that obtained by the simulation of the effect of considering the soil parameter frequency characteristics of the transient voltage curve.

And then continue to observe and analyze Fig. 13.6, it can be known that: respectively, without considering and considering the effect of soil soil parameter frequency characteristics, the time domain of the various algorithms simulated by the substation grounding grid lightning current injection point of the transient voltage curve basically meets the same, and at the same time, it can be seen, do not take into account the effect of soil parameter frequency characteristics of the transient voltage curve obtained by the simulation is higher than that obtained by the simulation of the effect of soil parameter frequency characteristics of the transient voltage curve obtained by the simulation of the soil parameter frequency characteristics of the effect of the soil parameter is taken into account.

Finally, observing and analyzing Fig. 13.7, it can be known that: under the premise of considering the effect of soil ionization, the transient voltage curves of the lightning current injection point of the substation grounding grid simulated by various algorithms in the time domain basically conform to the same when the effects of soil parameter frequency characteristics are not considered and considered respectively, and at the same time, it can be seen that the transient voltage curves obtained by simulating the effect of soil parameter frequency characteristics are higher than those obtained by simulating the effect of soil parameter frequency characteristics are taken into account.

13.6.3 Time-Domain Analysis of the 3-d Distribution of Transient Voltages, Leakage Currents, Branch Currents and Virtual Conductor Radii on the Grounding Grid (Taking into Account Soil Ionization Effects and Frequency Characteristics of Soil Parameters)

In this section, the 3-d distributions of transient voltages, leakage and branch currents, and virtual conductor radii at critical moments obtained from numerical simulations based on a hybrid algorithm in the time domain of the PI 2 I case under a large lightning current impact are given.

Observing the nine 3-d transient voltage distribution subplots in Fig. 13.8, it can be seen that the temporal and spatial distribution of the transient voltage on the whole grounding grid is basically closely related to the excitation source, i.e., the lightning current. First of all, the transient voltage induced on the whole grounding grid starts from the lightning current injection and gradually increases from near zero value, and the overall transient voltage reaches the maximum at the moment of 3.07 μs. Referring to the 2-d Fig. 13.7b in the previous section, it can be seen that the transient voltage curve at the lightning current injection point of the grounding

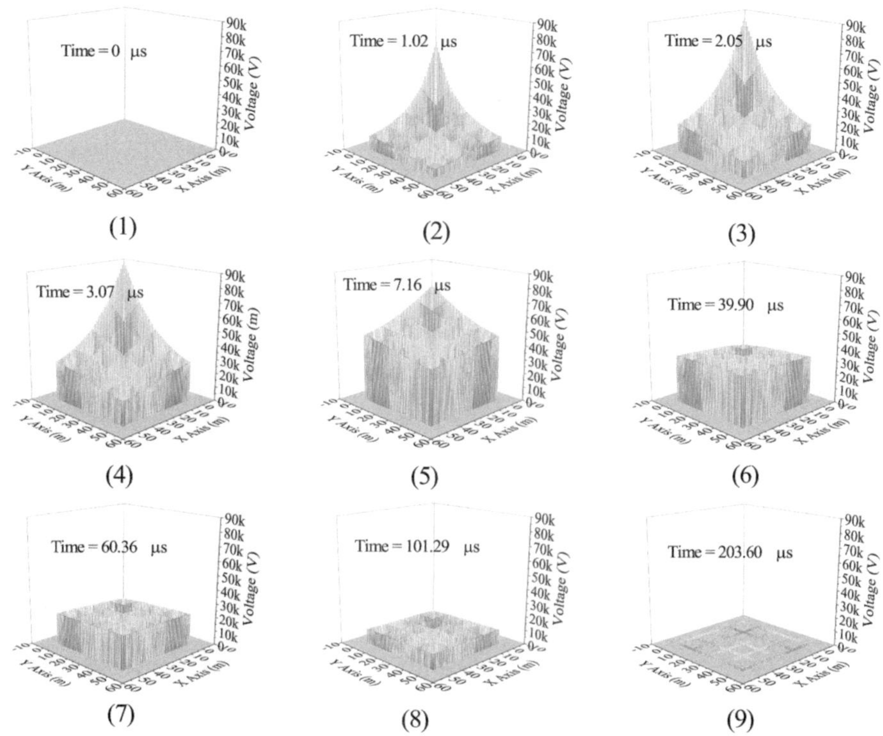

Fig. 13.8 3-d distribution of voltage along the grid (Time domain with soil ionization effect and soil frequency characteristic case)

grid reaches the maximum value at this time. In the 4th (moment of 3.07 μs) 3-d transient voltage distribution subplot, it also shows the maximum value of transient voltage at the injection point of the grounding grid, while the transient voltage at the diagonal point of the injection point is minimum at that moment. After 3.07 μs, the overall transient voltage induced on the grounding grid gradually decreases from its peak value until it disappears. It is basically the same as that of the lightning current, which starts from zero and increases gradually, reaches the peak and then gradually decreases until it disappears, except that the peak occurs at a different moment. The peak of the lightning current occurs at 10.23 μs, while the peak of the induced transient voltage occurs at 3.07 μs.

Here observe the nine small diagrams in Fig. 13.9, the leakage current on the grounding grid and the temporal and spatial distribution of the transient voltage induced on the grounding grid are basically the same, and the change rule of the lightning current of the excitation source is the same. At each critical moment, the largest value of leakage current occurs at the injection point of lightning current on the grounding grid; at the same time, in the whole time period, as in the case of

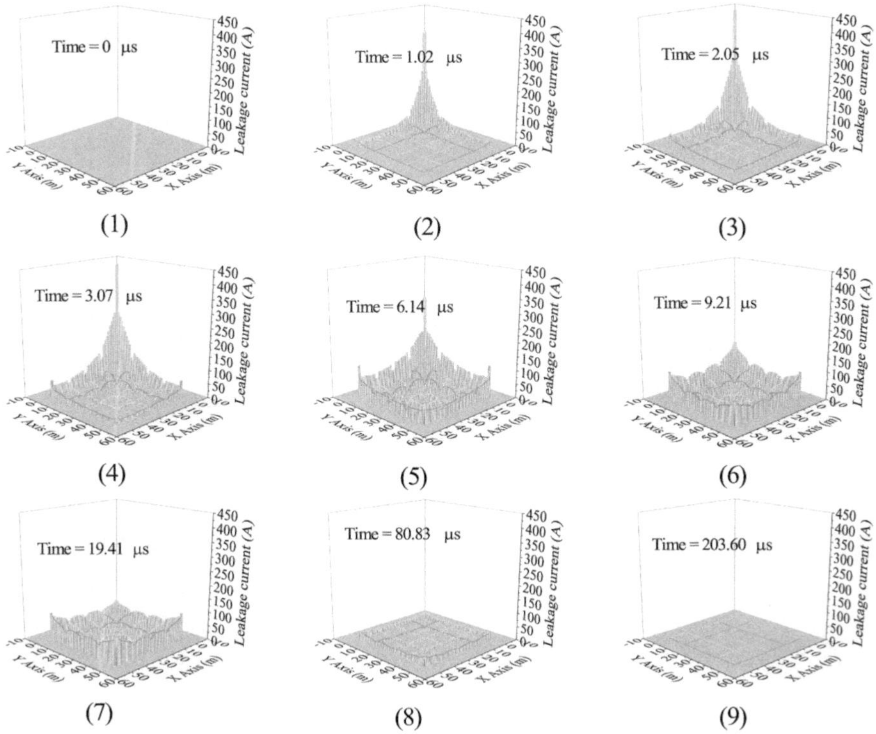

Fig. 13.9 3-d distribution of leakage current along the grid (Time domain with soil ionization effect and soil frequency characteristic case)

transient voltage induced on the grounding grid, the largest value of leakage current also occurs at the moment of 3.07 μs.

Next, observing the nine subplots in Fig. 13.10, it can be known that the overall temporal and spatial distribution of the transient branch currents on the grounding grid is basically the same as that of the transient voltages and leakage currents. It is also the same as the development of the excitation source lightning current, and the largest value of the branch current in each moment occurs at the point of lightning current injection on the grounding grid, only that the moment of the largest value of the branch current occurs in the whole time period is slightly delayed to 4.09 μs.

The time-domain algorithm for the soil ionization phenomenon occurring under the action of large-value lightning currents is realized by numerical simulation of the increasing virtual conductor radius approach. In the absence of soil ionization, the virtual conductor half disappears and recedes to the actual radius of the conductor. The nine subplots in Fig. 13.11 reflect the process of development of the virtual radius of the conductor of the grounding grid at the critical moment. It also starts from nothing and gradually increases until the maximum virtual radius appears, and then gradually decreases until it disappears. The maximum virtual radius occurs at

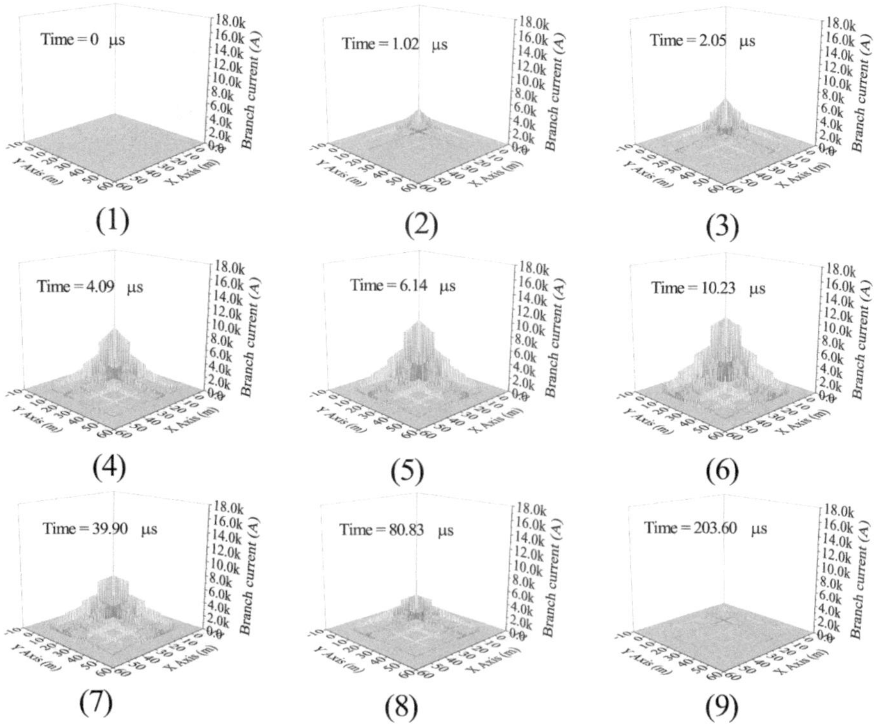

Fig. 13.10 3-d distribution of branch current along the grid (Time domain with soil ionization effect and soil frequency characteristic case)

the moment of 3.07 μs during the whole period of soil ionization phenomenon. This moment is the same moment when the maximum values of transient voltage and leakage current on the grounding grid occur.

13.7 Summary

In this section, the correctness and validity of the simulation results obtained by various frequency and time domain algorithms for simulation calculations with or without consideration of the frequency characteristics of soil parameters and soil ionization effects are discussed. Through the analysis and discussion of numerical examples, the following conclusions are obtained: (1) After considering the frequency characteristics of the soil parameter, regardless of the frequency domain or time domain numerical algorithms, the simulated lightning transient response of the substation grounding grid is lower than that simulated when the frequency characteristics of the soil parameter are not considered. (2) Regardless of whether the frequency characteristics of the soil parameter are taken into account, the lightning transient response

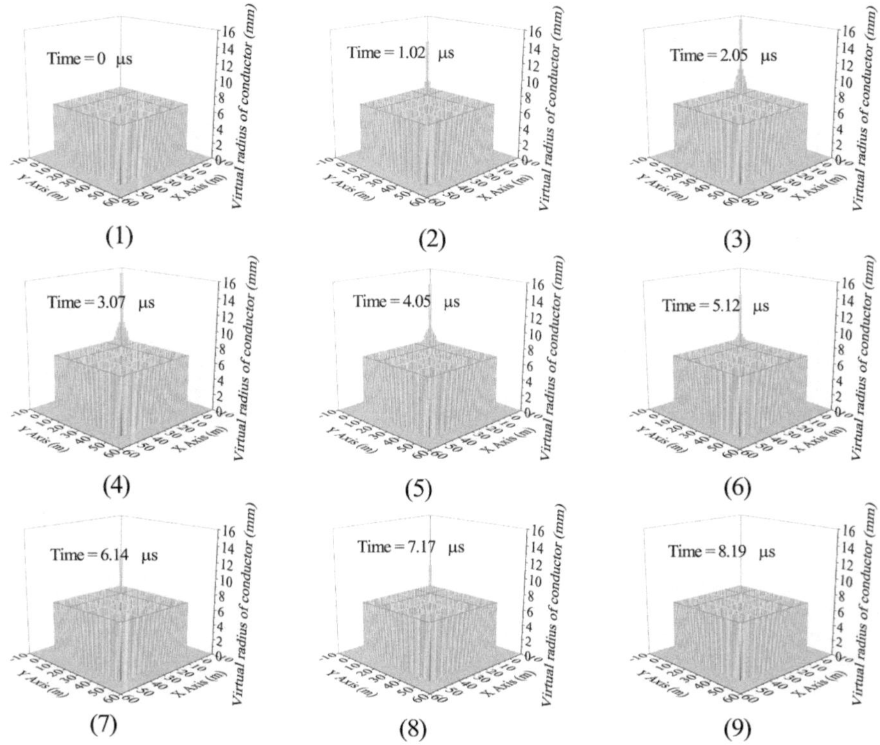

Fig. 13.11 3-d distribution of virtual radius of conductor along the grid (Time domain with soil ionization effect and soil frequency characteristic case)

of the substation grounding grid simulated by the time-domain algorithm without taking into account the influence of soil ionization effect is higher than that simulated by the time-domain algorithm with taking into account the influence of soil ionization effect.

References

1. J.G. Safar, R. Shariatinasab, etc., Comprehensive modeling of grounding electrode buried in ionized soil based on mom-hbm approach. IEEE Trans. Power Delivery **35**(3), 1390–1398 (2020)
2. M. Moradi, Analysis of transient performance of grounding system considering frequency-dependent soil parameters and ionization. IEEE Trans. Electromagn. Compat. **62**(3), 785–797 (2020)
3. M. Moradi, Comparison of different approaches for incorporating frequency dependence of soil parameters in time domain simulation of grounding system. IEEE Trans. Power Delivery **39**(4), 2028–2040 (2024)
4. J.-B. Fan, Z.-X. Li, Lightning transient impulse problem of substation grounding grid in multilayer horizontal layered soil considering soil ionization effect and frequency characteristics. IEEE Trans. Electromagn. Compat., 1–10 (2024)
5. L. Grcev, Modeling of grounding electrodes under lightning currents. IEEE Trans. Electromagn. Compat. **51**(3), 559–571 (2009)

Chapter 14
Summary for Numerical Method in the Book

14.1 Perface

This is the final chapter of the book, and it is the content that summarizes all the numerical methods for electromagnetic fields developed in this book.

14.2 The Basic Theory of Grounding Problem of Power System Substation Grounding Grids

Various numerical methods of electromagnetic field developed in this book are aimed at the grounding problem of substation grounding grid in multilayered horizontal stratified soil. Since all the numerical methods of electromagnetic field developed in this book are aimed at the grounding problem of substation grounding grid in mutilayered horizontal layered soil, all the numerical calculation methods developed in this book are based on the "source domain problem". The "source domain problem" here is to first solve the distributed current source of the grounding grid in the grounding problem, and then, the electromagnetic field quantity of any concerned field point is solved by a known current source. Therefore, it is necessary to solve the Green function of the point source in the multilayered horizontal stratified soil. The second chapter of this book is devoted to the study of frequency domain and time domain Green's function of point current source in multilayered horizontal stratified soil by using matrix pencil method, and related articles [1, 2] are published.

Z.-X. Li, *Numerical Simulation Algorithm of Electromagnetic Field for Grounding Problems in Power System Substation Grounding Grids*,
https://doi.org/10.1007/978-981-96-0904-8_14

14.3 Methods of Substation Grounding Steady-State Problems

The electromagnetic field numerical methods related to the steady-state grounding problem of power system substation grounding grids have basically been theoretically developed relatively maturely.

The fifth chapter of this book presents an in-depth study of the steady-state grounding problem of substation grounding grids in power systems and develops the spatial 1-d scalar Galerkin BEM and Galerkin MM, and publishes a number of algorithmic papers [3–8] on the subject. Articles [3, 4] in these literature all developed numerical methods for equipotential electromagnetic fields, namely the 1-d scalar Galerkin MM in space. Especially in Articles [5–8], the spatial 1-d Galerkin scalar MM for high-order basis functions of equipotential electromagnetic field numerical methods is further developed.

In addition to the equal potential electromagnetic field numerical algorithm, this book also develops the unequal potential electromagnetic field numerical algorithm. The sixth chapter of this book is devoted to the study of the vector Galerkin BEM, the vector Galerkin MM and the vector Galerkin hybrid algorithm based on various basis functions used to simulate the single frequency grounding of substation grounding grid in multilayered horizontal stratification soil. In this way, several articles [9–15] based on PI basis function vector Galerkin hybrid algorithm are published for numerical simulation of steady-state single-frequency grounding of multilayer horizontal stratified ground substations.

14.4 Methods of Substation Grounding Transient Problems

In this book, the lightning transient fault grounding problem of power substation grounding grids is deeply discussed and studied from the frequency domain and time domain electromagnetic field numerical method.

Chapter eighth of this book focuses on spatial 1-d vector BEM, vector Galerkin MM and vector Galerkin hybrid algorithm based on various basis functions in frequency domain. Therefore, corresponding algorithms are published in articles [16–20], of which article [20] is based on vector Galerkin BEM to simulate and calculate the lightning transient response grounding problem of substation grounding grids in multilayered horizontal stratified soil. In Article [19], a vector Galerkin hybrid algorithm based on PEEC basis function is used to simulate and calculate the grounding problem of lightning transient fault response of substation grounding grids in multilayered horizontal stratified soil. While articles [16, 17] study the problem of simulating the lightning transient impact grounding process of the grounding grid buried in multilayer horizontal stratified soil based on the vector Galerkin hybrid algorithm based on PI basis function, and reference [18] uses the vector Galerkin hybrid algorithm based on the T basis function to numerically simulate the lightning transient impact grounding problem of the grounding grid buried in multilayer horizontal stratified soil.

Similar to frequency domain algorithms, Chapter tenth of this book focuses on the spatial 1-d vector Galerkin BEM, vector Galerkin MM, and vector Galerkin hybrid algorithm based on various basis functions in the time domain, And published multiple scientific articles [21–26] on corresponding algorithms. These several technological articles are all numerical simulations of the lightning transient impulse grounding problem of the substation grounding grids in multi-layered horizontal layered soil. Articles [21, 26] introduce the time-domain Galerkin BEM. And articles [22–24] are about the vector Galerkin hybrid algorithm based on PI basis functions, while Article [25] is about the vector Galerkin hybrid algorithm based on PEEC basis functions. These articles are all developing time-domain electromagnetic field numerical algorithms, therefore, they all discuss the phenomenon of soil ionization around the grounding grid.

14.5 Prospects for the Research Direction of Transient Fault Grounding Problem in the Grounding Grid of Power System Substations

The lightning transient impulse grounding problem of the power system substation grounding grid is a problem in the time domain. Therefore, the development and research of time-domain electromagnetic field numerical algorithms are undoubtedly the optimal solution. This book discusses the spatial 1-d vector Galerkin BEM, vector Galerkin MM, and vector Galerkin hybrid algorithm based on various basis functions (PI and PEEC) in the time domain space, which can numerically simulate the lightning transient impulse grounding problem of the substation grounding grid in multi-layer horizontal layered soil. However, all these electromagnetic field algorithms are developed based on the quasi-static electromagnetic field theory, ignoring the delay effect of time-domain current on the substation grounding grid conductor. Although the simulated impulse grounding impedance and the data obtained by considering this delay effect algorithm are used in most substation lightning impulse grounding problems. The difference in error is not significant, however, considering the delayed effect, developing all corresponding algorithms based on the theory of dynamic electromagnetic fields is an upgrade and development of the theory in this book.

References

1. Z.-X. Li, G.-F. Li, J.-B. Fan, Quasi-static complex image method for a point source buried in horizontal multilayered earth model. Progress Electromagn. Res. B **34**, 187–204 (2011)
2. Z.-X. Li, G.-F. Li, J.-B. Fan, Y. Yu, Quasi-static complex image method with matrix pencil method for a scalar monopole buried in horizontal multilayered earth model for grounding system problem. Adv. Develop. Mathem. Sci. **4**(1 and 2), 30–49 (2013)
3. L.-P. Zhang, J.-S. Yuan, Z.-X. Li, The complex image method and its application in numerical simulation of substation grounding grids. Commun. Numer. Method Eng. **15**(11), 835–839 (1999)
4. Z.-X. Li, J.-B. Fan, W.-J. Chen, Numerical simulation grounding grids buried in both horizontal and vertical multilayer earth model. Int. J. Numer. Methods Eng. **69**(11), 2359–2380 (2007)
5. Z.-X. Li, G.-F. Li, J.-B. Fan, Numerical simulation of substation grounding system in low frequency domain basing on the boundary element method with linear basis function. Int. J. Numer. Methods Biomed. Eng. **26**(12), 1819–1914 (2010)
6. Z.X. Li, J.-B. Fan, Y. Yu, Numerical simulation of substation grounding grids buried in vertical earth model based on the thin-wire approximation with linear basis functions. Appl. Comput. Electromagn. Soc. J. **26**(4), 353–366 (2011)
7. Z.-X. Li, G.-F. Li, J.-B. Fan, Y. Yu, Numerical calculation of ac substation grounding systems buried in a vertical multilayered earth model by the boundary element method with higher order basis function. Int. J. Numer. Model. Electron. Network Dev. Fields **25**(3), 258–274 (2012)
8. Z.-X. Li, G.-F. Li, J.-B. Fan, C.-X. Zhang, Numerical calculation of ac substation grounding systems buried in a horizontal multilayered earth by higher order basis function. Int. J. Power Energy Syst. **32**(1), 21–30 (2012)
9. Z.-X. Li, W.-J. Chen, J.-B. Fan, A novel mathematical modelling of grounding system buried in multilayer earth. IEEE Trans. Power Delivery **21**(3), 1267–1272 (2006)
10. Z.-X. Li, W.-J. Chen, Numerical simulation grounding system buried within multilayer earth based on quasi-static complex image method in frequency domain. Commun. Numer. Method Eng. **23**(1), 11–27 (2007)
11. Z.-X. Li, J.-B. Fan, Numerical calculation of grounding system in low frequency domain based on the boundary element method. Int. J. Numer. Methods Eng. **73**(5), 685–705 (2008)
12. Z.-X. Li, G.-F. Li, J.-B. Fan, C.-X. Zhang, Numerical calculation of grounding system buried in vertical earth model in low frequency domain based on the boundary element method. Europ. Trans. Electr. Power **19**(8), 1177–1190 (2009)
13. Z.-X. Li, Y. Yin, C.-X. Zhang, C.-X. Zhang, Frequency dependent characteristics of grounding systems buried in multilayered earth model based on quasi-static electromagnetic field theory. Progress Electromagn. Res. M **33**, 185–196 (2013)
14. P.D. Rancic, L.V. Stefanovic, D.R. Djordjevic, Numerical simulation of grounding grid buried in horizontal multilayered earth based on “t” typical basic elements in frequency domain. IET Sci. Measur. Technol. **9**(6), 645–653 (2015)
15. Z.-X. Li, Y. Yin, C.-X. Zhang, L.-C. Zhang, Numerical simulation of currents distribution along grounding grid buried in horizontal multilayered earth in frequency domain based dynamic state complex image method. Electr. Power Components Syst. **43**(14), 1573–1582 (2015)
16. Z.-X. Li, G.-F. Li, J.-B. Fan, Y. Yin, A novel mathematical model for the lightning response of a grounding system buried in multilayered earth based on the quasi-static complex image method. Europ. Trans. Electr. Power **22**(6), 831–845 (2012)
17. Z.X. Li, Y. Yin, C.-X. Zhang, L.-C. Zhang, A mathematical model for the transient lightning response from grounding systems. Progress Electromagn. Res. B **57**, 47–61 (2014)
18. Z.-X. Li, K.-L. Gao, Y. Yin, D. Ge,Transient lightning responses of grounding systems buried in horizontal multilayered earth with a hybrid method. J. Electrost. **72**(5) (2014)
19. Z.-X. Li, K.-L. Gao, S.-W. Rao, Lightning response of a grounding system buried in multiple layers in earth using the peec method based on the quasi-static complex image method. Electr. Power Components Syst. **48**(3), 291–303 (2020)

20. Z.-X. Li, Y. Yin, K.-C. Wang, Lightning response of a grounding system buried in multilayered earth based on the galerkin's boundary element and quasi-static complex image methods. IEEE Trans. Electromagn. Compat. **63**(4), 1118–1127 (2021)
21. Z.-X. Li, P. Li, X. Zhao, Galerkin boundary element method for simulating lightning response of grounding grid in horizontal multilayered soil model considering soil ionization effect. Int. J. Numer. Model. Electron. Network Dev. Fields **37**(5), e3285 (2024)
22. Z.-X. Li, P. Li, K.-C. Wang, Transient lighting response to grounding grids buried in horizontal multilayered earth model considering time domain quasi-static complex image method and soil ionization effect. Int. J. Computation Mathem. Electr. Electron. Eng. **40**(3), 516–534 (2021)
23. Z.-X. Li, S.-W. Rao, Transient lightning response of grounding grids in a horizontal, multilayered soil model that considers soil ionization effects with time-domain quasi-static complex images. Int. J. Numer. Model. Electron. Network Dev. Fields **32**(1), e2463 (2019)
24. Z.-X. Li, W.-J. Chen, K.-C. Wang, Frequency domain and time domain analysis of the transient behavior of buried grounding grids in horizontal multilayered earth model. Electr. Eng. **104**(4), 2512–2529 (2022)
25. Z.-X. Li, Lightning response transient characteristics of substation grounding grid in multilayer horizontal stratified earth model considering soil ionization effect. IEEE Trans. Electromag. Compat. **65**(6), 1665–1673 (2023)
26. J.-B. Fan, Z.-X. Li, Lightning transient impulse problem of substation grounding grid in multilayer horizontal layered soil considering soil ionization effect and frequency characteristics. IEEE Trans. Electromagn. Compat., 1–10 (2024)

MIX
Papier aus verantwortungsvollen Quellen
Paper from responsible sources
FSC® C105338

If you have any concerns about our products,
you can contact us on
ProductSafety@springernature.com

In case Publisher is established outside the EU,
the EU authorized representative is:
Springer Nature Customer Service Center GmbH
Europaplatz 3, 69115 Heidelberg, Germany

Printed by Libri Plureos GmbH
in Hamburg, Germany